SKELETAL MUSCLE REPAIR AND REGENERATION

Advances in Muscle Research

Volume 3

Series Editor

G.J.M. Stienen, *Vrije Universiteit, Amsterdam, The Netherlands*

Skeletal Muscle Repair and Regeneration

Edited by

Stefano Schiaffino

University of Padova and Venetian Institute of Molecular Medicine (VIMM), Padova, Italy

and

Terence Partridge

Center for Genetic Medicine, Washington, DC, USA

Library of Congress Control Number: 2008920295

ISBN 978-1-4020-6767-9 (HB)
ISBN 978-1-4020-6768-6 (e-book)

Published by Springer,
P.O. Box 17, 3300 AA Dordrecht, The Netherlands.

www.springer.com

Printed on acid-free paper

CONTENTS

PREFACE

Barring misfortune, there is a general expectation that our skeletal muscles will remain in effective service for our allotted 'three-score- years-and-ten' (Psalm 90, king James Bible, 1611) despite daily exposure to a variety of traumatic insults. It is now clear that this maintenance entails efficient operation of appropriate repair processes. To cover the whole spectrum of severity of muscle injury ranging from i) focal intracellular damage, such as plasma membrane rupture or myofibril disruption, to ii) segmental damage or necrosis involving the whole length of individual muscle fibers, to iii) injuries affecting whole muscle bundles, including blood vessels and interstitial tissues, requires that these repair processes be subtly tailored to cope appropriately (Sloper et al., 1978). So ingrained, currently, is the notion that skeletal muscle can repair itself, that it comes as somewhat of a shock to be reminded (Chapter 1) how recently this idea has come into general acceptance and of how hot was the debate as to the fundamental mechanisms behind this repair process.

By virtue of its mechanical function skeletal muscle is subject to continual mechanical stress caused by muscle contractions while its largely superficial distribution places it as the primary cushion against external physical impact. In consequence, some form of muscle damage and repair must occur on a continuous day-to-day basis. Muscle fibres are especially sensitive to eccentric (lengthening) contractions, such as downhill running, that produce focal membrane disruptions leading to influx of plasma proteins or tracers into the damaged fibers and efflux of muscle proteins, such as creatine kinase, into the blood (McNeil and Khakee, 1992). Such eccentric contractions appear to be especially damaging in muscular dystrophies caused by defects in components of the muscle fibre surface membrane. This is most conspicuous in Duchenne muscular dystrophy, the archetype of a number of inherited defects of the complex of proteins linking the internal cytoskeleton to the sarcolemma-extracellular matrix, absence or disturbance of which results in an increased membrane fragility (Rowland, 1980).

An important insight to emerge during the past few years, is that muscular dystrophy can result not only from diminished membrane resilience but also from deficiencies in the mechanisms that undertake repair of damaged muscle membranes. This notion is based on the discovery that a membrane protein called dysferlin is a major component of a specific membrane repair mechanism responsible for the maintenance of plasmalemmal integrity in many cells including skeletal muscle fibres (Bansal et al., 2003; Lennon et al., 2003). Following membrane damage dysferlin, together with other proteins, promotes a rapid membrane resealing through the Ca2+-dependent fusion of subsarcolemmal vescicles to the plasma membrane. In humans, mutations in dysferlin result in limb-girdle muscular dystrophy 2B, as well as in other clinically distinct muscular dystrophies. Myofibrillar disruption and Z-disk streaming are also frequent sequelae of mechanical stress induced by eccentric contractions. Though generally considered as an expression of muscle damage and responsible for the force drop caused by eccentric contractions (Lieber and Friden, 2002), these changes have also been interpreted as an expression of myofibrillar remodeling, namely as part of the repair process that leads to formation of new myofibrils and to muscle hypertrophy (Yu et al., 2004).

More severe muscle damage characteristically provokes segmental or global muscle fiber necrosis which in turn triggers the typical response of muscle regeneration, mediated largely, if not entirely under normal circumstances, by a specific type of stem cell, the muscle satellite cell (Sherwood et al., 2004; Collins et al., 2005). Satellite cells undergo proliferation after muscle injury, followed by differentiation accompanied by fusion with one another into myotubes as well as with undamaged portions of the fibre, prior to their transformation into mature myofibre. Even when the injury is restricted to the muscle fibre, with maintenance of an intact basal lamina, this process requires the active participation of other cell types. In particular, macrophages, derived from blood monocytes, are attracted to the site of muscle injury and play an important role not only by clearing the necrotic debris, but also by releasing growth factors that promote myogenic cell proliferation and differentiation (Arnold et al., 2007). Finally, functional recovery requires re-establishment of any lost motor innervation to the regenerated fibers as well as formation of new myotendinous junctions to restitute mechanical linkage.

A number of articles in this volume identify and characterize the players in this regeneration process; Chapter 2 describing the origin of the satellite cell and Chapter 3 giving a detailed description of it, insofar as that is possible with what appears increasingly to be a heterogeneous category united principally by its anatomical definition. This reassessment of the satellite cell has been provoked in large part by evidence from a variety of studies as presented in Chapter 4, that cells from other sources can, under some circumstances, participate in regeneration of skeletal muscle (Ferrari et al., 1998; Gussoni et al., 1999). The mechanisms that control the activities of myogenic cells during regeneration must be finely tuned if we are to end up with appropriate amounts of muscle and of reserve myogenic cells to cope with future bouts of degeneration. This issue is considered both from the viewpoint of the regulation of the activation and proliferation of the myogenic cells themselves (Chapter 6) and of the factors influencing their entry into the actual formation of new muscle fibres (Chapter 7). Both of these would appear to play important roles in regulating the amount of muscle formed and the size of the regenerated fibres, which, although more variable than that of uninjured muscle is still remarkably constant. Increasingly too, we have come to understand that the inflammatory cells that enter muscle lesions do not simply clean up the debris but also play a central role in coordinating the various stages of the regeneration process (Chapter 12) and are heavily implicated in determining the balance between muscle fibres and interstitial connective tissue.

While muscle fibre necrosis that leaves the basal lamina intact can be repaired by muscle regeneration with complete restitutio ad integrum, the problem becomes more complicated when entire muscle fibre bundles are damaged by major mechanical traumas or by muscle strain injuries in athletes. In this case, more extensive remodeling of muscle ensues (Chapter 8), often accompanied by formation of a scar of fibrous tissue. This latter tends to disturb muscle micro-anatomy, an effect which is to some extent coped with by development of insertions of the newly regenerated muscle fibers via new-formed myotendinous junctions onto disruptive collagenous sheets (see Chapter 11 Kalimo). An often unrecognized participant in this process is the intercellular matrix (Chapter 13) whose prompt and apposite restitution is clearly important in a tissue whose chief function is mechanical, but which also plays an integral part in the complex of cell signaling events that underlie control of the restoration of this tissue. Similarly, the matter of effectively re-innervating a muscle that is regenerating is essential for restoration of function (Chapter 14). This interaction between muscle and motor nerve is one of the most extensively and intensively studied archetypes of ectodermal/mesodermal interaction and is a clear example of the need for dynamic control mechanisms in so responsive a system as that which provides our locomotion. Much of our practical interest in muscle regeneration is focused on this more severe end of the spectrum of damage, in good part in the belief that a better understanding of the sequence of events involved in this complex process is likely to improve our prospects of optimizing its outcome. This partiality of interest is strongly reflected in the balance of research and hence in the selection of articles in this volume, which are concerned predominantly with regeneration on this larger scale.

Over the years, histopathological diagnostics has, by analysis of muscle obtained from various human diseases, evolved a sophisticated classification of the patterns of necrosis and regeneration associated with individual disease conditions and this (Chapter 10) remains as a critical basis of reference for all of the animal models of muscle disease and regeneration that have been developed. In recent years, it has become possible to view these histopathological processes from a quite different angle, in terms of the patterns of gene expression that characterize the variety of disease phenotypes (Chapter 5). Combination of these two descriptive dimensions looks set to provide a more coherent and comprehensive picture of the mechanisms involved and of the influence on overall pathology of changes in balance between the cellular and biochemical components. Of the animal models, perhaps the most informative have been based on muscle transplantation or other forms of major physical traumatic damage in mammalian muscle (Chapter 8). These were among the first to indicate the robustness of muscle regeneration and to give some hope that it was a process that might be managed to advantage. A still more startling instance of the astonishing potential for regeneration is that seen in urodele amphibians, where there is complete restitution of entire limbs, including the muscle tissue (Chapter 9). As an extreme example of the possibilities of reorganizing tissues in the adult, this phenomenon has long-attracted those seeking to elucidate the fundamental mechanisms that might, one day, be exploited in man.

Our ultimate aim in medical research is to develop ideas that may become of clinical use. In the case of skeletal muscle, the main interest is in conserving muscle mass and strength in the faceof affliction by genetic or acquired disease of skeletal muscle and in combating the profoundly debilitating loss of muscle mass and strength that accompanies disease of other systems, e.g. cachexia associated with cancer or chronic cardio-pulmonary disease. This search has led to a particular interest in the mechanisms of action of a variety of factors that have been shown to possess potent growth-promoting or growth-inhibiting activities on skeletal muscle (Chapter 15).

Finally, the fantasy of eternal youth leads us to wonder why only 'three-score-years-and-ten'? What goes wrong to prevent us maintaining our muscle bulk and strength (Chapter 16). It is with some surprise and stirrings of hope that we learn that muscle itself is probably not the main culprit.

Stefano Schiaffino
University of Padova, Italy

Partridge Terence
Center for Genetic Medicine, Washington, DC, USA

REFERENCES

Arnold L, Henry A, Poron F, Baba-Amer Y, van Rooijen N, Plonquet A, Gherardi RK, Chazaud B (2007) Inflammatory monocytes recruited after skeletal muscle injury switch into antiinflammatory macrophages to support myogenesis. J Exp Med 204:1057–1069

Bansal D et al (2003) Defective membrane repair in dysferlin-deficient muscular dystrophy. Nature 423:168–172

Collins CA, Olsen I, Zammit PS, Heslop L, Petrie A, Partridge TA, Morgan JE (2005) Stem cell function, self-renewal, and behavioral heterogeneity of cells from the adult muscle satellite cell niche. Cell 122:289–301

Ferrari G, Angelis C, Coletta M, Paolucci E, Stornaiuolo A, Cossu G, Mavilio F (1998) Muscle Regeneration by bone marrow derived myogenic progenitors. Science 279:1528–1530

Gussoni E, Soneoka Y, Strickland CD, Buzney EA, Khan MK, Flint AF, Kunkel LM, Mulligan RC (1999) Dystrophin expression in the mdx mouse restored by stem cell transplantation. Nature 401:390–394

Lennon NJ, Kho A, Bacskai BJ, Perlmutter SL, Hyman BT, Brown RH Jr (2003) Dysferlin interacts with annexins A1 and A2 and mediates sarcolemmal wound-healing. J Biol Chem 278:50466–50473

Lieber RL, Friden J (2002) Mechanisms of muscle injury gleaned from animal models. Am J Phys Med Rehabil 81(11 Suppl):S70–S79

McNeil PL, Khakee R (1992) Disruptions of muscle fiber plasma membranes. Role in exercise-induced damage. Am J Pathol 140:1097–1109

Rowland LP (1980) Biochemistry of muscle membranes in Duchenne muscular dystrophy. Muscle Nerve 3:3–20

Sherwood RI, Christensen JL, Conboy IM, Conboy MJ, Rando TA, Weissman IL,Wagers AJ (2004) Isolation of adult mouse myogenic progenitors: functional heterogeneity of cells within and engrafting skeletal muscle. Cell 119:543–554

Sloper JC, Barrett MC, Partridge TA (1978) The muscle cell. J Clin Pathol Suppl R Coll Pathol 12:25–43

Yu JG, Carlsson L, Thornell LE (2004) Evidence for myofibril remodeling as opposed to myofibril damage in human muscle with DOMS: an ultrastructural and immunoelectron microscopic study. Histochem Cell Biol 121(3):219–227

CONTRIBUTORS

Ville Äärimaa Department of Orthopedic Surgery, University and University Central Hospital of Turku, Turku, Finland
E-mail: ville.aarimaa@utu.fi

Andrew S. Brack Department of Neurology and Neurological Sciences, Stanford University School of Medicine, Stanford, CA 94305-5235, USA
E-mail: asbrack@stanford.edu

Margaret Buckingham Department of Developmental Biology, URA CNRS 2578, Institut Pasteur, 25 rue du Dr. Roux, 75015 Paris, France
E-mail: margab@pasteur.fr

Bruce M. Carlson Institute of Gerontology, 300 North Ingalls Building, University of Michigan, Ann Arbor MI 48109, USA
E-mail: brcarl@umich.edu

Giulio Cossu Stem Cell Research Institute, Dibit, H. San Raffaele, via Olgettina 58, 20132, Milan, Italy
E-mail: cossu.giulio@hsr.it

Graziella Messina Stem Cell Research Institute, Dibit, H. San Raffaele, via Olgettina 58, 20132, Milan, Italy
E-mail: messina.graziella@hsr.it

Stefano Biressi Department of Neurology and Neurological Sciences, Stanford University School of Medicine, Stanford, CA 94305-5235, USA
E-mail: sbiressi@stanford.edu

Miranda Grounds School of Anatomy & Human Biology, The University of Western Australia, 35 Stirling Hwy Crawley, Western Australia, Australia 6009
E-mail: mgrounds@anhb.uwa.edu.au

Eric Hoffman Research Center for Genetic Medicine, Children's National Medical Center, 111 Michigan Ave NW, Washington DC 20010, USA
E-mail: EHoffman@cnmcresearch.org

Markku Järvinen Medical School, University of Tampere, Tampere, Finland; and Departments of Orthopaedic and Plastic Surgery, Tampere University Hospital, Tampere, Finland
E-mail: Markku.Jarvinen@uta.fi

Tero A.H. Järvinen Medical School, University of Tampere, Tampere, Finland; and Departments of Orthopaedic and Plastic Surgery, Tampere University Hospital, Tampere, Finland
E-mmail: Tero.AH.Jarvinen@uta.fi

George Karpati Neuromuscular Research Group, Montreal Neurological Institute, McGill University, Montreal, Quebec, Canada, H3A 2B4
E-mail: George.Karpati@mcgill.ca

Hannu Kalimo, MD Department of Pathology, University and University Hospital of Helsinki, P.O.Box 21, FI-00014, Helsinki, Finland
E-mail: hannu.kalimo@helsinki.fi

Minna Kääriäinen Medical School, University of Tampere, Tampere, Finland; and Departments of Orthopaedic and Plastic Surgery, Tampere University Hospital, Tampere, Finland
E-mail: minna.t.kaariainen@gmail.com

Maria J. Molnar Neuromuscular Research Group, Montreal Neurological Institute, McGill University, Montreal, Quebec, Canada; Molecular Medicine Division, National Institute of Psychiatry and Neurology, Budapest
E.mail: molnarm@opni.hu

Didier Montarras Department of Developmental Biology, URA CNRS 2578, Institut Pasteur, 25 rue du Dr. Roux, 75015 Paris, France
E-mail: dmontarr@pasteur.fr

Tommaso Nastasi EMBL – Mouse Biology Unit, Campus A. Buzzati-Traverso, Via Ramarini, 32, 00015, Monterotondo Scalo (RM), Italy
E-mail: nastasi@embl.it

Roddy S. O'Connor Graduate Program in Molecular and Systems Pharmacology, Department of Pharmacology, Emory University, School of Medicine, Atlanta, GA 30322-4218, USA
E-mail: rsoconn@emory.edu

Terence Partridge Center for Genetic Medicine Research, Children's National Medical Center, 111 Michigan Ave NW, Washington DC 20010, USA
E-mail: tpartridge@cnmcresearch.org

Grace K. Pavlath Emory University School of Medicine, Deptartment of pharmacology, Room 5027, O.W. Rollins Research Building, Atlanta, GA 30322-3090, USA
E-mail: gpavlat@emory.edu

Tom Rando Department of Neurology and Neurological Sciences, Stanford University School of Medicine, Stanford, CA 94305-5235, USA
E-mail: rando@stanford.edu

Nadia Rosenthal EMBL-Monterotondo Outstation, European Molecular Biology Laboratory, Campus "A. Buzzati-Traverso", Italy
E-mail: rosenthal@embl-monterotondo.it

Stefano Schiaffino Department of Biomedical Sciences, University of Padova, and Venetian Institute of Molecular Medicine (VIMM), Padova, Italy
E-mail: stefano.schiaffino@unipd.it

Gabi Shefer Department Cell and Developmental Biology, Sackler School of Medicine, Tel Aviv University, Tel Aviv, Israel
E-mail: gabise@post.tau.ac.il

Clarke R. Slater School of Neurology, Neurobiology & Psychiatry, Faculty of Medical Sciences, University of Newcastle upon Tyne, Framlington Place, Newcastle upon Tyne, NE2 4HH, UK
E-mail: C.R.Slater@newcastle.ac.uk

Frank E. Stockdale Maureen Lyles D'Ambrogio Professor of Medicine, Emeritus, Stanford University, Stanford School of Medicine, Cancer Center Room 2238 Stanford, CA 94305-5826
E-mail: Stockdale@Stanford.edu

Elly M. Tanaka Max-Planck Institute of Molecular Cell Biology and Genetics, Pfotenhauerstrasse 108, 01307 Dresden, Germany
E-mail: tanaka@mpi-cbg.de

James G. Tidball Department of Physiological Science, University of California. Los Angeles, Los Angeles CA 90095-1606, USA
E-mail: jtidball@physci.ucla.edu

Zipora Yablonka-Reuveni Deptartment of Biological Structure, Box 357420, Warren B. Magneson HSB-G514, University of Washington, Seattle, WA, 98195, USA
E-mail: reuveni@u.washington.edu

Peter S. Zammit Randall Division of Cell and Molecular Biophysics, King's College London, New Hunt's House, Guy's Campus, London, SE1 1UL, England
E-mail: peter.zammit@kcl.ac.uk

Po Zhao Research Center for Genetic Medicine, Children's National Medical Center, 111 Michigan Ave NW, Washington DC 20010, USA
E-mail: pzhao@cnmcresearch.org

CHAPTER 1

MYOGENESIS – THE EARLY YEARS

FRANK E. STOCKDALE

Maureen Lyles D'Ambrogio Professor of Medicine, Emeritus

Over a hundred years has passed between the first publication on skeletal muscle regeneration and when a mechanistic understanding of the biology of regeneration was realized. The reason for the delay is rooted in the pace of change in the technology at hand to investigate this process. While the microscope was invented in the 18th century, the first description of the response of skeletal muscle to injury did not appear until the middle of the 19th century (Waldeyer, 1865). It was believed that muscles basically did not regenerate, however, Waldeyer found that human muscle regenerated in tissue isolated from patients who survived a typhoid epidemic. Several decades earlier Schwann (Schwann, 1839) observed that muscles of embryos were multinucleated and proposed they formed by a process of coalescences, though he probably did not propose cell fusion, but merely aggregation. It became accepted that skeletal muscle was multinucleated and that many nuclei appeared in the region of muscle damage. The question was how could regenerating fiber become nucleated? It was commonly thought for nearly the next 100 years that cytoplasmic buds protruded from the broken ends of fibers to close the gap between damaged fiber ends. But it was difficult to explain the ribbons of nuclei that were so apparent to all that observed the process under the microscope. A number of theories were proposed – that muscle fiber nuclei were extruded from the damage fiber into the buds, that they came from the proliferating cells surrounding the fiber, or that they were from cells that migrated from other sites. Of course, the latter two hypotheses alone could not explain multinucleation of the fibers, although they were prophetic in recognizing some century and a half later that blood borne stem cells could be shown to contribute to the nuclear population of damaged muscle. Left unsettled and a source for much speculation was the mechanism of nucleation.

S. Schiaffino and T. Partridge (eds.), Skeletal Muscle Repair and Regeneration, 1–17.

From the very outset embryonic muscle development, which was thought to occur de novo in mesenchyme, and skeletal muscle regeneration in the adult, were intertwined conceptually. It was recognized that both the embryo and the adult skeletal muscle fiber had to become nucleated by some mechanism. While in the case of regeneration it was possible to postulate that a fiber itself could be the source of nuclei in a regenerate fiber, how could one explain the many nuclei in an embryonic fiber where initially there were no fibers? This difficulty was acknowledged by observers for over century who never reported seeing a mitotic figure within an embryonic, or for that matter, regenerating muscle fiber. Thus arose the expedient hypothesis that there must be a cryptic form of nuclear division, dubbed amitosis. This hypothesis emerged long before replication of DNA was understood. Thus it was proposed that this replication occurred within embryonic and regenerated nuclei, but it was not accompanied by mitosis (a term that was not coined until 1882). A major problem for the early investigators was that they had no method for directly observing the process of fiber formation in either the embryonic or the regenerate muscle context.

1. THE BLOSSOMING OF EXPERIMENTAL MYOGENESIS

Well into the 20th century it was quite broadly held that both embryonic and regenerate muscle became nucleated by amitosis or nuclear splitting. The reason that the concept of amitosis persisted from the turn of the 19th century until the 1960's was because no observers ever saw mitotic figures within developing or regenerating muscle, yet there was rapid accumulation of nuclei within muscle fibers. This all changed with the introduction of tissue and cell culture and the development of cytological analytic techniques. There were few contributions of greater importance to the study of myogenesis, and for that matter, biological sciences than those that permitted in vitro cultivation of tissue and cells. While the microscope had permitted a detailed view of the appearance of the muscle fiber in both embryonic and the adult muscle, opportunities to revisit muscle regeneration and fiber nucleation in both adults and embryos languished until Ross Harrison (Harrison, 1907) reported observations on living embryonic tissue under the microscope in 1907. Following his initial reports there was a rapid emergence of publications as techniques for direct observation of embryonic tissue in culture. It is reported that just before Harrison published his observations on tissue culture, at age 37 he was lured from Johns Hopkins Medical School to a professorship at Yale. The inducement was a zoology department in a new biological sciences building that was completed in 1907. It was in these laboratories that Harrison first described "tissue culture" – a technique for direct observation of living tissue behavior outside the animal.

Harrison was not focused on muscle formation, but on another developmental question for which the technique was well suited – how do the axons of neurons form. Were these very long processes the outgrowth of a single cell, a neuron, or did they coalesce from the environment of the cell? By being the first to incubate pieces of spinal cord of the developing frog in a lymph fluid clot on a glass coverslip suspended over a slide with a depression, he could observe that the axon actually grew from the cell body of the neuron. While Harrison deserves credit for the new techniques of tissue culture, there is little in sciences for which there are not earlier contributions important to the new observations. A renowned physician working at University College, London, Sydney Ringer (Ringer, 1882) had earlier developed a method of the incubation of a beating frog heart in total isolation from the animal, by the development of a suitable salt solution – Ringer's solution – subsequent modifications include Tyrode's, Locke's (Locke, 1895), and Hank's solution. Locke had added dextrose to Ringer's salt solution because he noted that the isolated heart stopped beating unless he added dextrose to Ringer's solution. These solutions, particularly Tyrode's' solution, became the preferred salt solution in early cell culture. Ringer died of a stroke at age 75 just 3 years after Ross Harrison developed the method for cultivation of pieces of organs rather than the organ itself. Judith Schiff (Schiff, 2002) reports that "Twice Harrison was seriously considered for the Nobel Prize. In 1917 the Nobel committee recommended him for science's greatest honor, but due to the World War, a prize was not awarded in his field. In 1933 he was one of two finalists, but because the full value of tissue culture was not yet appreciated, the Nobel went to geneticist Thomas Hunt Morgan."

Thus it was tissue culture, most often spoken of as organ culture in embryonic parlance, which provided a method for the study of living skeletal muscle tissue. While many quickly applied and modified the technique, it was Alex Carrel at the Rockefeller Institute, who for decades most influenced the field of tissue culture (Carrel and Burrows, 1911). However, it was Margaret and Warren Lewis in 1917 who first extensively observed the behavior of muscle tissue in vitro using the techniques of Carrel (Lewis and Lewis, 1917). For a number of years Carrel and Burrows focused on technical aspects to improve tissue culture demonstrating that a plasma clot permitted outgrowth of cells that could be mechanically serially transferred to other culture dishes – the Carrel flask. The plasma clot was the centerpiece of this method. It may have been Burrows who was responsible for the use of a plasma clot for he had gone to Yale to observe Harrison's tissue culture method. Harrison used the lymph of a frog, a cold blooded animal, to make a clot in which to place the tissue, and Burrows needed something that would clot for tissues from warm blooded animals. When he returned to the Carrel laboratory plasma clots became the substrate for cell culture. Placing a piece of embryonic chick muscle in Locke's solution on a glass substrate coated with a plasma clot by the method of Carrel and Burrows the Lewises observed what we observe today – fibers appear to extend from the tissue mass. They were particularly impressed that the tissue culture model strikingly resembled muscle regeneration seen in vivo, suggesting that the tissue cultures displayed essentially the same process as seen in regeneration. As expected, they saw no mitotic figures within the muscle buds concluding that it was "probable that nuclear division does take place." They reported that there was "Abundant outgrowth of skeletal muscles of chick embryos...by means of tissue cultures in Locke's solution...". They postulated a process of dedifferentiation of the old fibers leading to the formation of muscle buds. Thus their observations tended to support the long held belief that it was the cut ends of fibers that sent out processes that become multinucleated. While these direct observations did not appear to move beyond what had been observed in fixed materials at earlier times, the Lewises surmised that either amitosis or cell fusion could explain myogenesis.

It was the extension of the tissue culture approach to cell or monolayer culture that was central to our current understanding of the biology of myogenesis. Over several decades Strangeways (Strangeways and Fell, 1926), and subsequently Fell (Fell, 1972), perfected organ culture of embryonic tissues using the techniques developed by Carrel. But it was another thirty years before individual cells or monolayer culture techniques were introduced. The latter refinements in tissue culture techniques mostly came from investigators who were interested in cancer and virology, not from those interested in embryogenesis. Foremost among the investigators that focused on cell culture rather than tissue culture were Earle and Sanford (Earle et al., 1951), Puck and Marcus (Puck and Marcus, 1955), and Gey (Gey, 1954). Near the end of the second World War, using mechanical methods to dissociate minced tissue to release single cells, and other approaches such as removing an explant and leaving behind outgrowth from the explant, as Carrel had done, the first cell culture were performed using embryonic chick tissues and tumor tissue. Contamination was a common problem with tissue culture, as no antibiotics were available. Whether related to the ending of the Second World War, the discovery of antibiotic, or something else, there was a virtual explosion in techniques of cell culture in the 1940's and 1950's. At that time dissociation of tissues for cell culture most commonly employed mechanical dissociation of tissue, or use of low calcium salts or chelating agents (Zwilling, 1954), or high pH (Holtfreter, 1943). These techniques were harsh, so a method was needed that produced high cellular yields but did not damage cells. The answer lay in proteolytic enzymatic digestion of minced embryonic tissues. It is not entirely clear who should receive credit for developing this technique. Rous and Jones (Rous and Jones, 1916) were at the Rockefeller Institute working with the first tissue culture chambers that their colleague, Carrel had developed. They wanted a method for transferring the outgrowth of tissue explants from one Carrel flask to another. They showed that this could be done with trypsin dissociation of the plasma clot that freed up these cells for transfer. They did not proposed dissociating tissues or embryos to introduce them into culture as a monolayer of cells but they did develop the ideas of "subculture." It appears to me that is was Aaron Moscona (Moscona, 1951) working in the Strangeways Laboratories as a post doctoral fellow with Honor Fell in 1951 who proposed using proteolytic enzymes to make a single cell suspension of tissues and organ rudiments for cell culture. He used trypsin and hyluronidase to digest chick embryonic limb buds and obtain a suspension of single cells. Moscona was a young embryologist whose intention was not to develop a method of cell culture, but to obtain cell suspensions of various embryonic organs as a means to studying cell reaggregation and tissue formation in vitro. However, in his first publication of this method he plated cell suspensions derived from limb buds on a plasma clot attached to a cover slip building upon the approach Honor Fell had perfected from

Harrison for organ culture or tissue culture of embryonic rudiments. The photos he published in 1951 undoubtedly were of myoblasts admixed with chondrocytes of the limb bud. Thus the technique of monolayer cell culture used for nearly of all cell types today was first demonstrated on muscle of the chick embryo.

There was divergence in the nature of work of those interested in cancer and those interested in embryonic development, with the cancer biologist developing new techniques of culture of tumor cells and the developmental biologist refining the techniques introduced by Harrison and Carrel to investigate embryonic tissues. Dulbecco (Dulbecco, 1952) citing Evans and colleagues (Evans et al., 1951), wished to have a monolayer of animal cells as a quantitative basis for his pioneering work on viral replication. He sought a monolayer of susceptible cells to demonstrated plaque formation by viruses. Dulbecco published a paper within a year of Moscona's where he mechanically shredded 10-day chick embryo and then trypsinized the debris to form 'chick embryonic fibroblast' monolayers. Neither author cited the other. Little did Delbecco know that by dissociating 10-day chick embryos and plating them on Pyrex glass, that the suspension of cells contained mostly myoblasts. Gey had established a cancer cell line, the HeLa cell line, in 1943 using mechanical dissociation and this cell line quickly became widely used in cell culture experiments. Earle and Puck and others were performing nutritional and radiobiological studies on these and other cancer cell lines before trypsinization was used. Tissue culture (cell culture) had become such a prominent part of biological experimental work at that time that the American Tissue Culture Association was founded in 1946. By 1949 there were commercial sources of tissue culture media.

The focus of the cancer cell culturists was working out conditions of optimal cell proliferation whereas the embryologists were interested in culture techniques that fostered cell differentiation. In fact most embryologists thought cell proliferation and differentiation were antagonistic processes. Earle recognized that complex media consisting of serum, cell extracts and salt solutions needed to be "conditioned" by cells to optimally support growth (Earle et al., 1951). Sanford, Earle and their colleagues (Sanford et al., 1948) first cloned a single mammalian cell in vitro but it was a complex process, requiring placement of a single cancer cell within a small capillary tube containing conditioned medium. It was not possible to easily clone cells until Puck and colleagues (Puck et al., 1956) recognized that the problem of conditioning could be circumvented by using irradiated monolayers of cells as a substrate on which to plate small numbers of cells – thus his group was the first to readily establish clonal growth albeit with the HeLa cancer cell line developed by Gey in 1943. Another decade was to pass before Richard Ham (Ham, 1965) developed completely defined media that are still used today and are a component of muscle cell culture medium, although serum is still required for myogenesis in cell culture and in myogenic cell cloning. Ham was also first to clone a cell using a completely defined medium without the addition of serum. Konigsberg (Konigsberg, 1963) using these newer techniques of cloning was the first to show that a single myoblast could produce a muscle fiber, though the process required the addition of serum and conditioned medium. Konigsberg had had an appointment as an assistant professor at the University of Colorado until 1958 where undoubtedly he knew of Puck's successful work done at that time on cloning cancer cells because Puck was also at the University of Colorado. Konigsberg moved to the National Institutes of Health in 1958 where Earle and his colleagues were using conditioned media to clone tumor cells. As with Earle it was the conditioned medium that permitted Konigsberg to clone an embryonic myoblast.

Those interested in embryological tissue culture continued to use modifications of Alex Carrel's methods well into 1950's, perhaps because cells differentiated well in such cultures. On the other had the cancer cell tissue culturists were critical of Carrel's methods because there was limited growth of tumor tissue as an explant and quantitation was difficult with this technique. Carrel had added embryo extract to what previously was a salt solution such as Locke's or Tyrode's solution and noted that the cells which otherwise did not proliferate began to do so and that embryo extract also enhanced viability. Embryo extract has remained a common component of tissue culture media since that time. Carrel and Burrows use of the plasma clots and embryo extract plus a salt solution became the basis for nearly all embryological tissue culture in the 1920's through the early 60's. This is illustrated by the early work from Holtzer's (Holtzer et al., 1960), Konigsberg's (Konigsberg et al., 1960) and Grobstein's (Grobstein, 1953) laboratories where the media developed by Carrel and the use of trypsin were employed. Grobstein immediately picked up on the report by Moscona and using "very early rudiments of the mouse submandibular salivary gland, kidney, and lung"... And demonstrating "normal morphogenesis at the glass-clot interface in Carrel flasks." He remarked, "that each of these rudiments can be separated into viable epithelial and mesenchymal components by exposure to trypsin as has been described for chick rudiments by Moscona" and by so doing Grobstein initiated a re-emergence of studies on embryonic induction.

2. THE "PRE-MODERN" ERA

Within a 3 to 4-year period, from 1958 to 1961 there was unprecedented progress and excitement in unraveling the questions posed so long ago in the field of myogenesis, particularly in the study of myogenesis in the embryo. Surprisingly, only slowly did the new concepts and observation that were swiftly forthcoming from studies of embryonic myogenesis get transferred to ongoing work in muscle regeneration. However, a pivotal discovery was the observation by Hewson Swift (Lash et al., 1957) that the nuclei of regenerating anterior tibialis muscle of young adult mouse muscle were always diploid. Swift was an expert and developer of microspectrophotometric measurement of DNA using Fuelgen staining, a technique that made it possible to measure DNA content within a single nucleus. Teaming with Jay Lash and Howard Holtzer, who had been exploring the question of nucleation of muscle fibers in regeneration, Swift found that the nuclei in regenerated fibers were always diploid. This seemed to make untenable any explanation of muscle fiber nucleation during regeneration that required DNA replication. If amitosis or splitting of nuclei were the explanation these would have required DNA synthesis, and as there was relative uniformity in the signal measured in each nucleus, this was unlikely. They found no evidence of amitosis. Never the less, in their publication they still concluded "Since we ascribe a key role to the mononucleated basophilic cells during regeneration, it is necessary to consider their mode of origin. The injury clearly caused early degeneration of some muscle nuclei, and the enlargement, and apparent activation of others. Although critical evidence is lacking, it is possible that these activated muscle nuclei with their surrounding cytoplasm became mobilized into the regenerating area." "It is tentatively proposed that the accumulation of the centrally placed nuclei is the result of mobilization, not extensive proliferation within the regenerating myotube." They did not anticipate that the nuclei did in fact come from fibers, that is, from satellite cells on the surface of the fiber rather than from nuclei within the fiber.

An important technique introduced in the mid 1950's was that of autoradiography. Herbert Taylor was aware that thymidine had been shown to be a precursor for DNA synthesis. He reasoned that if he could covalently link a radioactive isotope to thymidine he would be able to investigate mitosis and DNA synthesis at the single cell level in tissue culture (Taylor, 1953). While first he used P^{32} to label the thymidine, he realized that a less energetic isotope was needed. So he synthesized tritiated thymidine in his laboratory and exposed English broad bean seedlings to tritiated thymidine of reasonable high specific activity (Taylor et al., 1957). He showed that he could identify incorporation into single chromosomes with excellent resolution. Within a year of this report tritiated thymidine was commercially available.

The experiments that clinched the conclusion that neither amitosis, nor some cryptic forms of nuclear splitting occur in myogenesis, were to those that demonstrated that DNA synthesis did not occur within any nuclei of a embryonic skeletal muscle fiber. I arrived as a graduate student at the University of Pennsylvania in 1958 and began to work with Howard Holtzer and Jay Lash who were experimenting with chick embryonic myogenic cell culture using techniques modified from Carrel and Fell. While Lash worked principally with organ culture, Abbott and Holtzer (Abbott and Holtzer, 1966; Holtzer et al., 1960) had implemented the technique of Carrel of forming a plasma clot on cover slips for study of muscle cell and chondrocytes in cell rather than tissue or explant culture. Adult chickens were bled; their plasma was isolated, and if one were quick it was possible to get a clot to form evenly on the surface of a coverslip by mixing a drop plasma with a drop of extract of chick embryo. The plasma clot formed a semi-ridged substrate on which we could plate a suspension of cells derived by trypsinization of 11-day chick breast muscle. The medium was two parts salt solution, two parts serum, and one part extract of 11-day chick embryo. The most robust myotubes one has ever seen form on the surface of a plasma clot. Using autoradiography with tritiated thymidine I was able to show that the nuclei of mononucleated myoblasts incorporated the DNA precursor, while nuclei within myotubes were never labeled (Stockdale and Holtzer, 1961). On the other hand, the nuclei of myoblasts labeled in the culture prior to the appearance of myotubes, readily appeared several days later within the myotubes that subsequently became plentiful in the cultures. At this same time Irv Konigsberg was working on monolayer cultures of myoblasts. Using nitrogen mustard to treat cultures of embryonic avian myotubes, Konigsberg (Konigsberg et al., 1960) reasoned that this antiproliferative agent should kill myotubes if their nuclei were proliferating. He found that the mononucleated cells were killed by nitrogen mustard, whereas the myotubes in the culture persisted suggesting that mononucleated cell nuclei proliferated, but myotube nuclei did not. Thus it should have been apparent to all that DNA synthesis did not occur in muscle cell nuclei either in vivo or in vitro and that cell fusion was the most likely explanation. A short time later Cooper and Konigsberg (Cooper and Konigsberg, 1961) used time-lapse photography of cultures of myoblasts, a technique introduced by Hughes and Hughes in 1948, to show that one could observe cell fusion, clearly bringing to a close the question of how multinucleated fibers form in embryonic muscle cell culture. Though they did not observe cell fusion, Mintz and Baker (Mintz and Baker, 1967), using the allophenic mice technique Mintz had developed, deduced that cell fusion is also the most likely mechanism for multinucleation of muscle fibers formed in vivo. They mixed the blastomeres from blastulae of mice with allelic variants of isozymes of isocitrate dehydrogenase that differed in electrophoretic mobility. When reconstituted blastulae consisting of blastomeres of the two different strains were transplanted in surrogate mothers, the muscles of the new born mice contained hybrid enzymes composed of the subunits from both alleles, a result explained best by two different nuclei within a common cytoplasm. They reported that muscle was the only tissue in which an exchange of subunits could be detected – thus cell fusion must have occurred in the developing muscle among the cellular progeny of the two different types of blastomeres.

In the same year that we reported that fusion must be responsible for myogenesis, Alex Mauro published a one and a half page, one figure paper (Mauro, 1961) demonstrating cells that lay on the surface of adult frog muscle fibers beneath the basal lamina. These he called satellite cells. This discovery should have turned the field of muscle regeneration around. Mauro postulated that such cells could be responsible for muscle regeneration. For reasons that are not clear, but perhaps because satellite cells were few in number on the surface of muscle fibers, or could only be reliably demonstrated using an electron microscope, this discovery was not recognized for the importance it merited in the field. Though the discovery of satellite cells was reported in 1961 at the same year as fusion in embryonic muscle had been demonstrated, there was skepticism of the role of satellite cells in muscle regeneration for nearly a decade. In a report in 1965 (Gilbert and Hazard, 1965) Gilbert and Hazard said, “In particular it is clear that many new muscle fibers are formed in injured muscle and that these can be derived from mononuclear precursor cells. These multiply by mitosis and fuse to yield myotubes which then develop into mature muscle fibers. The source of the mononuclear muscle cell precursors is still uncertain. The idea that an undifferentiated, as it were embryonic, satellite precursor cell can persist through adult life, lying between plasma and basement membranes of the muscle fiber, has gained wide acceptance; but it has not entirely superseded the view that myoblasts can arise by segregation of differentiated myonuclei. It remains possible too that local connective tissue cells and, again, circulating cells may have accessory roles in myogenesis." It was not until nine years after Mauro published his observation, that the words "satellite cell" can be found in the title of any published paper, while for the last decade there is on average at least one such publication every month. Some 17 years later, in a review of muscle regeneration Sloper and Partridge (Partridge et al., 1978) concluded that "that satellite cells in young rats can differentiate into multinucleated myotubes following muscle injury." In large measure a work that firmly established the role of satellite cells in the regenerative process was published by 1978 (Snow, 1978). Using autoradiography with tritiated thymidine, satellite cells were traced during the regeneration of skeletal muscle in young Sprague-Dawley rats. They state, "Satellite cells in uninjured muscles appeared labeled after injections of tritiated thymidine; none of the myonuclei were labeled in the same muscles. Four to six days after transplanting the radioactive muscles to non-radioactive littermates, regenerating myotube nuclei in the host appeared labeled." Thus, this study demonstrated that satellite cells could differentiate into multinucleated myotubes following muscle injury. Some might point out that there could have been cells in the interstitium of the transplants that became labeled and contributed to fiber nuclei. However, it is unlikely that many such cells would have been replicating DNA within uninjured muscle and, therefore, would have become labeled. So, until about 1975–1980 it really was not broadly accepted that satellite cells were the sole source of nuclei in regenerated muscle fibers.

Two notable achievements for understanding myogenesis and muscle regeneration that emerged from cell culture were the development of single fiber culture and muscle cell lines. Richard Bischoff (Bischoff, 1975) and Ursula Kongisberg and colleagues (Konigsberg et al., 1975) developed a method for the isolation of intact single fibers from the rat or quail which when placed in culture permitted these and many other investigators to dissect the function of satellite cells that lay on the surfaces of these isolated fibers. They demonstrated that satellite cells could be activated by injury in vitro just as they could by injury in vivo. The technique is widely used today. Earlier David Yaffe (Yaffe, 1968) used cell culture techniques to develop the first myogenic cell lines – these were from the rat, the so called "L" lines and subsequently from the mouse – the C2 line. The "C2" meaning these were derived from the second control mouse in a set of experiments the goal of which was to develop a cell line of dystrophic muscle from the a mouse strain carrying the *dy* gene a recessive gene for a dystrophy similar to human muscular dystrophy (Yaffe, 1977). Though thwarted in this goal, the result was extremely important as the control cell line, the C2 myogenic cell line, was to become central in studies of biochemical and molecular genetic aspects of myogenesis in vitro.

As the mechanisms for embryonic myogenesis and regeneration became clear, reports appeared that myosin was not a single protein, but probably existed in several isoforms in nearly every species. Perry and Chappell (Perry and Chappell, 1957) reported that the physical properties of this contractile protein, isolated from adult muscles of differing functional type, were not the same. Using purified myosin extracted from a variety of muscles and species as diverse as the sloth and the squid, Bárány (Bárány, 1967) showed muscles were of "slow" and "fast" contractile types based upon corresponding "fast" and "slow" myosin-ATPase activity of the myosins they contained. Likewise histochemical staining initially perfected by Helen Padykula and Edith Herman (1954), from earlier approaches (Glick and Fisher, 1945), showed that cross sections of muscle had varying amount of ATPase staining over a range of pH indicating that the myosins of differing types were present within adjacent fibers. Thus the concept of fiber type emerged. John Marshall is extremely important in developing a technique that refined our understanding of fiber type. John Marshall (Marshall, 1951) developed a method for linking antisera to fluorescein isocyanate for the purpose of identifying native antigens within cells, which he first applied to identifying adrenocorticotropic hormone within pituitary cells. The approach was based on the earlier work of Coons and colleagues on using fluorescent conjugated antibodies to identify foreign antigens, such as products of infectious agents of bacteria or viruses within cells (Coons and Kaplan, 1950). Holtzer, and Finck produced an antiserum to myosin from extracts of chicken muscle without considerations of fiber type and in collaboration with Marshall (Finck et al., 1956; Holtzer et al., 1957) they stained the muscle fibers of the chick myotome using this antiserum conjugated with a fluorescent dye. Thus began immunohistochemistry of muscle. The contribution of this group and Coons and Marshall in particular, can not be over stated because immunohistochemistry has been instrumental for progress not only in analyses of myogenesis, but also biomedical science and medical practice in general.

With the success of muscle cell culture there was rapid application of emerging molecular approaches to the study myogenesis. Patterson and Strohman (Paterson and Strohman, 1970; Paterson and Strohman, 1972), Yaffe (Shainberg et al., 1971), and Stockdale and O'Neill (Stockdale and O'Neill, 1972) were among the first to use cultures to study the synthesis of myosin and its relationship to the process of cell fusion. In addition Coleman (Coleman and Coleman, 1968), Emerson (Emerson and Beckner, 1975), Nadal-Ginard (Nadal-Ginard, 1979), and Bandman (Bandman et al., 1981) published biochemical methods studying muscle protein expression within myogenic cells differentiating in cell culture. Introduction of hybridoma technology to produce monoclonal antibodies to myosin occurred in the laboratories of Fischman (Bader et al., 1982), Stockdale (Crow and Stockdale, 1986a) and Bandman (Bandman, 1985a). Coupled with John Marshall's technique these were to provide a more precise assessment of the diversity among muscle fibers than had previous been possible and to become the basis for recognition of cell lineages in myogenesis. Based on biochemical evidence Hoh (Hoh et al., 1976), Whalen (Whalen et al., 1979; Whalen et al., 1981), Rushbrook and Stracher (Rushbrook and Stracher, 1979) proposed that there were not only different isoforms of myosin heavy chain, but specific developmental patterns to the expression of these isoforms as well. It soon became apparent that there was also a temporal pattern to the expression of myosins in regenerating muscle of adults as well. Work in the

Strohman (Matsuda et al., 1983) and the Schiaffino (Gorza et al., 1983; Sartore et al., 1982) laboratories demonstrated patterns of expression in regenerating muscle of adult animals that included "embryonic", "fetal", as well as "adult" forms of myosin, reinforcing the conceptual connection between embryonic development and regeneration of muscle. In chickens the patterns of myosin expression were even more complex if one analyzed myosin expression in specific anatomic muscles using immunopeptide mapping with monospecific antibodies to fasts and slow myosin heavy chain (Bandman, 1985b; Crow and Stockdale, 1986a). Adult chicken muscles differ markedly from one another in the pattern of myosin expression that cannot be explained by sequential transitions from embryonic to fetal, to adult myosins, for there are adult muscles in which the embryonic form is the predominate "adult" form. There is no single pattern to the temporal expression of the myosin heavy chain within the avian muscles.

That there was a temporal sequence of myosin expression that was similar in both the embryo and in the regenerating muscle was also demonstrated in embryonic myoblast and satellite cell culture. Miller and Stockdale (Miller et al., 1985; Miller and Stockdale, 1986a; Miller and Stockdale, 1986b) and Feldman and Stockdale (Feldman and Stockdale, 1988) showed that myoblasts of the chick embryo (day 3 to 8 of chick development); fetus (day 8 to day 12 of development) as well at adult (day 12 of fetal development through adulthood) formed muscle fibers that expressed different isoforms of myosin. The work was based on the earlier observations of the Hauschka laboratory (Hauschka, 1974; White et al., 1975; White and Hauschka, 1971), where extending the cloning work of Kongsberg (Konigsberg, 1963), they described morphologically distinctive types of clones that formed in culture from cell suspensions of limb muscles of the chick, mouse, and human at specific stages of development. The early clones, those from embryonic stages, formed fibers with few nuclei while clones from fetal and adult muscles formed fibers with many nuclei. Some types required conditioned medium for formation. Miller et al. (1985) found that clones from embryonic muscle were of two predominant myosin-expressing types – one expressing fast and slow isoforms and one expressing only fast isoforms. Clones from the fetal stages formed fibers of single type – only fast myosin expressing fibers. Satellite cell clones appeared between day 12 and 14 of chick development and depending upon whether they were derived from fast or slow muscle rudiments, formed fibers of either the fast or fast/slow myosin-expressing type (Feldman et al., 1992; Feldman and Stockdale, 1988). Satellite cells from adult rabbit fast or slow muscles were also shown to form fibers in culture in the absence of innervation that correspondingly expressed fast or slow isoforms of MyHC (Düsterhoft et al., 1990). Schiaffino and his colleagues (Kalhovde et al., 2005) extended these observations on satellite cells to the in vivo setting. They observed that regenerating muscle in the presence or absence of innervation expressed myosins characteristic of the muscle that was injured, suggesting as had Hoh and Hughes (Hoh and Hughes, 1988) that in vivo satellite cells are autonomous or committed to form fibers of specific types. In single-fiber cultures performed as Bischoff had done, where satellite cells are derived from a specific fiber type, Rosenblatt and colleagues (Rosenblatt et al., 1996) showed as well that in such cultures the satellite cells that migrated from slow muscle fibers formed predominately slow muscle fibers in culture. There have been a number of other investigators who have confirmed these observations in birds, fish, mammals, and humans (Devoto et al., 1996; Ghosh and Dhoot, 1998a; Ghosh and Dhoot, 1998b; Pin et al., 2002; Pin and Merrifield, 1993). Thus, whether in the avian embryo or in the avian and mammalian adult, mononucleated muscle-fiber precursor cells are of differing differentiate types. While there appears to be widespread confirmation that satellite cells in birds and mammals can have autonomous functions when forming fibers, there is less agreement that this is the case with mammalian embryonic and fetal myoblasts (Blau and Hughes, 1990; Rubinstein and Kelley, 1978). To some degree this may reflect difference of technique and approaches to the question of muscle lineages.

These studies in birds and in mammals in vivo and in vitro are refinements in understanding of the earlier findings on embryonic myogenesis and muscle regeneration. They suggest that myogenesis in these two contexts is more complex than merely understanding multinucleation by cell fusion because intrinsic differences exist in myoblasts and satellite cells in the setting of normal development and muscle regeneration. Because studies in cell culture by definition exclude effects of innervation, it was assumed that innervation played no important role in influencing myosin heavy chain expression in fibers in culture or regeneration. This conclusion was supported by the observation that muscles of differing myosin expressing types form normally in embryos in vivo where from the onset of embryonic development functional innervation of developing muscles is completely blocked (Crow and Stockdale, 1986b), an observation supported by Duxson and Harris in mammals (Duxson et al., 1989). Specific anatomic muscles and their constituent fibers types formed normally when curare was injected into fertilized eggs during the earliest stage of embryonic development. Thus it appeared that the embryonic stages of fiber-type determination did not required innervation. But it was noted that at fetal stages of development the fibers did not grow when innervation was blocked. So innervation appeared to be important once fetal stages of avian development were reached. It had been noted earlier (Miller et al., 1985; Miller and Stockdale, 1986a) that fetal muscles contained myoblast that when cloned expressed only fast isoforms of myosin whether or not the myoblasts were isolated from fetal fast or slow type muscles. But this was confusing because at later stages of fetal development (Feldman and Stockdale, 1990) there were differences in myosin expressed in fetal myoblast clones. These differences in the fate of myoblasts in the early as opposed to the late fetus were thought to be based on satellite cells becoming the predominate mononucleated myogenic population in mid- to late fetal development. However, DiMario and Stockdale (DiMario and Stockdale, 1997) found it was not true that early fetal stage myoblasts formed only fast-myosin expressing fibers. If co-cultured with a source of motor neurons (spinal cord), neural outgrowth as described by Harrison occurred, and as myotubes formed in vitro from early fetal myoblasts these axons formed motor endplates on the surface of these newly formed fibers.

Dependent on the anatomic source of the early fetal myoblasts, the innervated myotubes expressed either fast or slow myosin heavy chains (MyHC) just as the fibers would have had they developed from these myoblasts in vivo (DiMario and Stockdale, 1997). Thus early fetal myoblasts from the fast pectoralis major formed fibers that expressed only fast myosin isoforms when formed in cell culture whether or not they were innervated, whereas those myoblasts from a slow muscle such as the medial adductor express slow isoforms of myosin heavy chain as well when innervated, but when not innervated, only expressed fast myosin heavy chains. In the embryo myogenesis proceeds independently of innervation, whereas in fetal development myogenesis is innervation dependent (Stockdale and Miller, 1987).

This work was extended to the fibers of the myotome (Sacks et al., 2003). Myotomal muscle fibers appear to be autonomous with regard to myosin expression because myosin begins to be expressed before innervation occurs in the myotome of the intact developing embryo. As the myotome develops and innervation occurs both slow and fast isoforms are expressed. However, this expression occurs even if the functional innervation of the myotome is blocked. But as with fetal myogenesis there are aspects that are innervation dependent. Only embryonic slow MyHC 1 is expressed in the absence of innervation where as the transition to expression of slow MyHC 2 only occurs if innervation of the myotomal fibers is permitted. Thus the generalization that emerges is that myoblasts as well as satellite cells are cells committed to express certain isoforms to myosin heavy chain, expression that earlier investigators have established is responsible for fiber type and the speed of contraction. This expression occurs independently of environmental factors, but among the repertoire of MyHC genes there are some that required motor innervation of a fiber for transcription and specific myogenic cell lineages in which innervation can activate these MyHC genes.

It has been recognized for many years that innervation is not required for muscle regeneration to occur (Kirby, 1892; Clark, 1946; Saunders and Sissons, 1953), but as in the embryo, when it is blocked the regenerate fibers are atrophic. As stated above the regenerate muscle expresses myosins characteristic of the muscle in which regeneration is occurring. There is probably no better example of this than found in the work of Hoh and colleagues (Hoh and Hughes, 1988, 1991) where they showed that regenerated temporalis muscle in ectopic sites within the cat, expressed the unique superfast myosin heavy chain characteristic of the temporalis muscle, suggesting autonomous function specific to temporalis satellite cells. However, the pattern of expression in the regenerate muscle is dependent on the pattern of impulses delivered by the nerve or artificially supplied by electrical stimulation. Thus in all stages of development from the time of neural tube induction of myogenesis within the somite, to innervation of newly formed muscle fibers within the embryo and fetus, to regeneration of adult muscle fibers, the central nervous system influences muscle fiber functional properties which are rooted in difference in fates of the myoblasts and the satellite cells.

The relationship of embryonic development of muscle and muscle regeneration remains as close today as it has been over the last 150 years. While today we converse in terms of gene activation, studies show common regulatory mechanisms for development and regeneration indicating that regeneration and embryonic myogenesis are intertwined and successfully studied together. Current work in both fields is dependent on flow cytometry, the ability to construct transgenic mice, the ability to stain with monospecific antibodies and gene specific probes, transfect cells, silence gene expression, and perform microarrays analysis. However, cell and tissue culture remain central to many current studies on myogenesis. The concepts and many of the systems developed in the last century provide the biological basis for the application of these approaches.

REFERENCES

Abbott J, Holtzer H (1966) The loss of phenotypic traits by differentiated cells. 3. The reversible behavior of chondrocytes in primary cultures. J Cell Biol 28:473–487

Bader D, Masaki T, Fischman DA (1982) Immunochemical analysis of myosin heavy chain during avian myogenesis in vivo and in vitro. J Cell Biol 95:763–770

Bandman E (1985a) Continued expression of neonatal myosin heavy chain in adult dystrophic skeletal muscle. Science 227: 780–782

Bandman E (1985b) Myosin isoenzyme transitions in muscle development, maturation, and disease. Int Rev Cytol 97:97–131

Bandman E, Matsuda R, Micou-Eastwood J, Strohman R (1981) In vitro translation of RNA from embryonic and from adult chicken pectoralis muscle produces different myosin heavy chains. FEBS Lett 136:301–305

Bárány M (1967) ATPase activity of myosin correlated with speed of muscle shortening. J Gen Physiol Suppl 50:197–218

Bischoff R (1975) Regeneration of single skeletal muscle fibers in vitro. Anat Rec 182:215–35

Blau HM, Hughes SM (1990) Cell lineage in vertebrate development. Curr Opin Cell Biol 2:981–985

Carrel A, Burrows M (1911) Cultivation of tissues in vitro and its technique. J Exp Med 13:387–396

Clark WE (1946) An experimental study of the regeneration of mammalian striped muscle. J Anat 80:24–36

Coleman JR, Coleman AW (1968) Muscle differentiation and macromolecular synthesis. J Cell Physiol 72:19–34

Cooper G, Konigsberg I (1961) Dynamics of myogenesis in vitro. Anat Rec 140:195–205

Coons AH, Kaplan MH (1950) Localization of antigen in tissue cells; improvements in a method for the detection of antigen by means of fluorescent antibody. J Exp Med 91:1–13

Crow MT, Stockdale FE (1986a) Myosin expression and specialization among the earliest muscle fibers of the developing avian limb. Dev Biol 113:238–254

Crow MT, Stockdale FE (1986b) The developmental program of fast myosin heavy chain expression in avian skeletal muscles. Dev Biol 118:333–342

Devoto SH, Melançon E, Eisen JS, Westerfield M (1996) Identification of separate slow and fast muscle precursor cells in vivo, prior to somite formation. Development 122:3371–3380

DiMario JX, Stockdale FE (1997) Both myoblast lineage and innervation determine fiber type and are required for expression of the slow myosin heavy chain 2 gene. Dev Biol 188:167–80

Dulbecco R (1952) Production of plagues in monolayer tissue culture by single particles of an animal virus. Proc Nat Acad Sci 38:747–752

Düsterhoft S, Yablonka-Reuveni Z, Pette D (1990) Characterization of myosin isoforms in satellite cell cultures from adult rat diaphragm, soleus and tibialis anterior muscles. Differentiation 45:185–191

Duxson MJ, Usson Y, Harris AJ (1989) The origin of secondary myotubes in mammalian skeletal muscles: ultrastructural studies. Development 107:743–750

Earle WR, Sanford KK, Evans VJ, Waltz HK, Shannon JE Jr (1951) The influence of inoculum size on proliferation in tissue cultures. J Natl Cancer Inst 12:133–53

Emerson CP Jr, Beckner SK (1975) Activation of myosin synthesis in fusing and mononucleated myoblasts. J Mol Biol 93: 431–447

Evans V, Earle W, Sanford KK, Shannon JE, Waltz HK (1951) The preparation and handling of replicate tissue cultures for quantitative studies. J Natl Cancer Inst 11:907–927

Feldman JL, DiMario JX, Stockdale FE (1992) Developmental appearance of adult myoblasts (satellite cells): studies of adult myoblasts in culture and adult myoblast transfer into embryonic avian limbs. In: Limb Development & Regeneration, Part B. (Ed. J.A. Fallon, P.F. Goetinck, R.O. Kelly and D.L. Stocum). John Wiley & Sons, Inc., New York, NY 563–574

Feldman JL, Stockdale FE (1988) Commitment to formation of distinct myotube types in chicken satellite cells. J Cell Bioch 12C:325

Feldman JL, Stockdale FE (1990) Skeletal muscle satellite cell diversity: satellite cells form fibers of different types in cell culture. Dev Biol 143:320–334

Fell H (1972) Tissue culture and its contribution to biology and medicine. J Exp Biol 57:1–13

Finck H, Holtzer H, Marshall JM, Jr. (1956) An immunochemical study of the distribution of myosin in glycerol extracted muscle. J Biophys Biochem Cytol 2(Suppl 4):175–178

Gey GO (1954) Some aspects of the constitution and behavior of normal and malignant cells maintained in continuous culture. Harvey Lect 50:154–229

Ghosh S, Dhoot GK (1998a) Both avian and mammalian embryonic myoblasts are intrinsically heterogeneous. J Muscle Res Cell Motil 19:787–795

Ghosh S, Dhoot GK (1998b) Evidence for distinct fast and slow myogenic cell lineages in human foetal skeletal muscle. J Muscle Res Cell Motil 19:431–441

Gilbert RK, Hazard JB (1965) Regeneration in human skeletal muscle. J Pathol Bacteriol 89:503–512

Glick D and Fisher EE (1945) Scientific apparatus and laboratory methods the histochemical localization of adenosinetriphosphatase in plant and animal tissues. Science 102:429–430

Gorza L, Sartore S, Triban C, Schiaffino S (1983) Embryonic-like myosin heavy chains in regenerating chicken muscle. Exp Cell Res 143:395–403

Grobstein C (1953) Morphogenetic interaction between embryonic mouse tissues separated by a membrane filter. Nature 172:869–870

Ham RG (1965) Clonal growth of mammalian cells in a chemically defined, synthetic medium. Proc Natl Acad Sci U S A 53:288–293

Harrison R (1907) Observations on the living developing nerve fibre. Proc Soc Exp Biol (N.Y.) 4:140–143

Hauschka SD (1974) Clonal analysis of vertebrate myogenesis: III. Developmental changes in muscle-colony forming cells of the human fetal limb. Dev Biol 37:345–368

Hoh JF, Hughes S (1988) Myogenic and neurogenic regulation of myosin gene expression in cat jaw-closing muscles regenerating in fast and slow limb muscle beds. J Muscle Res Cell Motil 9:59–72

Hoh JFY, Hughes S (1991) Expression of superfast myosin in aneural regenerates of cat jaw muscle. Muscle Nerve 14:316–325

Hoh JFY, McGrath PA, White RI (1976) Electrophoretic analysis of multiple forms of myosin in fast-twitch and slow-twitch muscles of the chick. Biochem J 157:87–95

Holtfreter J (1943) Properties and functions of the surface coat in amphibian embryos. J Exp Zool 93:251–323

Holtzer H, Abbott J, Lash J, Holtzer S (1960) The loss of phenotypic traits by differentiated cells in Vitro, I. Dedifferentiation of Cartilage Cells. Proc Natl Acad Sci U S A 46:1533–1542

Holtzer H, Marshall JM, Finck H (1957) An analysis of myogenesis by the use of fluorescent antimyosin. J Biophysic Biochem Cytol 3:705–729

Kalhovde JM, Jerkovic R, Sefland I, Cordonnier C, Calabria E, Schiaffino S, Lomo T (2005) "Fast" and "slow" muscle fibres in hindlimb muscles of adult rats regenerate from intrinsically different satellite cells. J Physiol 562:847–57

Kirby E (1892) Experimentelle Untersuchungen uber die Regeneration des quergestreiftem Muskelgewebes. Beit zur Pathol Anat und Allegem Pathol 11:302–119

Konigsberg IR (1963) Clonal analysis of myogenesis. Science 140:1273–1284

Konigsberg IR, McElvain N, Tootle M, Herrmann H (1960) The dissociability of deoxyribonucleic acid synthesis from the development of multinuclearity of muscle cells in culture. J Biophys Biochem Cytolo 8:333–343
Konigsberg U, Lipton H, Konigsberg IR (1975) The regenerative response of single mature muscle fibers isolated in vitro. Dev Biol 45:260–275
Lash J, Holtzer H, Swift H (1957) Regeneration of mature skeletal muscle. Anat Rec 128:679–697
Lewis W, Lewis M. (1917) Behavior of cross striated muscle in tissue cultures. Amer J Anat 22:169–194
Locke FS (1895) Towards the ideal artificial circulating fluid for the isolated frog's heart: preliminary communication. J Physiol 5: 332–333
Marshall JM (1951) Localization of adrenocorticotropic hormone by histochemical and immunochemical methods. J Exp Med 94:21–30
Matsuda R, Spector DH, Strohman RC (1983) Regenerating adult chicken skeletal muscle and satellite cell cultures express embryonic patterns of myosin and tropomysin isoforms. Dev Biol 100:478–488
Mauro A (1961) Satellite cell of skeletal muscle. J Biophys Biochem Cytolo 9:493–495
Miller JB, Crow MT, Stockdale FE (1985) Slow and fast myosin heavy chain content defines three types of myotubes in early muscle cell cultures. J Cell Biol 101:1643–1650
Miller JB, Stockdale FE. (1986a) Developmental origins of skeletal muscle fibers: clonal analysis of myogenic cell lineages based on expression of fast and slow myosin heavy chains. Proc Natl Acad Sci U S A 83:3860–3864
Miller JB, Stockdale FE (1986b) Developmental regulation of the multiple myogenic cell lineages of the avian embryo. J Cell Biol 103:2197–2208
Mintz B, Baker WW (1967) Normal mammalian muscle differentiation and gene control of isocitrate dehydrogenase synthesis. Proc Natl Acad Sci U S A 58:592–598
Moscona A (1951) Tissues from dissociated cells. Exp Cell Res 3:535–539
Nadal-Ginard B (1979) Most myosin heavy chain mRNA in L6E9 rat myotubes has a short poly (A) tail. Proc Natl Acad Sci U S A 76:1853–1857
Partridge TA, Grounds M, Sloper JC (1978) Evidence of fusion between host and donor myoblasts in skeletal muscle grafts. Nature 273:306–308
Paterson B, Strohman RC (1970) Myosin structure as revealed by simultaneous electrophoresis of heavy and light subunits. Biochemistry 9:4094–4105
Paterson B, Strohman RC (1972) Myosin synthesis in cultures of differentiating chicken embryo skeletal muscle. Dev Biol 29:113–138
Perry SV, Chappell JB (1957) The action of 2:4-dinitrophenol on myosin and mitochondrial adenosine triphosphatase systems. Biochem J 65:469–476
Pin CL, Hrycyshyn AW, Rogers KA, Rushlow WJ, Merrifield PA (2002) Embryonic and fetal rat myoblasts form different muscle fiber types in an ectopic in vivo environment. Dev Dyn 224:253–266
Pin CL, Merrifield PA (1993) Embryonic and fetal rat myoblasts express different phenotypes following differentiation in vitro. Dev Genet 14:356–368
Puck T, Marcus P (1955) A rapid method for viable cell titration and clone production with HELA cells in tissue culture: The use of X-irradiated cells to supply conditioning factors. Proc Natl Acad Sci U S A 41:432–437
Puck TT, Marcus PI, Cieciura SJ (1956) Clonal growth of mammalian cells in vitro; growth characteristics of colonies from single HeLa cells with and without a feeder layer. J Exp Med 103:273–283
Ringer S (1882) Regarding the action of hydrate of soda, hydrate of ammonia, and hydrate of potash on the ventricle of the Frog's Heart. J Physiol 3:195–202
Rosenblatt J, Parry D, Partridge T (1996) Phenotype of adult mouse muscle myoblasts reflects their fiber type of origin. Differentiation 60:39–45
Rous P, Jones FS (1916) A method for obtaining suspensions of living cells from the fixed tissues, and for the plating out of individual cells. J Exp Med 23: 549–555
Rubinstein NA, Kelley AM (1978) Myogenic and neurogenic contributions to the development of fast and slow twitch muscles in rat. Dev Biol 62:473–485

Rushbrook JI, Stracher A (1979) Comparison of adult, embryonic, and dystrophic myosin heavy chains form chicken muscle by sodium dodecyl sulfate/polyacrylamide gel electrophoresis and peptide mapping. Proc Natl Acad Sci U S A 76:4331–4334

Sacks L, Cann G, Nikovits W, Conlon S, Espinoza N, Stockdale F (2003) Regulation of myosin expression during myotome formation. Development 130:3391–3402

Sanford K, Earle W, Likely G (1948) The growth in vitro of single Tissue cells. J Natl Cancer Inst 9:229–246

Sartore S, Gorza L, Schiaffino S (1982) Fetal myosin heavy chains in regenerating muscle. Nature 298:294–296

Saunders JH, Sissons HA (1953) The effect of denervation on the regeneration of skeletal muscle after injury. J Bone Joint Surg Br 35-B:113–24

Schiff J (2002) Old Yale – An Unsung Hero of Medical Research. Yale Alumni Magazine Vol. 64, February

Schwann T (1839) Mikroskopische Untersuchungen über die Übereinstimmung in der Struktur und dem Wachsthum der Thiere und Pflanzen. *Berlin: Verlag der Sander'schen Buchhandlung*

Shainberg A, Yagil G, Yaffe D (1971) Alterations of enzymatic activities during muscle differentiation in vitro. Dev Biol 25:1–29

Snow MH (1978) An autoradiographic study of satellite cell differentiation into regenerating myotubes following transplantation of muscles in young rats. Cell Tissue Res 186:535–540

Stockdale FE, Holtzer H (1961) DNA synthesis and myogenesis. Exp Cell Res 24:508–520

Stockdale FE, Miller JB (1987) The cellular basis of myosin heavy chain isoform expression during development of avian skeletal muscles. Dev Biol 123:1–9

Stockdale FE, O'Neill MC (1972) DNA synthesis, mitosis, and skeletal muscle differentiation. In Vitro 8:212–227

Strangeways T, Fell H (1926) Experimental studies on the differentiation of embryonic tissues growing in vivo and in vitro. Proc Royal Soc, London 99:340–366

Taylor JH (1953) Intracellular localization of labeled nucleic acid determined with autoradiography. Science 118:555–557

Taylor J, Woods P, Hughes W (1957) The organization and duplication of chromosomes as revealed by autoradiographic studies using tritium-labeled thymidine. Proc Natl Acad Sci U S A 43:122–128

Waldeyer W (1865) Über die Veränderungen der quergestreiften Muskeln bei der Entzündung und dem Typhus – Prozess, sowie über die Regeneration derselben nach Substanzdefecten. Vircheress Arch path Anat Physiol 34:473–514

Whalen RG, Schwartz K, Bouveret P, Sell SM, Gros F (1979) Contractile protein synthesis in muscle development: Identification of an embryonic form of myosin heavy chain. Proc Natl Acad Sci U S A 76:5197–5201

Whalen RG, Sell SM, Butler-Browne GS, Schwartz K, Bouveret P, Pinset-Harstrom I (1981) Three myosin heavy-chain isozymes appear sequentially in rat muscle development. Nature 292:805–809

White NK, Bonner PH, Nelson DR, Hauschka SD (1975) Clonal analysis of vertebrate myogenesis. IV. Medium-dependent classification of colony-forming cells. Dev Biol 44:346–361

White NK, Hauschka SD (1971) Muscle development in vitro: A new conditioned medium effect on colony differentiation. Exp Cell Res 67:479–482

Yaffe D (1968) Retention of differentiation potentialities during prolonged cultivation of myogenic cells. Proc Natl Acad Sci U S A 61:477–483

Yaffe D (1977) Serial passaging and differentiation of myogenic cells isolated from dystrophic mouse muscle. Nature 270:725–727

Zwilling E (1954) Dissociation of chick embryo cells by means of a chelating compound. Science 120:219

CHAPTER 2

THE ORIGIN AND GENETIC REGULATION OF MYOGENIC CELLS: FROM THE EMBRYO TO THE ADULT

MARGARET BUCKINGHAM AND DIDIER MONTARRAS
Département de Biologie du Développement, CNRS URA 2578, Institut Pasteur, 25 rue du Dr Roux, 75015 Paris, France

In this chapter, we shall discuss the satellite cells of post-natal muscle and their contribution to growth and regeneration from the perspective of myogenesis during development. The cellular and molecular strategies employed during the formation of skeletal muscle in the embryo can provide insight into the mechanisms underlying post-natal myogenesis, notably into the origin and genetic regulation of satellite cells. In the molecular context, we shall mainly focus on the myogenic regulatory factors and on the Pax transcription factors, Pax3 and Pax7.

1. THE ORIGIN OF MYOGENIC CELLS

Much of our understanding of the embryonic origin of myogenic cells comes from lineage studies in the chick embryo. The overall conclusions apply also to mammals and since mammalian muscle has tended to be the object of studies on regeneration, we shall take the mouse, in which most genetic studies have been performed, as the model of reference.

Skeletal muscles of the trunk and limbs derive from somites, epithelial structures that form by segmentation of paraxial mesoderm on either side of the neural tube and notochord that run along the back of the embryo (see Tajbakhsh & Buckingham, 2000). This process takes place progressivly from about mouse embryonic day (E)8, following an anterior to posterior developmental gradient, such that the most posterior somites are the least mature. As somites mature, they acquire a mesenchymal compartment, located ventrally, which contributes the cartilage and

S. Schiaffino and T. Partridge (eds.), Skeletal Muscle Repair and Regeneration, 19–44.

bone of the vertebral column and ribs while a sub-compartment, the syndetome, will form tendons (Brent et al., 2003). The dorsal part of the somite remains epithelial. This epithelium, known as the dermomyotome, will give rise, as its name implies, to the skeletel muscle of the myotome (and all other myogenic cells in the body and limbs of the embryo) and also to the derm (skin) of the back. It is also a source of endothelial and smooth muscle cells that contribute to blood vessels. These different cell types derive from common multipotent progenitor cells present in the dermomyotome. The existence of a common progenitor has been shown by labelling experiments in the chick embryo, for derm and skeletal muscle (Ben-Yair & Kalcheim, 2005) and for endothelial and skeletal muscle cells that migrate to the limb (Kardon et al., 2002). Retrospective clonal analysis in the mouse embryo demonstrates that smooth muscle cells of the dorsal aorta also share a common progenitor with skeletal muscle (Esner et al., 2006). Such multipotent progenitor cells express the dermomyotomal markers Pax3/7 which are downregulated as they enter a non-myogenic programme (Cinnamon et al., 2006; Esner et al., 2006). In the case of muscle and derm it has been shown that asymmetric cell division accompanies this cell fate choice, with retention of N-cadherin, another dermomyotomal marker, in the Pax positive cell which requires this cell surface molecule in order to contribute to the skeletal muscle of the myotome at later stages (Cinnamon et al., 2006).

1.1 The Onset of Myogenesis

Specification of skeletal muscle cells depends on signalling molecules such as Wnts or Sonic Hedgehog that are produced by tissues adjacent to the somite, notably the neural tube, notochord and dorsal ectoderm (see Tajbakhsh & Buckingham, 2000). These inductive signals are accompanied by the production of modulator molecules, such as Noggin that counteracts repression exerted by BMPs. There is a complex balance of inductive and repressive signals that leads to the activation of myogenic determination genes in the correct spatio-temporal pattern in the somite. The first skeletal muscle to form is the myotome, as a result of the delamination of cells from the edges of the dermomyotome (Fig. 1a) (Gros et al., 2004; see also Kalcheim & Ben-Yair, 2005). This is initiated from the epaxial lip, (adajent to the axis, medial) before extending to the other lips, notably that of the hypaxial dermomyotome (furthest from the axis, ventro-lateral), which is a major source of myogenic progenitor cells. These cells, while still present at the edges of the dermomyotome, have already begun to express myogenic regulatory genes which determine their correct localisation under this epihelium as well as their myogenic cell fate (Tajbakhsh et al., 1996a). This notion of position is very important during embryogenesis, determining which signals a cell will receive, both from other cell types and cells of the same type which can generate a community effect, as shown at the onset of mouse myogenesis (Cossu et al., 1995). The initial positioning of cells in the epaxial myotome depends on $\alpha6\beta1$integrin interaction with laminin, which

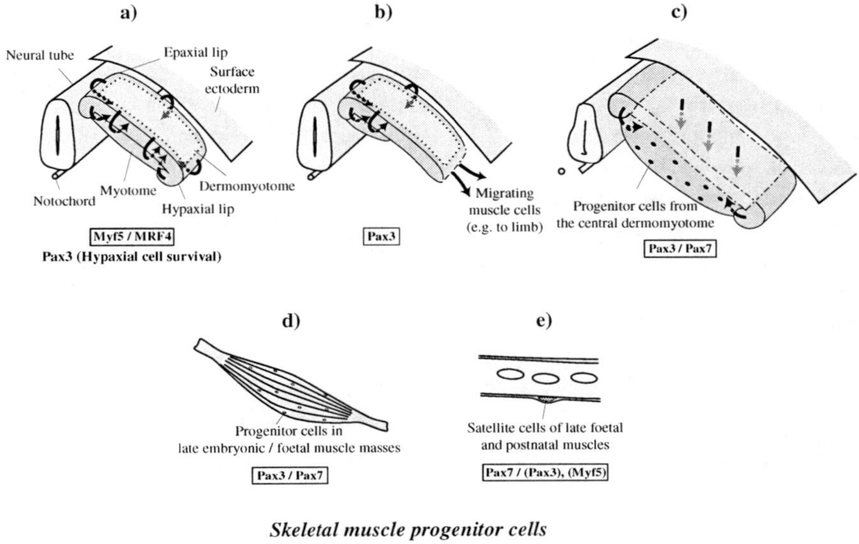

Figure 1. Myogenic progenitor cells at different stages of development. (a) early myotome formation by delamination of Myf5/Mrf4 positive myogenic cells from the edges of the dermomyotome (from E8.0). (b) delamination of Pax3 positive migrating progenitors from the hypaxial dermomyotome (from E9.5). (c) Pax3/Pax7 positive cells enter the myotome as the central dermomyotome disaggregates (from E10.5). (d) Pax3/Pax7 positive progenitors in later muscle masses. (e) Pax7 positive satellite cells most of which also transcribe *Pax3* and *Myf5*, located under the basal lamina of muscle fibres (from E16)

also plays a role in limiting the myogenic potential of cells in the dermomyotome, at the onset of myogenesis (Bajanca et al., 2006).

Other myogenic progenitor cells, which express *Pax3*, but have not yet activated myogenic regulatory genes, will delaminate from the hypaxial dermomyotome and migrate away from the somite to form muscle masses elsewhere (Fig. 1b) (see Tajbakhsh & Buckingham, 2000). This process, which is Pax3 dependent, occurs at certain axial levels, notably at the level of the limb buds where these cells will form the muscle masses of the limbs or, more anteriorly, at the level of the branchial arches where myogenic progenitor cells, that migrate into these transitory embryonic structures, will contribute to muscles of the neck and jaws, for example. Cells that will form the skeletal muscle of the diaphragm also migrate from somites close to the anterior level of the fore-limb (Dietrich et al., 1999; F. Relaix & M. Buckingham, unpublished observations). Delamination and migration of cells from the hypaxial somite also depend on signalling molecules (see Birchmeier & Brohmann, 2000). The c-Met receptor is expressed on these cells, while its ligand, HGF/scatter factor, lines the route of migration. Both receptor and ligand are essential for the movement of muscle progenitor cells from the somite. The cytokine receptor, CXCR4, with its ligand, SDF1, has also been shown recently to be implicated in this process (Vasyutina et al., 2005). This receptor is also expressed on migrating cells, and is essential for the localisation of a subpopulation of them; in *CXCR4* mutant embryos some limb muscles, for example, are missing, indicating heterogeneity among the cells that leave the hypaxial somite. Heterogeneity in myogenic progenitor cells that participate in limb myogenesis is also indicated by the distinct phenotypes of *Lbx1* (see Birchmeier & Brohmann, 2000), *Mox2* (Mankoo et al., 1999) and *Six1* (Laclef et al., 2003) mutants, where only some skeletal muscles of the limb are affected. Once migrating myogenic progenitor cells have reached their destination, they are again subject to inductive signals, such as Wnts from the limb ectoderm (Geetha-Loganathan et al., 2005) resulting in the activation of myogenic regulatory genes which are not expressed by the migrating cells (Tajbakhsh & Buckingham, 1994). In addition to a mesenchymal mode of cell migration, some myogenic progenitor cells leave the hypaxial somite as a coherent sheet of cells which have already begun expressing myogenic factors (see Noden & Francis-West, 2006). This mode of movement characterises cells in the hypoglossal chord which contribute the muscles of the larynx or part of the tongue. Again c-Met is implicated in the delamination and directed migration of these cells.

1.2 Later Muscle Growth and the Origin of Satellite Cells

The myogenic cells that form the early myotome differentiate into skeletal muscle rapidly and the myotome continues to grow as further cells feed in from the edges of the dermomyotome. However this cannot account for later myogenesis, after the somite structure begins to break down and the early myotome expands into distinct muscle masses. The central region of the dermomyotome loses its epithelail structure from about E10.5 in the mouse embryo. Recent results in both mouse (Relaix et al., 2005; Kassar-Duchossoy et al., 2005) and chicken (Gros et al., 2005; Ben-Yair & Kalcheim, 2005) have shown that cells from this region move down into the developing muscle mass of the myotome (Fig. 1c). These cells are characterised by the expression of both Pax3 and Pax7, and do not express myogenic regulatory factors. In avian embryos, they have been shown to express the Fgf receptor, FREK (Ben-Yair & Kalcheim, 2005) and their myogenic contribution depends on N-cadherin (Cinnamon et al., 2006). These Pax3/7 positive cells constitute a source of replicating myogenic progenitors that can activate the myogenic regulatory factor genes, move into the skeletal muscle differentiation programme and contribute to muscle growth. As development proceeds all muscle masses contain these reserve cells (Fig. 1d), including those of the limb (Schienda et al., 2006), where, presumably, some Pax3 positive cells which have migrated from the somites do not enter the myogenic programme, but retain their progenitor cell status. In the mouse embryo, *Pax7* is only activated in these cells once they have reached the limb (Relaix et al., 2004), in contrast to the situation in the chick embryo, where Pax7 is already present in the hypaxial dermomyotome.

During later foetal development, Pax3/7 positive cells begin to locate under the developing basal lamina which forms around muscle fibres, thus acquiring the characteristic position of satellite cells (Relaix et al., 2005) (Fig. 1e). The somitic origin of satellite cells had initially been suggested by chick/quail grafting experiments (Armand et al., 1983). This approach, together with dye labelling of dermomyotomal cells in the chick embryo, now demonstrate that satellite cells derive from the Pax3/7 expressing population of the central dermomyotome (Gros et al., 2005). Similar experiments in avian embryos show a somitic origin for the satellite cells of limb muscles (Schienda et al., 2006). Examination of chick/quail grafting experiments is limited to the post-natal period, when at least 90% of satellite cells are shown to be of somitic origin (Gros et al., 2005). However cell labelling experiments using a *Pax3-Cre* on a conditional *GFP* reporter line led to the conclusion that in the hindlimb muscles of adult (4–6 weeks old) mice, all or almost all satellite cells are derived from *Pax3* expressing progenitor cells (Schienda et al., 2006). Satellite cells of the limb muscles examined do not express *Pax3* (Relaix et al., 2006), and do not activate the gene during regeneration (Montarras et al., 2005) so that this result reflects their embryonic origin. The implication is that the Pax3/7 positive reserve cells, derived from the somite, gradually take up a satellite cell position prior to birth and constitute the satellite cells of adult skeletal muscle. It is not possible to exclude that a small proportion of satellite cells come from another source, but this, if it exists, is probably a minor phenomenon.

However, there may be other somite derived cells that have the potential to contribute to muscle. After sorting of limb muscle cells by flow cytometry, a side population (SP), that excludes Hoechst dye, was also shown to be derived from the dermomyotome (Schienda et al., 2006). SP cells isolated from skeletal muscle, can display myogenic potential and have been implicated in muscle regeneration (Gussoni et al., 1999; Seale et al., 2000; Asakura et al., 2002), although it is not clear that they make a significant contribution under normal conditions. The mesoangioblast mesodermal stem cell, which is derived from the walls of blood vessels (Minasi et al., 2002) may also have a somitic origin. The dorsal aorta is a source of these cells. In this embryonic blood vessel both endothelial and smooth muscle cells come from the paraxial mesoderm of the somite (Pardanaud et al., 1996; Pouget et al., 2006; Esner et al., 2006) and indeed in the mouse embryo it has been shown that the smooth muscle component, which includes pericytes, derives from *Pax3* expressing cells, that have probably migrated from the hypaxial dermomyotome of adjacent somites (Esner et al., 2006). It is therefore possible that the mesoangioblast, which contains *Pax3* transcripts (G. Cossu, personal communication), is a rare multipotent dermomyotomal cell and that such cells have been sequestered in the walls of blood vessels during development. Again it is not clear that mesoangioblasts, or indeed pericytes, normally contribute to the satellite cell population of postnatal muscle, but they do have regenerative potential demonstrated to be of therapeutic interest (Sampaolesi et al., 2003; 2006).

1.3 Head Muscles

It is interesting to consider head muscles in a developmental context. In the adult, some of these anterior muscles show increased resistance to degenerative diseases, such as Duchenne Muscular Dystrophy, which affect muscles of the body and limbs (see Porter et al., 2003), whereas in other myopathies, such as Myesthenia Gravis, eye muscles are particularly affected. There are interesting indications that satelllite cells in these muscles may behave differently. This is especially notable in extra-ocular muscles where they appear to be chronically activated, with constant muscle fiber turnover (McLoon et al., 2004). The molecular regulation of head myogenesis also differs from that in the trunk.

Head muscles have a varied developmental origin (see Noden & Francis-West, 2006). Those of the lower face, notably the jaw, derive from the paraxial mesoderm of the branchial arches. Unlike myogenic progenitor cells derived from the somites of the trunk, these cells do not express *Pax3*. In *Pax3/Myf5/Mrf4* mutants, *MyoD* activation and skeletal myogenesis is blocked in the trunk, but not in the head (Tajbakhsh et al., 1997). In the head, Pax7 does not appear to mark progenitor cells since it is expressed after the myogenic determination factors Myf5 and MyoD (Horst et al., 2006). Although some upstream regulators of myogenesis, such as Six1/4 (Grifone et al., 2005), play a role in both trunk and head myogenesis, other transcription factors, such as MyoR and Capsulin (Lu et al., 2002) or Tbx1 (Kelly et al., 2004), are specifically required for the formation of muscles derived from the branchial arches. Different head muscles show distinct progenitor cell behaviour and regulation. This is exemplified by the extra-ocular muscles that are derived from prechordal mesoderm. Their gene expression patterns differ from that of other muscles (Porter et al., 2001) and in the *Pitx2* mutant mouse, it is only these head muscles that are compromised (Kitamura et al., 1999). During head morphogenesis, anterior muscle progenitors undergo extensive displacement, usually moving as a coherent mass. Myogenic cells in the head are also in very close contact with neural crest cells, which potentially influence myogenesis (see Noden & Francis-West, 2006). They secrete antagonists to Wnts and Bmps, signalling pathways which can negatively affect the onset of cranial myogenesis (Tzahor et al., 2003). However transplantation of paraxial mesoderm from head and trunk, along the body axis, indicates that it has similar response capabilities (Borue & Noden, 2004). Once, however, the myogenic response is initiated, it is clear that, in the context of the head, different regulatory strategies operate. Little is known about the origin or properties of satellite cells of head muscles, although, as mentioned, there are intriguing indications that they have distinct features.

2. MYOGENIC REGULATORY FACTORS IN THE DETERMINATION AND DIFFERENTIATION OF SKELETAL MUSCLE CELLS

The synchronised activation of muscle specific genes, observed when a cultured myoblast differentiates into a myotube, had led to the proposition that a master switch regulated this process. This idea received experimental support in the mid-eighties when it was shown that in heterokaryons between a muscle cell and a non-muscle cell of a different species, the non-muscle cell nucleus expressed muscle specific genes (Blau et al., 1983; Wright, 1984). The model system provided by the C3H 10T1/2 mesenchymal cell line resulted in the demonstration that such a master switch existed. These cells undergo myogenic conversion, when treated with the DNA hypo-methylating agent, 5-azacytidine, at very high frequency (c. 25%), suggesting that the activation of very few genes or a single gene was required (Konieczny & Emerson, 1984), and indeed it was shown that a myogenic phenotype could be transmitted by transfection of limiting amounts of genomic DNA into non-muscle cells (Lassar et al., 1986). Myogenic conversion of C3H 10T1/2 cells was obtained with cloned cDNAs from myogenic cells, leading to the isolation of MyoD (Davis et al., 1987). Myogenin was subsequently isolated independently as a sequence that activated myoblast differentiation in a muscle cell line (Wright et al., 1989). *MyoD* and *Myogenin* encode a sub-class of basic-helix-loop-helix transcription factors and subsequently two other members of this gene family – *Myf5* and *MRF4* (also named *herculin* or *Myf6*) were characterized on the basis of sequence homology (see Buckingham & Tajbakhsh, 1999). All four myogenic regulatory factors (MRFs) are capable of converting C3H 10T1/2 cells to myogenesis and furthermore demonstrate this remarkable property in differentiated cell types also, such that a cell expressing retinal markers, for example, will activate skeletal muscle genes (Choi et al., 1990). Thus they do indeed function as effectors of a master switch (Weintraub et al., 1991). More recently it has been shown that this is related to their capacity to re-model chromatin, a function that myogenin performs less well (Gerber et al., 1997; Bergstrom & Tapscott, 2001; see Tapscott, 2005 for a discussion of the master switch and transcriptional mechanisms).

2.1 In the Embryo

During embryonic development, each myogenic regulatory factor gene has a characteristic pattern of expression (see Tajbakhsh & Buckingham, 2000).

2.1.1 Myogenic cell determination

Myf5, MyoD and Mrf4 can act as myogenic determination factors, directing progenitor cells into the myogenic programme (Fig. 2). *Myf5* is transcribed prior to the onset of myogenesis, in cells located at the edges of the dermomyotome (Fig. 1a), which will subsequently delaminate to form the skeletal muscle of the myotome. *MyoD* is activated later in hypaxial and then in epaxial progenitors which contribute to the mature myotome. *Myf5* mutant mice lack the early myotome (Braun et al., 1992) and myogenic progenitor cells, in which the gene has been activated, but the protein is absent, leave the dermomyotome but fail to locate correctly (Tajbakhsh et al., 1996a). These cells may die or become incorporated into other tissues, such as the cartilage and bone that is derived from the sclerotomal compartment of the somite, if they have mis-located to this region. *MyoD* activation

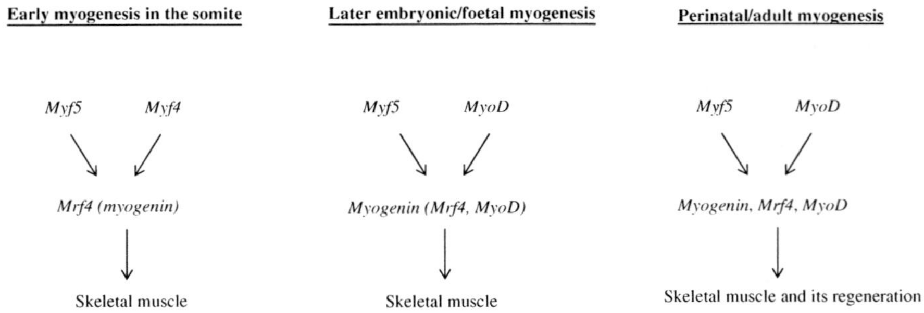

Figure 2. The role of myogenic regulatory factors in muscle cell determination and differentiation. This depends on where and when the genes are activated, and on their relative levels of expression. In some cases, ex. Myf5/MyoD, feedback compensation can occur at the transcriptional level. In the absence of one determination or differentiation factor myogenesis can still take place, unless the other factor(s) are insufficiently expressed (indicated by brackets)

is delayed in *Myf5* mutant embryos, showing that it is initially under Myf5 control (Tajbakhsh et al., 1997). However *MyoD* is subsequently activated independently and myogenesis then takes place with the correct localisation of cells that would now normally co-express *Myf5* and *MyoD* (Tajbakhsh et al., 1996b).

In the absence of MyoD, myogenesis takes place apparently normally (Rudnicki et al., 1992), although in the limb, for example, there is a delay, probably because initial levels of Myf5 are insufficient to trigger differentiation (Kablar et al., 1999). In the *MyoD* mutant, *Myf5* expression, which normally decreases after activation of *MyoD*, remains high. In the *Myf5/MyoD* double mutant (Rudnicki et al., 1993), myogenesis does not take place and myogenic cells, marked by desmin expression for example, are absent. Cells that have activated the genes, but do not express the proteins, undergo cell death or assume other cell fates in the absence of these myogenic determination factors. These mutants are characterised by oedema and by excess fat, suggesting that this may be a default pathway for myogenic progenitor cells.

Myf5 and *Mrf4* genes are within 6 kb of each other in the same chromosomal locus and it had been suggested that mutating one gene could interfere with the other, such that *Mrf4* expression is affected in the original *Myf5* mutants (see Olson et al., 1996). It has now been shown, as a result of a new series of *Myf5* mutants, that Mrf4 can also function as a myogenic determination factor at the onset of myogenesis (Kassar-Duchossoy et al., 2004). This gene is also expressed in the lips of the dermomyotome (Summerbell et al., 2002) and when only *Myf5* expression is affected, the early myotome forms. Furthermore, in a double *Myf5/MyoD* mutant, in which *Mrf4* is expressed correctly, myogenesis takes place. The function of Mrf4 as a myogenic determination factor is consistent with its role in efficiently re-modelling chromatin (Gerber et al., 1997; Bergstrom & Tapscott, 2001). After the formation of the early myotome, *Mrf4* is no longer expressed in myogenic progenitor cells and its function is restricted to myogenic differentiation (Kassar-Duchossoy et al., 2004).

In the Pax3/7 positive reserve cells, it is *Myf5* and *MyoD* that are activated as these cells enter the myogenic programme (Relaix et al., 2005; Kassar-Duchossoy et al., 2005). In the limb, also, these two factors regulate myogenic determination.

2.1.2 *Myogenic cell differentiation*

Myogenin, Mrf4 and MyoD can control myogenic cell differentiation (Fig. 2), which is characterised by the activation of downstream muscle genes and the formation of muscle fibres. Initially in the early myotome, differentiated cells are present as myocytes and muscle cell fusion only occurs in the mature myotome, at the time when MyoD is expressed (from E10.5) and M-cadherin begins to accumulate (Rose et al., 1994). Intitially *Mrf4* and *myogenin* are transcribed in differentiating muscle of the myotome. However myogenin protein only accumulates in all myotomal cells later, from about E10.5 (Cusella-De Angelis et al., 1992). The main factor responsible for the onset of differentiation in the myotome is Mrf4; in *Mrf4* mutant embryos, differentiated markers are not expressed initially (see Buckingham, 1994). Conversely, in the *myogenin* mutant, where there is a major defect in muscle differentiation (Hasty et al., 1993; Nabeshima et al., 1993), the early myotome forms correctly. After its early expression, *Mrf4* is downregulated and is transcribed again at a high level during later foetal muscle development (see Tajbakhsh & Buckingham, 2000), whereas myogenin is present in differentiating muscle cells throughout later embryonic and foetal development. In the *myogenin* mutant, later myogenesis, characterized by the formation of secondary fibres (Ontell & Kozeka, 1984) from about E14 when innervation is initiated, is severely affected (Venuti et al., 1995) and the later phase of *Mrf4* expression is compromised. Myoblasts are present, but they fail to differentiate in the absence of myogenin and the mice are not viable at birth. Analysis of a hypomorphic *myogenin* allele has shown the sensitivity of skeletal muscle formation to quantitative differences in this factor (Vivian et al., 1999). MyoD and Mrf4 may partially compensate for a reduction in myogenin, however in cultured myoblasts from the *myogenin* mutant, differentiation takes place, but Mrf4 is not expressed, although MyoD is present (Nabeshima et al., 1993). The role of these two factors is demonstrated by the phenotype of *MyoD/Mrf4* double mutants (Rawls et al., 1998) which show a striking deficit in differentiated muscle, suggesting that myogenin alone is not sufficient. The levels of MyoD and Mrf4, as well as myogenin, are probably more or less limiting in this later developmental context, when these genes tend to be co-expressed in differentiating muscle. The effects on individual muscles in the different mutants provide interesting pointers to heterogeneity between muscles.

2.2 In Perinatal and Adult Skeletal Muscle

In later foetal and adult muscle, conditional mutants have not yet been described and therefore the effects of mutation of the myogenic regulatory genes (Fig. 2) at these later stages also reflects their earlier impact on muscle development.

2.2.1 *Myogenin*

One exception is myogenin where a conditional mutation activated in newborn mice has been described recently (Knapp et al., 2006). *Myogenin* is expressed as satellite cells begin to differentiate. In muscle cell cultures it has been shown to be important as a transcriptional activator of genes, such as that encoding P21, that assure cell cycle with-drawal, as well as of skeletal muscle genes (see Buckingham and Tajbakhsh, 1998). Surprisingly, in mice that lack myogenin postnatally, satellite cell differentiation appears to occur normally and the muscle masses of the mice grow during the post-natal period. Expression of Myf5, MyoD and Mrf4 is not affected and one or more of these factors may therefore compensate for the lack of myogenin. The mutant mice are, however, generally smaller and their viability is compromised, perhaps suggesting that some autocrine function of the muscle is affected.

2.2.2 *Mrf4*

Mrf4 is also expressed in adult muscle fibres, but not in satellite cells, and is present with MyoD and myogenin in newly formed myotubes in regenerating muscle (Zhou & Bornemann, 2001). Although *Mrf4* mutant mice are viable, they do have muscle defects. In addition to developmental defects in some trunk muscles derived from the early myotome, including deep back and thoracic muscles (see Vivian et al., 2000), they also show defects in adult innervated muscle. Expression of the Na^+ channel gene *Na(V)1.4*, is selectively downregulated, both in the surface membrane and at neuromuscular junctions, despite increased expression of the other myogenic regulatory factors (Thompson et al., 2005).

2.2.3 *MyoD*

MyoD mutant mice are viable, however regeneration is affected and muscle recovery occurs more slowly after injury. This is particularly evident in $mdx : MyoD^{-/-}$ mice (Megeney et al., 1996). *MyoD* is not expressed in quiescent satellite cells, but is normally activated when these cells leave the basal lamina of the muscle fibre and begin to proliferate. In the absence of MyoD, the defect in regeneration is thought to result from the reduced capacity of activated muscle satellite cells to cease proliferation and differentiate into muscle fibres (Sabourin et al., 1999; Cornelison et al., 2000). Differentiation takes place, but is delayed (Yablonka-Reuveni et al., 1999). This function of MyoD appears to be only partially compensated by the upregulation of *Myf5* in the mutant (Braun et al., 1992). Although MyoD deficient myoblasts proliferate normally, after serial passaging, cells appear to undergo senescence (Montarras et al., 2000), which is probably another indication of perturbation in the balance between proliferation and differentiation. The $mdx :$ $MyoD^{-/-}$ double mutant reveals a deficit in skeletal muscle fibres in the diaphragm which can lead to perinatal lethality (Inanlou et al., 2003). Furthermore, in this muscle, in the absence of MyoD, correct maturation of acetylcholine receptor clusters does not occur and innvervation is abnormal (Wang et al., 2003). As in the case of the *Mrf4* mutant (Thompson et al., 2005), this indicates that myogenic regulatory factors have specific effects on the neuromuscular junction.

2.2.4 Myf5

Myf5, like MyoD, is expressed in replicating satellite cells. However β-galactosidase activity is also detected in the majority (about 80%) of quiescent satellite cells on adult muscle fibres of *Myf5* $^{nlacZ/+}$ mice, in which one allele of the *Myf5* gene has been targeted with the *nlacZ* reporter (Beauchamp et al., 2000). Such mice have proved to be valuable for providing marked satellite cells. The Myf5 protein is not detectable in quiescent satellite cells and the reporter labelling probably corresponds to residual transcription of the gene. This may reflect a choice for the progeny of activated satellite cells either to go on to differentiate, adding to existing fibres and forming new ones, or to revert to a progenitor state as quiescent satellite cells under the basal lamina. This capacity of satellite cells to self-renew (Zammit et al., 2004a; Collins et al., 2005; Montarras et al., 2005) is now clearly demonstrated. Such cells, that have passed through an activation phase, may retain residual *Myf5* transcription. During peri-natal muscle growth, this will apply to the majority of satellite cells. The presence of a number of satellite cells on each fibre that do not express Myf5-β–gal demonstrates satellite cell heterogeneity and may point to a populaton of Pax positive satellite cells that still belong to the progenitor cell population observed during development (Fig. 3). This population self-renews, prior to activation of *Myf5* and *MyoD* in the proportion of the population which will go on to differentiate. Such "upstream" cells might be expected to have more stem-like characteristics of the type recently demonstrated for some satellite cells which show asymmetric segregation of maternal DNA strands (Shinin et al., 2006). It remains to be seen how the Myf5-β-gal negative satellite cells participate in regeneration.

The original *Myf5* mutants, in which *Mrf4* was also affected, died at birth due to a rib defect, an indirect effect of myogenic perturbations in the early somite. Muscle cells isolated from late foetal stages show growth defects in the absence of Myf5, leading to premature differentiation (Montarras et al., 2000). Viable *Myf5* mutants now exist (Kaul et al., 2000; Kassar-Duchossoy et al., 2004) so that analysis of adult satellite cells and their regenerative capacity in the absence of Myf5 has become possible.

2.3 Transcriptional Control of Myogenic Regulatory Factor Genes

Study of the regulation of myogenic factor genes can provide insight into how myogenesis is controlled.

2.3.1 MyoD

MyoD has an enhancer element at −20 kb from the gene that is sufficient to direct transgene expression to early sites of *MyoD* expression in the embryo (Goldhamer et al., 1995). Despite extensive functional analysis, it is still not clear what regulatory

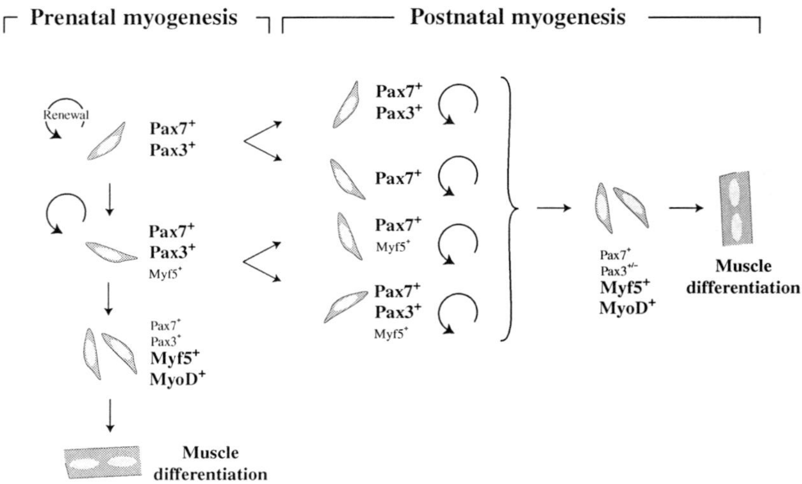

Figure 3. This represents Pax dependent progenitor cells in prenatal myogenesis (late embryonic and fœtal muscle growth) and in postnatal myogenesis (perinatal muscle growth and adult regeneration). In the first case, Pax3/7 positive cells self-renew and those that activate myogenic regulatory genes go on to differentiate. During the late fœtal period, when progenitor cells begin to take up a satellite cell position, they lose *Pax3* expression in some muscles, such as those of the hind limb. At this time some proliferating cells that have just activated *Myf5*, probably return to the quiescent state accounting for satellite cells (the majority) that show low level transcription of the gene ($Myf5^{nlacZ/+}$, β-galactosidase positive). Self-renewal is indicated by a blue circular arrow

factors directly control this sequence. A second region within 6 kb of the transcriptional start site contains enhancers that control later expression of *MyoD* in vivo and also in dividing or differentiating cultured muscle cells (Asakura et al., 1995). Here again the molecular control of these sequences is not fully understood. However it is clear that distinct regulatory strategies are involved in the activation of the gene in the somite and its later activation in Pax positive cells of developing or mature muscle, as well as in differentiated myotubes.

2.3.2 Mrf4-Myf5

Regulation of the *Mrf4-Myf5* locus has been extensively studied, with an emphasis on *Myf5* expression. This myogenic determination gene, like other genes that act upstream in a developmental hierarchy, has a complex regulation, no doubt reflecting the fact that it is the target of many different developmental signals. Different regulatory sequences direct its spatio-temporal expression in different populations of myogenic cells during development. The earliest expression of *Myf5*, in the epaxial dermomyotome, depends on an early epaxial enhancer (Summerbell et al., 2000; Teboul et al., 2003) that is regulated by sites that bind Gli (McDermott et al., 2005; see also Gustafsson et al., 2002; Teboul et al., 2003) and Tcf (Borello et al., 2006) transcription factors, the read-out of Hedgehog and canonical

Wnt signalling respectively, in keeping with the initial activation of this gene by such signals from the neural tube and notochord (Cossu et al., 1996; Tajbakhsh et al., 1998). Subsequently much of its embryonic expression depends on a region at −48/−58 kb from the gene which contains elements that direct transcription to sites in mature somites and in the limb buds (Hadchouel et al., 2000, 2003; Buchberger et al., 2003). The onset of *Myf5* expression in the limbs is controlled by a short sequence within this region; the activity of this sequence depends on Pax3, demonstrating the direct regulation of a myogenic determination gene by Pax transcription factors (Bajard, Relaix et al., 2006). This element is also regulated by Six1/4 transcription factors (Giordani, Bajard et al., 2007). Another transcription factor implicated in the regulation of the early epaxial enhancer (Teboul et al., 2002) and of another enhancer at −17 kb from *Myf5* (Chang et al., 2004), is USF, a member of the basic-helix-loop-helix super family of transcription factors. Other sequences are involved in expression of *Myf5* in the branchial arches, in keeping with the distinct properties of head myogenesis (Summerbell et al., 2000). Regulation of the gene in satellite cells is not yet understood at a molecular level, but upstream regions have been shown to direct such expression (Zammit et al., 2004b), and indeed the −17 kb sequence can show such activity. This sequence can also direct transgene expression in myonuclei instead of satellite cells with an *Mrf4* promoter (Chang, Vincent et al., 2007). *Mrf4* regulation has been shown to depend on regulatory sequences within 8.5 kb 5′ of the gene both in the embryo (ex. Pin et al., 1997; Fomin et al., 2004; Chang et al., 2004) and in the myonuclei of adult fast fibres (Pin & Konieczny, 2002). However more distal sequences are probably also involved (see Carvajal et al., 2001). How regulatory elements in this locus interact with either the *Myf5* or *Mrf4* promoter, to direct the different expression patterns of these two genes, poses a complex molecular problem which may depend on three dimensional constraints imposed by chromatin configurations at the locus.

2.3.3 Myogenin

Embryonic expression of the *myogenin* gene is efficiently recapitulated by 1 kb of upstream sequence. This depends on sites present in the proximal promoter region that bind Mef2, and myogenic regulatory factors, such as Myf5, with varying effects on somite versus limb expression (Cheng et al., 1993; Yee & Rigby, 1993). Six homeobox proteins, acting through a Mef3 binding site in the promoter, are also implicated in the transcription of the gene (Spitz et al., 1998). Regulation of the *myogenin* gene in satellite cells has not been examined, however it is likely to be myogenic factor dependent.

3. PAX3 AND PAX7 IN MYOGENIC PROGENITOR CELLS

These *Pax* genes belong to a family of transcription factors characterised by the presence of a paired domain DNA binding motif and in many cases, including Pax3 and Pax7, of a homeodomain (see Buckingham & Relaix 2007). Pax factors play important roles in tissue specification and organogenesis during embryogenesis.

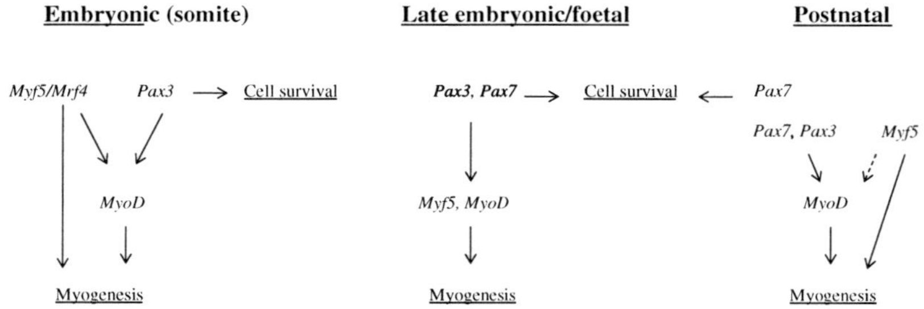

Figure 4. Pax3/7 regulation of progenitor cell survival and entry into the myogenic programme, based on genetic experiments. At the onset of myogenesis in the embryo Pax3 assures cell survival and also delamination and migration of progenitor cells from the hypaxial somite

Pax5, for example, is required for the emergence of the B lymphocyte cell lineage (Busslinger, 2004), whereas Pax6 is required for eye formation (see Relaix & Buckingham, 1999). Neither Pax3 nor Pax7 are skeletal muscle specific (see Relaix et al., 2004). They are also expressed in regions of the central nervous system and Pax3 plays an important role in neural crest. However it has become increasingly clear that these *Pax* genes are of major importance in myogenic progenitor cells during development and in the adult (Fig. 4).

3.1 In the Embryo

Pax3 is expressed in pre-somitic mesoderm and then in newly formed somites where it subsequently becomes restricted to the dermomyotome, which is the source of myogenic progenitor cells. *Pax7* transcripts are first detected slightly later, concentrated in the central region of the dermomyotome, in the mouse embryo, where, in contrast to the chick, *Pax7* is not expressed in the hypaxial dermomyotome nor initially in myogenic cells that migrate to the limb (see Relaix et al., 2004).

Pax3 mutant mice, either engineered or as the spontaneously occuring mutant, *splotch*, are characterised by the loss of the hypaxial dermomyotome and its myogenic derivatives (see Tajbakhsh & Buckingham, 2000). These cells progressively undergo apoptosis (Borycki et al., 1999). The delamination and subsequent migration of hypaxial myogenic cells does not occur, so that muscle masses are not formed in the limbs, for example, or the diaphragm.The *c-Met* gene is probably a direct target of Pax3 (Epstein et al., 1996; Relaix et al., 2003) and this aspect of the *Pax3* phenotype ressembles that of *c-Me*t mutants (Bladt et al., 1995). In *Pax3* mutants, the epaxial extremity of the dermomyotome is also affected, with some loss of cells; however this myogenic phenotype is less severe (Tajbakhsh et al., 1997), probably because early epaxial myogenesis occurs before the progressive cell death seen in the absence of Pax3.

Pax7 mutants do not have a detectable embryonic phenotype (Mansouri et al., 1996), probably due to overlapping expression of *Pax3* in the Pax7

positive cells. The myogenic progenitor cells of the central dermomyotome are not affected in single *Pax3* or *Pax7* mutants. Indeed the two factors perform similar functions during myogenesis in the trunk. This is demonstrated by a genetic experiment in which a Pax7 coding sequence was targeted to an allele of *Pax3* (Relaix et al., 2004). In homozygote embryos in which both alleles of *Pax3* encode Pax7 there is no developmental defect in the trunk. Interestingly myogenesis in the limbs is partially compromised, indicating that the fuction of these two Pax factors has diverged where appendicular muscle formation is concerned. Pax3 appears to have acquired additional functions required for the efficient migration and proliferation of myogenic progenitor cells in the limbs. The progressively more severe phenotypes seen when one or two alleles of *Pax3* are replaced by *Pax7* are more pronounced in the forelimbs and in distal limb muscles.

When both *Pax3* and *Pax7* genes are mutated, there is a major deficit in skeletal muscle. The Pax3/7 positive progenitor cell population, present in developing muscles, fails to activate *Myf5* or *MyoD* and does not enter the myogenic programme (Relaix et al., 2005). Furthermore as in the case of *Myf5* mutants, cells that should express Pax factors mis-locate and become associated with other tissues such as the ribs, or undergo apoptosis. This phenotype underlies the role of these Pax factors in directing a progenitor cell into the myogenic programme and also, as in the case of Pax3 in the hypaxial dermomyotome, in assuring progenitor cell survival (Fig. 4). *Pax3* mutants tend to die from E13.5 due to non-myogenic defects so that it has not been possible to examine this double mutant phenotype at later developmental stages. However the lack of a diaphragm, seen already in *Pax3* mutants, means that the myogenic phenotype is lethal at birth and the expectation is that there will be no muscle progenitor cells and only vestiges of skeletal muscle in the trunk.

Two other observations on *Pax* mutants require comment. *Pax3/Myf5(and Mrf4)* triple mutants showed no skeletal muscle formation, with no later activation of *MyoD* (Tajbakhsh et al., 1997) (Fig. 4). This phenotype is surprising in that Pax3/Pax7 positive progenitor cells in the central dermomyotome might be expected to retain their myogenic potential in the presence of Pax7, as in the case of the single *Pax3* mutant. In the absence of Myf5 and Mrf4, the early myotome does not form (Tajbakhsh et al., 1996a; Kassar-Duchossoy et al., 2004), and, furthermore, in the absence of Pax3, myogenic cells, including those that should activate *MyoD* later, are lost from the hypaxial and epaxial extremities of the dermomyotome. There is therefore no myotome to function as a myogenic scaffold to receive the Pax positive cells of the central dermomyotome. It would appear that they require the presence of myogenic cells, perhaps as a source of survival/myogenic signals, in order to avoid cell death and participate in myogenesis. A second observation concerns *Pax7/Myf5/Mrf4* triple mutants (Kassar-Duchossoy et al., 2005). Again, in this case, the early myotome does not form completely, although *Pax3* expressing cells are present at the epaxial and hypaxial extremities of the dermomyotome so that *MyoD* activation in these domains can occur. Skeletal muscle does form, but the Pax positive population of progenitor cells tends to undergo apoptosis, despite the presence of Pax3. This is not observed when *Pax7* alone is mutated and suggests that in the absence of part of the early skeletal musculature, insufficient survival signals are received to adequately protect the Pax positive progenitor population from cell death.

3.2 In Perinatal and Adult Muscle

The important observation that satellite cells are missing and that regeneration is compromised in the *Pax7* mutant mouse led to the conclusion that Pax7 is required for the specification of muscle satellite cells (Seale et al., 2000). However since then it has become clear that satellite cells are present, although in reduced numbers, in the absence of Pax7 and that these cells retain their myogenic potential (Oustanina et al., 2004). Immediately after birth the numbers of satellite cells are close to normal, suggesting that they have been "specified" correctly (Relaix, Montarras et al., 2006). This is in keeping with what we now know about the origin of satellite cells from the Pax3/7 expressing cells of the somite. These myogenic reserve cells depend on either Pax3 or Pax7 under normal conditions and therefore one can conclude that Pax3 alone is responsible for their initial presence in the *Pax7* mutant after birth. Their subsequent loss may be partly due to cell cycle defects and also to apoptosis observed immediately after birth, principally in desmin positive satelite cells in vivo (Relaix, Montarras et al., 2006). This would suggest that apoptosis occurs when satellite cells are activated, during the extensive muscle growth that takes place postnatally. *Pax7* mutant pups are smaller and tend to die within the first week or two after birth. Their survival is variable and appears to depend on the genetic background (Kuang et al., 2006), pointing to the intervention of modifier genes. Experiments on cultured satellite cells, infected with dominant negative Pax-En constructs, where the DNA binding domain of Pax is fused to the repression domain of Engrailed, confirm that Pax7 plays a crucial anti-apoptotic role after birth (Relaix, Montarras et al., 2006) (Fig. 4). Similarly, in amphibia, where Pax7 positive cells have also now been described in skeletal muscle (Chen et al., 2006); Morrison & Brickman, 2006), Pax7-En constructs introduced in vivo antagonise Pax7 function, leading to apoptosis (Chen et al., 2006).

This raises the question of the presence and role of Pax3 after birth. In the culture experiments, dominant negative Pax3-En did not produce as severe a cell death phenotype as Pax7-En (Relaix, Montarras et al., 2006). This suggests that their anti-apoptotic functions, which are interchangeable in the embryo, as shown by the *Pax7* knock-in into *Pax3* (Relaix et al., 2004), have diverged in postnatal satellite cells. *Pax7* is expressed in the satellite cells of all muscles examined and is the best marker of these cells (Seale et al., 2000). Pax protein is present and transcriptionally active, as shown by a Pax reporter mouse (Relaix, Montarras et al., 2006; Zammit et al., 2006). *Pax3* expression, on the other hand, is not detected in the satellite cells of all adult muscles. It is notably absent in lower and most upper hindlimb muscles (Relaix, Montarras et al., 2006). There is no obvious reason, either in terms of embryological origin or function, as to why it is completely down-regulated before birth in some muscles and not others. Furthermore the muscles in which Pax3 is down-regulated are juxtaposed to those in which it is retained, as seen for intercostal muscles or the *Gracilis* muscle of the hind-limb which remains Pax3 positive in a Pax3 negative environment. In Pax3 positive adult muscles such as the diaphragm, many trunk muscles and fore-limb muscles, quiescent satellite cells, visualised histologically, are marked by expression of nlacZ or GFP reporters targeted to the *Pax3* gene. Furthermore *Pax3* transcripts and protein are accumulated in these adult muscle masses (Relaix, Montarras et al., 2006). There are, however, variable reports on this point, with one paper showing general down-regulation of Pax3 protein levels prior to birth (Horst et al., 2006) ; reporter gene expression is probably a sensitive read-out of reduced levels of expression. *Pax3* expression appears to be an intrinsic property of satellite cells ; Pax3-GFP positive satellite cells grafted into a regenerating *Tibialis Anterior* muscle, which is usually Pax3 negative, retain *Pax3* expression in the regenerated muscle whereas Pax3 negative cells remain negative (Montarras et al., 2005). In contrast, it has been proposed that Pax3 protein is present in all satellite cells after activation, based on observations of cultured cells from normal and injured *Tibialis Anterior* muscle (Conboy & Rando 2002). However, in a recent report on cultured satellite cells, this was not observed and Pax3 negative cells did not show activation of *Pax3* even on the basis of reporter expression (Relaix, Montarras et al., 2006). Antibodies to Pax3 with improved specificity and sensitivity should help resolve some of these expression issues.

As in the embryo, Pax proteins play an important role in regulating the entry of satellite cells into the myogenic programme (Fig. 4). This is illustrated by the manipulation of dominant negative Pax3 or Pax7 constructs in satellite cell cultures (Relaix, Montarras et al., 2006). In both cases, the expression of MyoD is prevented. However in cultures of wild type satellite cells, activation of *myogenin* and differentiation still take place, whereas this is not seen in satellite cell cultures from *Myf5* mutant mice. In this adult situation, therefore, *Myf5* and *Pax/MyoD* are on parallel genetic pathways (Fig. 4). This would explain why the remaining satellite cells in *Pax7* mutant mice, whether from Pax3 positive or Pax3 negative muscles, can differentiate into skeletal muscle, reflecting the fact that a low level of *Myf5* expression is detectable in the majority of quiescent satellite cells (Beauchamp et al., 2000), which have therefore already entered the myogenic programme. As satellite cells differentiate, they down-regulate *Pax* gene expression and those cells which retain high levels of Pax7 probably reconstitute the satellite cell population (Zammit et al., 2004a). Recently it has been shown that satellite cells can differentiate in the presence of Pax proteins (Relaix, Montarras et al., 2006; Zammit et al., 2006), whereas a previous report had claimed that forced expression of Pax7 inhibited differentiation (Olguin & Olwin, 2004). This may depend on the level of expression. Myogenic cell lines are also more sensitive to forced expression of Pax7, which downregulates MyoD and affects proliferation (D.M., unpublished observations).

This discussion on *Pax* gene expression in skeletal muscle has concentrated on satellite cells, which have been shown, as pure populations (Montarras et al., 2005; Collins et al., 2005), to efficiently undergo myogenic differentiation in vitro and to efficiently repair injured skeletal muscle when grafted in vivo. Such cells, in addition to their myogenic properties, have characteristic cell surface markers and granulosity which permits their isolation by flow cytometry (Montarras et al., 2005). However other cell types are present in the skeletal muscle mass and some of these may express *Pax3*. This has been reported for a small proportion (< 1%) of interstitial cells which show myogenic properties (Kuang et al., 2006). It is possible that they are related to somite derived SP cells (Schienda et al., 2006), although they are not detectable as a Pax3-GFP population with different flow cytometry parameters (Montarras et al., 2005; D.M. & A. Cumano, unpublished observations). They may also be related to positive cells observed in the walls of adult blood vessels, a phenomenon seen with *Pax3* reporter mice (M. Esner, F. Relaix, D.M., M.B., unpublished observations). This may reflect a somitic origin of some blood vessel cells, as described for the embryonic dorsal aorta (Esner et al., 2006; Pouget et al., 2006), and may be related to the blood vessel derived mesoangioblast stem cells (Minasi et al., 2002) which expresses *Pax3* (G. Cossu, personal communication). One can speculate that vestiges of Pax3 expression confer myogenic potential.

3.3 Molecular Aspects of Pax3/7 Function

Pax3 and Pax7 are poor transcription factors on their own and appear to need co-factors to function efficiently. In cell systems which are far removed from a myogenic context, Pax3 co-repressors such as Hira or Dax have been isolated (see Relaix et al., 2003). It was not clear whether Pax3 and Pax7 acted as transcriptional activators or repressors during skeletal myogenesis. Pediatric rhabdomyosarcomas, muscle tumours found in children, are often caused by chromosomal translocations that result in a fusion protein in which the DNA binding domain of PAX3 or PAX7 is fused to the transcriptional activation domain of FOXO1a, also known as FKHR. This results in a transcription factor that acts as a strong transcriptional activator. Targeting of *Pax3* by a PAX3-FKHR coding sequence (Relaix et al., 2003) showed that in the absence of Pax3 this fusion protein replaced its myogenic function, so that hypaxial myogenesis took place, for example. This therefore establishes that Pax3 acts as a transcriptional activator during myogenesis. Since Pax3 function in the trunk can be replaced by Pax7 (Relaix et al., 2004) this applies to Pax7 also. In the presence of PAX3-FKHR, Pax3 targets tend to be overactivated. In the embryo, this is seen for *c-Met* or for *MyoD* and *Myf5* in certain contexts. As mentioned in the section on myogenic factors, *Myf5* has now been shown to be a direct target of Pax3 in the myogenic cells of the limb and in the mature hypaxial somite (Bajard, Relaix et al., 2006). There is an extensive literature on rhabdomyosarcoma cells (see Barr, 2001) and indeed the reason why these become tumorgenic is particularly interesting in the context of Pax3/7 control of the cell cycle. Microarray analyses of these cells have identified potential Pax targets. Discussion of rhabdomyosarcomas is outside the scope of this review. However the identification of more Pax3/7 targets is clearly of future importance for the molecular characterisation of the myogenic progenitor populations that are regulated by these factors. Microarrays on skeletal muscle cells are already uncovering potential regulators, like the FgfR4 receptor (Zhao et al., 2006). In the future, the characterisation of Pax3 and Pax7 co-factors should also add another dimension to our understanding of myogenesis.

The regulation of *Pax3* and *Pax7* gene expression during myogenesis is not well understood at the molecular level. Their gene structure is complex with large introns with evidence for multiple splice forms for Pax7 (Vorobyov & Horst, 2004). Wnt signalling from the dorsal ectoderm influences Pax3 both at the level of maintaining its expression (Fan et al., 1997) and of regulating its transcriptional activity (Brunelli et al., 2007). Non-canonical signalling through Wnt6 or Wnt7a is probably involved and indeed the planar cell polarity pathway acting through PKC has been shown to be important in the latter. Six homeoproteins also affect *Pax3* in the embryo (Grifone et al., 2005), possibly acting through the hypaxial enhancer element which lies upstream of the gene and has been shown to direct transgene expression to the hypaxial dermomyotome and domains of myogenesis that derive from this region (Brown et al., 2005). Deciphering the transcriptional regulation of *Pax* genes will be important in understanding the upstream events that preceed myogenesis.

4. CONCLUDING REMARKS

In this review we have presented cellular aspects of myogenesis in the embryo and new insights into the origin of the satellite cell. At the molecular level, we have focussed on the role of the myogenic regulatory factors in the determination and differentiation of embryonic and foetal skeletal muscle and discussed how they affect muscle regeneration. The Pax3 and Pax7 transcription factors, which have more recently emerged as major players in pre- and post-natal myogenesis, are also a focus of this review. We have underlined their role in orchestrating the entry of cells into the myogenic programme and in ensuring cell survival, dual functions that control stem cell fate. Other transcriptional regulators that affect skeletal muscle formation are mentioned briefly. Of these, homeodomain factors, such as Mox2, Lbx1, Msx1 and Six1/4 are particularly interesting in the context of myogenic progenitor cell behaviour. Six proteins and their co-factors Eya and Dach intervene at multiple levels (see Grifone et al., 2005), through feedback loops with Pax3, reminiscent of the Pax6 paradigm for eye formation (Relaix & Buckingham, 1999), on the myogenic regulatory factors, with direct control of the *myogenin* gene (Spitz et al., 1998) and on fibre type determination in adult muscle (Grifone et al., 2004). It will be important to investigate the potential role of these and the other homeodomain factors in satellite cells and during muscle regeneration. Another very important aspect, only touched on in this review, concerns the signalling molecules that orchestrate myogenesis. In the embryo, canonical and non-canonical Wnts are positive effectors, whereas Bmps antagonise the onset of myogenesis. Fgfs are also implicated in myogenic regulation. As discussed in the context of progenitor cell migration to the limb c-Met and CXCR4 signaling also play a role. Mutation of *MAML1*, which encodes a Notch co-activator also interacting with the muscle differentiation factor, Mef2, provides a recent example of a myogenic phenotype observed in the embryo, probably resulting from interference with cell fate decisions (Shen et al., 2006). There are indications that asymmetric cell decisions, probably involving Notch, take place when cells leave the dermomyotome to enter the myogenic pathway (Cinnamon et al., 2006; see Buckingham 2006). These different signalling pathways also potentially affect satellite cell behaviour. Indeed Notch signalling has been shown to be important in the regenerative response (Conboy et al., 2003) and asymmetric cell division is also potentially involved in the decision to self renew or differentiate (see also Zammit et al., 2004a; Shinin et al., 2006). Clearly regulation of myogenesis during development provides important leads for the study of postnatal muscle growth and regeneration and conversely aspects of adult myogenesis repercute on the study of muscle development.

ACKNOWLEDGEMENTS

Skeletal muscle research in M.B.'s laboratory is supported by the Pasteur Institute, the CNRS, and by grants from the AFM, the EU Integrated Project "EuroStemCell", and the EU networks of Excellence "MYORES" and "Cells into Organs".

REFERENCES

Armand O, Boutineau AM, Mauger A, Pautou MP, Kieny M (1983) Origin of satellite cells in avian skeletal muscles. Arch Anat Microsc Morphol Exp 72:163–181

Asakura A, Lyons GE, Tapscott, SJ (1995) The regulation of MyoD gene expression: conserved elements mediate expression in embryonic axial muscle. Dev Biol 171:386–398

Asakura A, Seale P, Girgis-Gabardo A, Rudnicki MA (2002) Myogenic specification of side population cells in skeletal muscle. J Cell Biol 159:123–134

Bajanca F, Luz M, Raymond K, Martins GG, Sonnenberg A, Tajbakhsh S, Buckingham M, Thorsteinsdottir S (2006) Integrin alpha6beta1-laminin interactions regulate early myotome formation in the mouse embryo. Development 133:1635–1644

Bajard L, Relaix F, Lagha M, Rocancourt D, Daubas P, Buckingham ME (2006) A distinct genetic hierarchy controls hypaxial myogenesis: Pax3 directly activates Myf5 in muscle progenitor cells in the limb. Genes & Dev 20:2450–2464

Barr FG (2001) Gene fusions involving PAX and FOX family members in alveolar rhabdomyosarcoma. Oncogene 20:5736–5746

Beauchamp JR, Heslop L, Yu DSW, Kelly RG, Tajbakhsh T, Buckingham ME, Partridge TA, Zammit PS (2000) Expression of CD34 and Myf5 defines the majority of quiescent adult skeletal muscle satellite cells. J Cell Biol 151:1221–1233

Ben-Yair R, Kalcheim C (2005) Lineage analysis of the avian dermomyotome sheet reveals the existence of single cells with both dermal and muscle progenitor fates. Development 132:689–701

Bergstrom DA, Tapscott SJ (2001) Molecular distinction between specification and differentiation in the myogenic basic helix-loop-helix transcription factor family. Mol Cell Biol 21:2404–2412

Birchmeier C, Brohmann H (2000) Genes that control the development of migrating muscle precursor cells. Curr Opin Cell Biol 12:725–730

Bladt F, Riethmacher D, Isenmann S, Aguzzi A, Birchmeier C (1995) Essential role for the c-met receptor in the migration of myogenic precursor cells into the limb bud. Nature 376:768–771

Blau H, Chiu C-P, Webster C (1983) Cytoplasmic activation of human nuclear genes in stable heterocaryons. Cell 32:1171–1180

Borello U, Berarducci B, Murphy P, Bajard L, Buffa V, Piccolo S, Buckingham M, Cossu G (2006) The Wnt/β-catenin pathway regulates the Shh-mediated *Myf5* activation during somitogenesis. Development 133:3723–3732

Borue X, Noden DM (2004) Normal and aberrant craniofacial myogenesis by grafted trunk somitic and segmental plate mesoderm. Development 131:3967–3980

Borycki AG, Li J, Jin F, Emerson CP, Epstein JA (1999) Pax3 functions in cell survival and in Pax7 regulation. Development 126:1665–1674

Braun T, Rudnicki MA, Arnold HH, Jaenisch R (1992) Targeted inactivation of the muscle regulatory gene Myf-5 results in abnormal rib development and perinatal death. Cell 71:369–382

Brent AE, Schweitzer R, Tabin CJ (2003) A somitic compartment of tendon progenitors. Cell 113: 235–248

Brown CB, Engleka KA, Wenning J, Lu MM, Epstein JA (2005) Identification of a hypaxial somite enhancer element regulating *Pax3* expression in migrating myoblasts and characterization of hypaxial muscle Cre transgenic mice. Genesis 41:202–209

Brunelli S, Relaix F, Baesso S, Buckingham M, Cossu G (2007) Beta catenin-independent activation of MyoD by Wnt7a requires PKC and depends on Pax3 transcriptional activity. Dev Biol. 304:604-614

Buchberger A, Nomokonova N, Arnold HH (2003) *Myf5* expression in somites and limb buds of mouse embryos is controlled by two distinct distal enhancer activities. Development 130, 3297–3307

Buckingham M (1994) Which myogenic factors make muscle? Curr Biol 4(1):61–63

Buckingham M, Tajbakhsh S (1999) Myogenic cell specification during somitogenesis. In: Moody SA (ed) Cell lineage and fate determination. Academic Press, Chapter 41, pp 617–633

Buckingham M (2006) Myogenic progenitor cells and skeletal myogenesis in vertebrates. Curr Opin Genet Dev 16:525–532

Buckingham M, Relaix F (2007) The role of *Pax* genes in the deveopment of tissues and organs: *Pax3* and *Pax7* regulate muscle progenitor cell functions. Ann. Rev. Cell Dev. Biol. 23:645–673.

Busslinger M (2004) Transcriptional control of early B cell development. Annu Rev Immunol 22:55–79

Carvajal JJ, Cox D, Summerbell D, Rigby PW (2001) A BAC transgenic analysis of the Mrf4/Myf5 locus reveals interdigitated elements that control activation and maintenance of gene expression during muscle development. Development 128:1857–1868

Chang TH-T, Primig M, Hadchouel J, Tajbakhsh, S, Rocancourt D, Fernandez A, Kappler R, Scherthan H, Buckingham M. (2004) An enhancer directs differential expression of the linked *Mrf4* and *Myf5* myogenic regulatory genes in the mouse. Dev Biol 269:595–608

Chang TH-T, Vincent SD, Buckingham ME, Zammit PS. (2007) The A17 enhancer directs expression of *Myf*5 to satellite cells but *Myf*4 to myonudei. Dev. Dynamics 236, in press

Cheng TC, Wallace MC, Merlie JP, Olson EN (1993) Separable regulatory elements governing myogenin transcription in mouse embryogenesis. Science 261:215–218

Chen Y, Lin G, Slack JMW (2006) Control of muscle regeneration in the *Xenopus* tadpole tail by Pax7. Development 133:2303–2313

Choi J, Costa ML, Mermelstein CS, Chagas C, Holtzer S (1990) MyoD converts primary dermal fibroblasts, chondroblasts, smooth muscle, and retinal pigmented epithelial cells into striated mononucleated myoblasts and multinucleated myotubes. Proc Natl Acad Sci USA 87:7988–7992

Cinnamon Y, Ben-Yair R, Kalcheim C (2006) Differential effects of N-cadherin-mediated adhesion on the development of myotomal waves. Development 133:1101–1112

Collins CA, Olsen I, Zammit PS, Heslop L, Petrie A, Partridge TA, Morgan JE (2005) Stem cell function, self-renewal, and behavioral heterogeneity of cells from the adult muscle satellite cell niche. Cell 122:289–301

Conboy IM, Rando TA (2002) The regulation of Notch signaling controls satellite cell activation and cell fate determination in postnatal myogenesis. Dev Cell 3:397–409

Conboy IM, Conboy MJ, Smythe GM, Rando TA (2003) Notch-mediated restoration of regenerative potential to aged muscle. Science 302:1575–1577

Cornelison DD, Olwin BB, Rudnicki MA, Wold BJ (2000) MyoD(-/-) satellite cells in single-fiber culture are differentiation defective and MRF4 deficient. Dev Biol 224:122–137

Cossu G, Kelly R, Di Donna S, Vivarelli E, Buckingham M (1995) Myoblast differentiation during mammalian somitogenesis is dependent upon a community effect. Proc Natl Acad Sci USA 92: 2254–2258

Cossu G, Kelly R, Tajbakhsh S, Di Donna S, Vivarelli E, Buckingham M. (1996) Activation of different myogenic pathways: myf-5 is induced by the neural tube and MyoD by the dorsal ectoderm in mouse paraxial mesoderm. Development 122:429–437

Cusella-De Angelis MG, Lyons G, Sonnino C, De Angelis L, Vivarelli E, Farmer K, Wright WE, Molinaro M, Bouche M, Buckingham M et al. (1992) MyoD, myogenin independent differentiation of primordial myoblasts in mouse somites. J Cell Biol 116:1243–1255

Davis RL, Weintraub H, Lassar AB (1987) Expression of a single transfected cDNA converts fibroblasts to myoblasts. Cell 51:987–1000

Dietrich S, Abou-Rebyeh F, Brohmann H, Bladt F, Sonnenberg-Riethmacher E, Yamaai T, Lumsden A, Brand-Saberi B, Birchmeier C (1999) The role of SF/HGF and c-Met in the development of skeletal muscle. Development 126:1621–1629

Epstein JA, Shapiro DN, Cheng J, Lam PY, Maas RL (1996) Pax3 modulates expression of the c-Met receptor during limb muscle development. Proc Natl Acad Sci USA 93:4213–4218

Esner M, Meilhac SM, Relaix F, Nicolas J-F, Cossu G, Buckingham ME (2006) Smooth muscle of the dorsal aorta shares a common clonal origin with skeletal muscle of the myotome: a model for precursor cell migration from the somite. Development 133:737–749

Fan CM, Lee CS, Tessier-Lavigne M (1997) A role for WNT proteins in induction of dermomyotome. Dev Biol 191:160–165

Fomin M, Nomokonova N, Arnold HH (2004) Identification of a critical control element directing expression of the muscle-specific transcription factor MRF4 in the mouse embryo. Dev Biol 272:498–509

Geetha-Loganathan P, Nimmagadda S, Prols F, Patel K, Scaal M, Huan R, Christ B (2005) Ectodermal Wnt-6 promotes Myf5-dependent avian limb myogenesis. Dev Biol 288:221–233

Gerber AN, Klesert TR, Bergstrom DA, Tapscott SJ (1997) Two domains of MyoD mediate transcriptional activation of genes in repressive chromatin: a mechanism for lineage determination in myogenesis. Genes Dev 11:436–450

Giordani J, Bajard L, Demignon J, Buckingham M, Maire P (2007) Six proteins regulate the activation of Myf5 expression in embryonic mouse limbs. PNAS 104:11310–11315

Goldhamer DJ, Brunk BP, Faerman A, King A, Shani M, Emerson CP Jr (1995) Embryonic activation of the myoD gene is regulated by a highly conserved distal control element. Development 121:637–649

Grifone R, Laclef C, Spitz F, Lopez S, Demignon J, Guidotti JE, Kawakami K, Xu PX, Kelly R, Petrof BJ, Daegelen D, Concordet JP, Maire P (2004) Six1 and Eya1 expression can reprogram adult muscle from the slow-twitch phenotype into the fast-twitch phenotype. Mol Cell Biol 24:6253–6267

Grifone R, Demignon J, Houbron C, Souil E, Niro C, Seller MJ, Hamard G, Maire P (2005) Six1 and Six4 homeoproteins are required for Pax3 and Mrf expression during myogenesis in the mouse embryo. Development 132:2235–2249

Gros J, Scaal M, Marcelle C (2004) A two-step mechanism for myotome formation in chick. Dev Cell 6:875–882

Gros J, Manceau M, Thome V, Marcelle C (2005) A common somitic origin for embyronic muscle progenitors and satellite cells. Nature 435:954–958

Gustafsson MK, Pan H, Pinney DF, Liu Y, Lewandowski A, Epstein DJ, Emerson CP Jr (2002) Myf5 is a direct target of long-range shh signaling and Gli regulation for muscle specification. Genes Dev 16:114–126

Gussoni E, Soneoka Y, Strickland CD, Buzney EA, Khan MK, Flint AF, Kunkel LM, Mulligan RC (1999) Dystrophin expression in the mdx mouse restored by stem cell transplantation. Nature 401: 390–394

Hadchouel J, Tajbakhsh S, Primig M, Chang THT, Daubas P, Rocancourt D, Buckingham M (2000) Modular long-range regulation of Myf5 reveals unexpected heterogeneity between skeletal muscles in the mouse embryo. Development 127:4455–4467

Hadchouel J, Carvajal JJ, Daubas P, Bajard L, Chang T, Rocancourt D, Cox D, Summerbell D, Tajbakhsh S, Rigby PWJ, Buckingham M (2003) Analysis of a key regulatory region upstream of the *Myf5* gene reveals multiple phases of myogenesis, orchestrated at each site by a combination of elements dispersed throughout the locus. Development 130:3415–3426

Hasty P, Bradley A, Morris JH, Edmondson DG, Venuti JM, Olson EN, Klein WH (1993) Muscle deficiency and neonatal death in mice with a targeted mutation in the myogenin gene. Nature 364:501–506

Horst D, Ustanina S, Sergi C, Mikuz G, Juergens H, Braun T, Vorobyov E (2006) Comparative expression analysis of Pax3 and Pax7 during mouse myogenesis. Int J Dev Biol 50:47–54

Inanlou MR, Dhillon GS, Belliveau AC, Reid AM, Ying C, Rudnicki MA, Kablar B (2003) A significant reduction of the diaphragm in mdx: MyoD-/- 9th embryos suggests a role for MyoD in the diaphragm development. Dev. Biol 261:324–336

Kablar B, Krastel K, Ying C, Tapscott SJ, Goldhamer DJ, Rudnicki MA (1999) Myogenic determination occurs independently in somites and limb buds. Dev. Biol 206:219–231

Kalcheim C, Ben-Yair R (2005) Cell rearrangements during development of the somite and its derivatives. Curr Opin Genet Dev 15:371–380

Kardon G, Campbell JK, Tabin CJ (2002) Local extrinsic signals determine muscle and endothelial cell fate and patterning in the vertebrate limb. Dev Cell 3:533–545

Kassar-Duchossoy L, Gayraud-Morel B, Gomès D, Rocancourt D, Buckingham M, Shinin V, Tajbakhsh S (2004) Mrf4 directs skeletal muscle indentity in *Myf5: MyoD* double mutant mice. Nature 431: 466–471

Kassar-Duchossoy L, Giacone E, Gayraud-Morel B, Jory A, Gomes D, Tajbakhsh S (2005) Pax3/Pax7 mark a novel population of primitive myogenic cells during development. Genes Dev 19:1426–1431

Kaul A, Köster M, Neuhaus H, Braun T (2000) Myf-5 revisited: loss of early myotome formation does not lead to a rib phenotype in homozygous Myf-5 mutant mice. Cell 102:17–19

Kelly RG, Jerome-Majewska LA, Papaioannou VE (2004) The del22q11.2 candidate gene *Tbx1* regulates branchiomeric myogenesis. Hum Mol Genet 13:2829–2840

Kitamura K, Miura H, Miyagawa-Tomita S, Yanazawa M, Katoh-Fukui Y, Suzuki R, Ohuchi H, Suehiro A, Motegi Y, Nakahara Y, Kondo S, Yokoyama M (1999) Mouse Pitx2 deficiency leads to anomalies of the ventral body wall, heart, extra- and periocular mesoderm and right pulmonary isomerism. Development 126:5749–5758

Knapp JR, Davie JK, Myer A, Meadows E, Olson EN, Klein WH (2006) Loss of myogenin in postnatal life leads to normal skeletal muscle but reduced body size. Development 133:601–610

Konieczny SF, Emerson SP Jr (1984) 5-Azacytidine induction of stable mesodermal stem cell lineages from 10T1/2 cells: evidence for regularoty genes controlling determination. Cell 38:791–800

Kuang S, Charge SB, Seale P, Huh M, Rudnicki MA (2006) Distinct roles for Pax7 and Pax3 in adult regenerative myogenesis. J. Cell Biol 172:103–113

Laclef C, Hamard G, Demignon J, Souil E, Houbron C, Maire P (2003) Altered myogenesis in Six1-deficient mice. Development 130:2239–2252

Lassar AB, Paterson BM, Weintraub H (1986) Transfection of a DNA locus that mediates the conversion of 10T1/2 fibroblasts to myoblasts. Cell 47:649–656

Lu J-R, Bassel-Duby R, Hawkins A, Chang P, Valdez R, Wu H, Gan L, Shelton JM, Richardson JA, Olson EN (2002) Control of facial muscle development by MyoR and Capsulin. Science 298:2378–2381

Mankoo BS, Collins NS, Ashby P, Grigorieva E, Pevny LH, Candia A, Wright CV, Rigby PW, Pachnis V (1999) Mox2 is a component of the genetic hierarchy controlling limb muscle development. Nature 400:69–73

Mansouri A, Stoykova A, Torres M, Gruss P (1996) Dysgenesis of cephalic neural crest derivatives in $Pax7^{-/-}$ mutant mice. Development 122:831–838

McDermott A, Gustafsson M, Elsam T, Hui CC, Emerson CP Jr, Borycki AG (2005) Gli2 and Gli3 have redundant and context-dependent function in skeletal muscle formation. Development 132:345–357

McLoon LK, Rowe J, Wirtschafter J, McCormick KM (2004) Continuous myofiber remodeling in uninjured extraocular myofibers: myonuclear turnover and evidence for apoptosis. Muscle Nerve 29:707–715

Megeney LA, Kablar B, Garrett K, Anderson JE, Rudnicki, MA (1996) MyoD is required for myogenic stem cell function in adult skeletal muscle. Genes & Dev 10:1173–1183

Minasi MG, Riminucci M, De Angelis L, Borello U, Berarducci B, Innocenzi A, Caprioli A, Sirabella D, Baiocchi M, De Maria R, Boratto R, Jaffredo T, Broccoli V, Bianco P, Cossu G (2002) The meso-angioblast: a multipotent, self-renewing cell that originates from the dorsal aorta and differentiates into most mesodermal tissues. Development 129:2773–2783

Montarras D, Lindon C, Pinset C, Domeyne P (2000) Cultured myf5 null and myoD null muscle precursor cells display distinct growth defects. Biol. Cell 92:565–572

Montarras D, Morgan J, Collins C, Relaix F, Cumano A, Partridge T, Buckingham M (2005) Direct isolation of muscle satellite cells demonstrates their major role in skeletal muscle self renewal. Science 309:2064–2067

Morrison GM, Brickman JM (2006) Conserved roles for Oct4 homologues in maintaining multipotency during early vertebrate development. Development 133:2011–2022

Nabeshima Y, Hanaoka K, Hayasaka M, Esumi E, Li S, Nonaka I, Nabeshima Y (1993) Myogenin gene disruption results in perinatal lethality because of severe muscle defect. Nature 364:532–535

Noden DM, Francis-West P (2006) The differentiation and morphogenesis of craniofacial muscles. Dev Dyn 235:1194–1218

Olguin HC, Olwin BB (2004) Pax-7 up-regulation inhibits myogenesis and cell cycle progression in satellite cells: a potential mechanism for self-renewal. Dev Biol 275:375–388

Olson E, Arnold HH, Rigby PWJ, Wold BJ (1996) Know your neighbours: three phenotypes in null mutants of the myogenic bHLH gene *Mrf4*. Cell 85:1–4

Ontell M, Kozeka K (1984) Organogenesis of the mouse extensor digitorum logus muscle: a quantitative study. Am J Anat 171:149–161

Oustanina S, Hause G, Braun T (2004) Pax7 directs postnatal renewal and propagation of myogenic satellite cells but not their specification. EMBO J 23:3430–3439

Pardanaud L, Luton D, Prigent M, Bourcheix LM, Catala M, Dieterlen-Lievre F (1996) Two distinct endothelial lineages in ontogeny, one of them related to hemopoiesis. Development 122:1363–1371

Pin CL, Ludolph DC, Cooper ST, Klocke BJ, Merlie JP, Konieczny, SF (1997) Distal regulatory elements control MRF4 gene expression in early and late myogenic cell populations. Dev Dyn 208:299–312

Pin CL, Konieczny, SF (2002) A fast fiber enhancer exists in the muscle regulatory factor 4 gene promoter. Biochem Biophys Res Commun 299:7–13

Porter JD, Khanna S, Kaminski HJ, Rao JS, Merriam AP, Richmonds CR, Leahy P, Li J, Andrade FH (2001) Extraocular muscle is defined by a fundamentally distinct gene expression profile. Proc Natl Acad Sci USA 98:12062–12067

Porter JD, Merriam AP, Khanna S, Andrade FH, Richmonds CR, Leahy P, Cheng G, Karathanasis P, Zhou X, Kusner LL, Adams ME, Willem M, Mayer U, Kaminski HJ (2003) Constitutive properties, not molecular adaptations, mediate extraocular muscle sparing in dystrophic mdx mice. FASEB J 17:893–895

Pouget C, Gautier R, Teillet MA, Jaffredo T (2006) Somite-derived cells replace ventral aortic heman-gioblasts and provide aortic smooth muscle cells of the trunk. Development 133:1013–1022

Rawls A, Valdez MR, Zhang W, Richardson J, Klein WH, Olson EN (1998) Overlapping functions of the myogenic bHLH genes MRF4 and MyoD revealed in double mutant mice. Development 125:2349–2358

Relaix F, Buckingham M (1999) From insect eye to vertebrate muscle: redeployment of a regulatory network. Genes & Dev 13:3171–3178

Relaix F, Polimeni M, Rocancourt D, Ponzetto C, Schäfer BW, Buckingham M (2003) The transcriptional activator PAX3-FKHR rescues the *Pax3* mutant phenotype and induces a gain of function phenotype with ligand-independent activation of Met signaling. Genes & Dev 17:2950–2965

Relaix F, Rocancourt D, Mansouri M, Buckingham M. (2004) Divergent functions of murine Pax3 and Pax7 in limb muscle development. Genes and Dev 18:1088–1105

Relaix F, Rocancourt D, Mansouri A, Buckingham MA (2005) A Pax3/Pax7-dependent population of skeletal muscle progenitor cells. Nature 435:948–953

Relaix F, Montarras D, Zaffran S, Gayraud-Morel B, Rocancourt D, Tajbakhsh S, Mansouri A, Cumano A, Buckingham M (2006) Pax3 and Pax7 have distinct and overlapping functions in adult muscle progenitor cells. J Cell Biol 172:91–102

Rose O, Rohwedel J, Reinhardt S, Bachmann M, Cramer M, Rotter M, Wobus A, Starzinski-Powitz A. (1994) Expression of M-cadherin protein in myogenic cells during prenatal mouse development and differentiation of embryonic stem cells in culture. Dev Dyn 201:245–259

Rudnicki MA, Braun T, Hinuma S, Jaenisch R (1992) Inactivation of MyoD in mice leads to up-regulation of the myogenic HLH gene Myf-5 and results in apparently normal muscle development. Cell 71:383–390

Rudnicki MA, Schnegelsberg PN, Stead RH, Braun T, Arnold HH, Jaenisch R (1993) MyoD or Myf-5 is required for the formation of skeletal muscle. Cell 75:1351–1359

Sabourin LA, Girgis-Gabardo A, Seale P, Asakura A, Rudnicki MA (1999) Reduced differentiation potential of primary MyoD-/- myogenic cells derived from adult skeletal muscle. J Cell Biol 144:631–643

Sampaolesi M, Torrente Y, Innocenzi A, Tonlorenzi R, D'Antona G, Pellegrino MA, Barresi R, Bresolin N, De Angelis MG, Campbell KP, Bottinelli R, Cossu G (2003) Cell therapy of alpha-sarcoglycan null dystrophic mice through intra-arterial delivery of mesoangioblasts. Science 301:487–492

Sampaolesi M, Blot S, D'Antona G, Granger N, Tonlorenzi R, Innocenzi A, Mognol P, Thibaud JL, Galvez BG, Barthelemy I, Perani L, Mantero S, Guttinger M, Pansarasa O, Rinaldi C, Cusella De Angelis MG, Torrente Y, Bordignon C, Bottinelli R, Cossu G (2006) Mesoangioblast stem cells ameliorate muscle function in dystrophic dogs. Nature 444:574–579

Schienda J, Engleka KA, Jun S, Hansen MS, Epstein JA, Tabin CJ, Kunkel LM, Kardon G (2006) Somitic origin of limb muscle satellite cell and side population cells. Proc Natl Acad Sci USA 103:945–950

Seale P, Sabourin LA, Girgis-Gabardo A, Mansouri A, Gruss P, Rudnicki MA (2000) Pax7 is required for the specification of myogenic satellite cells. Cell 102:777–786

Shen H, McElhinny AS, Cao Y, Gao P, Liu J, Bronson R, Griffin JD, Wu L (2006) The Notch coactivator, MAML1, functions as a novel coactivator for MEF2-C-mediated transcription and is required for normal myogenesis. Genes Dev 20:675–688

Shinin V, Gayraud-Morel B, Gomes D, Tabjakhsh S (2006) Asymmetric division and cosegregation of template DNA strands in adult muscle satellite cells. Nat Cell Biol 8:677–687

Spitz F, Demignon J, Porteu A, Kahn A, Concordet J-P, Daegelen D, Maire P (1998) Expression of myogenin during embryogenesis is controlled by Six/sine oculis homeoproteins through a conserved MEF3 binding site. Proc Natl Acad Sci USA 95:14220–14225

Summerbell D, Ashby PR, Coutelle O, Cox D, Yee S, Rigby PW (2000) The expression of Myf5 in the developing mouse embryo is controlled by discrete and dispersed enhancers specific for particular populations of skeletal muscle precursors. Development 127:3745–3757

Summerbell D, Halai C, Rigby PW (2002) Expression of the myogenic regulatory factor Mrf4 precedes or is contemporaenous with that of Myf5 in the somitic bud. Mech. Dev 117:331–335

Tajbakhsh S, Buckingham ME (1994) Mouse limb muscle is determined in the absence of the earliest myogenic factor myf-5. Proc Natl Acad Sci USA 91:747–751

Tajbakhsh S, Rocancourt D, Buckingham M (1996a) Muscle progenitor cells failing to respond to positional cues adopt non-myogenic fates in *myf-5* null mice. Nature 384:266–270

Tajbakhsh S, Bober E, Babinet C, Pournin S, Arnold H, Buckingham M (1996b) Gene targeting the *myf-5* locus with *nlacZ* reveals expression of this myogenic factor in mature skeletal muscle fibres as well as early embryonic muscle. Dev Dynamics 206:291–300

Tajbakhsh S, Rocancourt D, Cossu G, Buckingham M (1997) Redefining the genetic hierarchies controling skeletal myogenesis: *Pax3* and *Myf-5* act upstream of MyoD. Cell 89:127–138

Tajbakhsh S, Borello U, Vivarelli E, Kelly R, Papkoff J, Duprez D, Buckingham M, Cossu G (1998) Differential activation of Myf5 and MyoD by different Wnts in explants of mouse paraxial mesoderm and the later activation of myogenesis in the absence of Myf5. Development 125:4155–4162

Tajbakhsh S, Buckingham M (2000) The birth of muscle progenitor cells in the mouse: spatiotemporal considerations. In: Ordahl CP (ed) Current topics in developmental biology: Somitogenesis, vol. 47. Academic Press, pp 225–268

Tapscott SJ (2005) The circuitry of a master switch: MyoD and the regulation of skeletal muscle gene transcription. Development 132:2685–2695

Teboul L, Hadchouel J, Daubas P, Summerbell D, Buckingham M, Rigby PWJ (2002) The early epaxial enhancer of Myf5 is essential for the initial transcription of this myogenic determination gene in the somite but not for subsequent expression in the myotome. Development 129:4571–4580

Teboul L, Summerbell D, Rigby PW (2003) The initial somitic phase of Myf5 expression requires neither Shh signaling nor Gli regulation. Genes Dev 17:2870–2874

Thompson AL, Filatov G, Chen C, Porter I, Li Y, Rich MM, Kraner SD (2005) A selective role for MRF4 ininnervated adult skeletal muscle: Na(V) 1.4 Na+ channel expression is reduced in MRF4-null mice. Gene Expr 12:289–303

Tzahor E, Kempf H, Mootoosamy RC, Poon AC, Abzhanov A, Tabin CJ, Dietrich S, Lassar AB (2003) Antagonists of Wnt and BMP signaling promote the formation of vertebrate head muscle. Genes Dev 17:3087–3099

Vasyutina E, Stebler J, Brand-Saberi B, Schulz S, Raz E, Birchmeier C (2005) CXCR4 and Gab1 cooperate to control the development of migrating muscle progenitor cells. Genes Dev 19:2187–2198

Venuti JM, Morris JH, Vivian JL, Olson EN, Klein, WH (1995) Myogenin is required for late but not early aspects of myogenesis during mouse development. J Cell Biol 128:563–576

Vivian JL, Gan L, Olson EN, Klein WH (1999) A hypomorphic *myogenin* allele reveals distinct *myogenin* expression levels required for viability, skeletal muscle development and sternum formation. Dev Biol 208:44–55

Vivian JL, Olson EN, Klein WH (2000) Thoracic skeletal defects in myogenin- and MRF4-deficient mice correlate with early defects in myotome and intercostal musculature. Dev Biol 224:29–41

Vorobyov E, Horst J (2004) Expression of two protein isoforms of PAX7 is controlled by competing cleavage-polyadenylation and splicing. Gene 342:107–112

Wang ZZ, Washabaugh CH, Yao Y, Wang JM, Zhang L, Ontell MP, Watkins SC, Rudnicki MA, Ontell, M (2003) Aberrant development of motor axons and neuromuscular synapses in MyoD-null mice. J Neurosci 23:5161–5169

Weintraub H, Davis R, Tapscott S, Thayer M, Krause M, Benezra R, Blackwell TK, Turner D, Rupp R, Hollenberg S et al (1991) The MyoD gene family: nodal point during specification of the muscle cell lineage. Science 251:761–766

Wright WE (1984) Induction of muscle genes in neural cells. J Cell Biol 98:427–435

Wright WE, Sassoon DA, Lin VK (1989) Myogenin, a factor regulating myogenesis, has a domain homologous to MyoD. Cell 56:607–617

Yablonka-Reuveni Z, Rudnicki MA, Rivera AJ, Primig M, Anderson JE, Natanson P (1999) The transition from proliferation to differentiation is delayed in satellite cells from mice lacking MyoD. Dev Biol 210:440–455

Yee SP, Rigby PW (1993) The regulation of myogenin gene expression during the embryonic development of the mouse. Genes & Dev 7:1277–1289

Zammit PS, Golding JP, Nagata Y, Hudon V, Partridge TA, Beauchamp JR (2004a) Muscle satellite cells adopt divergent fates: a mechanism for self-renewal? J Cell Biol 166:347–357

Zammit PS, Carvajal JJ, Golding JP, Morgan JE, Summerbell D, Zolnerciks J, Partridge TA, Rigby PW, Beauchamp JR (2004b) Myf5 expression in satellite cells and spindles in adult muscle is controlled by separate genetic elements. Dev Biol 273:454–465

Zammit PS, Relaix F, Nagata Y, Ruiz AP, Collins CA, Partridge TA, Beauchamp JR (2006) Pax7 and myogenic progression in skeletal muscle satellite cells. J Cell Sci 119:1824–1832

Zhao P, Caretti G, Mitchell S, McKeehan WL, Boskey AL, Pachman LM, Sartorelli V, Hoffman EP (2006) Fgfr4 is required for effective muscle regeneration in vivo. Delineation of a MyoD-Tead2-Fgfr4 transcriptional pathway. J Biol Chem 281:429–438

Zhou Z, Bornemann A (2001) MRF4 protein expression in regenerating rat muscle. J Muscle Res Cell Motil 22:311–316

CHAPTER 3

THE MUSCLE SATELLITE CELL: THE STORY OF A CELL ON THE EDGE!

PETER S. ZAMMIT

King's College London, Randall Division of Cell and Molecular Biophysics, New Hunt's House, Guy's Campus, London, SE1 1UL, UK

1. INTRODUCTION

We have the Romans to thank for the word muscle, which is derived from the latin "musculus" (little mouse), since they thought that muscle movement resembled tiny mice running around under the skin. Skeletal muscle is composed of multiple muscle fibres (myofibres) that are responsible for producing force by contraction. The myofibre is a highly structured syncytium consisting of repeated arrays of contractile elements controlled by a reticular network, and sustained by hundreds of peripherally located post-mitotic myonuclei. Residing between the plasmalemma of the myofibre and the surrounding basal lamina are satellite cells (anatomical name: myosatellitocytus). These mono-nucleated cells are responsible for supplying myonuclei to muscle for both the routine needs of growth and turnover and the more sporadic demands for repair and regeneration.

2. TIME FOR A BRIEF HISTORY

Since many of the early studies on muscle regeneration and satellite cells can now be readily accessed online, it might be useful to start with a personal view of some of the seminal discoveries in the field. Early observations on muscle regeneration appeared in the mid to late 19th Century, reporting that under certain circumstances, muscle damage resulted in nuclear proliferation and formation of new fibres (references included in Mazanet and Franzini-Armstrong, 1986). Highlights from the first half of the 20th century include Lewis and Lewis's examination of myogenesis in culture (Lewis and Lewis, 1917), Speidel's elegant hand illustrated

S. Schiaffino and T. Partridge (eds.), Skeletal Muscle Repair and Regeneration, 45–64.

study of single tadpole myofibre regeneration (Speidel, 1938) and Le Gros Clark's detailed observations on muscle regeneration in rabbit following ischemia induced degeneration (Le Gros Clark, 1946). A series of studies in Russia by Studitsky in the 1950s and 1960s showed that both rodent and bird muscle can be removed, minced and replaced and will still regenerate to form a new functional muscle (e.g. Studitsky, 1964, reviewed in Carlson, 1968). The presence of mitotic figures (Lash et al., 1957) and incorporation of tritiated thymidine (Bintliff and Walker, 1960) in mono-nucleated cells in regenerating muscle, combined with electron microscopic examination (Allbrook, 1962), revealed that regeneration was not an amitotic event as had been suggested. Myoblast culture experiments were then used to demonstrate that large numbers of myoblasts fused together to form multi-nucleated skeletal myofibers (Cooper and Konigsberg, 1961; Stockdale and Holtzer, 1961). These studies however, did not resolve the question of whether mammalian myoblasts arose from dedifferentiated myonuclei, as suggested in newt (e.g. Hay, 1959), or arose from mono-nucleated myogenic precursor cells.

While exploring the innervation of the frog muscle spindle using electron microscopy, Katz mentions that "the surface of many muscle fibres is invested here and there with hypectolemmal satellite cells" and speculated that they may be associated with development and growth of the fibre (Katz, 1961). In the same year, Mauro described a cell "wedged between the plasma membrane of the muscle fibre and basement membrane" in extrafusal myofibres of frog tibialis anticus muscle (Mauro, 1961). "Alerting" other investigators resulted in similar cells being found in other frog muscles and in rat tongue and sartorius (illustrated in the Mauro manuscript with an image supplied by G. Palade). Mauro states that these cells intimately associated with muscle fibres "we have chosen to call satellite cells", thus marrying the defining anatomical location with the name (Mauro, 1961). A few years later, more detailed morphological descriptions of mouse and bat (possibly the species finest hour in muscle biology!) satellite cells appeared (Muir et al., 1965; Venable, 1966). Satellite cells were also found to be osmotically independent of the underlying myofibre, confirming the electron microscopic observations of no continuity between the satellite cell cytoplasm and that of the muscle fibre (Muir et al., 1965).

The satellite cell was quickly assumed to be the major source of myonuclei for muscle growth (Macconnachie et al., 1964; Shafiq et al., 1968) and regeneration (Price et al., 1964; Shafiq and Gorycki, 1965; Church et al., 1966). In growing muscle, satellite cells were shown to be able to undergo mitosis (Shafiq et al., 1968) before Moss and Leblond established that satellite cells but not myonuclei, were able to incorporate tritiated thymidine when analysed shortly after the pulse (3, 6 and 10 hrs). The appearance of label in myonuclei did not occur until at least 24 hrs later, implying that the labelled satellite cells had given rise to myonuclei following cell division. Thus, satellite cells were attributed as the main source of myonuclei during post natal muscle growth (Moss and Leblond, 1970, 1971). Debate continued though, about the source of new myonuclei during muscle regeneration (see Carlson, 1973 for a topical discussion). The connection between satellite cells and proliferating myoblasts was effectively proved by culture of adult myofibres, isolated complete with their associated satellite cells still resident in their niche on the myofiber surface. Such cultures demonstrated that intact myofibers were a source of proliferative myoblasts, that could subsequently fuse into myotubes (Bischoff, 1975; Konigsberg et al., 1975). Since the myofiber nuclei are permanently withdrawn from the cell cycle (e.g. Bischoff, 1986), it was reasoned that the only cells that could give rise to the myogenic progeny were satellite cells. Changes in morphology during the weeks immediately following birth were deemed consistent with the satellite cell becoming less metabolically active (Schultz, 1976), with low levels of incorporated tritiated thymidine in adult, showing that they were also mitotically quiescent (Schultz et al. 1978). Cell or muscle transplantation assays were then employed to show that satellite cells re-entered the cell cycle and gave rise to new myonuclei following damage to adult muscle in vivo (Snow, 1977, 1977a, 1978; Lipton and Schultz, 1979).

Work by Christ and others in the early 1970s established that myogenic cells arise from somites, transitory mesoderm-derived structures formed in pairs on either side of the neural tube during embryonic development in vertebrates (reviewed in Stockdale et al., 2000). Somites differentiate into the dermomyotome and sclerotome with mesodermal cells specified as muscle precursors in the nascent myotome. Cells migrating from the somites were demonstrated to also be responsible for providing myogenic precursors to populate the muscle fields of the limbs (Stockdale et al., 2000). Transplantation experiments in chick were used to then demonstrate that the somite was also the origin of most satellite cells (Armand et al., 1983).

Different characteristics of embryonic and foetal myoblasts provided evidence that distinct myogenic lineages arose during development (e.g. Bonner and Hauschka, 1974). Response to a tumour promoter was an early indication that satellite cells too composed a separate lineage (Cossu et al., 1983) with criteria such as myosin heavy chain content after differentiation, further supporting the notion (Feldman and Stockdale, 1992; Hartley et al., 1992). Thus satellite cells are specified before they can first be distinguished on anatomical criteria after formation of the basal lamina in late foetal development in mouse (Ontell and Kozeka, 1984).

The description of the myogenic regulatory factor family (Myf5, MyoD, myogenin and MRF4) in the late 1980s was a major breakthrough in myogenesis (reviewed in Weintraub et al., 1991). Since satellite cell-derived myoblasts also express these transcription factors (Grounds et al., 1992; Fuchtbauer and Westphal, 1992; Yablonka-Reuveni and Rivera, 1994), developmental and post-natal myogenesis were confirmed as related processes also on the molecular level. Related yet not identical, because these transcription factors also highlighted differences: Lack of MyoD is not detrimental to muscle development but muscle regeneration is severely affected (Megeney et al., 1996), while Mrf4 acts as a myogenic determination factor only in embryonic cells (Kassar-Duchossoy et al., 2004) and myogenin is not required for differentiation after birth (Knapp et al., 2006). This dichotomy was most strikingly revealed by the lack of another transcription factor Pax7, in whose absence only post-natal myogenesis is obviously perturbed (Seale et al., 2000).

It was not until the description of the adhesion molecule M-Cadherin in 1994 though, that rodent quiescent satellite cells finally got a useful molecular marker, allowing their identification at the light microscope level (Irintchev et al., 1994). The lack of a reliable and widely available antibody to M-Cadherin though, meant that progress remained slow until several years later, when further reagents were described. The observation that satellite cells express Pax7 was crucial, since a good Pax7 antibody was commercially available (Seale et al., 2000). In the same year, the Myf5/β–gal fusion protein from the targeted allele of the $Myf5^{nlacZ/+}$ mouse was shown to be expressed in quiescent satellite cells, as was the saliomucin CD34 (Beauchamp et al., 2000).

Moss and Leblond proposed that in growing muscle, dividing satellite cells gave rise to both myonuclei and more satellite cells, leading to the idea that satellite cells maintained their population by self-renewal (Moss and Leblond, 1971). In vitro observations on myogenic cell proliferation supported the notion of asymmetric divisions and the presence of a myogenic stem cell (Quinn et al., 1985; 1984; Baroffio et al., 1996). This assumption that a viable satellite cell pool was maintained by self-renewal was brought into question by observations that bone marrow cells could be incorporated into myofibres (Ferrari et al., 1998), which initiated a series of investigations into the provenance of satellite cells (reviewed in Zammit et al., 2006a). Transplantation studies using the Myf5nlacZ/+ mouse revealed that donor cells could contribute functional satellite cells to the host (Heslop et al., 2001). Grafting of single myofibres with their associated satellite cells then established that satellite cells could self-renew (Collins et al., 2005 – discussed in detail in Section 5). This brings us up to date (February 2007), with the predominant opinion being that some satellite cells at least (Shinin et al., 2006), can be considered as adult stem cells.

3. IDENTIFICATION OF SATELLITE CELLS

3.1 Morphology and Anatomical Location of Satellite Cells

A generalised morphological description of a mammalian quiescent satellite cell is a bipolar cell with long processes. The nucleus contains abundant heterochromatin and there is a paucity of cytoplasm containing few mitrochondria, little rough enoplasmic reticulum, undeveloped Golgi apparatus and no myofilaments (reviewed in Bischoff, 1994). Despite their intimate association, satellite cells exhibit no specialised junctional complexes with the underlying myofibre, and are not electrically coupled (Bader et al., 1988), but do exhibit numerous caveolae on both surfaces of the cell. The absolute definition of a satellite cell remains anatomical though, i.e. a cell that resides between the basal lamina and plasmalemma of a myofibre (Mauro, 1961). Thus technically, satellite cells can only be identified on this anatomical criterion after the basal lamina has formed in late foetal development (~E18 in mouse -Ontell and Kozeka, 1984).

Satellite cells in adult rodent muscle comprise between 5–10% of myofibre nuclei (Zammit et al., 2002; Halevy et al., 2004, reviewed in Hawke and Garry, 2001).

Where myonuclei are also located on the periphery of muscle fibres as occurs in mammals, it is necessary to distinguish between the sub-plasmalemmal myonucleus and the supra-plasmalemmal satellite cell. Combined use of markers of the basal lamina (e.g. laminin) and the plasmalemma (e.g. dystrophin) on tissue sections has been used to detect satellite cells. However, clearly resolving between these two structures at the light level is difficult, thus essentially limiting their unambiguous identification on anatomical criteria to electron microscopy. Cells in this location have been described in most vertebrates including mammals, birds, amphibia and fish (references included in Mazanet and Franzini-Armstrong, 1986). Since any cell in this location is by definition, a satellite cell, irrespective of antigen expression, or technically, even function, this criterium alone has limitations. Thus the definition of a satellite cell really needs to be broadened to maybe also encompass a defined gene expression profile (Pax7 expression?) and have some relation to myogenic capacity, in addition to this traditional anatomical definition (as commented on in Muir et al. as far back as 1965!).

3.2 Molecular Markers of Satellite Cells

The advent of molecular markers of quiescent satellite cells has radically changed the field, and the most widely used in mouse are probably Pax7 (Seale et al., 2000), M-Cadherin (Irintchev et al., 1994), CD34 and the Myf5/β-gal fusion protein from the targeted allele of the $Myf5^{nlacZ/+}$ mouse (Beauchamp et al., 2000). Recently we have shown that satellite cells also have high levels of sphingomyelin in their plasma membranes, and this sphingolipid can be detected using the protein lysenin (Nagata et al., 2006a). Other reported markers of quiescent satellite cells now include caveolin-1 (Volonte et al., 2005), the cell-surface heparin sulphate proteoglycans syndecan 3 and 4 (Cornelison et al., 2001), FoxK1 (formerly myocyte nuclear factor) (Garry et al., 1997), Sox 8 (Schmidt et al., 2003), Sox 15 (Lee et al., 2004), a nestin transgene (Day et al., 2007) and the antibody SM/C2.6 (Fukada et al., 2004). In our hands, it is clear that the majority of quiescent satellite cells in mice of reproductive age express M-Cadherin, Pax7, CD34, Myf5/β-gal, caveloin-1 and have high levels of sphingomyelin in their plasma membranes (Fig. 1). It should be noted though that in "old" mice, cells in the satellite cell position can lose expression of markers such as Pax7 (Shefer et al., 2006; Collins et al., 2007).

Molecular markers that recognise antigens on the cell surface, or genetic modifications that result in production of fluorescent proteins are particularly useful since they allow satellite cells to be purified using fluorescently activated cell sorting techniques. This has now successfully been carried out with the $Pax3^{eGFP/+}$ mouse, in which the *Pax3* locus has been targeted with *eGFP* (Montarras et al., 2005) and a nestin transgenic mouse (Day et al., 2007). Some of these antibodies can be used for identifying satellite cells in other species, e.g. Pax7 in chicken (Halevy et al., 2004), salamander (Morrison et al., 2006) and Xenopus (Chen et al., 2006), while both Pax7 (Reimann et al., 2004) and N-CAM are useful in man (Illa et al., 1992; Dreyer et al., 2006). It is important to note though, that while these markers are all effective

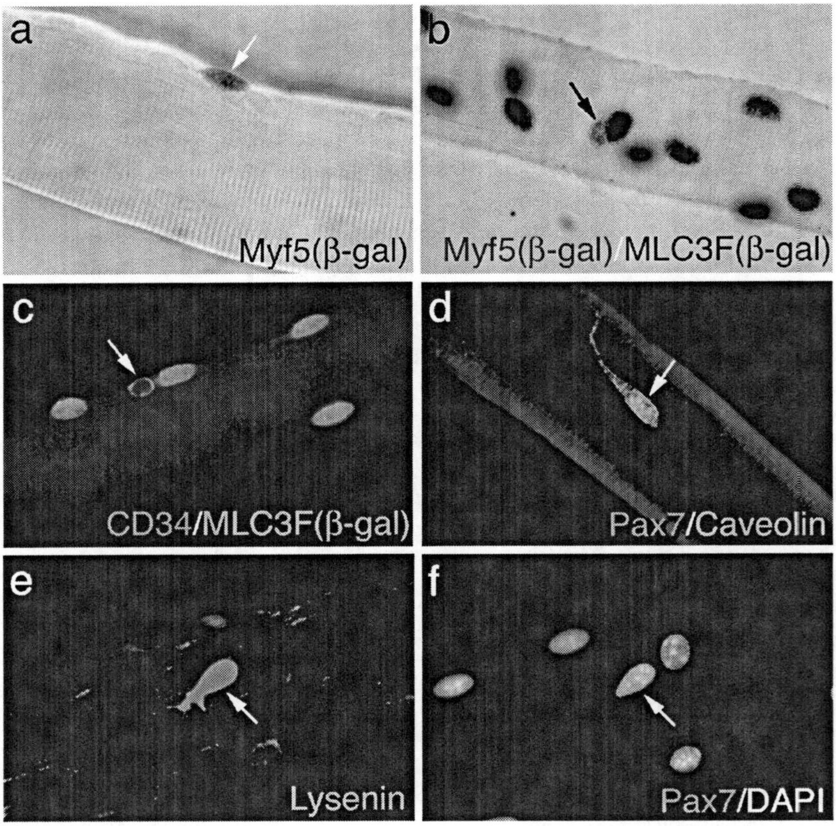

Figure 1. **Molecular markers of quiescent satellite cells** An isolated EDL myofibre from a $Myf5^{nlacZ/+}$ mouse (a) incubated in X-gal to reveal the presence of the Myf5/β-gal fusion protein in the nucleus of a quiescent satellite (arrow in a). The *3F-nlacZ-E* transgene marks all fast myonuclei and so provides an easy method to identify the whole satellite cell population on a myofibre, independently of antigen expression. Crossing $Myf5^{nlacZ/+}$ and *3F-nlacZ-E* mice allows myonuclei and satellite cells to be discriminated (b), since incubation in X-gal can be used to reveal myonuclei and salmon-gal to detect the Myf5/β-gal fusion protein in satellite cells (arrow in b). Immunostaining a *3F-nlacZ-E* myofibre with CD34 (c – red) identifies a satellite cell, distinguished from the surrounding myonuclei by lack of *3F-nlacZ-E* transgene expression, as revealed with an anti-β-gal antibody (c – green). Pax7 is a useful marker of satellite cells since there is a good commercially available antibody. Here an isolated EDL myofibre is immunostained with Pax7 (d – red) together with caveolin-1 (d – green) to show the satellite cell surface (arrow in d). Quiescent satellite cells have high levels of sphingomyelin in their plasma membrane (arrow in e and f). Co-immunostaining for the sphingomyelin binding protein lysenin (e – green) and Pax7 (f – red) identifies a satellite cell, with 4,6-diamidino-2-phenylindole (DAPI) used to counterstain all nuclei present

for identifying satellite cells on isolated myofibres, most of them are not specific to satellite cells (e.g. CD34, lysenin, syndecan 3 and 4). For applications such as detecting satellite cells on muscle sections, Pax7, M-Cadherin and, where possible, the Myf5/β-gal fusion protein, are probably the most useful.

3.3 Do Molecular Markers Reveal Satellite Cell Heterogeneity?

The *3F-nlacZ-E* transgene is composed of regulatory regions from the locus of the mouse *myosin light chain 1/3F* gene (Kelly et al., 1995), whose protein products are integral components of the contractile apparatus of fast myofibres. Since this transgene marks all myonuclei in fast myofibers, any nucleus under the basal lamina of an isolated myofibre that does not contain β-galactosidase is a satellite cell. This allows the unbiased identification of the total population of satellite cells on a myofibre, irrespective of antigen expression (Beauchamp et al., 2000). By use of this transgene, together with CD34 and the $Myf5^{nlacZ/+}$ mouse, we have shown that some molecular markers do not label all quiescent satellite cells, showing heterogeneity within the satellite cell pool of young adult mice (Beauchamp et al., 2000). More recently, it has been shown that a small percentage of satellite cells identified with Pax7 do not contain the Myf5/β-gal fusion protein (Day et al., 2007).

Satellite cell heterogeneity is most vividly illustrated in the $Pax3^{eGFP/+}$ mutant mouse. This heterogeneity is manifested not only between different muscles, but also within a particular muscle. For example, of the hind limb musculature, only the gracilus muscle contains significant numbers of *eGFP (Pax3)* expressing satellite cells, while in contrast, the forelimbs and upper body contain many muscles with large numbers of eGFP + ve satellite cells (Montarras et al., 2005; Relaix et al., 2006). However, even within a muscle, expression of *eGFP* is not uniform. While the vast majority of eGFP + ve satellite cells also express Pax7 in the diaphragm, other satellite cells only express *Pax7* or *eGFP* (Relaix et al., 2006). This heterogeneity does not appear to be related to either ontogeny of the muscle, or muscle fibre type composition. When transplanted, cells maintain *eGFP* expression in a host muscle environment that does not normally contain such cells, indicating that this is probably lineage based (Montarras et al., 2005). Whether this targeted allele reflects endogenous *Pax3* expression in adult muscle though, is in debate (Horst et al., 2006; Day et al., 2007). It has been reported that activated satellite cells transiently contain Pax3 protein (Conboy and Rando, 2002; Shinin et al., 2006) but the targeted *Pax3* allele is not up-regulated in injured muscle (Montarras et al., 2005).

4. REGULATION OF SATELLITE CELL ACTIVATION

4.1 Satellite Cell Activation

In rat, satellite cells form ~32% of sublaminal nuclei at birth, which then drops to the adult value of ~5% by 2 months of age (Bischoff, 1994), by which time most have become mitotically quiescent (Schultz et al., 1978). In the haematopoetic system, quiescence is an active transcriptional state (Yusuf and Fruman, 2003), and although little is known about quiescence in myogenic cells, it seems probable that, rather than constituting a passive "default" position, satellite cell quiescence is also actively maintained (reviewed in Dhawan and Rando, 2005).

To be able to contribute myonuclei to skeletal muscle for hypertrophy, repair and regeneration, satellite cells must be activated from this normal quiescent state to overcome the G0-G1 block and proliferate. There is currently a paucity of markers that can be used to partition the stages of activation, before the first cell division occurs at around 30hr after muscle injury (McGeachie and Grounds, 1987). One of the earliest indications that a satellite cell has been activated is the co-expression of MyoD with Pax7, which occurs within hours of stimulation in vitro (Yablonka-Reuveni and Rivera, 1994; Zammit et al., 2004) or in vivo (Grounds et al., 1992). Other landmark events of activation include isoform switches of CD34 (Beauchamp et al., 2000) and FoxK1, with mice null for FoxK1 exhibiting defective satellite cell activation (Garry et al., 2000). Since most means of obtaining satellite cells lead to activation as shown by MyoD expression, methods of induced "quiescence" in myogenic cells such as the reserve cell model (Baroffio et al., 1996; Kitzmann et al., 1998; Yoshida et al., 1998) or maintenance of myogenic cells in non-adherent cultures are also employed (Milasincic et al., 1996; Sachidanandan et al., 2002). In contrast, the vast majority of satellite cells isolated together with a myofibre do not express MyoD (Zammit et al., 2002), indicating that even if the activation process has been initiated, it is not advanced.

4.2 What Triggers Satellite Cell Activation?

The stimuli that trigger satellite cell activation for growth and routine myonuclear turnover are likely to be different from those that elicit a more widespread and synchronous response for muscle regeneration. It is not surprising therefore, that diverse biochemical and biomechanical stimuli including ischemia, stretch, nitric oxide and denervation have all been implicated. These various stimuli however, may well all act locally by inducing release of hepatocyte growth factor (HGF). Satellite cells express Cmet, the HGF receptor, while HGF is present in muscle in an active form in the extracellular matrix (Tatsumi and Allen, 2004). Exogenous HGF has been shown to elicit satellite cell activation in vivo, which is prevented by the application of blocking antibodies (Tatsumi et al., 1998). It is crucial to distinguish satellite cell activation and subsequent entry into the cell cycle from the maintenance of proliferation once a permissive environment is attained. While HGF, fibroblast growth factors (FGF), and insulin-like growth factors have all been shown to promote proliferation, HGF alone has been suggested to be involved in activation of satellite cells from quiescence (Tatsumi et al., 1998). Other studies however, have demonstrated similar effects of FGF and HGF on recruitment of satellite cells to proliferation (Yablonka-Reuveni et al., 1999b; Kastner et al., 2000). Several FGF receptor isoforms (particularly FGF-R1 and FGF-R4), together with syndecan-3 and syndecan-4 that are required for signaling through such tyrosine kinase type receptors, are present on satellite cells (Cornelison et al., 2001; Kastner et al., 2000). Disruption of these signalling pathways, for example in syndecan 4 null mice, produces defective satellite cell activation (Cornelison et al., 2004).

4.3 What Signalling Pathways Control Satellite Cell Activation and Proliferation?

Since Cmet and FGF-R are tyrosine kinase type receptors, the role of kinase-mediated signalling in the control of satellite cell activation has been investigated. Such analysis has revealed that quiescent and activated satellite cells immunostain for the mitogen activated protein kinase (MAPK) ERK1/2 (Yablonka-Reuveni et al., 1999b; Shefer et al., 2001) and that HGF can lead to ERK phosphorylation in myogenic cells (Volonte et al., 2005). Activation of the p38β/β family of MAPK occurs concomitantly with satellite cell activation, while inhibition of p38β/β allows satellite cells to remain quiescent by insulating them from external stimuli (Jones et al., 2005). These observations raise the question of what links the initiation signal for satellite cell activation to these protein kinases?

We have recently shown that the bioactive lipid, sphingosine-1-phosphate (S1P), has an important role as a regulator of satellite cells and muscle regeneration (Nagata et al., 2006a; 2006b). S1P and homologous phosphorylated long-chain sphingoids act in diverse organisms, from mammals through to worms, flies, slime mould, yeast and plants (reviewed in Spiegel and Milstien, 2003). In mammals, S1P is mitogenic for several cell types including fibroblasts and endothelial cells (Olivera et al., 1999; Olivera and Spiegel, 1993; Zhang et al., 1991). While SIP can initiate entry of myogenic cells into the cell cycle, it does not appear to be able to maintain proliferation in the absence of mitogens, unlike lipids such as lysophosphatidic acid (Nagata et al., 2006b; Yoshida et al., 1998). Upon activation signals, sphingomyelin located in the inner leaflet of the plasma membrane is hydrolysed by neutral sphingomyelinase to initiate a signalling cascade that results in generation of S1P. Targeting sphingosine kinase (which generates S1P from sphingosine) and neutral sphingomyelinase with pharmacological inhibitors reduces the number of satellite cells able to divide in response to mitogen stimulation (Nagata et al., 2006b). As would be expected from these observations, inhibiting S1P production after muscle damage perturbs subsequent regeneration. It remains to be determined what initiates this sphingolipid signalling cascade, and whether the mitogenic action is elicited by S1P acting directly as a second messenger, or signalling through any of the five known S1P ($S1P_{1-5}$) receptors. Crosstalk between sphingolipid and kinase signalling has been shown in certain cell types. VEGF binding to VEGF-R2 stimulates endothelial cell growth through protein kinase C, which leads to the activation of sphingosine kinase 1 and generation of S1P that in turn activates MAPK signalling resulting in cell division (Shu et al., 2002). We are currently investigating whether sphingolipid signaling interacts with MAPK pathways during satellite cell activation and proliferation.

5. MAINTENANCE OF THE SATELLITE CELL POOL

Satellite cell-derived myoblasts proliferate and differentiate rapidly after injury (McGeachie and Grounds, 1987) so that the generation of large numbers of new myotubes can occur within as little as 3–4 days (Whalen et al., 1990). This rate and extent of repair is particularly striking considering how few satellite cells are associated with each myofibre. For example, a myofibre from the mouse extensor digitorum longus muscle has only 5–7 associated satellite cells, requiring a rapid expansion in their number to replace the mean 274 myofibre nuclei in a biological time frame (Zammit et al., 2002). More remarkably, this regenerative ability is retained even after repeated cycles of degeneration/regeneration following extensive injury using myotoxins (Basson and Carlson, 1980; Luz et al., 2002; Sadeh et al., 1985), which in the case of the Luz study, consisted of 50 consecutive weekly injections of toxin! Therefore an effective mechanism must exist to maintain a viable satellite cell pool.

5.1 Is the Satellite Cell Pool Maintained by Non-satellite Cell Progenitors?

The established dogma is that satellite cells maintain their population by self-renewal. In the last few years however, several reports have emerged challenging this paradigm and suggesting that satellite cells could be replenished from non-satellite progenitors. These non-mutually exclusive populations could either reside within muscle, or be able to "home" to muscle via the circulation. Candidates for the resident satellite cell progenitor include endothelial-associated cells (De Angelis et al., 1999), interstitial cells (Kuang et al., 2006; Polesskaya et al., 2003; Tamaki et al., 2002) and side population cells (Asakura et al., 2002; Gussoni et al., 1999). Some of these cell types have been shown, under certain circumstances, to be able to occupy the satellite cell niche, but with very low efficiency. Whether this is a biologically relevant source of satellite cells though, remains unproven (reviewed in Zammit et al., 2006a).

Of the proposed satellite cell progenitors located outside muscle tissue, bone marrow-derived cells are the best studied. The seminal discovery that bone marrow cells could be incorporated into myofibres and express a muscle specific transgene (Ferrari et al., 1998) initiated a series of investigations into non-satellite cell-derived muscle. It also prompted the question of whether these cells needed to first contribute to the satellite cell pool or could become directly incorporated into myofibres. Certainly following bone marrow transplant, donor cells can be observed in association with myofibers and can express some markers of bona fide satellite cells (Dreyfus et al., 2004; LaBarge and Blau, 2002). In general though, occupation of the satellite cell niche by non-satellite cells is a rare event, and the niche seems unable to instruct bone marrow-derived cells to function as myogenic precursors in a manner similar to satellite cells (Sherwood et al., 2004). Indeed, a large proportion of bone marrow-derived cells incorporating into the syncitial myofibre itself, and so exposed to the common cytoplasm, fail to activate the myogenic programme (Lapidos et al., 2004; Wernig et al., 2005).

5.2 Is the Satellite Cell Pool Maintained by Self-renewal

During postnatal growth, the evidence indicates that satellite cells are maintained by self-renewal, with an asymmetric cell division resulting in one daughter cell fusing with the myofibre while the second continues to proliferate or becomes quiescent (Moss and Leblond, 1971). Diversification in satellite cell fate can indeed be observed using molecular markers in growing (Baroffio et al., 1996; Halevy et al., 2004; Schultz et al., 2006) and adult muscle (Kitzmann et al., 1998; Shefer et al., 2006; Zammit et al., 2004). Satellite cells activate to co-express Pax7 and MyoD before undergoing their first division between 24–48 hrs of culture. After further proliferation, the majority then suppress Pax7, maintain MyoD, express myogenin and differentiate. Others though, down-regulate MyoD and maintain Pax7 (Halevy et al., 2004; Zammit et al., 2004), which remains transcriptionally active as shown by expression of a transgenic construct containing concatermerised Pax3/7 binding sites controlling a minimal promoter (Zammit et al., 2006b). These cells eventually exit the cell cycle and enter a state resembling quiescence, expressing Pax7 with high levels of sphingomyelin restored to their plasma membranes (Nagata et al., 2006a) and re-expression of a nestin transgene associated with myogenic quiescence (Day et al., 2007). This is consistent with observations in the *MyoD*$^{-/-}$ mouse, where the absence of MyoD delays differentiation in satellite cells (Sabourin et al., 1999; Yablonka-Reuveni et al., 1999a). Importantly, these Pax7 + ve/MyoD-ve cells can be re-stimulated and will again up-regulate MyoD, divide and differentiate (Zammit et al., 2004). Together these observations provide a mechanism whereby some satellite cells "opt out" of differentiation and so could replenish the satellite cell pool.

5.3 Role of Pax and Notch Genes in Determining Satellite Cell Fate

Satellite cells are severely depleted during the early postnatal period in mice lacking Pax7, with those from muscles in which the *Pax3* locus remains active, suffering the same fate (Oustanina et al., 2004; Relaix et al., 2006). Although Pax7 null mice also experience perturbed neural development, it is likely that Pax7 has a direct effect on satellite cell function. Pax7 is transcriptionally active in quiescent satellite cells and Pax7 + ve/MyoD-ve satellite cell-derived myoblasts, indicating that it may have a role in maintaining the quiescent state (Zammit et al., 2006b). Whether the absence of Pax7 causes a failure of self-renewal however, is debatable. Constitutively expressed Pax7 can up-regulate MyoD in myogenic cells and is compatible with MyoD and cell proliferation in satellite cells, although it does delay myogenic differentiation (Relaix et al., 2006; Zammit et al., 2006b but see Olguin and Olwin, 2004). Since suppression of Pax7, but not Pax3, transcriptional activity leads to apoptosis in the vast majority of myoblasts, Pax7 may instead control maintenance and survival rather than self-renewal (Relaix et al., 2006), with similar observations made in lower vertebrates (Chen et al., 2006). It should be remembered though, that Pax7 exists in at least 4 isoforms generated by alternative splicing, and it is currently unknown whether any functional differences exist between these proteins (Lamey et al., 2004).

Notch signalling is also implicated in satellite cell fate (Conboy and Rando, 2002). Activation of the Notch receptor by its ligands Delta and Jagged results in cleavage of the intracellular region, which then translocates to the nucleus to activate various transcription factors. Activation of Notch-1 in satellite cells promotes proliferation and prevents differentiation, and Wnt family members may anatagonise the Notch-1 pathway to control this process (Dhawan and Rando, 2005). Notch signalling is also antagonised by Numb, which interacts with the cleaved intracellular region thus preventing its nuclear translocation. When Notch-1 signalling is inhibited by Numb, satellite cells exit the cell cycle. Numb can be asymmetrically inherited during cell division in some satellite cell progeny and appears to segregate with Pax7 (Conboy and Rando, 2002; Shinin et al., 2006). These data indicate that inhibition of Notch-1 in certain satellite cell progeny can influence cell fate and may be associated with self-renewal.

5.4 Satellite Cell Self-renewal Operates in vivo

Transplantation of cells into muscle provides a useful assay for their function and fate in vivo. Early experiments showed that grafted myogenic cells can differentiate and contribute functional myonuclei to host muscle (Lipton and Schultz, 1979; Partridge et al., 1989; Watt et al., 1982). Such transplanted myogenic cells can also provide viable myogenic precursors, which can be re-activated by muscle damage, will proliferate and differentiate ex vivo, and form new muscle after serial transplantation (Cousins et al., 2004; Gross and Morgan, 1999; Morgan et al., 1994; Yao and Kurachi, 1993). Importantly, donor-derived myogenic precursors can occupy the satellite cell niche and remain undifferentiated for weeks, but are still able to activate and proliferate in response to appropriate stimuli (Blaveri et al., 1999; Heslop et al., 2001). These studies however, failed to resolve two important points. Firstly, since the source of the donor cells was whole muscle tissue, the actual cells responsible for occupation of the satellite niche was not defined. Secondly, since thousands of muscle tissue-derived cells were grafted, typically in the order of $\sim 2\text{–}5 \times 10^5$ in our studies (Heslop et al., 2001), it was unknown whether the cells first underwent proliferation before entering the satellite cell niche; a pre-requisite for self-renewal.

To be able to address these points, we grafted a single isolated myofiber into an irradiated host muscle, thus the donor cells were anatomically defined as satellite cells and the average number per myofibre was known (Collins et al., 2005). Considerable amounts of new muscle were produced from a single myofiber with a mean of only ~ 7 associated satellite cells. In addition, many new satellite cells were also found, in some cases in excess of 10 times the input number, which could remain dormant for several weeks until stimulated by muscle damage to activate, proliferate and regenerate muscle. Since the number of donor-derived myonuclei and satellite cells far exceeds the number implanted on the single myofiber, satellite cells must have undergone extensive proliferation before either differentiating or becoming quiescent (Collins et al., 2005), a property that is maintained throughout life (Collins et al., 2007). Thus satellite cells can be considered as adult stem cells, since they fulfil the basic criteria, i.e. they proliferate to give rise to progeny that can differentiate and maintain their own population by self-renewal.

5.5 Do all Satellite Cells Self-renew?

A question that remains outstanding is whether all satellite cells on an isolated myofibre are able to self-renew, or whether this property is limited to a subpopulation. The vast majority of transplanted myogenic cells die within hours, but the few survivors begin to proliferate and eventually produce significant amounts of new muscle (Beauchamp et al., 1999). These observations imply that there are some myoblasts with more "stem cell-like" properties. Certainly functional heterogeneity exists: satellite cells can be separated into two typified categories by their rate of cell division during postnatal muscle growth (Schultz, 1996). During regeneration, some myoblasts express myogenin within 8 hours of damage and so presumably commit to differentiation with little or no proliferation, while most myoblasts do not even divide much before 24 hours (McGeachie and Grounds, 1987; Rantanen et al., 1995). These observations are reflected in culture, where myogenic progenitors vary in proliferation rate and clonogenic capacity (Molnar et al., 1996). Furthermore, a population of myogenic precursor cells in muscle appear resistant to the irradiation which ablates most satellite cells (Wakeford et al., 1991) and can still regenerate muscle following damage (Gross, 1999; Heslop et al., 2000).

Demonstration of functional heterogeneity has now been combined with direct identification of satellite cells. A limited number of satellite cells were found in adult muscle that retain incorporated BrdU from a pulse administered peri-natally. When stimulated to divide, some of these adult satellite cells did not then segregate the incorporated BrdU label (Shinin et al., 2006). These observations can be interpreted as showing that some satellite cells have more stem cell-like characteristics, with non-equivalent genomic DNA strands. It has been proposed that stem cells protect the template strand from being copied and so safeguard the DNA from replication errors (Cairns, 1975). Therefore, evidence is emerging that within the satellite cell pool are some satellite cells with a more stem cell-like characteristics, able to both produce myonuclei and also self-renew indefinitely.

6. CONCLUSIONS

For most of the time since their discovery, satellite cells remained difficult to identify and largely inaccessible. It is only in the first years of this new century that the tools have become available to make these tasks more routine and progress has been rapid. Questions outstanding include the role of the niche, what controls quiescence, triggers activation and influences the choice between differentiation or self-renewal. These questions are more than merely academic though, since it is now accepted that adult skeletal muscle provides an accessible stem cell paradigm, in which post-mitotic functionally specialised cells are maintained and repaired by resident stem cells, which sustain their own population by self-renewal. As such, unravelling how this system is regulated will give insights into adult stem cell control in general. It will also aid understanding of disease progression in muscle conditions, from the ~34 disorders currently designated as muscular dystrophies, to secondary muscle loss that contributes to morbidity in cancer and AIDS.

ACKNOWLEDGEMENTS

I would like to thank Zipora Reuveni-Yablonka (University of Washington School of Medicine, Seattle, USA), Jennifer Morgan (Imperial College, London, UK) and Charlotte Collins (Wellcome Trust Centre for Stem Cell Research, University of Cambridge, Cambridge, UK) for comments on the manuscript. I also gratefully acknowledge The Medical Research Council, Muscular Dystrophy Campaign, Association of International Cancer Research and the MYORES Network of Excellence, contract 511978, from the European Commission 6th Framework Programme for support.

REFERENCES

Allbrook D (1962) An electron microscopic study of regenerating skeletal muscle. J Anat 96:137–152

Armand O, Boutineau AM, Mauger A, Pautou MP, Kieny M (1983) Origin of satellite cells in avian skeletal muscles. Arch Anat Microsc Morphol Exp 72:163–181

Asakura A, Seale P, Girgis-Gabardo A, Rudnicki MA (2002) Myogenic specification of side population cells in skeletal muscle. J Cell Biol 159:123–134

Bader CR, Bertrand D, Cooper E, Mauro A (1988) Membrane currents of rat satellite cells attached to intact skeletal muscle fibers. Neuron 1:237–240

Baroffio A, Hamann M, Bernheim L, Bochaton-Piallat ML, Gabbiani G, Bader CR (1996) Identification of self-renewing myoblasts in the progeny of single human muscle satellite cells. Differentiation 60:47–57

Basson MD, Carlson BM (1980) Myotoxicity of single and repeated injections of mepivacaine (Carbocaine) in the rat. Anesth Analg 59:275–282

Beauchamp JR, Heslop L, Yu DS, Tajbakhsh S, Kelly RG, Wernig A, Buckingham ME, Partridge TA, Zammit PS (2000) Expression of CD34 and Myf5 defines the majority of quiescent adult skeletal muscle satellite cells. J Cell Biol 151:1221–1234

Beauchamp JR, Morgan JE, Pagel CN, Partridge TA (1999) Dynamics of myoblast transplantation reveal a discrete minority of precursors with stem cell-like properties as the myogenic source. J Cell Biol 144:1113–1122

Bintliff S, Walker BE (1960) Radioautographic study of skeletal muscle regeneration. Am J Anat 106:233–245

Bischoff R (1986) Proliferation of muscle satellite cells on intact myofibers in culture. Dev Biol 115:129–139

Bischoff R (1975) Regeneration of single skeletal muscle fibers in vitro. Anat Rec 182:215–235

Bischoff R (1994). The satellite cell and muscle regeneration. In: Engel AG, Franzini-Armstrong C (eds) Myology, vol. 1. McGraw-Hill, Inc., New York

Blaveri K, Heslop L, Yu DS, Rosenblatt JD, Gross JG, Partridge TA, Morgan JE (1999) Patterns of repair of dystrophic mouse muscle: studies on isolated fibers. Dev Dyn 216:244–256

Bonner PH, Hauschka SD (1974) Clonal analysis of vertebrate myogenesis. I. Early developmental events in the chick limb. Dev Biol 37:317–328

Cairns J (1975) Mutation selection and the natural history of cancer. Nature 255:197–200

Carlson BM (1973) The regeneration of skeletal muscle. A review. Am J Anat 137:119–149

Carlson BM (1968) Regeneration research in the Soviet Union. Anat Rec 160:665–674

Chen Y, Lin G, Slack JM (2006) Control of muscle regeneration in the Xenopus tadpole tail by Pax7. Development 133:2303–2313

Church JCT, Noronha RFX, Allbrook DB (1966) Satellite cells and skeletal muscle regeneration. Br J Surg 53:638–642

Collins CA, Olsen I, Zammit PS, Heslop L, Petrie A, Partridge TA, Morgan JE (2005) Stem cell function, self-renewal, and behavioral heterogeneity of cells from the adult muscle satellite cell niche. Cell 122:289–301

Collins CA, Zammit PS, Ruiz AP, Morgan JE, Partridge TA (2007) A population of myogenic stem cells that survives skeletal muscle aging. Stem Cells 25:885–894

Conboy IM, Rando TA (2002) The regulation of Notch signaling controls satellite cell activation and cell fate determination in postnatal myogenesis. Dev Cell 3:397–409

Cooper WG, Konigsberg IR (1961) Dynamics of myogenesis in vitro. Anat Rec 140:195–205

Cornelison DD, Filla MS, Stanley HM, Rapraeger AC, Olwin BB (2001) Syndecan-3 and syndecan-4 specifically mark skeletal muscle satellite cells and are implicated in satellite cell maintenance and muscle regeneration. Dev Biol 239:79–94

Cornelison DD, Wilcox-Adelman SA, Goetinck PF, Rauvala H, Rapraeger AC, Olwin BB (2004) Essential and separable roles for Syndecan-3 and Syndecan-4 in skeletal muscle development and regeneration. Genes Dev 18:2231–2236

Cossu G, Molinaro M, Pacifici M (1983) Differential response of satellite cells and embryonic myoblasts to a tumor promoter. Dev Biol 98:520–524

Cousins JC, Woodward KJ, Gross JG, Partridge TA, Morgan JE (2004) Regeneration of skeletal muscle from transplanted immortalised myoblasts is oligoclonal. J Cell Sci 117:3259–3269

Day K, Shefer G, Richardson JB, Enikolopov G, Yablonka-Reuveni, Z (2007) Nestin-GFP reporter expression defines the quiescent state of skeletal muscle satellite cells. Dev Biol 304:246–259

De Angelis L, Berghella L, Coletta M, Lattanzi L, Zanchi M, Cusella-De Angelis MG, Ponzetto C, Cossu G (1999) Skeletal myogenic progenitors originating from embryonic dorsal aorta coexpress endothelial and myogenic markers and contribute to postnatal muscle growth and regeneration. J Cell Biol 147:869–878

Dhawan J, Rando TA (2005) Stem cells in postnatal myogenesis: molecular mechanisms of satellite cell quiescence, activation and replenishment. Trends Cell Biol 15:666–673

Dreyer HC, Blanco CE, Sattler FR, Schroeder ET, Wiswell RA (2006) Satellite cell numbers in young and older men 24 hours after eccentric exercise. Muscle Nerve 33:242–253

Dreyfus PA, Chretien F, Chazaud B, Kirova Y, Caramelle P, Garcia L, Butler-Browne G, Gherardi RK (2004) Adult bone marrow-derived stem cells in muscle connective tissue and satellite cell niches. Am J Pathol 164:773–779

Feldman JL, Stockdale FE (1992) Temporal appearance of satellite cells during myogenesis. Dev Biol 153:217–226

Ferrari G, Cusella-De Angelis G, Coletta M, Paolucci E, Stornaiuolo A, Cossu G, Mavilio F (1998) Muscle regeneration by bone marrow-derived myogenic progenitors. Science 279:1528–1530

Fuchtbauer EM, Westphal H (1992) MyoD and myogenin are coexpressed in regenerating skeletal muscle of the mouse. Dev Dyn 193:34–39

Fukada S, Higuchi S, Segawa M, Koda K, Yamamoto Y, Tsujikawa K, Kohama Y, Uezumi A, Imamura M, Miyagoe-Suzuki Y, Takeda S, Yamamoto H (2004) Purification and cell-surface marker characterization of quiescent satellite cells from murine skeletal muscle by a novel monoclonal antibody. Exp Cell Res 296:245–255

Garry DJ, Meeson A, Elterman J, Zhao Y, Yang P, Bassel-Duby R, Williams RS (2000) Myogenic stem cell function is impaired in mice lacking the forkhead/winged helix protein MNF. Proc Natl Acad Sci U S A 97:5416–5421

Garry DJ, Yang Q, Bassel-Duby R, Williams RS (1997) Persistent expression of MNF identifies myogenic stem cells in postnatal muscles. Dev Biol 188:280–294

Gross JG, Bou-Gharios G, Morgan JE (1999) Potentiation of myoblast transplantation by host muscle irradiation is dependent on the rate of radiation delivery. Cell Tissue Res 298:371–375

Gross JG, Morgan JE (1999) Muscle precursor cells injected into irradiated mdx mouse muscle persist after serial injury. Muscle Nerve 22:174–185

Grounds MD, Garrett KL, Lai MC, Wright WE, Beilharz MW (1992) Identification of skeletal muscle precursor cells in vivo by use of MyoD1 and myogenin probes. Cell Tissue Res 267:99–104

Gussoni E, Soneoka Y, Strickland CD, Buzney EA, Khan MK, Flint AF, Kunkel LM, Mulligan RC (1999) Dystrophin expression in the mdx mouse restored by stem cell transplantation. Nature 401:390–394

Halevy O, Piestun Y, Allouh MZ, Rosser BW, Rinkevich Y, Reshef R, Rozenboim I, Wleklinski-Lee M, Yablonka-Reuveni Z (2004) Pattern of Pax7 expression during myogenesis in the posthatch chicken establishes a model for satellite cell differentiation and renewal. Dev Dyn 231:489–502

Hartley RS, Bandman E, Yablonka-Reuveni Z (1992) Skeletal muscle satellite cells appear during late chicken embryogenesis. Dev Biol 153:206–216

Hawke TJ, Garry DJ (2001) Myogenic satellite cells: physiology to molecular biology. J Appl Physiol 91:534–551

Hay ED (1959) Electron microscopic observations of muscle dedifferentiation in regenerating ambystoma limbs. Dev Biol 1:555–585

Heslop L, Beauchamp JR, Tajbakhsh S, Buckingham ME, Partridge TA, Zammit PS (2001) Transplanted primary neonatal myoblasts can give rise to functional satellite cells as identified using the Myf5nlacZl+ mouse. Gene Ther 8:778–783

Heslop L, Morgan JE, Partridge TA (2000) Evidence for a myogenic stem cell that is exhausted in dystrophic muscle. J Cell Sci 113(Pt 12):2299–2308

Horst D, Ustanina S, Sergi C, Mikuz G, Juergens H, Braun T, Vorobyov E (2006) Comparative expression analysis of Pax3 and Pax7 during mouse myogenesis. Int J Dev Biol 50:47–54

Illa I, Leon-Monzon M, Dalakas MC (1992) Regenerating and denervated human muscle fibers and satellite cells express neural cell adhesion molecule recognized by monoclonal antibodies to natural killer cells. Ann Neurol 31:46–52

Irintchev A, Zeschnigk M, Starzinski-Powitz A, Wernig A (1994) Expression pattern of M-cadherin in normal, denervated, and regenerating mouse muscles. Dev Dyn 199:326–337

Jones NC, Tyner KJ, Nibarger L, Stanley HM, Cornelison DD, Fedorov YV, Olwin BB (2005) The p38alpha/beta MAPK functions as a molecular switch to activate the quiescent satellite cell. J Cell Biol 169:105–116

Kassar-Duchossoy L, Gayraud-Morel B, Gomes D, Rocancourt D, Buckingham M, Shinin V, Tajbakhsh S (2004) Mrf4 determines skeletal muscle identity in Myf5:Myod double-mutant mice. Nature 431:466–471

Kastner S, Elias MC, Rivera AJ, Yablonka-Reuveni Z (2000) Gene expression patterns of the fibroblast growth factors and their receptors during myogenesis of rat satellite cells. J Histochem Cytochem 48:1079–1096

Katz B (1961) The terminations of the afferent nerve fibre in the muscle spindle of the frog. Philos Trans Royal Soc Lond [Biol] 243:221–240

Kelly R, Alonso S, Tajbakhsh S, Cossu G, Buckingham M (1995) Myosin light chain 3F regulatory sequences confer regionalized cardiac and skeletal muscle expression in transgenic mice. J Cell Biol 129:383–396

Kitzmann M, Carnac G, Vandromme M, Primig M, Lamb NJ, Fernandez A (1998) The muscle regulatory factors MyoD and myf-5 undergo distinct cell cycle-specific expression in muscle cells. J Cell Biol 142:1447–1459

Konigsberg UR, Lipton BH, Konigsberg IR (1975) The regenerative response of single mature muscle fibers isolated in vitro. Dev Biol 45:260–275

Knapp JR, Davie JK, Myer A, Meadows E, Olson EN, Klein WH (2006) Loss of myogenin in postnatal life leads to normal skeletal muscle but reduced body size. Development 133:601–610

Kuang S, Charge SB, Seale P, Huh M, Rudnicki MA (2006) Distinct roles for Pax7 and Pax3 in adult regenerative myogenesis. J Cell Biol 172:103–113

LaBarge MA, Blau HM (2002) Biological progression from adult bone marrow to mononucleate muscle stem cell to multinucleate muscle fiber in response to injury. Cell 111:589–601

Lamey TM, Koenders A, Ziman M (2004) Pax genes in myogenesis: alternate transcripts add complexity. Histol Histopathol 19:1289–1300

Lapidos KA, Chen YE, Earley JU, Heydemann A, Huber JM, Chien M, Ma A, McNally EM (2004) Transplanted hematopoietic stem cells demonstrate impaired sarcoglycan expression after engraftment into cardiac and skeletal muscle. J Clin Invest 114:1577–1585

Lash JW, Holtzer H, Swift H (1957) Regeneration of mature skeletal muscle. Anat Rec 128:679–697

Le Gros Clark WE (1946) An experimental study of the regeneration of mammalian striped muscle. J Anat 80:24–36

Lee HJ, Goring W, Ochs M, Muhlfeld C, Steding G, Paprotta I, Engel W, Adham IM (2004) Sox15 is required for skeletal muscle regeneration. Mol Cell Biol 24:8428–8436

Lewis WH, Lewis MR (1917) Behaviour of cross striated muscle in tissue cultures. Am J Anat 22: 169–194

Lipton BH, Schultz E (1979) Developmental fate of skeletal muscle satellite cells. Science 205: 1292–1294

Luz MA, Marques MJ, Santo Neto H (2002) Impaired regeneration of dystrophin-deficient muscle fibers is caused by exhaustion of myogenic cells. Braz J Med Biol Res 35:691–695

Macconnachie HF, Enesco M, Leblond CP (1964) The mode of increase in the number of skeletal muscle nuclei in the postnatal rat. Am J Anat 114:245–253

Mauro A (1961) Satellite cell of skeletal muscle fibers. J Biophys Biochem Cytol 9:493–495

Mazanet R, Franzini-Armstrong C (1986) The Satellite Cell. In: Engel RG, Banker BQ (eds) Myology. McGraw-Hill, New York, pp. 285–307

McGeachie JK, Grounds MD (1987) Initiation and duration of muscle precursor replication after mild and severe injury to skeletal muscle of mice. An autoradiographic study. Cell Tissue Res 248:125–130

Megeney LA, Kablar B, Garrett K, Anderson JE, Rudnicki MA (1996) MyoD is required for myogenic stem cell function in adult skeletal muscle. Genes Dev 10:1173–1183

Milasincic DJ, Dhawan J, Farmer SR (1996) Anchorage-dependent control of muscle-specific gene expression in C2C12 mouse myoblasts. In Vitro Cell Dev Biol Anim 32:90–99

Molnar G, Ho ML, Schroedl NA (1996) Evidence for multiple satellite cell populations and a non-myogenic cell type that is regulated differently in regenerating and growing skeletal muscle. Tissue Cell 28:547–556

Montarras D, Morgan J, Collins C, Relaix F, Zaffran S, Cumano A, Partridge T, Buckingham M (2005) Direct isolation of satellite cells for skeletal muscle regeneration. Science 309:2064–2067

Morgan JE, Beauchamp JR, Pagel CN, Peckham M, Ataliotis P, Jat PS, Noble MD, Farmer K, Partridge TA (1994) Myogenic cell lines derived from transgenic mice carrying a thermolabile T antigen: a model system for the derivation of tissue-specific and mutation-specific cell lines. Dev Biol 162: 486–498

Morrison JI, Loof S, He P, Simon A (2006) Salamander limb regeneration involves the activation of a multipotent skeletal muscle satellite cell population. J Cell Biol 172:433–440

Moss FP, Leblond CP (1971) Satellite cells as the source of nuclei in muscles of growing rats. Anat Rec 170:421–435

Moss FP, Leblond CP (1970) Nature of dividing nuclei in skeletal muscle of growing rats. J Cell Biol 44:459–462

Muir AR, Kanji AH, Allbrook D (1965) The structure of the satellite cells in skeletal muscle. J Anat 99:435–444

Nagata Y, Kobayashi H, Umeda M, Ohta N, Kawashima S, Zammit PS, Matsuda R (2006a) Sphingomyelin levels in the plasma membrane correlate with the activation state of muscle satellite cells. J Histochem Cytochem 54:375–384

Nagata Y, Partridge TA, Matsuda R, Zammit PS (2006b) Entry of muscle satellite cells into the cell cycle requires sphingolipid signaling. J Cell Biol 174:245–253

Olguin HC, Olwin BB (2004) Pax-7 up-regulation inhibits myogenesis and cell cycle progression in satellite cells: a potential mechanism for self-renewal. Dev Biol 275:375–388

Olivera A, Kohama T, Edsall L, Nava V, Cuvillier O, Poulton S, Spiegel S (1999) Sphingosine kinase expression increases intracellular sphingosine-1-phosphate and promotes cell growth and survival. J Cell Biol 147:545–558

Olivera A, Spiegel S (1993) Sphingosine-1-phosphate as second messenger in cell proliferation induced by PDGF and FCS mitogens. Nature 365:557–560

Ontell M, Kozeka K (1984) The organogenesis of murine striated muscle: a cytoarchitectural study. Am J Anat 171:133–148

Oustanina S, Hause G, Braun T (2004) Pax7 directs postnatal renewal and propagation of myogenic satellite cells but not their specification. Embo J 23:3430–3439

Partridge TA, Morgan JE, Coulton GR, Hoffman EP, Kunkel LM (1989) Conversion of mdx myofibres from dystrophin-negative to -positive by injection of normal myoblasts. Nature 337:176–179

Polesskaya A, Seale P, Rudnicki MA (2003) Wnt signaling induces the myogenic specification of resident CD45+ adult stem cells during muscle regeneration. Cell 113:841–852

Price HM, Howes EL, Jr., Blumberg JM (1964) Ultrastructural alterations in skeletal muscle fibers injured by cold. Ii. Cells on the sarcolemmal tube: Observations on "Discontinuous" regeneration and myofibril formation. Lab Invest 13:1279–1302

Quinn LS, Holtzer H, Nameroff M (1985) Generation of chick skeletal muscle cells in groups of 16 from stem cells. Nature 313:692–694

Quinn LS, Nameroff M, Holtzer H (1984) Age-dependent changes in myogenic precursor cell compartment sizes. Evidence for the existence of a stem cell. Exp Cell Res 154:65–82

Rantanen J, Hurme T, Lukka R, Heino J, Kalimo H (1995) Satellite cell proliferation and the expression of myogenin and desmin in regenerating skeletal muscle: evidence for two different populations of satellite cells. Lab Invest 72:341–347

Reimann J, Brimah K, Schroder R, Wernig A, Beauchamp JR, Partridge TA (2004) Pax7 distribution in human skeletal muscle biopsies and myogenic tissue cultures. Cell Tissue Res 315:233–242

Relaix F, Montarras D, Zaffran S, Gayraud-Morel B, Rocancourt D, Tajbakhsh S, Mansouri A, Cumano A, Buckingham M (2006) Pax3 and Pax7 have distinct and overlapping functions in adult muscle progenitor cells. J Cell Biol 172:91–102

Sabourin LA, Girgis-Gabardo A, Seale P, Asakura A, Rudnicki MA (1999) Reduced differentiation potential of primary MyoD-/- myogenic cells derived from adult skeletal muscle. J Cell Biol 144: 631–643

Sachidanandan C, Sambasivan R, Dhawan J (2002) Tristetraprolin and LPS-inducible CXC chemokine are rapidly induced in presumptive satellite cells in response to skeletal muscle injury. J Cell Sci 115:2701–2712

Sadeh M, Czyewski K, Stern LZ (1985) Chronic myopathy induced by repeated bupivacaine injections. J Neurol Sci 67:229–238

Schmidt K, Glaser G, Wernig A, Wegner M, Rosorius O (2003) Sox8 is a specific marker for muscle satellite cells and inhibits myogenesis. J Biol Chem 278:29769–29775

Schultz E (1996) Satellite cell proliferative compartments in growing skeletal muscles. Dev Biol 175:84–94

Schultz E (1976) Fine structure of satellite cells in growing skeletal muscle. Am J Anat 147:49–70

Schultz E, Chamberlain C, McCormick KM, Mozdziak PE (2006) Satellite cells express distinct patterns of myogenic proteins in immature skeletal muscle. Dev Dyn 235:3230–3239

Schultz E, Gibson MC, Champion T (1978) Satellite cells are mitotically quiescent in mature mouse muscle: an EM and radioautographic study. J Exp Zool 206:451–456

Seale P, Sabourin LA, Girgis-Gabardo A, Mansouri A, Gruss P, Rudnicki MA (2000) Pax7 is required for the specification of myogenic satellite cells. Cell 102:777–786

Shafiq SA, Gorycki MA (1965) Regeneration in skeletal muscle of mouse: some electron-microscope observations. J Pathol Bacteriol 90:123–127

Shafiq SA, Gorycki MA, Mauro A (1968) Mitosis during postnatal growth in skeletal and cardiac muscle of the rat. J Anat 103:135–141

Shefer G, Oron U, Irintchev A, Wernig A, Halevy O (2001) Skeletal muscle cell activation by low-energy laser irradiation: a role for the MAPK/ERK pathway. J Cell Physiol 187:73–80

Shefer G, Van de Mark DP, Richardson JB, Yablonka-Reuveni Z (2006) Satellite-cell pool size does matter: defining the myogenic potency of aging skeletal muscle. Dev Biol 294:50–66

Sherwood RI, Christensen JL, Conboy IM, Conboy MJ, Rando TA, Weissman IL, Wagers AJ (2004) Isolation of adult mouse myogenic progenitors: functional heterogeneity of cells within and engrafting skeletal muscle. Cell 119:543–554

Shinin V, Gayraud-Morel B, Gomes D, Tajbakhsh S (2006) Asymmetric division and cosegregation of template DNA strands in adult muscle satellite cells. Nat Cell Biol 8:677–687

Shu X, Wu W, Mosteller RD, Broek D (2002) Sphingosine kinase mediates vascular endothelial growth factor-induced activation of ras and mitogen-activated protein kinases. Mol Cell Biol 22:7758–7768

Snow MH (1978) An autoradiographic study of satellite cell differentiation into regenerating myotubes following transplantation of muscles in young rats. Cell Tissue Res 186:535–540

Snow MH (1977) Myogenic cell formation in regenerating rat skeletal muscle injured by mincing. I. A fine structural study. Anat Rec 188:181–200

Snow MH (1977a) Myogenic cell formation in regenerating rat skeletal muscle injured by mincing. II. An autoradiographic study. Anat Rec 188:201–217

Speidel CC (1938) Studies in living muscles 1. Growth, injury and repair of striated muscle, as revealed by prolonged observations of individual fibers in living frog tadpoles. Am J Anat 62:179–235

Spiegel S, Milstien S (2003) Sphingosine-1-phosphate: an enigmatic signalling lipid. Nat Rev Mol Cell Biol 4:397–407

Stockdale FE, Holtzer H (1961) DNA synthesis and myogenesis. Exp Cell Res 24:508–520

Stockdale FE, Nikovits W, Jr., Christ B (2000) Molecular and cellular biology of avian somite development. Dev Dyn 219:304–321

Studitsky AN (1964) Free Auto- and Homografts of Muscle Tissue in Experiments on Animals. Ann N Y Acad Sci 120:789–801

Tamaki T, Akatsuka A, Ando K, Nakamura Y, Matsuzawa H, Hotta T, Roy RR, Edgerton VR (2002) Identification of myogenic-endothelial progenitor cells in the interstitial spaces of skeletal muscle. J Cell Biol 157:571–577

Tatsumi R, Allen RE (2004) Active hepatocyte growth factor is present in skeletal muscle extracellular matrix. Muscle Nerve 30:654–658

Tatsumi R, Anderson JE, Nevoret CJ, Halevy O, Allen RE (1998) HGF/SF is present in normal adult skeletal muscle and is capable of activating satellite cells. Dev Biol 194:114–128

Venable JH (1966) Morphology of the cells of normal, testosterone deprived and testosterone-stimulated levator ani muscles. Am J Anat 119:271–302

Volonte D, Liu Y, Galbiati F (2005) The modulation of caveolin-1 expression controls satellite cell activation during muscle repair. Faseb J 19:237–239

Wakeford S, Watt DJ, Partridge TA (1991) X-irradiation improves mdx mouse muscle as a model of myofiber loss in DMD. Muscle Nerve 14:42–50

Watt DJ, Lambert K, Morgan JE, Partridge TA, Sloper JC (1982) Incorporation of donor muscle precursor cells into an area of muscle regeneration in the host mouse. J Neurol Sci 57:319–331

Weintraub H, Davis R, Tapscott S, Thayer M, Krause M, Benezra R, Blackwell TK, Turner D, Rupp R, Hollenberg S, et al. (1991) The myoD gene family: nodal point during specification of the muscle cell lineage. Science 251:761–766

Wernig G, Janzen V, Schafer R, Zweyer M, Knauf U, Hoegemeier O, Mundegar RR, Garbe S, Stier S, Franz T, Wernig M, Wernig A (2005) The vast majority of bone-marrow-derived cells integrated into mdx muscle fibers are silent despite long-term engraftment. Proc Natl Acad Sci U S A 102: 11852–11857

Whalen RG, Harris JB, Butler-Browne GS, Sesodia S (1990) Expression of myosin isoforms during notexin-induced regeneration of rat soleus muscles. Dev Biol 141:24–40

Yablonka-Reuveni Z, Rivera AJ (1994) Temporal expression of regulatory and structural muscle proteins during myogenesis of satellite cells on isolated adult rat fibers. Dev Biol 164:588–603

Yablonka-Reuveni Z, Rudnicki MA, Rivera AJ, Primig M, Anderson JE, Natanson P (1999a) The transition from proliferation to differentiation is delayed in satellite cells from mice lacking MyoD. Dev Biol 210:440–455

Yablonka-Reuveni Z, Seger R, Rivera AJ (1999b) Fibroblast growth factor promotes recruitment of skeletal muscle satellite cells in young and old rats. J Histochem Cytochem 47:23–42

Yao SN, Kurachi K (1993) Implanted myoblasts not only fuse with myofibers but also survive as muscle precursor cells. J Cell Sci 105 (Pt 4):957–963

Yoshida N, Yoshida S, Koishi K, Masuda K, Nabeshima Y (1998) Cell heterogeneity upon myogenic differentiation: down-regulation of MyoD and Myf-5 generates "reserve cells". J Cell Sci 111 (Pt 6):769–779

Yusuf I, Fruman DA (2003) Regulation of quiescence in lymphocytes. Trends Immunol 24:380–386

Zammit PS, Golding JP, Nagata Y, Hudon V, Partridge TA, Beauchamp JR (2004) Muscle satellite cells adopt divergent fates: a mechanism for self-renewal? J Cell Biol 166:347–357

Zammit PS, Heslop L, Hudon V, Rosenblatt JD, Tajbakhsh S, Buckingham ME, Beauchamp JR, Partridge TA (2002) Kinetics of myoblast proliferation show that resident satellite cells are competent to fully regenerate skeletal muscle fibers. Exp Cell Res 281:39–49

Zammit PS, Partridge TA, Yablonka-Reuveni Z (2006a) The skeletal muscle satellite cell: the stem cell that came in from the cold. J Histochem Cytochem 54:1177–1191

Zammit PS, Relaix F, Nagata Y, Ruiz AP, Collins CA, Partridge TA, Beauchamp JR (2006b) Pax7 and myogenic progression in skeletal muscle satellite cells. J Cell Sci 119:1824–1832

Zhang H, Desai NN, Olivera A, Seki T, Brooker G, Spiegel S (1991) Sphingosine-1-phosphate, a novel lipid, involved in cellular proliferation. J Cell Biol 114:155–167

CHAPTER 4

NON MUSCLE STEM CELLS AND MUSCLE REGENERATION

GRAZIELLA MESSINA[1], STEFANO BIRESSI[1] AND GIULIO COSSU[1,2,3]

[1] *Stem Cell Research Institute, Dibit, H. San Raffaele, via Olgettina 58, 20132 Milan, Italy*

[2] *Institute of Cell Biology and Tissue Engineering. San Raffaele Biomedical Science Park of Rome* II°, *via Castel Romano 100/2, 00128 Rome, Italy*

[3] *Department of Biology, University of Milan, Via Celoria 28, 20129, Milan, Italy*

Abstract: Skeletal muscle of the vertebrate embryo originates from paraxial mesoderm (somites, somitomers and prechordal cephalic mesoderm) (Christ and Ordahl, 1995) and is formed in discrete steps by different classes of myogenic progenitor cells (Cossu and Biressi, 2005). After myotome formation, embryonic myoblasts give rise to primary fibers in the embryo, while fetal myoblasts give rise to secondary fibers, initially smaller and surrounding primary fibers. Satellite cells appear underneath the newly formed basal lamina that develops around each muscle fiber, and contribute to their post-natal growth and regeneration (Bischoff, 1994). In addition to canonical progenitors, evidence accumulated through the years that cells cultured from tissues that do not derive from paraxial mesoderm and do not contain skeletal muscle such as thymus, brain or kidney may differentiate at low frequency into skeletal muscle. Initially dismissed as a tissue culture artifact, the phenomenon came under closer scrutiny when it was unequivocally demonstrated that the bone marrow of adult normal mice contain cells capable of contributing to skeletal muscle regeneration in vivo (Ferrari et al., 1998). In the following years, different types of non-somitic stem-progenitor cells have been shown to contribute to muscle regeneration. The origin of these different cell types and their possible lineage relationships with other myogenic cells as well as their possible role in muscle regeneration is actively studied in these years and will be the subject of this chapter. Finally, the possible use of different non-canonic myogenic cells in experimental protocols of cell therapy will be briefly outlined.

Keywords: Skeletal myogenesis; muscle satellite cells; skeletal myoblasts; mesoangioblasts; muscle regeneration.

S. Schiaffino and T. Partridge (eds.), Skeletal Muscle Repair and Regeneration, 65–83.

Abbreviations: BMP2: Bone morphogenetic protein 2; GFP: green fluorescent protein; HSC: hematopoietic stem cell; MSC: mesoderm stem cell (referred to as non hematopoietic); PKC: protein kinase C; Shh: Sonic hedgehog; SP: side population; TGF β: transforming growth factor β.

1. A BRIEF HISTORY OF UNORTHODOX MYOGENESIS AND OF ITS POSSIBLE SIGNIFICANCE IN REGENERATION

Myogenic progenitor cells, termed myoblasts, have been isolated and cultured since the early 60' of the last century. Originally isolated from the muscle anlagen of avian embryos, myoblasts were later cultured from muscles of virtually all vertebrates, both embryonic and adult. Removal or consumption of growth factors (often provided as serum or embryo extracts) induces irreversible withdrawal from the cell cycle and terminal differentiation of myoblasts that fuse into multinucleated myotubes. During further maturation, which occurs only partially in vitro, myotubes complete sarcomerogenesis, assemble a functional excitation-contraction coupling system and contract in response to appropriate stimuli (Okazaki and Holtzer, 1966).

Because they are easily recognized morphologically in living cultures, myotubes were occasionally observed in cultures of cells that were not myogenic nor derived from tissues that in vivo contain skeletal muscle. These observations remained anecdotic and largely unpublished, also because they lacked a rational explanation. "Contamination with myogenic cells during isolation" or "tissue culture artifact" represented the easiest interpretations of these data (Cossu, 1997).

Nevertheless papers accumulated through the years, some of which reporting solid and unquestionable data. Perhaps the most striking example is represented by the thymus that is derived from pharyngeal pouches and does not contain any skeletal muscle fiber. In 1975, it was reported the occurrence of striated muscle fiber differentiation in monolayer cultures of adult thymus reticulum (Wekerle et al., 1975). Later it was reported that in the thymus from adult but not neonatal mice, MyoD or myogenin-positive cells are concentrated in the medullary region but do not differentiate within the normal murine thymic environment. However, myogenesis takes place both in vitro, as demonstrated in the original paper, and in vivo, upon transplantation into regenerating muscle (Grounds et al., 1992).

Another example is represented by the so called "myogenic conversion of fibroblasts" originating from dermis and, to different extent, other mesoderm tissues. The first example of this phenomenon was the correction by fibroblast-myoblast fusion of the genetic defect of the *mdg* mouse mutant muscle fibers (Chaudhary et al., 1989; Courbin et al., 1989). Subsequently, several groups reported that genetically labeled dermal fibroblasts could be incorporated into differentiated myotubes both in vitro and in vivo (Gibson et al., 1995; Breton et al., 1995; Salvatori et al., 1995). These studies showed evidence for fusion of fibroblasts with myogenic cells. In these cases myogenesis could be activated as it occurs in heterokaryons where the fibroblast nucleus is exposed to muscle transcription factors. Interestingly however, this myogenic potency is not restricted to dermis, but is present in virtually all organs containing a significant mesoderm component, such as smooth and skeletal muscle or kidney and also the central nervous system. At variance with cells from the thymus, these other cells require signals from differentiating myogenic cells, possibly related to a "community effect" (Gurdon et al., 1993; Cossu et al., 1995) present during skeletal muscle histogenesis and possibly regeneration. Moreover, normal murine dermal fibroblasts implanted into the muscles of the mdx mouse participate in new myofiber formation and direct the expression of the protein dystrophin, deficient in these mice (Gibson et al., 1995). Interestingly, the lectin galectin-1, expressed and secreted by the myoblasts, induces the conversion of dermal but not of muscle fibroblasts to skeletal muscle (Goldring et al., 2002).

Two additional examples are represented by in vitro myogenic differentiation of neuro-ectoderm cells from the developing central nervous system and by BHK (Baby Hamster Kidney) cells.

Spontaneous myogenic differentiation of cells from the brain was documented a number of times (examples quoted in Tajbakhsh et al., 1994) but it was only through insertion of the reporter gene *LacZ* into the *myf-5* locus that it was possible to unequivocally identify Myf-5 expressing cells in the neural tube and to show that these cells co-express neural and muscle markers (Tajbakhsh et al., 1994). Once explanted in cultures, some of *myf-5* expressing cells will differentiate into skeletal myocytes, thus suggesting escape from a community-induced inhibition. A similar situation was observed in a specific areas of the brain of the same mice: Myf-5 expression begins to be detected at embryonic day 8 (E8) in the mesencephalon and coincides with the appearance of the first differentiated neurons; expression in the secondary prosencephalon initiates at E10 and is confined to the ventral domain of prosomere p4, later becoming restricted to the posterior hypothalamus (Tajbakhsh and Buckingham, 1995).

BHK cells are derived from proteolitic digestion of newborn kidney and have been widely used as fibroblasts. More careful analysis revealed that these cells express MyoD and myogenin and can be induced to differentiate into skeletal muscle cells (Mayer and Leinwand, 1997).

All these cases of unorthodox myogenesis are conceptually distinct from trans-differentiation, a phenomenon by which an already differentiated cell can be induced the change the repertoire of gene expressed and to express genes typical of a different tissue. In higher vertebrates, this situation is mainly related to pathology (metaplasia), although trans-differentiation from smooth to skeletal muscle has been demonstrated at the single cell level in the post-natal mammalian esophagus (Patapoutian et al., 1995). Trans-differentiation is not discussed in this chapter.

2. A CURRENT CLASSIFICATION OF NON MUSCLE STEM CELLS POSSIBLY INVOLVED IN MUSCLE REGENERATION

2.1 Non Muscle Stem Cells From the Ectoderm and Endoderm

2.1.1 Neural stem cells as a source of myogenic cells

To date, neural stem cells are the only ectoderm-derived stem cells that have been shown to differentiate into skeletal muscle when co-cultured with skeletal myoblasts (Galli et al., 2000). Both acutely isolated cells and clonally expanded neurospheres of both murine and human origin could be induced to undergo myogenesis in vitro and in vivo, upon injection into regenerating muscle. Interestingly direct contact was shown to be required between myogenic cells and neural stem cells, as only the cells at the border of the neurosphere could be converted to myogenesis. Although the possible practical exploitation of these results is not immediate, they nevertheless represent unequivocal evidence of myogenesis arising from cells of a germ layer different from mesoderm. No evidence of skeletal muscle differentiation has so far been reported for stem cells from ectoderm or endoderm derived epithelia, suggesting that, if attempts have been made, they have been unsuccessful.

2.2 Non Muscle Stem Cells from the Hematopoietic System

2.2.1 Total bone marrow as a source of myogenic cells

The first evidence of in vivo development of skeletal muscle from cells of the hematopoietic system was reported in 1998, thanks to the use of a transgenic mouse expressing a nuclear *lacZ* under the control of muscle-specific regulatory elements (MLC3F-n*lacZ*) only in striated muscle (Kelly et al., 1995). Bone marrow-derived cells from these mice were transplanted into lethally-irradiated mice and, when reconstitution by donor bone marrow had occurred, muscle regeneration was induced by cardiotoxin injection into a leg muscle (*tibialis anterior*). Histochemical analysis unequivocally showed the presence of ß-gal positive nuclei at the center of regenerated fibers, demonstrating for the first time that murine bone marrow contains transplantable progenitors that can be recruited to an injured muscle through the peripheral circulation, and participate to muscle repair by undergoing differentiation into mature muscle fibers (Ferrari et al., 1998). The publication of this report raised new interest in myogenic progenitors and in their possible clinical use. It was reasoned that, although the frequency of the phenomenon was very low, in a chronically regenerating, dystrophic muscle myogenic progenitors would have found a favorable environment and consequently would have contributed significantly to regeneration of dystrophin positive, normal fibers.

2.2.2 SP cells as a source of myogenic cells

This, however, turned out not to be the case. In the following year the groups of Kunkel and Mulligan showed that *mdx* mice transplanted with the bone marrow side population, or SP (a fraction of total cells that is separated by die exclusion and contains stem/progenitor cells able to repopulate the hematopoietic system upon transplantation (Goodell et al., 2005)) from syngeneic C57BL/10 mice develop, within several weeks, a small number of dystrophin-positive fibers containing genetically marked (Y chromosome) donor nuclei (Gussoni et al., 1999) Even after many months from the transplantation, the number of fibers carrying both dystrophin and the Y chromosome never exceeded 1% of the total fibers in the average muscle, thus precluding a direct clinical translation for this protocol. Similar results were later obtained in a slightly different animal model, the *mdx4cv* mutant (Ferrari et al., 2001) Together these data indicate that myogenic differentiation from bone marrow occurs but a frequency discouraging low in order to predict possible clinical benefit.

Following these initial observations, experiments were conducted to identify the cell type within the heterogeneous bone marrow cells which may give rise to skeletal muscle upon transplantation. When bone marrow was fractionated into CD45 positive and negative fractions, the muscle forming activity was associated with the CD45+ fraction (McKinney-Freeman et al., 2002); retrospective analysis in a Duchenne patient that had undergone bone marrow transplantation confirmed persistence of donor derived skeletal muscle cells over a periods of many years, again at very low frequency (Gussoni et al., 2002). Together these data suggested that a myogenic potential is present in the hematopoietic stem cell itself or in a yet to be identified cell that expresses several markers in common with true HSC. More recent and sophisticated approaches confirmed these first observations but disagreed on the underlying mechanism: it was reported that the progeny of a single cell can both reconstitute the hematopoietic system and contribute to muscle regeneration (Corbel et al., 2003). Integration of bone marrow cells into myofibers was shown to occur spontaneously at low frequency and to increase with muscle damage. It was concluded that classically defined single hematopoietic stem cells can give rise to both blood and muscle. A similar study showed that, although myogenic activity in bone marrow is derived from HSCs and their hematopoietic progeny, contribution to regenerating skeletal muscle does not occur through a myogenic stem cell intermediate. Evidence was presented through a lineage tracing strategy, that myofibers were derived from fusion of mature myeloid cells in response to injury (Camargo et al., 2003).

SP cells are not exclusively present in bone marrow, but rather can be isolated from most tissues (for a review see Challen et al., 2006). It became thus obvious to search for myogenic potency of SP derived form skeletal muscle itself (Asakura et al., 2002). Indeed it was shown that freshly isolated progenitors contained within the adult skeletal muscle side population (SP) can engraft into muscle fibers of dystrophic mice after intravenous or intra-arterial transplantation (Bachrach et al., 2004 and 2006). Engraftment rate was however quite low, ranging from 1% of skeletal muscle fibers expressing donor-derived gene products for intra-venous to 8% for intra-arterial delivery.

2.2.3 *AC133^{+} cells as a source of myogenic cells*

As another example of non-muscle stem cells arising from the hematopoietic system, a subpopulation of circulating cells expressing AC133, a well-characterized marker of hematopoietic stem cells, also expresses early myogenic markers (Torrente et al., 2004). It was shown that freshly isolated, circulating AC133+ cells are able to undergo myogenesis when cocultured with myogenic cells or when transplanted in vivo into the muscles of transgenic *scid/mdx* mice (which allow survival of human cells). Injected cells also localized under the basal lamina of host muscle fibers and expressed satellite cell markers such as M-cadherin and Myf5. Furthermore, functional tests of injected muscles revealed a substantial recovery of force after treatment. As these cells can be isolated from the blood, manipulated in vitro, and delivered through the circulation, they represent a possible tool for future cell therapy applications in DMD disease or other muscular dystrophies: current limit of this approach is related to the difficulty of expanding in vitro this rare cell population to numbers that would be suitable to treat systemically a pediatric patient.

2.3 Non Muscle Stem Cells from Solid Mesoderm

2.3.1 Mesenchymal stem cells

Mesenchymal stem cells, mainly originate from perycytes, are located in the perivascular district of the bone marrow stroma and are the natural precursors of bone, cartilage and fat, the constituent tissues of the bone (Bianco and Gehron Robey, 2000). Although MSC were reported to give rise to myotubes in culture upon induction with 5-azacytidine (Wakitani et al., 1995) they do not differentiate into muscle under normal conditions (Bianco and Cossu, 1999). When transplanted in sheep fetus *in utero*, human MSC colonized most tissues, including skeletal muscle, although their effective muscle differentiation was not demonstrated (Liechty et al., 2000).

Recently it was reported that MSC expressing a truncated form of Notch and exposed to certain cytokines were able to differentiate into skeletal muscle in vitro with high efficiency (Dezawa et al., 2005). Induced cells differentiated into muscle fibers upon transplantation into degenerated muscles of mdx-nude mice. The induced population contained Pax7-positive cells that contributed to subsequent regeneration of muscle upon repetitive damage without additional transplantation of cells. These MSCs may represent a more ready supply of myogenic cells than other rare myogenic stem cells found in other tissues, but the underlying molecular mechanism needs to be fully elucidated and the risks related to the expression of an oncogenic protein need to be carefully evaluated.

2.3.2 Multipotent adult progenitors

The group of Verfaillie (Reyes et al., 2001) identified a rare cell, within adherent cells cultured from human or rodent bone marrow, which was termed multipotent adult progenitor cell (MAPC). This cell can be expanded for greater than 70 to 150 population doublings (PDs) and differentiates not only into mesenchymal lineage cells but also into endothelium, neuroectoderm, and endoderm. Similar cells can be selected from mouse muscle and brain, suggesting that they may be associated with the microvascular niche of probably many if not all tissues of the mammalian body (Jiang et al., 2002a). Furthermore, when injected into a blastocyst, MAP cells colonize all the tissues of the embryo, with a frequency comparable with ES cells (Jiang et al., 2002b). Because of their apparently unlimited lifespan and multipotency, MAP cells appear as obvious candidates for many cell replacement therapies, although complete differentiation into the desired cell type still needs to be optimized. For what concerns skeletal muscle, neither the frequency at which MAP differentiate into skeletal muscle cells after 5-azacytidine treatment, not their ability to rescue dystrophic muscle have been investigated. In addition the ability of MAP to travel through the body using the circulatory route has not been formally demonstrated, although the general features of these cells strongly suggest this to be the case.

2.3.3 *Muscle derived stem cells (MDSC)*

Cells that adhere late to the culture dish after proteolytic digestion of adult skeletal muscle were isolated though differential pre-plating and shown to retain their phenotype for more than 30 passages with normal karyotype, ability to differentiate into muscle, neural, and endothelial lineages both in vitro and in vivo. These cells that co-express CD34 and Sca-1 like mesoangioblasts (see below) are clearly different from resident satellite cells and were termed "muscle derived stem cells" (MDSC). Transplantation of MDSC improved the efficiency of muscle regeneration and dystrophin delivery to dystrophic muscle (Qu et al., 1998). The ability to proliferate in vivo for an extended period of time, combined with their strong capacity for self-renewal, their multipotent differentiation, and their immune-privileged behavior, suggested that these cells may be very efficient for future cell transplantation experiments.

More recently it was reported that freshly isolated MDSC are potentially useful for reconstitution therapy of the vascular, muscular, and peripheral nervous systems. These results provide new insights into somatic stem and/or progenitor cells with regard to vasculogenesis, myogenesis, and neurogenesis (Tamaki et al., 2005).

2.3.4 *Mesoangioblasts*

Searching for the origin of the bone marrow cells that contribute to muscle regeneration (Ferrari et al., 1998) we identified, by clonal analysis, a progenitor cell derived from the embryonic aorta (De Angelis et al., 1999). When expanded on a feeder layer of embryonic fibroblasts, the clonal progeny of a single cell from the mouse dorsal aorta acquires unlimited life-span, expresses angioblastic markers (CD34, Sca1 and Flk1) and maintains multipotency in culture or when transplanted into a chick embryo. We proposed that these newly identified, vessel associated stem cells, the mesoangioblasts, participate in post-embryonic development of the mesoderm and speculated that postnatal mesodermal stem cells may be rooted in a vascular developmental origin (Minasi et al., 2002).

In as much as mesoangioblasts can be expanded indefinitely, are able to circulate and are easily transduced with lentiviral vectors, they appeared as a potential novel strategy for the cell therapy of genetic diseases. Recently we have succeeded in isolating mesoangioblast-like cells also from post-natal mouse, dog and human tissues. When injected into the blood circulation, mesoangioblasts accumulate in the first capillary filter they encounter and are able to migrate outside the vessel, but only in the presence of inflammation, as in the case of dystrophic muscle. We thus reasoned that if these cells were injected into an artery, they would accumulate into the capillary filter and from there into the interstitial tissue of downstream muscles. Indeed, intra-arterial delivery of wild type mesoangioblasts in the α-sarcoglycan KO mouse, a model for limb girdle muscular dystrophy, corrects morphologically and functionally the dystrophic phenotype of all the muscles downstream of the injected vessel Furthermore, mesoangioblasts isolated from α-sarcoglycan null mice and transduced with a lentiviral vector expressing α-sarcoglycan, reconstituted skeletal muscle similarly to wild type cells (Sampaolesi et al., 2003). These data represented the first successful attempt to treat a murine model of muscular dystrophy with a novel class of mesoderm stem cells. In order to move towards clinical experimentation, we have recently isolated canine mesoangioblasts. Indeed, the only animal model specifically reproducing the full spectrum of human pathology is the golden retriever dog model. Affected animals present a single mutation in intron 6, resulting in complete absence of the dystrophin protein, and early and severe muscle degeneration with nearly complete loss of motility and walking ability. Intra-arterial delivery of wild-type canine mesoangioblasts (vessel-associated stem cells) results in an extensive recovery of dystrophin expression, normal muscle morphology and function (confirmed by measurement of contraction force on single fibres). The outcome was a remarkable clinical amelioration and preservation of active motility (Sampaolesi et al., 2006). Overall the data so far accumulated qualify the mesoangioblasts as candidates for future stem cell therapy for Duchenne patients.

2.3.5 Endothelial progenitor cells (EPC) and other endothelia

Initially identified as CD34+, FlK-1+ circulating cells (Asahara et al., 1997), EPC were shown to be transplantable and to participate actively to angiogenesis in a variety of physiological and pathological conditions. *In vitro* expansion of EPC is still problematic and few laboratories have succeeded in optimizing this process. The clear advantage of EPC would be their natural homing to site of angiogenesis that would target them to site of muscle regeneration. It is known that human umbilical cord blood (UCB) contains high numbers of endothelial progenitors cells (EPCs) characterized by co-expression of CD34, CD133, Flk1 and VE-Cadherin (Murohara et al., 2000) and several studies have shown that these CD34+/CD133+ EPCs from the cord or peripheral blood (PB) can give rise to endothelial cells and induce angiogenesis in ischemic tissues (Takahashi et al., 1999; Kocher et al., 2001). Recently, it has been shown that freshly isolated human cord blood CD34+ cells injected into ischemic adductor muscles give rise to endothelial but also to skeletal muscle cells in mice (Pesce et al., 2003). In fact, the treated limbs exhibited enhanced arteriole length density and regenerating muscle fiber density. Under similar experimental conditions, CD34− cells did not enhance the formation of new arterioles and regenerating muscle fibers. These results support the notion that also endothelial cells, either resident inside adult skeletal muscle (Tamaki et al., 2002) or isolated from fetal lung and yolk sac (Cusella De Angelis et al., 2003) have the ability to participate to muscle regeneration.

2.3.6 *Stem cells from adipose tissue*

Several studies have recently reported the isolation of a human multipotent adipose-derived stem (hMADS) cell population from adipose tissue of young donors (Rodriguez et al., 2005). hMADS cells display normal karyotype, have active telomerase, proliferate over 200 population doublings and differentiate into adipocytes, osteoblasts and myoblasts. Flow cytometry analysis indicates that hMADS cells are positive for CD44 and other mesenchymal markers but negative for CD34, c-Kit, Flk-1, CD133. Transplantation of hMADS cells into the mdx mouse, an animal model of Duchenne muscular dystrophy, resulted in substantial expression of human dystrophin in the injected tibialis anterior and the adjacent gastrocnemius muscle (Rodriguez et al., 2005). Surprisingly, long-term engraftment of hMADS cells also takes place in non-immunocompromised animals, which may be due to the very low level of HLA expressed. It remains to be explained if hMADS-derived muscle fibers did not express high level of class I HLA as all muscle fibers do. Still, the easily available tissue source, their strong capacity for expansion ex vivo, their multipotent differentiation and their immune-privileged behavior, suggest that hMADS cells could be an important tool for muscle cell-mediated therapy.

2.3.7 *Stem cells from sinovium*

Several years ago mesenchymal stem cells were isolated and characterized from human synovial membrane (SM): it was shown that SM-derived MSCs have a multilineage differentiation potential in vitro (De Bari et al., 2001). The same group demonstrated later their myogenic differentiation in a nude mouse model of skeletal muscle regeneration providing proof of principle of their potential use for muscle repair in the mdx mouse model of Duchenne muscular dystrophy (De Bari et al., 2003). Indeed, when implanted into regenerating nude mouse muscle, hSM-MSCs contributed to myofibers and to long term persisting functional satellite cells. Interestingly no nuclear fusion hybrids were observed between donor human cells and host mouse muscle cells as the myogenic differentiation proceeded through a molecular cascade resembling embryonic muscle development. Moreover, the differentiation was sensitive to environmental cues, since hSM-MSCs injected into the bloodstream engrafted in several tissues, but acquired the muscle phenotype only within skeletal muscle. When administered into dystrophic muscles of immunosuppressed mdx mice, hSM-MSCs restored sarcolemmal expression of dystrophin and ameliorated muscle morphology.

All the examples of stem/progenitor cells that we have described above because of their myogenic potency, differ among themselves for a number of biological features (origin, proliferation and differentiation ability etc.) as well as for expression of myogenic and stem cell markers. These are summarized in Tables 1 and 2 respectively, that suffer of over-simplification but hopefully help the get a general view

Table 1. Features of different myogenic progenitor cells under various experimental conditions

Cell type	Origin	Proliferation	Systemic delivery	In vitro differentiation	Dystrophin expression in vivo
Satellite cells	Somite	High	No	Spontaneous	Yes
MSC	Vessel wall	High	ND	Induced by Aza-cytidine	Yes
EPC	Vessel wall	Low	Yes	Induced by muscle cells	ND
MAPC	Vessel wall	High	ND	Induced by Aza-cytidine	ND
MDSC	Skeletal muscle	High	ND	ND	Yes
MAB	Vessel wall	High	Yes	Induced by muscle cells	Yes
ADSC	Adipose tissue	High	ND	Spontaneous	Yes
SDSC	Synovium	High	ND	Induced by Aza-cytidine	Yes
HSC	Bone marrow	Low	Yes	Induced by muscle cells	Yes

Main biological features of satellite cells and other stem-progenitor cells endowed with myogenic potency. MSC: mesenchymal stem cells; EPC: endothelial progenitor cells; MAPC: multipotent adult progenitors; MDSC: muscle derived stem cells; MAB: mesoangioblasts; ADSC: adipose derived stem cells; SDSC: Synovium derived stem cells; HSC cells refer to hematopoietic stem cells, independently from the selection method (lineage negative, expression of markers such as c-Kit, CD34, Sca-1, dye exclusion – SP population).

of the current situation. It is likely that the list, admittedly incomplete, may still grow in the future, but it should considered that different source, age and species, different methods of isolation and culture may have led to rediscover several times the same cell types, differences among which may depend on these variables. Time will be needed to reach a clearer and more definitive picture.

Table 2. Expression of myogenic and stem cell markers in satellite cells and other stemprogenitor cells endowed with myogenic potency

Cell Type	MRF	Pax3/7	Sca-1	CD45	CD34	CD31
Satellite cells	Yes	Low/High	Yes	No	Yes	Yes
MSC	No	No	Yes	No	No	Yes
MAPC	No	No	Yes	No	No	ND
MDSC	No	No	Yes	No	Yes	Yes
MAB	No	High/No	Yes	No	Yes	Yes
ADSC	No	ND	Yes	No	No	Yes
SDSC	No	ND	Yes	No	No	Yes
HSC	No	No	Yes	Yes	Yes	Yes

3. THE POSSIBLE DEVELOPMENTAL ORIGIN OF NON MUSCLE STEM CELLS

At first sight the origin of non muscle-derived stem cells, able to make muscle, appears to be mainly restricted to the hemo-vascular system (hematopoietic, endothelial, pericytes) that derives from the splanchno pleura. Cells associated with developing vessels would be evenly distributed to developing tissues with fetal angiogenesis and thus allocated to the local pool of progenitors for further tissue growth or regeneration. Non muscle stem cells with similar myogenic potency are also present in the neural tissue, but it is possible that they ingress the nervous system with fetal angiogenesis. Although this has never been demonstrated, the reported association of neural stem cells (or possibly a subset of them) with the vasculature (Palmer et al., 2000) would be compatible with this hypothesis. Although all the above mentioned embryonic tissues are unrelated to somites and paraxial medoserm, the situation may be more complex.

3.1 Clonal Studies in Mouse and Chick Embryos

Canonic skeletal myogenic progenitors originate from the dorsal somite but several other cell types such as dermis fibroblasts, endothelial cells and smooth muscle also originate in part from the dermomyotome (Christ and Ordahl, 1995). Therefore detecting myogenesis arising from an endothelial or a smooth vascular progenitor would not necessarily imply that it is non somitic in origin.

An unbiased search for a skeletal myogenic progenitor outside the somite in the developing mouse embryo identified the dorsal aorta as a source of skeletal myogenic clones that could not be derived from other anlagen such as the heart, the ectoderm or the gut (De Angelis et al., 1999). Virtually all the cells of the clones derived from the dorsal aorta co-express early endothelial and myogenic markers such as VE-cadherin and MyoD as well as smooth alpha actin. Few years later, an elegant study identified a common progenitor that gives rise to endothelium and skeletal muscle. A library of replication-defective retroviral vectors was used to infect cells in the somite, from which both myogenic and endothelial progenitors migrate to the limb. Single cell PCR confirmed the clonal origin of differentiated cells that shared integration of the same proviral sequence: surprisingly, approximately one third of myogenic and endothelial cells were found to derive from a common somitic precursor.

In this context, a recent report clearly indicated a common clonal origin for cells in the myotome and in the dorsal aorta. A genetic approach that permits retrospective clonal analysis (Bonnerot and Nicolas, 1993) is based on a laacZ reporter that contains a duplication of the lacZ coding sequence under the control of regulatory sequences directing expression to the tissues of interest. In the embryo, a rare intragenic recombination event will remove the duplication to give lacZ, which encodes a functional β-galactosidase (β-gal) protein when the gene is expressed. A common progenitor cell that has undergone such a recombination event will give rise to β-gal+ cells that are clonally related. When the α-cardiac actin gene was targeted with a nlaacZ reporter it became possible to examine, in addition to the heart (Meilhac et al., 2004), also embryonic skeletal muscle and the dorsal aorta where this gene is also transiently expressed (Sassoon et al., 1988). This retrospective clonal analysis showed that cells in the dorsal aorta and in the myotome have a common clonal origin. Moreover, based on the long half life of the GFP protein, it was possible to follow the fate of Pax3GFP/+ progenitors in the paraxial mesoderm that appear to migrate from the somite to the dorsal aorta. Most of the clones contained smooth muscle cells, but occasional labeled endothelial cells were present in the clones, in keeping with the existence of a common vascular progenitor.

Thus the relationship among somitic and non somitic vascular progenitors may be complex: cells from the somite may migrate to the dorsal aorta and eventually be distributed to developing tissues with vessels branching from the aorta. If some of these branches reach developing skeletal muscle, these somitic derived vascular progenitors may be recruited to a myogenic fate by signals emanating from developing muscle fibers. Moreover, although experimentally not tested, somitic vascular progenitors may easily associate with inter-somitic arteries and thus be distributed to developing tissues with the same mechanism proposed for the dorsal aorta.

Therefore all the studies showing origin of myogenic cells from non somitic tissue, should be interpreted with the caveat that cells in vascular system may ultimately derive from somites through the developmental events described above. Since the vascular tree grows into virtually any tissue (excluding cartilage and epidermis) and it may be carrying along somite derived progenitors, a somitic origin for myogenic cells found in other tissues cannot be excluded. Indeed, to formally demonstrate a non somitic origin of at least some of these progenitors, we dissected the lateral mesoderm from mouse embryos at the stage of 3–5 somites, before a vascular connection between somites and lateral mesoderm is established. The embryos expressed the n-LacZ reporter gene under the transcriptional control of the Myosin light chain 1/3 fast promoter/enhancer, restricting transgene expression to striated muscle. As expected no transgene expression was observed in the lateral mesoderm explants, cultured in isolation or on a feeder layer of fibroblasts. However, when the same explants were co-cultured with differentiating C2C12 myogenic cells, many LacZ expressing nuclei were detected inside multinucleated myotubes, indicating that truly non somitic cells have at least the option of fusing in vitro into differentiated myotubes and trans-activate a skeletal muscle promoter (Fig. 1).

3.2 Studies on the Origin of Satellite Cells and of Non-muscle Stem Cells

Together these studies strongly argue in favor of a complex lineage relationship among early endothelial, smooth and skeletal myogenic progenitors, but the exact underlying mechanism remains elusive. Since most of these studies were limited to early post-somitic stages, none sheds light on the origin of later progenitors or

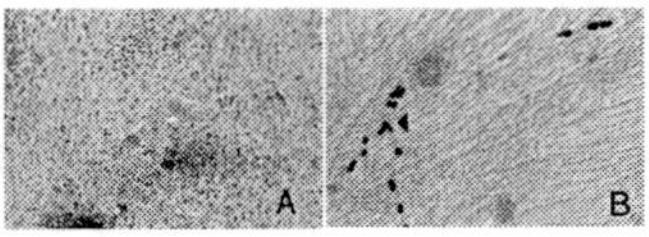

Figure 1. Skeletal myogenic differentiation in cells isolated from 3–5 somite stage mouse embryo lateral mesoderm. Lateral mesoderm was dissected from MLC1/3F-nLacZ embryos and cultured either on a feeder layer of 10T1/2 fibroblasts (A) or C2C12 myogenic cells (B). After 5 days, cultures were stained with X-Gal. β-gal positive nuclei are shown by arrowhead

post-natal stem cells. Recently however three studies agreed that also post-natal satellite cells and, in one case, muscle SP, are somite derived. A cell population that expresses the transcription factors Pax3 and Pax7 but no skeletal-muscle-specific markers was recently identified in the mouse. These cells are maintained as a proliferating population in embryonic and fetal muscles of the trunk and limbs throughout development and later adopt a satellite cell position characteristic of progenitor cells in postnatal muscle (Relaix et al., 2005).

In another study, electroporation of GFP in chick somites and quail-chick grafting experiments showed that the dorsal compartment of the somite, the dermomyotome, is the origin of a population of muscle progenitors that contribute to the growth of trunk muscles during embryonic and fetal life, including satellite cells (Gros et al., 2005). Finally it was shown, through different approaches (replication-defective retroviruses, quail/chick chimeras, and mouse Pax3-Cre lines) that the majority of limb muscle satellite cells arise from cells expressing Pax3 specifically in the hypaxial somite; moreover they show that a significant number of limb muscle SP cells are derived from the hypaxial somite (Schienda et al., 2006).

As for the origin of the other stem cells described above, not much is known at the moment. We can assume, based on previous embryological studies, that hematopoietic stem cells, pericytes, endothelial progenitors, mesoangioblasts, MAPs and mesenchymal stem cells are all associated to the hemo-vascualar system, which is derived, but not entirely (see above) from the ventral lateral mesoderm or splanchnopleura. Unfortunately, expression of a given repertoire of surface antigens may be useful to prospectively isolate these cells form adult or fetal tissues, but is not informative on their origin since the same cell lineage may change gene expression during development. Indeed, genetic labeling by the cre-lox system has been used so far to demonstrate that endothelial cells in the adult may derive from a common myeloid progenitors. In general these studies are limited by paucity of truly specific promoters, which are also expressed early during development, to allow tracing the developmental origin of a given stem/progenitor cell. In the past we used VE-Cadherin/cre and Tie2/cre mice crossed to floxed Rosa 26 mice aiming to detect β-gal+ cells, originating from the endothelium, inside smooth, skeletal or cardiac muscle. The results of these experiments showed that rare (less than 1%) smooth muscle cells are derived from founders that once expressed either VE-Cadherin or Tie2. However the frequency of cardiac or skeletal muscle derived from endothelial founders was extremely low (less than 0.01%) indicating that virtually no skeletal muscle is derived from an endothelial cell, at least at a stage when it already expressed VE-Cadherin or Tie2 (Berarducci et al. unpublished results). It remains possible that some muscle cells are derived from a more immature endothelial progenitor or angioblast but, by the time the cells has activated differentiated gene products such as VE-Cadherin its fate is restricted to mature endothelium and possibly rare smooth muscle cell. Here again, absence of a well characterized, truly "angioblast" specific promoter, prevents this kind of approach to be extended to a more immature and possibly still multipotent progenitor.

3.3 The Possible Lineage Relationship of Mesoderm Stem Cells with Satellite Cells

Mesoangioblasts are derived from the vessel wall and so are mesenchymal stem cells, EPC and multipotent adult progenitors: thus the vascular niche in the bone marrow and possibly in all mesoderm is a site where different types of multipotent (and potentially myogenic cells) are found in the adult. Furthermore, hematopoietic stem cells (HSC), which also show myogenic potency, are present in the same anatomical site, within the bone marrow and other hematopoietic tissues.

A question relevant to muscle regeneration is whether there is any lineage relationship between one or more types of mesoderm stem cells and muscle satellite cells. In other words it is possible that any of these cells may leave the vessel wall, enter the interstitial space, then cross the basal lamina of the muscle fiber and eventually adopt a satellite cell position, possibly expressing satellite cell specific genes. Evidence for this event has been claimed of the basis of co-expression of a satellite cell markers (M-Cadherin, CD34, Pax7) and a donor cell marker (GFP, LacZ etc.) in a cell located underneath the basal lamina but outside the sarcolemma, after either intra-muscular or intra-arterial injection or bone marrow transplantation. An example is shown in Fig. 2. Even though this event has been found to be rare when analyzed in vivo, a real possibility exists that it may occur constantly during late fetal and post-natal muscle growth, so that it may feed a significant proportion of cells into the satellite cell compartment and thus contribute indirectly to regenerating fibers. Obviously experiments carried out in a short period of time

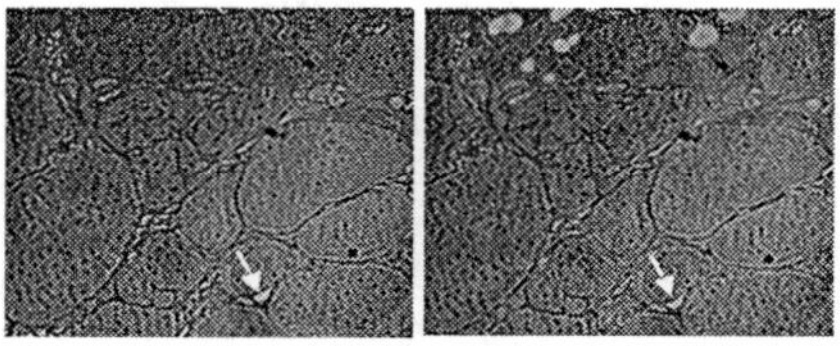

Figure 2. Human mesoangioblasts give rise to satellite cells after intra-arterial transplantation. Human cells identified and express satellite cell markers. A Myf5 (green) expressing cell (arrow), located at the periphery of a small fiber, also express Lamin A/C. Human nuclei appear violet (arrowhead), after co-staining with DAPI. Fluorescence is superimposed on the phase contrast image of the tissue. Bar = 20 μm

would miss the alternative origin of satellite cells that may have been derived from other mesoderm stem cells before the time of analysis. Importantly, in all these experiments a damage to skeletal muscle and often a depletion of the resident pool of myogenic cells are required to provide a selective advantage to donor cells. This means that it will be very difficult to know what is the turn-over of satellite cells and what part of this turn-over may be carried out by non resident progenitors cells in the healthy muscle of a normal mammal or in the course of a primary myopathy. The argument raised above of the somitic origin of most satellite cells does not contrast this possibility because of the somitic origin of endothelial and smooth muscle cells described above.

4. PERSPECTIVES FOR CELL THERAPY

The scenario described above is complex and likely will be expanded, refined and possibly modified by the rapidly accumulating data from the many laboratories involved in this area of research.

Nevertheless, a quest for a therapy for muscular dystrophy and other primary muscle diseases, raises the additional need to choose among these myogenic progenitors those which may best fit the requirements for a successful restoration of muscle morphology and function (Cossu and Sampaolesi, 2004).

To this aim selection of the appropriate cell type should meet the following criteria: (a) accessible source (e.g. blood, bone marrow, fat aspirate, muscle or skin biopsy); (b) ability to grow as a relatively homogeneous population in vitro for extended periods without loss of differentiation potency (since it appears currently unlikely that cells may be acutely isolated in numbers sufficient for therapeutic purposes); (c) susceptibility to in vitro transduction with vectors encoding therapeutic genes (these vectors should themselves meet criteria of efficiency, safety and long term expression); (d) ability to reach the sites of muscle degeneration/regeneration through a systemic route and in response to cytokines released by dystrophic muscle; (e) ability to differentiate in situ into new muscle fibers with high efficiency and to give rise to physiologically normal muscle cells.

Satellite cells that were considered as the first and most obvious candidate for the cell therapy of muscular dystrophy are not able to cross the vessel wall when delivered systemically and need to be locally injected into skeletal muscle at a distance of few mm from each other, since they cannot migrate extensively in the muscle. This fact alone limits the potential application of satellite cells, at least with current technology. Moreover, most of the injected cells die within the first day and this explains the failure of the first trials with satellite cell derived myoblasts in the early 90'.

Advantages and disadvantages of the other types of non muscle stem cells vary and are summarized in Table 1. Some are difficult to expand in vitro, others show inefficient myogenic differentiation while for others the ability to negotiate the vessel wall when systemically delivered has not been experimentally tested. Right now mesoangioblasts are the cell type for which most parameters have been tested in vitro and more importantly in vivo, first in a mouse model of muscular dystrophy (Sampaolesi et al., 2003) and more recently in the Golden Retriever dystrophic dog (Sampaolesi et al., 2006).

Hopefully in a few years time, phase I clinical trials with stem cells may start and set the stage for one more, and at least in part successful attack to defeat these genetic diseases.

ACKNOWLEDGEMENTS

Work in the Authors laboratory is supported by grants from from MDA, Telethon, AFM, Parent Project Onlus, CARIPLO, EC "Eurostemcell", "Cellsintoorgan" and "Myoamp", Italian Ministries of Health and Research.

REFERENCES

Asakura A, Seale P, Girgis-Gabardo A, Rudnicki MA, (2002) Myogenic specification of side population cells in skeletal muscle. J Cell Biol (2002) 159:123–134

Asahara T, Murohara T, Sullivan A, Silver M, van der Zee R, Li T, Witzenbichler B, Schatteman G, Isner JM (1997) Isolation of putative progenitor endothelial cells for angiogenesis. Science 275:964–967

Bachrach E, Perez AL, Choi YH, Illigens BM, Jun SJ, del Nido P, McGowan FX, Li S, Flint A, Chamberlain J, Kunkel LM (2006) Muscle engraftment of myogenic progenitor cells following intraarterial transplantation. Muscle Nerve 34(1):44–52

Bachrach E, Li S, Perez AL, Schienda J, Liadaki K, Volinski J, Flint A, Chamberlain J, Kunkel LM (2004) Systemic delivery of human microdystrophin to regenerating mouse dystrophic muscle by muscle progenitor cells. Proc Natl Acad Sci U S A 101(10):3581–3586

Bianco P, Cossu G (1999) Uno, nessuno e centomila: searching for the identity of mesodermal progenitors. Exp Cell Res 251(2):257–263

Bianco P, Gehron Robey P (2000) Marrow stromal stem cells. J Clin Invest 105:1663–1668

Bischoff R (1994) The satellite cell and muscle regeneration, In: Engel AG, Franzini-Armstrong C (eds) Myology 2nd ed., McGraw-Hill: New York pp. 97–133

Bonnerot C, Nicolas JF (1993) Application of LacZ gene fusions to postimplantation development. Methods Enzymol 225:451–469.

Breton M, Li Z, Paulin D, Harris JA, Rieger F, Pincon-Raymond M. Garcia L (1995). Myotube driven myogenic recruitment of cells during in vitro myogenesis. Develop Dynam 202:126–136

Camargo FD, Green R, Capetanaki Y, Jackson KA, Goodell MA (2003) Single hematopoietic stem cells generate skeletal muscle through myeloid intermediates. Nat Med 9:1520–1527

Challen GA, Little MH (2006) A side order of stem cells: the SP phenotype. Stem Cells 24(1):3–12

Chaudhary N, Delay R. & Beam KG (1989) Restoration of normal function in genetically defective myotubes by spontaneous fusion with fibroblasts. Nature 341:445–447

Christ B, Ordahl C (1995) Early stage of chick somite development. Anat Embryol 191:381–396

Corbel SY, Lee A, Yi L, Duenas J, Brazelton TR, Blau HM et al (2003) Contribution of hematopoietic stem cells to skeletal muscle. Nat Med 9:1528–1532

Courbin P, Koenig J, Ressouches A, Beam KG Powell JA (1989) Rescue of excitation-contraction coupling in disgenic muscle by addition of fibroblasts in vitro. Neuron 2:1341–1350

Cossu G, Kelly R, Di Donna S, Vivarelli E Buckingham M (1995) Myoblast differentiation during mammalian somitogenesis is dependent upon a community effect. Proc Natl Acad Sci USA 92:2254–2258

Cossu G (1997) Unorthodox myogenesis: possible developmental significance and implications for tissue histogenesis and regeneration. Histol Histopathol 12(3):755–760

Cossu G, Sampaolesi M. (2004) New therapies for muscular dystrophy: cautious optimism. Trends Mol Med 10:516–520

Cossu G, Biressi S. (2005) Satellite cells, myoblasts and other occasional myogenic progenitors: possible origin, phenotypic traits and role in muscle regeneration. Sem Cell Dev Biol Aug-Oct; 16(4–5):623–631.

Cusella De Angelis MG, Balconi G, Bernasconi S, Zanetta L, Boratto R, Galli D, Dejana E, Cossu G (2003) Skeletal myogenic progenitors in the endothelium of lung and yolk sac. Exptl Cell Res 290:207–216

De Angelis L, Berghella L, Coletta M, Lattanzi L, Zanchi M, Cusella-De Angelis MG et al (1999) Skeletal myogenic progenitors originating from embryonic dorsal aorta coexpress endothelial and myogenic markers and contribute to postnatal muscle growth and regeneration. J Cell Biol 147:869–878

De Bari C, Dell'Accio F, Vandenabeele F, Vermeesch JR, Raymackers JM, Luyten FP (2003) Skeletal muscle repair by adult human mesenchymal stem cells from synovial membrane. J Cell Biol 160(6):909–918

Dezawa M, Ishikawa H, Itokazu Y, Yoshihara T, Hoshino M, Takeda S, Ide C, Nabeshima Y (2005) Bone marrow stromal cells generate muscle cells and repair muscle degeneration. Science 309(5732):314–317

Ferrari G, Cusella-De Angelis MG, Coletta M, Paolucci E, Stornaiuolo A, Cossu G et al (1998) Muscle regeneration by bone marrow-derived myogenic progenitors. Science 279:1528–1530

Ferrari G, Stornaiuolo A, Mavilio F (2001) Failure to correct murine muscular dystrophy. Nature 411(6841):1014–1015

Galli R, Borello U, Gritti A, Minasi MG, Bjornson C, Coletta M et al (2000) Skeletal Myogenic Potential of Adult Neural Stem Cells. Nature Neurosci 3:986–991

Gibson AJ, Karasinski J, Relvas J, Moss J, Sherratt TG, Strong PN Watt DJ. (1995) Dermal fibroblasts convert to a myogenic lineage in *mdx* mouse muscle. J Cell Science 108:207–214

Goldring K, Jones GE, Thiagarajah R, Watt DJ (2002) The effect of galectin-1 on the differentiation of fibroblasts and myoblasts in vitro. J Cell Sci 115:355–366

Goodell MA, McKinney-Freeman S, Camargo FD (2005) Isolation and characterization of side population cells. Methods Mol Biol 290:343–352

Gros J, Manceau M, Thome V, Marcelle C. (2005) A common somitic origin for embryonic muscle progenitors and satellite cells. Nature 435(7044):954–958

Grounds MD, Garrett KL, Beilharz MW (1992) The transcription of MyoD1 and myogenin genes in thymic cells in vivo. Exp Cell Res 198(2):357–361

Gurdon JB. (1993) Community effect and related phenomena in development. Cell 75:501–506

Gussoni E, Soneoka Y, Strickland CD, Buzney EA, Khan MK, Flint AF, Kunkel LM, Mulligan RC (1999) Dystrophin expression in the mdx mouse restored by stem cell transplantation. Nature 401(6751):390–394

Gussoni E, Bennett RR, Muskiewicz KR, Meyerrose T, Nolta JA, Gilgoff I, Stein J, Chan YM, Lidov HG, Bonnemann CG, Von Moers A, Morris GE, Den Dunnen JT, Chamberlain JS, Kunkel LM, Weinberg K (2002) Long-term persistence of donor nuclei in a Duchenne muscular dystrophy patient receiving bone marrow transplantation. J Clin Invest 110(6):807–814

Jiang Y (2002a) Multipotent progenitor cells can be isolated from postnatal murine bone marrow, muscle, and brain. Exp Hematol 30(8):896–904. Erratum in: Exp Hematol 2006 Jun; 34(6):809

Jiang Y, Jahagirdar BN, Reinhardt RL, Schwartz RE, Keene CD, Ortiz-Gonzalez XR, Reyes M, Lenvik T, Lund T, Blackstad M, Du J, Aldrich S, Lisberg A, Low WC, Largaespada DA, Verfaillie CM (2002b) Pluripotency of mesenchymal stem cells derived from adult marrow. Nature 418(6893):41–49

Kelly R, Alonso S, Tajbakhsh S, Cossu G, Buckingham M (1995) Myosin light chain 3F regulatory sequences confer regionalized cardiac and skeletal muscle expression in transgenic mice. J Cell Biol 129(2):383–396

Kocher AA, Schuster MD, Szabolcs MJ, Takuma S, Burkhoff D, Wang J, Homma S, Edwards NM, Itescu S (2001) Neovascularization of ischemic myocardium by human bone-marrow-derived angioblasts prevents cardiomyocyte apoptosis, reduces remodeling and improves cardiac function. Nat Med 7(4):430–436

Liechty KW, MacKenzie TC, Shaaban AF, Radu A, Moseley AM, Deans R, Marshak DR, Flake AW (2000) Human mesenchymal stem cells engraft and demonstrate site-specific differentiation after in utero transplantation in sheep. Nat Med 6(11):1282–1286

Mayer DC, Leinwand LA (1997) Sarcomeric gene expression and contractility in myofibroblasts. J Cell Biol 139(6):1477–1484.

McKinney-Freeman SL, Jackson KA, Camargo FD, Ferrari G, Mavilio F, Goodell MA (2002) Muscle-derived hematopoietic stem cells are hematopoietic in origin. Proc Natl Acad Sci U S A 99(3):1341–1346

Meilhac SM, Esner M, Kelly RG, Nicolas JF, Buckingham ME (2004) The clonal origin of myocardial cells in different regions of the embryonic mouse heart. Dev Cell 6(5):685–698

Minasi MG, Riminucci M, De Angelis L, Borello U, Berarducci B, Innocenzi A et al (2002) The meso-angioblast: a multipotent, self-renewing cell that originates from the dorsal aorta and differentiates into most mesodermal tissues. Development 129:2773–2783

Murohara T, Ikeda H, Duan J, Shintani S, Sasaki K, Eguchi H, Onitsuka I, Matsui K, Imaizumi T (2000) Transplanted cord blood-derived endothelial precursor cells augment postnatal neovascularization. J Clin Invest 105(11):1527–1536

Okazaki K, Holtzer H (1996) Myogenesis: fusion, myosin synthesis, and the mitotic cycle. Proc Natl Acad Sci U S A 56(5):1484–1490

Palmer TD, Willhoite AR, Gage FH (2000) Vascular niche for adult hippocampal neurogenesis J Comp Neurol 425(4):479–494

Patapoutian A, Wold BJ, Wagner R (1995) Evidence for developmentally programmed transdifferentiation in mouse esophageal muscle. Science 270:1818–1821

Pesce M, Orlandi A, Iachininoto MG, Straino S, Torella AR, Rizzuti V, Pompilio G, Bonanno G, Scambia G, Capogrossi MC (2003) Myoendothelial differentiation of human umbilical cord blood-derived stem cells in ischemic limb tissues. Circ Res 93(5):e51–e62

Qu Z, Balkir L, van Deutekom JC, Robbins PD, Pruchnic R, Huard J. (1998). Development of approaches to improve cell survival in myoblast transfer therapy. J Cell Biol 142:1257–1267

Relaix F, Rocancourt D, Mansouri A, Buckingham, M (2005) A Pax3/Pax7-dependent population of skeletal muscle progenitor cells. Nature 435:948–953

Reyes M, Lund T, Lenvik T, Aguiar D, Koodie L. & Verfaillie CM (2001) Purification and ex vivo expansion of postnatal human marrow mesodermal progenitor cells. Blood 98:2615–2625

Rodriguez AM et al (2005) Transplantation of a multipotent cell population from human adipose tissue induces dystrophin expression in the immunocompetent mdx mouse. J Exp Med 201(9):1397–1405

Salvatori G, Lattanzi L, Coletta M, Aguanno S, Vivarelli E, Kelly R, Ferrari G, Harris AJ, Mavilio F, Molinaro M & Cossu G. (1995) Myogenic conversion of mammalian fibroblasts induced by differentiating muscle cells J Cell Science 108:2733–2739

Sampaolesi M, Torrente Y, Innocenzi A, Tonlorenzi R, D'Antona G, Pellegrino MA et al (2003) Cell therapy of alpha-sarcoglycan null dystrophic mice through intra-arterial delivery of mesoangioblasts. Science 301:487–492

Sampaolesi M, Blot S, D'Antona G, Granger N, Tonlorenzi R, Innocenzi A, Mognol P, Thibaud JL, Galvez BG, Barthélémy I, Perani L, Mantero S, Guttinger M, Pansarasa O, Rinaldi C, Cusella De Angelis MG, Torrente Y, Bordignon C, Bottinelli R. & Cossu G. (2006) Mesoangioblast stem cells ameliorate muscle function in dystrophic dogs. Nature 444(7119):574–579

Sassoon DA, Garner I, Buckingham M (1988) Transcripts of alpha-cardiac and alpha-skeletal actins are early markers for myogenesis in the mouse embryo. Development 104(1):155–164

Schienda J, Engleka KA, Jun S, Hansen MS, Epstein JA, Tabin CJ, Kunkel LM, Kardon G (2006) Somitic origin of limb muscle satellite and side population cells. Proc Natl Acad Sci U S A 103(4):945–950

Tajbakhsh S, Vivarelli G, Cusella-De Angelis G, Rocancourt D, Buckingham M, Cossu G (1994) A population of myogenic cells derived from the mouse neural tube. Neuron 13:813–821

Tajbakhsh S, Buckingham ME (1995) Lineage restriction of the myogenic conversion factor myf-5 in the brain. Development 121(12):4077–83

Takahashi T, Kalka C, Masuda H, Chen D, Silver M, Kearney M, Magner M, Isner JM, Asahara T (1999) Ischemia- and cytokine-induced mobilization of bone marrow-derived endothelial progenitor cells for neovascularization. Nat Med 5(4):434–438

Tamaki T, Akatsuka A, Ando K, Nakamura Y, Matsuzawa H, Hotta T, Roy RR, Edgerton VR (2002) Identification of myogenic-endothelial progenitor cells in the interstitial spaces of skeletal muscle. J Cell Biol 157:571–577

Tamaki T, Uchiyama Y, Okada Y, Ishikawa T, Sato M, Akatsuka M. & Asahara T. (2005) Functional recovery of damaged skeletal muscle through synchronized vasculogenesis, myogenesis, and neurogenesis by muscle-derived stem cells Circulation 112:2857–2866.

Torrente Y, Belicchi A, Sampaolesi M, Pisati F, Lestingi M, D'Antona G, Tonlorenzi R, Porretti L, Gavina M, Mamchaoui K, Pellegrino MA, Furling D, Mouly V, Butler-Browne GS, Bottinelli R, Cossu G. & Bresolin N (2004) Human circulating AC133+ stem cells replenish the satellite cell pool, restore dystrophin expression and ameliorate function upon transplantation in murine dystrophic skeletal muscle. J Clin Invest 2004;114:182–195

Wakitani S, Seito T, Caplan AI. (1995) Myogenic cells derived from rat bone marrow mesenchimal strem cells exposed to 5′-Azacytidine. Muscle & Nerve 18:1417–1426

Wekerle TH, Paterson B, Ketelsen U, Feldman M (1975) Striated muscle fibres differentiate in monolayer cultures of adult thymus reticulum. Nature. 256(5517):493–449

CHAPTER 5

TRANSCRIPTIONAL CASCADES IN MUSCLE REGENERATION

PO ZHAO AND ERIC HOFFMAN
Research Center for Genetic Medicine, Children's National Medical Center, 111 Michigan Ave NW, Washington DC 20010, USA

1. INTRODUCTION

1.1 Muscle as Paradigm for Molecular Networks in Cell Lineage Commitment and Differentiation

Muscle is the largest organ of the human body, and it contains the largest cells – the myofiber. The large syncytial multi-nucleated myofibers originate from mononuclear myogenic precursor cells; during development these arise in the embryonic mesoderm, and during muscle damage and repair they derive from muscle satellite cells. The transition of a myogenic cell from a proliferative undifferentiated state to a mature syncytial myofiber thousands or millions of times larger has been a research model for cell lineage commitment and differentiation. The molecular events defining myogenesis were given an initial boost in the 1980's, with the first identification of *MyoD* and other key transcriptional regulatory proteins (*Myf5, Mrf4,* Myogenin). Forced expression of these transcriptional regulators were able to convert non-myogenic cells into muscle cells in vitro (Davis et al., 1987; Braun et al., 1989; Edmondson and Olson, 1989; Wright et al., 1989). These proteins belong to the family of basic helix-loop-helix transcriptional factors, and are able to induce expression of muscle specific genes.

More recently, the development of genomics and microarray methods has enabled the study of the myogenic differentiation transitions downstream (transcriptional target genes) and upstream of these factors in a genome-wide manner. Key to new experimental approaches is 'time' as a controlled variable. By staging myogenesis in vitro, and muscle degeneration/regeneration in vivo, a temporal

S. Schiaffino and T. Partridge (eds.), Skeletal Muscle Repair and Regeneration, 85–106.

cascade of transcriptional factors can be defined. Through time series experiments, genome-wide cause-effect of gene-protein networks driving cell lineage and commitment can be modeled then validated. It is important to note that many aspects of gene-protein regulatory networks are *not* queried by microarrays. Protein phosphorylation and acetylation states, and subcellular localizations are known to be critical for molecular networks in myogenesis but are not assayed by mRNA profiling. Moreover, chromatin configuration, acetylation and methylation patterns of histones, methylation of DNA and promoter residency states are not queried by simply measuring steady state levels of RNA. Nevertheless, mRNA profiling of time series both in vitro and in vivo myogenesis provide one of the better molecular models of vertebrate cellular and tissue development in any organ system. Also, genome-wide assessments of DNA methylation, and transcription factor binding to chromatin promises a more complete picture of the regulatory networks involved in myogenesis.

Before describing the state-of-the-art in genome-wide studies of myogenesis, it is important to compare and contrast the three key experimental models used in most studies. In vitro studies typically utilize cultured myogenic cells (immortalized mouse C2C12 cells, or muscle tissue-derived primary myoblasts), with proliferation in high serum media (10–20% fetal calf serum), and initiation of myogenic differentiation via serum starvation (2% horse serum). While this important and highly utilized experimental system has provided enormous amounts of data on myogenesis, it is also quite different than the in vivo process of myogenesis. Commitment and differentiation in vivo is driven by positional cues (development), and interactions with basal lamina and other cell types (regeneration); these cues are lacking in the in vitro models used. Thus, the reduction of complexity in these classic in vitro systems enables the definition of clear cause/effect in myogenic molecular pathways, but it can be challenging to prove that these pathways are relevant to the in vivo process.

For in vivo studies, vertebrate myogenesis is typically studied during embryonic development or during adult muscle regeneration. These two processes share many molecular pathways, but they are also distinct in many ways. Embryonic myogenesis is discussed in depth in other chapters in this volume, and will not be discussed in detail here. As might be expected, positional cues in the embryonic mesoderm are critical for embryonic myogenesis, while less important to muscle regeneration within pre-existing myofiber basal lamina (Zhao and Hoffman, 2004).

In this current chapter we focus on staged muscle regeneration in adult muscle as a model for myogenic commitment and differentiation. An important characteristic of muscle is its capability for regeneration upon stimulation such as injury. This involves the activation of normally quiescent satellite cells in adult muscle. Satellite cells subsequently proliferate, exit the cell cycle, fuse into myotubes, and then differentiate into muscle fibers. It has been long recognized that developmentally regulated genes defined through in vitro studies are induced during muscle regeneration. Some examples are myogenic regulatory factors (MRFs, namely Myf5, Mrf4, MyoD, and myogenin), embryonic and neonatal myosin isoforms, acetylcholine receptors, α-cardiac actin, and cardiac isoforms of troponin T and I (d'Albis et al., 1988, 1989; Fuchtbauer and Westphal, 1992; Koishi et al., 1995; Rantanen et al., 1995; Franke et al., 1996; Creuzet et al., 1998; Nicolas et al., 1998; Messner et al., 2000; Launay et al., 2001). Recent expression profiling studies with microarrays have validated and expanded earlier findings (Zhao et al., 2002; Yan et al., 2003).

1.2 Status of Myogenesis Transcriptional Pathways 'Pre-genome-enabled'

Transcriptional factors, including Pax3, Pax7, and MRFs, have been implicated to be crucial for muscle development. Study of *Pax3* mutant mice (splotch, Sp), *Myf5* null mice and *Sp/Myf5* double mutant mice suggests that expression of MyoD depends on the presence of either Pax3 or Myf5, which regulate different pathways to activate *MyoD* in muscles derived from myotome or migrating muscle precursor cells, respectively (Kablar et al., 1997; Tajbakhsh et al., 1996, 1997).

The critical function of MRFs at different stages of embryonic myogenesis has been extensively evaluated in knockout mouse models. Although both *MyoD* null mice and *Myf5* null mice do indeed form skeletal muscle (Braun et al., 1992; Rudnicki et al., 1992), *MyoD/Myf5* double mutants fail to generate myoblasts and muscle fibers (Rudnicki et al., 1993). However, a recent study suggests that when the expression of *Mrf4*, which is genetically linked to *Myf5*, is intact when generating the *MyoD/Myf5* double mutants, skeletal muscle is still able to form (Kassar-Duchossoy et al., 2004). Furthermore, Mrf4, like Pax3 and Myf5, acts upstream of *MyoD*. Together, these studies suggest that MyoD, Myf5 and Mrf4 are essential for initial myoblast formation, and their functions compensate and/or are partially redundant with each other. Myogenin null mice do not form muscle fibers, although myoblasts did not seem to be affected (Hasty et al., 1993; Nabeshima et al., 1993), indicating that myogenin is critical for myoblast differentiation. When myogenin is disrupted after embryonic muscle formation, mice show normal muscle but reduced body size, suggesting myogenin plays a different role in postnatal muscle growth (Knapp et al., 2006).

Recent research has been focused on defining transcriptional cascades involving these important transcriptional factors. Traditional biochemistry-driven approaches (transfection of promoter-reporter constructs) are limited by the number of candidates studied, and such analyses are typically carried out in vitro, often in C2C12 cells.

Emerging high throughput technologies (microarray for steady state mRNA levels; ChIP-on-chip for promoter residency by transcription factors) provide powerful tools in the investigation of transcriptional cascades in the genome-wide scale. Applications of microarray on time series or inducible systems (cause/effect) are particularly useful in the study of transcription regulation. With the future advances to proteomic profiling, and integration of transcriptome/proteome data, transcriptional cascades in muscle regeneration may be completely defined.

2. GENOME-WIDE GENE EXPRESSION OF MUSCLE REGENERATION IN VIVO

2.1 Two High Throughput Platforms for Gene Expression Study: Affymetrix Oligonucleotide Array vs. Spotted cDNA or DNA Array

Microarrays are able to detect the expression of thousands of genes simultaneously. Two major platforms of microarrays have been developed: Affymetrix oligonucleotide microarrays and spotted cDNA or oligonucleotide microarrays (Fig. 1). Affymetrix oligonucleotide microarrays contain about 1 million features, where DNA fragments (probes) are chemically synthesized at specific addresses (features). Each feature contains millions of copies of a 25mer probe targeted to a specific gene sequence. Each mRNA transcript is queried by a set of probe pairs (probe set) consisting of both perfect match probes and mismatch probes. Mismatch probes can be used as an assessment of non-specific hybridization, and some probe set algorithms determine signal by assessing the intensity of the perfect match probes after subtraction of the intensity of mismatch probes. However, there are many methods for translation of hybridization image files into normalized

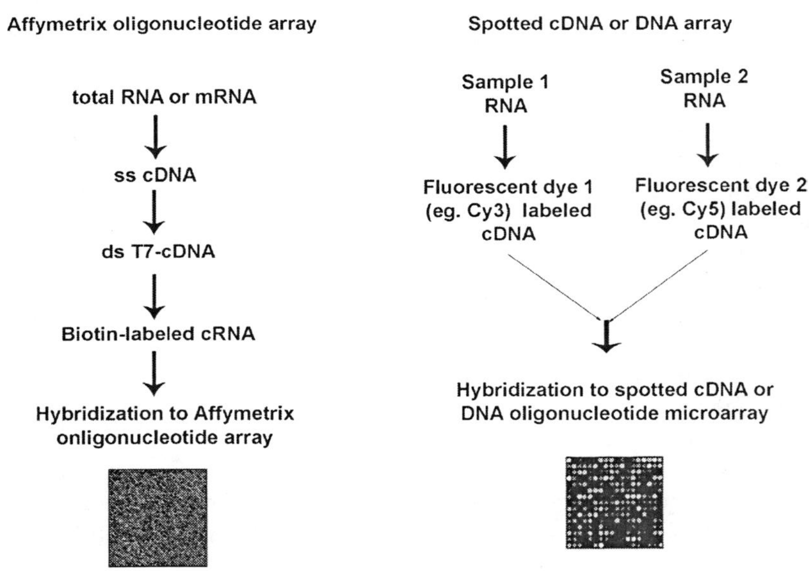

Figure 1. Flowchart of expression profiling analysis with Affymetrix oligonucleotide arrays and spotted cDNA or DNA arrays. Affymetrix array analysis uses biotin-labeled cRNA reverse transcribed from double-stranded cDNA linked to a T7 promoter. cRNA samples are subsequently fragmented, hybridized on microarrays, and detected with single-colored fluorescence (Streptavidin Phycoerythrin). Abundance of mRNA species is determined by the intensity of fluorescence. Spotted cDNA or DNA array analysis requires the experimental sample and the reference sample labeled with two different colors of fluorescent dyes of (red and green). Samples are then mixed in 1:1 ratio and hybridized on microarrays. Differential expression is determined by the ratio of two fluorescent dyes

transcriptional signals, and the reader is referred to recent reviews of this topic (Seo and Hoffman, 2006).

The gene capacity of an Affymetrix microarray has greatly increased with the reduced feature size. For example, the early version of murine genome U74Av2 array set consists of 3 arrays of 36,000 transcript units (12,000 probe sets/array, feature size 20 um, and 16 probe pairs/probe set), whereas the latest mouse genome 430 2.0 array has more than 45,000 probe sets on a single array (feature size 11 um, 11 probe pairs/transcript). Affymetrix oligonuleotide microarrays are single-colored, which means that samples are labeled with one fluorescent dye. The fluorescence intensity from each probe set is measured and calculated to determine the abundance of target sequence present in the sample. Samples and controls are hybridized to different individual microarrays, and compared to determine differentially expressed genes. DNA oligonucleotide microarrays are also available as spotted arrays, in which synthesized DNA oligos are printed on glass slides. Spotted DNA oligo microarrays often have longer probe lengths than Affymetrix arrays (e.g. 60-mer for Agilent gene expression DNA oligo arrays). The length of the probes is believed to be one of the factors affecting the sensitivity and specificity of the probes. cDNA microarrays are spotted arrays usually generated by PCR amplification of cDNA clones, and thus tend to have longer probes than oligonucleotide arrays. Both spotted DNA oligonucleotide arrays and cDNA arrays are typically two-colored, with samples and controls labeled with different fluorescent dyes (Cy3 or Cy5). Samples and controls are then mixed and hybridized to individual cDNA arrays. The differential expression of a gene is determined by the ratio of Cy3 to Cy5 fluorescence. A normalized ratio greater than 1 means higher expression in the sample than the control. Both Affymetrix oligonucleotide arrays and spotted arrays have been widely used for gene expression profiling study. The two platforms have also been used for gene expression study of myogenesis, both in vitro and in vivo (Bergstrom et al., 2002; Zhao et al., 2002; Yan et al., 2003).

2.2 Gene Expression Profiling of Muscle Regeneration in Vivo

2.2.1 Expression Profiling of Cardiotoxin Induced Muscle Regeneration in Normal Mice

Animal models of muscle regeneration can be generated by injection of snake venom cardiotoxin (CTX). We have generated a time course of muscle degeneration/regeneration in vivo in mice by injection of cardiotoxin into gastrocnemius muscles. Using Affymetrix U74Av2 mouse gene expression arrays (~12,000 probe sets), we profiled the gene expression of staged muscle degeneration/regeneration at 27 time points (0 to 40 day) (Zhao et al., 2002, 2003). A large number of differentially expressed genes were observed, including a series of myogenic transcription factors. Specifically, myogenic regulatory factors, *MyoD* and *myogenin* were induced at day 3 and day 3.5 post-injection (Fig. 2), which correlates the appearance of myotubes on muscle sections, suggesting this is a critical myoblast-myotube transition point. Clustering analysis of this data set with hierarchical

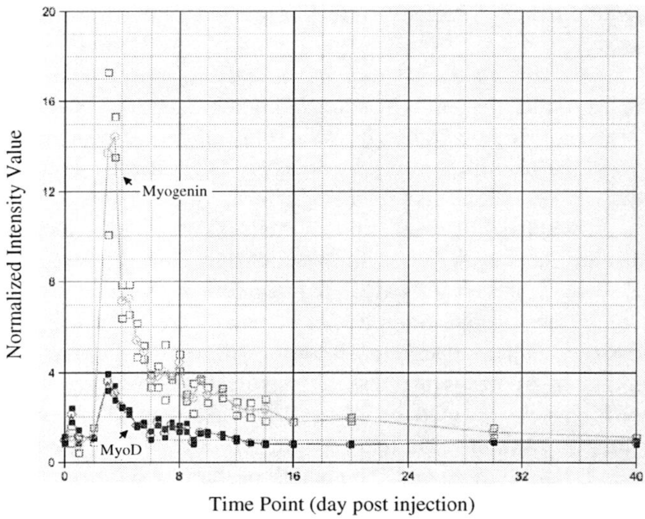

Figure 2. Temporal profile of MyoD and Myogenin in cardiotoxin induced muscle regeneration. Shown is the temporal profile of MyoD and Myogenin over the entire 27 time point series. MyoD and Myogenin were induced at day 3–3.5 post injection. Normalized (to day 0) expression values were graphed using a single gene query tool, which, together with the muscle regeneration data set, are accessible to the public through pepr.cnmcresearch.org. Error bars represent the two replicate values

clustering tool (GeneSpring and Hierarchical Clustering Explorer) and Soft Bayesian clustering tool (VISual Statistical Data Analyzer) revealed induction of genes involved three major processes of muscle degeneration/regeneration in sequential time windows (Fig. 3; Zhao et al., 2003). Inflammatory response genes are induced at early time points (day 0–2), when muscle structure genes were down-regulated. Myogenic and many developmentally regulated genes (*MyoD*, *myogenin, embryonic myosin heavy chain*) showed significant induction at day 3 to 3.5. At later time points, muscle structural genes (*actin, myosin*, others) gradually returned to normal expression level. This data set and the single gene query tool shown in Fig. 2 are openly accessible to the public (pepr.cnmcresearch.org; Chen et al., 2004).

Spotted cDNA arrays have also be used to expression profile of cardiotoxin induced muscle regeneration. Yan et al. (2003) studied 7 time points during a 14-day time course tibialis anterior (TA) muscle regeneration using NIA (National Institute on Aging) cDNA arrays, which contains 16,000 gene elements. Hierarchical clustering analysis also revealed highly coordinately regulated gene expression. Genes of cell cycle control and DNA replication were induced during the early phase (day 1 to day 3) of muscle regeneration, and genes for myogenic regulators and transition from proliferation to differentiation were induced during day 3 to day 5 after injury. Although the two expression profiling studies employed different platform (Affymetrix oligonucleotide arrays vs. NIA cDNA arrays) and different muscles (gastrocnemius vs. TA), both studies showed that gene expression in muscle

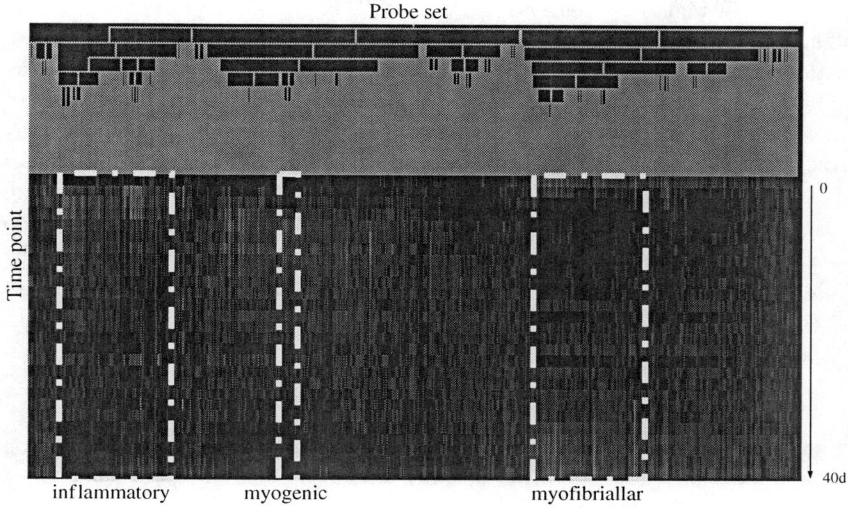

Figure 3. Supervised (time series) hierarchical clustering of genes over a 27 time-point temporal series of muscle degeneration/regeneration expression profiles. Shown is the dendrogram derived from the temporal hierarchical clustering algorithm (GeneSpring). Each row represents a time point of the time series averaged over two replicates at each time point (54 U74Av2 profiles). Each vertical colored bar (lower part of figure) represents a single probe set (gene) in the profile (4,687 total). Vertical bars in red color indicate over-expression relative to the reference value, which is the median of the expression levels of the corresponding gene in all 54 profiles. Blue color represents under-expression relative to the median. The intensity of the color represents the confidence of the data, which generally correlates with the fold changes relative to the reference value. This algorithm clusters genes with similar expression patterns based on correlation coefficients. The distance between two genes on the dendrogram reflects the temporal expression profile similarity. Specific functional clusters are indicated (early inflammation cluster containing macrophage marker genes, myogenesis cluster, and myofibrillogenesis cluster)

regeneration is temporally coordinately regulated, as has been shown previously for specific proteins (Swynghedauw, 1986; d'Albis et al., 1988, 1989). Both studies specifically observed upregulation of *MyoD* at day 3 after injury.

It should be noted that muscle cells do not behave entirely homogenously in cardiotoxin-induced in vivo muscle regeneration. Regenerating muscle cells at different stages, as well as different degrees of inflammation are often seen in the same muscle cross-section by histological examination. Expression profiling data reflects the average of the major events occurring at each time point. As indicated by microarray analysis, inflammatory response genes were markedly induced at early time points after muscle damage. Inflammation then subsided as satellite cells began to proliferate and differentiate, and expression of myogenic genes and muscle structure genes became the major event.

The above expression profiling studies of muscle regeneration also validated previous candidate gene studies showing that a subset of genes expressed in embryonic myogenesis was induced in adult muscle regeneration (e.g. *MyoD, myogenin, Myf5,* and *embryonic myosin heavy chain*) (d'Albis et al., 1988, 1989; Fuchtbauer and Westphal, 1992; Koishi et al., 1995; Rantanen et al., 1995; Franke et al., 1996; Creuzet et al., 1998; Nicolas et al., 1998; Messner et al., 2000; Launay et al., 2001). More importantly, the genome-wide assessment of transcription pattern provides large numbers of novel candidate genes and protein that could play key roles in myogenesis. For example, we have shown *Slug* and *Tead2* as direct targets of MyoD, and *Fgfr4* as a further downstream of Tead2. Furthermore, Slug knockout mice and Fgfr4 knockout mice show impaired muscle regeneration (Zhao et al., 2002, 2006). Another gene induced in muscle regeneration, *E2F1*, has also been shown critical for effective muscle regeneration through the examination of knockout mice (Yan et al., 2003).

2.2.2 Expression Profiling of Ischemia Induced Muscle Regeneration in Normal Mice

There are other models of muscle regeneration in addition to cardiotoxin. Expression profiling of muscle regeneration in ischemic animal models have been published (Paoni et al., 2002). Mouse calf muscle regeneration was induced by femoral artery ligation. Using Affymetix microarrays, eight time points after ligation were expression profiled during a 28-day time course. Similar to the regeneration induced by CTX, ischemia induced a large number of genes, including transcripts associated with inflammatory response, myogenic pathways, and muscle structural genes. While the CTX and ischemia data were qualitatively similar, they showed different temporal patterns. Specifically, in the ischemic mode, initial induction of *MyoD* was observed on day 1 following arterial ligation, and peak expression of embryonic myosin heavy chain (*MHC*) was seen on day 7. On the other hand, CTX induced muscle regeneration showed peak induction of *MyoD* much later (3.0 days), but embryonic *MHC* induction was earlier (3.5 days) (Zhao et al., 2002, 2003; Yan et al., 2003). This difference is probably due to the different severity of damage caused by CTX and femoral artery ligation. A more severe damage (CTX) may take a longer time for myogenic genes (*MyoD*) to become induced. The immediate induction of embryonic *MHC* following *MyoD* in CTX model suggests that activated myoblasts may go through a relatively short proliferation phase (a few hours) then began to differentiate. Whereas in the ischemic model, the delayed induction of embryonic *MHC* after *MyoD* activation suggests that myoblasts assume a longer activation and proliferation phase (6 days). It is not clear what factors cause this difference.

2.2.3 Expression Profiling of Muscle Regeneration in Muscular Dystrophy Mouse Models

Muscle regeneration is important during normal muscle repair after injury, and also is a prominent feature in muscle from muscular dystrophy patients or animal models. Muscle regeneration, and subsequent failure of regeneration have been long observed in a series of clinical muscular dystrophies, which cause disability, poor quality of life and shortened life span. While a failure of regeneration is widely accepted in Duchenne muscular dystrophy and many other progressive muscle disorders, the molecular basis for failed regeneration is not understood. More recently, it has been found that inappropriate muscle regeneration may be the primary cause of weakness in some types of muscular dystrophy, such as Emery Dreifuss and Facioscapulohumeral muscular dystrophy (FSHD) (Winokur et al., 2003; Bakay et al., 2006; Melcon et al., 2006). Transcriptional profiles in muscle tissue from human muscular dystrophy patients, and their mouse models, represent a 'snapshot' with many distinct cellular processes taking place simultaneously (inflammation, degeneration, regeneration). As such, it can be difficult to parse out the underpinnings of failed regeneration in 'mixed signals' in biopsies from Duchenne dystrophy, Emery Dreifuss, and FSHD. That said, recent publications have shown success in identifying specific defects in transcriptional pathways in muscle biopsies from patients with nuclear envelope defects (Emery Dreifuss muscular dystrophy) as the starting point.

Emery Dreifuss muscular dystrophy (EDMD) is characterized by early onset of severe contractures (particularly of the posterior cervical, elbows, and ankles), cardiac conduction defect (generally requiring a pace maker by age 20 yrs), cardiomyopathy, and slowly progressive weakness and wasting in a humero-peroneal distribution (Bonne et al., 1999; Brodsky et al., 2000). Emery Dreifuss muscular dystrophy can be caused by either the X-linked Emerin deficiency, or autosomal dominant missense mutations in the LMNA gene. Both Emerin and LMNA are nuclear envelope proteins. A 125 human muscle biopsy data set of mRNA profiles encompassing 13 diagnostic groups was generated, including patient biopsies from both X-linked and autosomal dominant Emery Dreifuss muscular dystrophy (250 Affymetrix U133 microarrays) (Bakay et al., 2006). Analysis of this data set showed that the muscle mRNA profiles of patients with LMNA and emerin mutations (EDMD) were very highly related to each other, despite the different genes and inheritance patterns, suggesting shared downstream molecular pathophysiology. A high proportion of top-ranked transcripts perturbed in both emerin and LMNA patient muscles were components of the MyoD-dependent transcriptional regulatory pathway during muscle regeneration (*Cri-1, CREBBP, Nap1L1, ECREBBP/p300*). A 27 time point in vivo murine muscle regeneration profiling data set was used to build a temporal model of transcriptional pathways involving MyoD and Rb at the time of exit from the cell cycle and terminal differentiation. A molecular model was evolved for transcriptional perturbations caused by LMNA and emerin mutations, where lack of appropriate LMNA–emerin–Rb–MyoD interaction and coordinated phosphorylation/acetylation states leads to incorrect timing of exit from cell cycle and poor terminal differentiation of myogenic cells in these two muscular dystrophies (Fig. 4).

The perturbations to the Rb/MyoD transcription pathways during muscle regeneration as a consequence of mutations in the *EMD* and *LMNA* genes were further tested in mice lacking a functional emerin gene or Lmna gene using expression profiling (Melcon et al., 2006). Emerin deficient mice show no overt pathology at any age. However, induction of muscle regeneration revealed abnormalities in

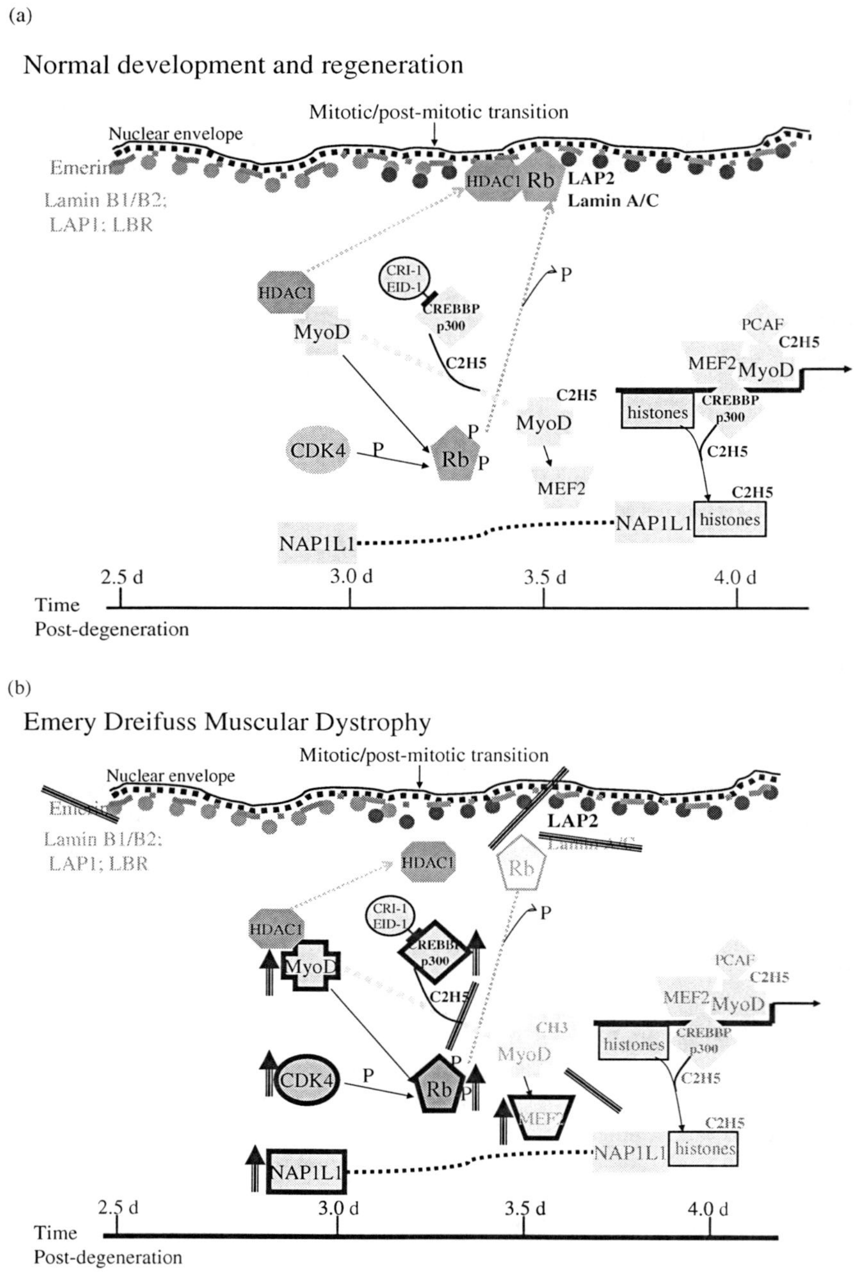
(a)
Normal development and regeneration
Mitotic/post-mitotic transition
Nuclear envelope
Emerin
Lamin B1/B2;
LAP1; LBR
HDAC1 Rb
LAP2
Lamin A/C
HDAC1
CRI-1
EID-1
CREBBP
p300
MyoD
C2H5
P
PCAF
C2H5
MEF2
MyoD
C2H5
histones
CREBBP
p300
MyoD
C2H5
CDK4
P
Rb
P
MEF2
C2H5
NAP1L1
histones
NAP1L1
2.5 d
3.0 d
3.5 d
4.0 d
Time
Post-degeneration
(b)
Emery Dreifuss Muscular Dystrophy
Mitotic/post-mitotic transition
Nuclear envelope
Lamin B1/B2;
LAP1; LBR
LAP2
HDAC1
Rb
P
HDAC1
CRI-1
EID-1
p300
MyoD
PCAF
C2H5
C2H5
MEF2
MyoD
CH3
histones
CREBBP
p300
MyoD
C2H5
CDK4
P
Rb
P
C2H5
NAP1L1
histones
NAP1L1
2.5 d
3.0 d
3.5 d
4.0 d
Time
Post-degeneration

cell-cycle parameters and delayed myogenic differentiation that were associated with perturbations to transcriptional pathways regulated by Rb and MyoD. Temporal activation of downstream targets of the transcriptional activator MyoD was significantly delayed while targets of the Rb1/E2F transcriptional repressor remained inappropriately active. The inappropriate modulation of Rb1/MyoD downstream transcriptional pathways was associated with up-regulation of Rb1, MyoD and their co-activators/repressors transcripts, suggesting a compensatory effort to overcome a biochemical block to differentiation at the myoblast/myotube transition (3rd day of staged regeneration). This compensation appeared effective for MyoD dependent pathways, although was less effective for Rb1 pathways. Analysis of Rb1 phosphorylation states showed an inappropriate hyperphosphorylation at key developmental stages in *Emd* –/Y myogenic cells, both in vivo and in vitro. *Lmna* null muscle did not show the same perturbations of Rb- and MyoD-dependent pathways, although it was observed both increased transcriptional expression of *LAP2α* and delayed expression of *Rb1*; both may serve to provide alternative biochemical pathways. Based on these data, it is suggested that Emery Dreifuss muscular dystrophy appears to be a block in appropriate Rb/nuclear envelope during muscle regeneration.

In the above example of EDMD, both human patients and mouse models were used in a complementary manner, despite the fact that EDMD mice show no disease symptoms, yet EDMD humans show progressive weakness and early death. While the EDMD mouse model shares the same genetic and biochemical lesion with patients with X linked EDMD, the discordance in clinical symptoms leads to questions regarding the 'appropriateness' of the mouse model for the human disease. The discrepancy between clinical symptoms of human dystrophies and their mouse models is shared with most types of muscular dystrophy. Mice with the same genetic problems as human muscular dystrophies often show much less severe symptoms than humans. In each case, the mouse model is an excellent genetic and biochemical model, but less appropriate for the study of failed regeneration and muscle wasting.

Figure 4. Disruptions of transcriptional pathways during muscle regeneration as a consequence of nuclear envelope defects (Emery Dreifuss muscular dystrophy). Panel A shows the temporal series of transcriptional events associated with MyoD and Rb at the 3.0–4.0 day time frame, as the myoblasts leave the cell cycle and begin to terminally differentiate as myotubes. Key to the process is de-phosphorylation of Rb, and subsequent association of Rb with the nuclear envelope. During this process, MyoD is acetylated, and begins activating downstream differentiation genes. Panel B shows the consequences of emerin or lamin A/C mutations in Emery Dreifuss muscular dystrophy. Rb fails to associate with the nuclear envelope, causing a block in the transcriptional pathways (red bars). Network members upstream of the block show upregulation in both human and mouse EDMD muscle, in an apparent attempt to compensate and overcome the biochemical block. Networks downstream of Rb/MyoD show transcriptional down-regulation, due to inappropriate failure of activation downstream of the biochemical block. Modified from Bakay et al., 2006

The most intensively studied mouse model for muscular dystrophy is the dystrophin-deficient *mdx* mouse for Duchenne dystrophy (DMD). DMD patients show progressive muscle weakness, with onset before age of 3, loss of the walking ability by age of 12, and death by the second or third decade of life due to respiratory failure. The *mdx* mouse model for DMD has a point mutation in exon 23 of the dystrophin gene, resulting in a premature stop codon. The *mdx* mice share the same genetic and biochemical pathophysiology with DMD, yet show a milder disease course. The *mdx* mice show widespread skeletal muscle necrosis and inflammation at 3–4 week of age, but regeneration is largely successful, with little progressive weakness subsequently. It has been hypothesized that muscle regeneration in the *mdx* mice may generate protective factors preventing further muscle damage. Expression profiling studies have been conducted to investigate such possibility. Tseng et al. (2002) studied the gastrocnemius muscles from 16-week-old *mdx* mice using Affymetrix murine U74Av2 arrays, and compared to human DMD muscle expression data (Chen et al., 2000). Differentially expressed genes in the *mdx* mice include decreased expression of *myostatin*, increased expression of *actin-related protein 2/3* (subunit 4), *beta-thymosin, calponin, mast cell chymase*, and *guanidinoacetate methyltransferase*. These genes could be the candidate protective factors in further investigation. Porter et al. (2002) studied 8-week-old gastrocnemius and soleus muscles from *mdx* mice using Affymetrix murine U74A arrays. Up-regulation of chronic inflammatory response genes were seen in *mdx* mice. Although *mdx* mice show less fibrosis than DMD, extracellular matrix genes (collagen) were up-regulated in *mdx* to levels similar to those in DMD. The authors suggested that collagen regulation at post-transcriptional stages may mediate extensive fibrosis in DMD. Turk et al. (2005) studied *mdx* hindlimb muscles at different ages from 1- to 20-week-old using spotted oligonucleotide arrays (65-mer). The authors further focused on genes possibly involved in regeneration. Most of such genes peaked at the age of 8 weeks, when maximal muscle regeneration was shown. Pathways induced in *mdx* mice include the activation of satellite cells (Notch-Delta pathway), and the proliferation and differentiation of satellite cells (Bmp15 ad Neuregulin 3). These genes were not seen up-regulated in DMD (Chen et al., 2000), which may account partially for the less efficient regeneration observed in humans. It is interesting that the three expression profiling studies of the *mdx* muscle revealed different pathophysiological aspects of muscle regeneration differentially regulated between *mdx* and DMD. All aspects could contribute the different clinical course observed between *mdx* and DMD.

Muscle regeneration as self-repairing process is also observed in other muscular dystrophies. Characterization of this process using expression profiling helps to better understand the disease process. Dysferlin deficiency, mostly due to non-sense or frame shift mutations, causes limb-girdle muscular dystrophy 2B (LGMD2B) and Miyoshi myopathy (MM). SJL mice, a model for human dysferlinopathy, show deficiency of dysferlin caused by an in-frame deletion, and a milder phenotype compared to human. Genes involved in muscle regeneration were found up-regulated in SJL mice, including cardiac ankyrin repeated protein

(*CARP*), *interleukin-6*, and a list of other genes (Suzuki et al., 2005). Similar to the *mdx* mice, *myostatin* was down-regulated in older SJL mice with increased muscle degeneration/regeneration and inflammation.

3. GENOME-WIDE GENE EXPRESSION PROFILING OF MYOGENIC CELLS

Muscle regeneration involves activation of quiescent satellite cells, which reenter into cell cycle, proliferate, and differentiate to fuse with existing myofibers. It has been thought that muscle regeneration recapitulates the basic events in embryonic myogenesis. Thus, it is hypothesized that study of myobast activation, proliferation and differentiation provides insight to the mechanism involved in muscle regeneration. However, it should be noted that there is also clear distinction between muscle regeneration and embryonic myogenesis, because the two occur in different environmental contexts. Not all cues from surrounding tissues are the same and unique regulatory pathways for each process likely exist. The ease of manipulation of cultured cells has made in vitro assays a popular approach to investigate and validate regulatory mechanisms in both myogenesis and muscle regeneration. A myoblast cell line, C2C12, has been widely used as a model to study myoblast proliferation and differentiation. C2C12 cells proliferate when growing in high serum medium, and differentiate and fuse into myotubes after serum withdrawal. Myoblast differentiation has been correlated with up-regulation of myogenin, myosin heavy chain, and some other muscle specific gene products.

By utilizing high throughput microarray technology, several studies have been conducted on C2C12 cells to characterize the global gene expression in this process (Moran et al., 2002; Shen et al., 2003; Delgado et al., 2003; Tomczak et al., 2004; Iezzi et al., 2004). These studies identified a number of gene clusters with distinct expression patterns that show temporal coordinate regulation when myoblasts proliferate, withdraw from cell cycle and undergo differentiation. Many of the differentially regulated genes are previously unknown to be involved in myoblast differentiation. Primary myoblast cells are believed to better reflect biological scenario in vivo. However, not much expression profiling data of primary myoblasts has been generated at this point. Available expression profiling data of human skeletal myoblast differentiation showed many differentially expressed genes not identified in mouse cell line (Sterrenburg et al., 2004). Many genes participating muscle regeneration were also found differentially expressed during human skeletal myoblast differentiation. Further construction and validation of regulatory models using gene expression information will help to provide novel insights into molecular and cellular events and elucidate the mechanisms of myoblast growth and differentiation. For example, using expression profiling method, it was found that C2C12 cell differentiation treated with deacetylase inhibitor (trichostatin A) showed induction of *follistatin*, which facilitates the recruitment and fusion of myblasts into preformed myotubes.

Animal muscles treated with trichostatin A showed induction of follistatin and improved muscle regeneration (Iezzi et al., 2004).

4. GENOME-WIDE TRANSCRIPTION TARGET ANALYSIS IN MYOGENIC CELLS

Expression profiling captures the abundance of transcripts at the cross-section of time. Abundance represents the dynamic balance between the production and degradation of the transcripts. Mechanisms of transcription regulation have been investigated in muscle regeneration and myogenic cells using expression profiling and the newly emerging ChIP-on-chip (chromatin immunoprecipitation coupled with microarray) analysis. It includes both transcriptional target analysis and posttranscriptional gene suppression analysis.

4.1 Expression Profiling of MyoD-inducible *MyoD-/-/Myf5-/-* Mouse Embryo Fibroblasts (MDER Cells)

MyoD is considered as a muscle determination factor induced in both muscle regeneration and embryonic myogenesis. MyoD then induces the transcription of muscle specific genes in the myogenic program mainly through binding to the E-box (CANNTG). An early study identified a large series of putative downstream targets of MyoD using embryonic fibroblast cultures transfected with an inducible MyoD expressing construct (Bergstrom et al., 2002). The study employed cell cultures knocked-out for endogenous MyoD and Myf5 expression, and also used cyclohexamide to block translation to limit the expression changes to those genes directly downstream of MyoD. Both spotted cDNA arrays (5103 elements) and Affymetrix murine mu6800 arrays (4144 probe sets) were used to identify more than 100 MyoD downstream targets, with about two-thirds induced and one-third repressed by MyoD. As stated above, it is clear that in vitro experimental systems have been critical to our knowledge of transcriptional cascades, and to our knowledge of biology in general. However, the relative simplicity and tightly controlled nature of in vitro experimental systems comes with the liability of unknown relevance to the in vivo state. Particularly worrying is the fact that the majority of in vitro systems utilize over-expression constructs, leading to non-physiological levels of specific proteins in the cultured cell under study. Many transcription factors function as heterodimers, as does MyoD, and non-physiological levels of the transcription factor is likely to alter affinity and binding to specific promoters. Thus, it is expected that supra-physiological levels of MyoD would activate transcription of target genes that are not similarly regulated by MyoD in the in vivo state (e.g. promiscuous, non-physiologically-relevant binding targets). To determine which of the potential downstream targets have relevance to either embryonic myogenesis, or post-natal muscle regeneration, the in vitro candidate gene list was filtered against a 27 time point series in vivo temporal series to identify those genes that were likely bone fide, biologically relevant downstream targets of MyoD (Zhao et al., 2003). A subset of the in vitro targets was strongly supported as biological relevant in vivo.

4.2 Genome-wide Location Analysis: ChIP-on-chip

Chromatin immunoprecipitation (ChIP) is an assay used to test the direct interaction between protein and DNA. Current ChIP assays are usually done on cultured cells as chromatin isolation is technically easier. However, a single report has studied ChIP on muscle tissues (Mejat et al., 2005). To perform ChIP, cells or tissues are treated with formaldehyde to cross-link DNA binding protein and chromatin. Chromatin is then extracted and fragmented by sonication to the size of ~200 bp to 1 kb. Specific DNA binding proteins with their tethered chromatin fragments are pulled down using specific antibodies. Purified DNA fragments are then studied by quantitative PCR to define transcription factor residency on specific gene promoter sequences.

ChIP requires antibodies for the transcription factor of interest, and knowledge of the specific gene promoter sequence that is a 'candidate' for binding by the transcription factor. A second generation method of scanning for all promoters bound by a specific transcription factor is called 'ChIP-on-chip' (chromatin immunoprecipitation coupled with microarray analysis of DNA pulled down) (Fig. 5). It has also been called genome-wide location analysis, because ChIP-on-chip allows transcriptional targets to be investigated in the genome wide scale. Two platforms have been developed for this approach—spotted DNA arrays and Affymetrix tiling arrays. Spotted DNA arrays generally use relatively long probes targeting the gene promoter region. For example, Agilent mouse promoter ChIP-on-chip microarray set covers −5.5 kb upstream to +2.5 kb downstream of the transcriptional start sites using 60-mer probes. Affymetrix tiling arrays are designed to comprehensively cover the whole genome. The probe length is 25-mer, with a gap of approximately 10 bp in between. Array sets for certain species (human and mouse) have been developed to cover either the whole genome or only the promoter region of about 6 kb upstream though 2.5 kb downstream of transcription start sites.

ChIP-on-chip with spotted DNA arrays has been used to study transcriptional cascades in the myogenesis process. Blais et al. (2005) investigated the targets of a few key myogenic transcriptional factors (MyoD, myogenin, and myocyte enhancer factor 2) during myoblast C2C12 cell differentiation. Total close to 200 genes were identified to be bound by the three transcriptional factors, with both overlapping and distinct targets. Novel downstream targets were identified, the role of which to be further defined in gene regulation in muscle development and repair. Using MyoD inducible *MyoD-/-/Myf5-/-* embryonic fibroblasts, Cao et al. (2006) investigated MyoD and myogenin downstream targets by ChIP-on-chip. A large overlap between MyoD and myogenin targets was also identified. By correlating to gene expression profiling data, it was proposed that MyoD and myogenin play different roles in regulating early and late expressed genes during myogenic differentiation. MyoD is able to fully activate early targets. Whereas transcription of late genes requires

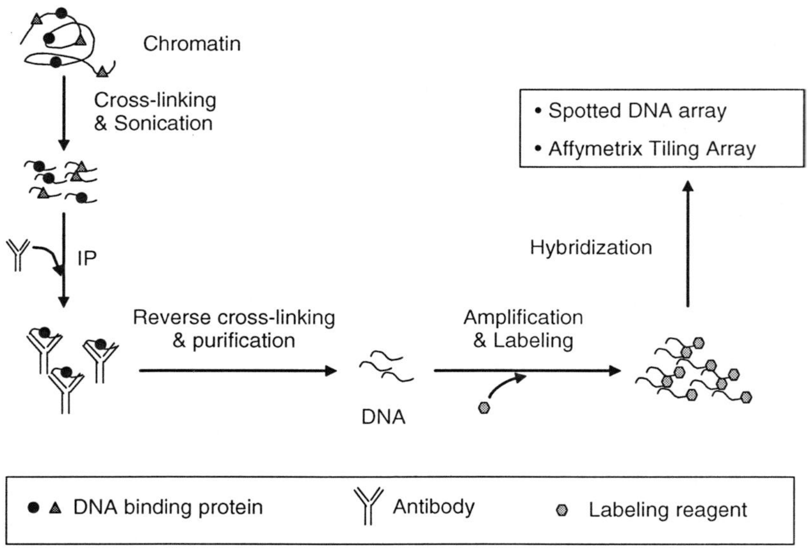

Figure 5. Flowchart of ChIP-on-chip analysis. In this analysis, antibodies to specific DNA binding proteins are used to pull down targeted DNA sequences, enrichment of which is compared to non-specific IgG or no antibody controls. For Affymetrix tilling array analysis, each DNA sample is labeled with biotin and detected with single-colored fluorescence (Streptavidin Phycoerythrin). Enrichment of DNA sequences is determined by the intensity of fluorescence. For spotted DNA arrays, experimental sample (specific antibody pull-down) and reference sample (IgG or no antibody pull-down) are labeled with of two different colors of fluorescent dyes (red and green). Samples are then mixed in 1:1 ratio and hybridized on microarrays. Enrichment of DNA sequences is determined by the ratio of two fluorescent dyes

coordinate action between MyoD and myogenin, with MyoD induction of histone acetylation to facilitate myogenin binding and transcription activation.

Affymetix tiling oligonucleotide arrays have been used to investigate gene regulatory mechanisms including histone modification, estrogen receptor binding in cancer cells, and developmental genes in embryonic stem cells (Bernstein et al., 2005, 2006; Carroll et al., 2005). Emerging data with tiling array application on myogenic cells during myogenesis and muscle regeneration are expected in the near future. ChIP-on-chip analysis of transcriptional factors implied in the pathogenesis of muscular dystrophies, such as Rb in EDMD, is of great interest as well.

The integration of ChIP-on-chip with expression profiling makes correlations between the promoter physical occupation information and transcription activation, and thus helps better understanding of gene regulatory mechanism. Recent findings also show that transcriptional factors binding to locations distant from transcription start sites, which may function as enhancers in the remodeling of chromatin and recruiting other transcription activators (Carroll et al., 2005). Whole genome tiling arrays make it possible to characterize the global DNA binding information, and reveal both promoter region and distant protein-DNA interactions.

4.3 Expression Profiling Study of mRNA Suppression by microRNA

Expression profiling also provides a powerful tool in the investigation of novel gene regulatory mechanism. microRNAs (miRNAs) are a recent discovered abundant class of RNAs with the length of ~22 nucleotides. miRNAs selectively repress gene expression by base pairing to specific 7-nucleotide sites on the mRNA sequences. Two miRNAs, miR-1 and miR-133, are muscle specific, and accumulate following cell-cycle arrest during myoblast differentiation. Study of expression profiles of myotube differentiation showed that mRNAs with miR-1 and miR-133 sites show reciprocal expression to miR-1 and miR-133 levels, with higher expression before miRNA expression and lower expression as miRNAs accumulate (Farh et al., 2005). This finding suggests the mechanism of miRNA controlled temporal regulation of gene expression, in which transition from one cellular process to the next cellular process is facilitated by miRNAs suppression of pre-existing transcripts.

5. CONSTRUCTION OF TRANSCRIPTIONAL CASCADES FROM HIGH THROUGHPUT DATA

5.1 Public Accessible Microarray Data Resources

Genome-wide mRNA profiling data sets contain very large amounts of information, and it becomes impossible for a single laboratory to fully characterize any particular data set. As the data sets increase in size, it becomes increasingly important to make the data sets public, so that research can be parallelized, with many laboratories mining the same data set. Internet accessible databases facilitate the public access process. There are two large mRNA profiling data resources that accept all or most data; GEO (NCBI, USA; www.ncbi.nlm.nih.gov/projects/geo/), and ArrayExpress (EBI, Europe; www.ebi.ac.uk/arrayexpress/). Most journals now require data to be submitted to one of these databases. Both databases store data as MIAME-compliant (Minimum Information About a Microarray Experiment), although this simply provides basic information fields about how the experiment was done. MIAME-compliant data does not imply that projects are directly comparable to each other with regards to mRNA levels.

An additional resource is heavily used by muscle and muscle disease researchers, namely PEPR (http://pepr.cnmcresearch.org) (Chen et al., 2004). PEPR contains data generated by only one site (Children's National Medical Center, Washington, DC), but contains a large amount of muscle data. PEPR differs from GEO and ArrayExpress in a number of ways. First, only Affymetrix arrays are included, and most adhere to quality control and standard operating procedures. Second, all types of raw data for over 2,000 mRNA profiles is provided, and projects are converted to five probe set algorithms. Finally, dynamic query tools are provided such that candidate genes can be quickly tested in different muscular dystrophies, or during muscle regeneration.

5.2 Strategies in Definition of Transcriptional Factor Downstream Targets

A major focus of the transcriptional cascade study is to identify direct targets regulated by specific transcriptional factors. Establishment of a direct regulator-target relationship is believed to require both physical interaction and transcriptional activation. Such information can be obtained from location analysis, expression studies, and functional assays. Temporal expression profiling has been proved very effective in defining coordinately regulated genes in muscle regeneration (Zhao et al., 2002; Yan et al., 2003). Mining temporally co-regulated gene clusters has successfully identified direct targets of transcriptional factors and novel transcriptional pathways (Zhao et al., 2002, 2006). It is assumed that a subset of the direct targets is commensurately expressed with their transcriptional factor in a temporal manner, and the targets bear binding sites for the transcriptional factor. Candidate downstream targets fulfill the assumption can be further confirmed by physical interaction with the transcriptional factor by in vitro gel shift assays and ChIP assays. One of the advantages of ChIP-on-chip is that protein-DNA physical interactions can be investigated without knowing the binding sites. Directly correlation of ChIP-on-chip data with expression profiling data provides better understanding of transcriptional factor actions (Blais et al., 2005; Cao et al., 2006).

The function of the transcriptional factor binding sequence element can be further evaluated by trans-activation assays, which could involve cloning sequence elements into reporter constructs and co-transfection with transcriptional factor constructs. Further function assays with mutagenesis of transcriptional factor binding sites could validate whether transcriptional activation is through the candidate binding sites. RNA interference assays and gene knock-out animals will help to evaluate whether the transcriptional factor is essential for target gene expression. Using such a strategy, we have identified *Slug* and *Tead2* as the direct targets of MyoD, and proposed a MyoD-Tead2-Fgfr4 pathway in muscle regeneration (Zhao et al., 2002, 2006).

5.3 Gene Regulatory Networks

The recent advance in high throughput microarray technology leads to the generation of a large amount of expression data, as well as the development of many powerful bioinformatic mining tools. Some of the tools are mainly focused on data manipulation and statistics, whereas a few others are designed to facilitate pathway/network generation by incorporating biological information derived from published literatures.

Among the widely used data analysis programs are GeneSpring (Silicon genetics), dchip, and BioConductor, which have been used in the aforementioned expression profiling studies of muscle regeneration as well. These programs usually provide tools for clustering analysis, as well as a variety of statistical packages, and are capable to manipulate large data sets.

Among commonly used pathway analysis programs are Ingenuity Pathway Analysis (IPA) and PathwayStudio (formerly known as PathwayAssist). IPA is considered as one of the best current pathway analysis programs. In IPA, input data (genes) are mapped to pre-existing knowledge done by manual searches of published scientific literature. The Ingenuity Pathways Knowledge Base includes any evidence for both physical and functional interactions between input genes and all other genes (and gene products). This interaction can be direct protein/protein (physical) interaction, protein/promoter interaction, or more indirect regulation (e.g. signal transduction cascades). However, ChIP-on-chip data is not currently included in Ingenuity networks. Additionally, this analysis does not take into consideration the tissue or cell types in which the interactions were defined. Thus, one should be cautious when interpreting the 'relevance' of network members to a specific system (e.g. myogenesis and muscle regeneration).

6. PROSPECT

Expression profiling and ChIP-on-chip capture the information of RNA abundance and protein-DNA physical interactions, from which gene transcription regulatory mechanisms can be derived. Most of the actions in biological processes, including muscle regeneration, are executed by proteins synthesized from mRNA. Construction of pathways from mRNA profiling indeed assumes that mRNA abundance well correlates with its protein abundance, which in turn correlates with protein activity in cells. However, there are also many cellular events regulated only at the protein level, such as signal transduction through protein modification (phosphorylation, acetylation, glycosylation, and etc.) and protein-protein interactions. To better understand muscle regeneration, it is necessary to characterize the protein level events in this process. The recent advance of high sensitivity and high throughput proteomics approaches (MALDI-TOF/TOF, and LC-MS/MS) has made it possible to investigate transcription regulation and signaling pathways at the protein level, which will be one of the future focuses in muscle regeneration study. The integration of proteomics discovery with expression profiling and ChIP-on-chip findings will help us to understand transcriptional and signaling pathways in muscle regeneration in a much greater detail.

Animal models of muscle regeneration truly reflects in vivo situation, but the information obtained usually represents the behavior of mixed cell populations (satellite cells, inflammatory cells, endothelial cells, and etc.). In vitro cultured myogenic cells are much purer, but lost the environmental context of regenerating muscle cells in vivo, which may result in induction of different pathways from in vivo process. One of the approaches to circumscribe the in vivo and in vitro short-comes is to isolate primary myogenic cells from regenerating muscles, and conduct expression analysis, protein-DNA binding analysis, and proteomics using these cells. This could be done by flow cytometry with specific satellite cell markers or labels (Montarras et al., 2005).

REFERENCES

Bakay M, Melcon G, Wang Z, Schiltz L, Xuan J, Zhao P, Sartorelli V, Seo J, Pegoraro E, Angelini C, Shneiderman B, Escolar D, Chen YW, Winokur ST, Fan C, Mandler R, Nevo Y, Gordon E, Zhu Y, Dong Y, Wang Y, Hoffman EP (2006) Large-scale microarray analysis of human monogenic defects: Nuclear envelope dystrophies show a transcriptional fingerprint suggesting disruption of Rb-MyoD pathways in muscle regeneration. Brain 129:996–1013

Bergstrom DA, Penn BH, Strand A, Perry RL, Rudnicki MA, Tapscott SJ (2002) Promoter-specific regulation of MyoD binding and signal transduction cooperate to pattern gene expression. Mol Cell 9:587–600

Bernstein BE, Kamal M, Lindblad-Toh K, Bekiranov S, Bailey DK, Huebert DJ, McMahon S, Karlsson EK, Kulbokas EJ 3rd, Gingeras TR, Schreiber SL, Lander ES (2005) Genomic maps and comparative analysis of histone modifications in human and mouse. Cell 120:169–181

Bernstein BE, Mikkelsen TS, Xie X, Kamal M, Huebert DJ, Cuff J, Fry B, Meissner A, Wernig M, Plath K, Jaenisch R, Wagschal A, Feil R, Schreiber SL, Lander ES (2006) A bivalent chromatin structure marks key developmental genes in embryonic stem cells. Cell 125:315–326

Blais A, Tsikitis M, Acosta-Alvear D, Sharan R, Kluger Y, Dynlacht BD (2005) An initial blueprint for myogenic differentiation. Genes Dev 19:553–569

Bonne G, Di Barletta MR, Varnous S, Becane HM, Hammouda EH, Merlini L, Muntoni F, Greenberg CR, Gary F, Urtizberea JA, Duboc D, Fardeau M, Toniolo D, Schwartz K (1999) Mutations in the gene encoding LMNA cause autosomal dominant Emery-Dreifuss muscular dystrophy. Nat Genet 21:285–288

Braun T, Buschhausen-Denker G, Bober E, Tannich E, Arnold HH (1989) A novel human muscle factor related to but distinct from MyoD1 induces myogenic conversion in 10T1/2 fibroblasts. EMBO J 8:701–709

Braun T, Rudnicki MA, Arnold HH, Jaenisch R (1992) Targeted inactivation of the muscle regulator gene Myf-5 results in abnormal rib development and perinatal death. Cell 71:369–382

Brodsky GL, Muntoni F, Miocic S, Sinagra G, Sewry C, Mestroni L (2000) LMNA gene mutation associated with dilated cardiomyopathy with variable skeletal muscle involvement. Circulation 101:473–476

Cao Y, Kumar RM, Penn BH, Berkes CA, Kooperberg C, Boyer LA, Young RA, Tapscott SJ (2006) Global and gene-specific analyses show distinct roles for Myod and Myog at a common set of promoters. EMBO J 25:502–511

Carroll JS, Liu XS, Brodsky AS, Li W, Meyer CA, Szary AJ, Eeckhoute J, Shao W, Hestermann EV, Geistlinger TR, Fox EA, Silver PA, Brown M (2005) Chromosome-wide mapping of estrogen receptor binding reveals long-range regulation requiring the forkhead protein FoxA1. Cell 122:33–43

Chen J, Zhao P, Massaro D, Clerch LB, Almon RR, DuBois DC, Jusko WJ, Hoffman EP (2004) The PEPR GeneChip data warehouse and implementation of a dynamic time series query tool (SGQT) with graphical interface. Nucleic Acids Res 32:D578–D581

Chen Y-W, Zhao P, Boroup R, Hoffman EP (2000) Expression profiling in muscular dystrophies: identification of novel aspects of molecular pathophysiology. J Cell Biol 151:1321–1336

Creuzet S, Lescaudron L, Li Z, Fontaine-Perus J (1998) MyoD, myogenin, and desmin-nls-lacZ transgene emphasize the distinct patterns of satellite cell activation in growth and regeneration. Exp Cell Res 243:241–253

d'Albis A, Couteaux R, Janmot C, Roulet A, Mira JC (1988) Regeneration after cardiotoxin injury of innervated and denervated slow and fast muscles of mammals. Eur J Biochem 174:103–110

d'Albis, Couteaux R, Janmot C, Mira JC (1989) Myosin isoform transitions in regeneration of fast and slow muscles during postnatal development of the rat. Dev Biol 135:320–325

Davis RL, Weintraub H, Lassar AB (1987) Expression of a single transfected cDNA converts fibroblasts to myoblasts. Cell 51:987–1000

Delgado I, Huang X, Jones S, Zhang L, Hatcher R, Gao B, Zhang P (2003) Dynamic gene expression during the onset of myoblast differentiation in vitro. Genomics 82:109–121

Edmondson DG, Olson EN (1989) A gene with homology to the myc similarity region of MyoD1 is expressed during myogenesis and is sufficient to activate the muscle differentiation program. Genes Dev 3:628–640

Farh KK, Grimson A, Jan C, Lewis BP, Johnston WK, Lim LP, Burge CB, Bartel DP (2005) The widespread impact of mammalian MicroRNAs on mRNA repression and evolution. Science 310: 1817–1821

Franke WW, Stehr S, Stumpp S, Kuhn C, Heid H, Rackwitz H-R, Schnolzer M, Baumann R, Holzhausen H-J, Moll R (1996) Specific immunohistochemical detection of cardiac/fetal α-actin in human cardiomyocytes and regenerating skeletal muscle cells. Differentiation 60:245–250

Fuchtbauer EM, Westphal H. (1992) MyoD and myogenin are coexpressed in regenerating skeletal muscle of the mouse. Dev Dyn 193:34–39

Hasty P, Bradley A, Morris JH, Edmondsnon DG, Venuti JM, Olson EN, Klein WH (1993) Muscle deficiency and neonatal death in mice with a targeted mutation in the myogenin gene. Nature 364: 501–506

Iezzi S, Di Padova M, Serra C, Caretti G, Simone C, Maklan E, Minetti G, Zhao P, Hoffman EP, Puri PL, Sartorelli V (2004) Deacetylase inhibitors increase muscle cell size by promoting myoblast recruitment and fusion through induction of follistatin. Dev Cell 6:673–684

Kablar B, Krastel K, Ying C, Asakura A, Tapscott SJ, Rudnicki MA (1997) MyoD and Myf5 differentially regulate the development of limb versus truck skeletal muscle. Development 124:4729–4738

Kassar-Duchossoy L, Gayraud-Morel B, Gomes D, Rocancourt D, Buckingham M, Shinin V, Tajbakhsh S (2004) Mrf4 determines skeletal muscle identity in Myf5:Myod double-mutant mice. Nature 431:466–471

Knapp JR, Davie JK, Myer A, Meadows E, Olson EN, Klein WH (2006) Loss of myogenin in postnatal life leads to normal skeletal muscle but reduced body size. Development 133:601–610

Koishi K, Zhang M, McLennan IS, Harris AJ (1995) MyoD protein accumulates in satellite cells and is neurally regulated in regenerating myotubes and skeletal muscle fibers. Dev Dyn 202:244–254

Launay T, Armand AS, Charbonnier F, Mira JC, Donsez E, Gallien CL, Chanoine C (2001) Expression and neural control of myogenic regulatory factor genes during regeneration of mouse soleus. J Histochem Cytochem 49:887–899

Mejat A, Ramond F, Bassel-Duby R, Khochbin S, Olson EN, Schaeffer L (2005) Histone deacetylase 9 couples neuronal activity to muscle chromatin acetylation and gene expression. Nat Neurosci 8:313–321

Melcon G, Kozlov S, Cutler DA, Sulivan T, Hernandez L, Zhao P, Mitchell S, Nader G, Bakay M, Rottman JN, Hoffman EP, Stewart CL (2006) Loss of emerin at the nuclear envelope disrupts the Rb1/E2F and MyoD developmental pathways during muscle regeneration. Hum Mol Genet 15: 637–651

Messner B, Baum H, Fischer P, Quasthoff S, Neumeier D (2000) Expression of messenger RNA of the cardiac isoforms of troponin T and I in myopathic skeletal muscle. Clin Chem 114:544–549

Montarras D, Morgan J, Collins C, Relaix F, Zaffran S, Cumano A, Partridge T, Buckingham M (2005) Direct isolation of satellite cells for skeletal muscle regeneration. Science 309:2064–2067

Moran JL, Li Y, Hill AA, Mounts WM, Miller CP (2002) Gene expression changes during mouse skeletal myoblast differentiation revealed by transcriptional profiling. Physiol Genomics 10:103–111

Nabeshima Y, Hanaoka K, Hayasaka M, Esumi E, Li S, Nonaka I, Nabeshima Y (1993) Myogenin gene disruption results in perinatal lethality because of severe muscle defect. Nature 364:532–535

Nicolas N, Mira J-C, Gallien CL, Chanoine C (1998) Localization of Myf-5, MRF4 and α cardiac actin mRNAs in regenerating Xenopus skeletal muscle. C R Acad Sic 321:355–364

Paoni NF, Peale F, Wang F, Errett-Baroncini C, Steinmetz H, Toy K, Bai W, Williams PM, Bunting S, Gerritsen ME, Powell-Braxton L (2002) Time course of skeletal muscle repair and gene expression following acute hind limb ischemia in mice. Physiol Genomics 11:263–272

Porter JD, Khanna S, Kaminski HJ, Rao JS, Merriam AP, Richmonds CR, Leahy P, Li J, Guo W, Andrade FH (2002) A chronic inflammatory response dominates the skeletal muscle molecular signature in dystrophin-deficient mdx mice Hum Mol Genet 11:263–272

Rantanen J, Hurme T, Lukka R, Heino J, Kalimo H (1995) Satellite cell proliferation and the expression of myogenin and desmin in regenerating skeletal muscle: evidence for two different populations of satellite cells. Lab Invest 72:341–347

Rudnicki MA, Braun T, Hinuma S, Jaenisch R (1992) Inactivation of MyoD in mice leads to up-regulation of the myogenic HLH gene Myf-5 and results in apparently normal muscle development. Cell 71:383–390

Rudnicki MA, Schneglesberg PNJ, Stead RH, Braun T, Arnold HH, Jaenisch R (1993) MyoD or Myf-5 is required for the formation of skeletal muscle. Cell 75:1351–1359

Seo J, Hoffman EP (2006) Probe set algorithms: is there a rational best bet? BMC Bioinformatics 7:395

Shen X, Collier JM, Hlaing M, Zhang L, Delshad EH, Bristow J, Bernstein HS (2003) Genome-wide examination of myoblast cell cycle withdrawal during differentiation. Dev Dyn 226:128–138

Sterrenburg E, Turk R, 't Hoen PA, van Deutekom JC, Boer JM, van Ommen GJ, den Dunnen JT (2004) Large-scale gene expression analysis of human skeletal myoblast differentiation. Neuromuscul Disord 14:507–518

Suzuki N, Aoki M, Hinuma Y, Takahashi T, Onodera Y, Ishigaki A, Kato M, Warita H, Tateyama M, Itoyama Y (2005) Expression profiling with progression of dystrophic change in dysferlin-deficient mice (SJL). Neurosci Res 52:47–60

Swynghedauw B (1986) Developmental and functional adaptation of contractile proteins in cardiac and skeletal muscles. Physiol Rev 66:710–771

Tajbakhsh S, Rocancort D, Buckingham M (1996) Muscle progenitor cells failing to respond to positional cues adopt non-myogenic fates in myf-5 null mice. Nature 384:266–270

Tajbakhsh S, Rocancourt D, Cossu G, Buckingham M (1997) Redefining the genetic hierarchies controlling skeletal myogenesis: Pax-3 and Myf5 act upstream of MyoD. Cell 89:127–138

Tomczak KK, Marinescu VD, Ramoni MF, Sanoudou D, Montanaro F, Han M, Kunkel LM, Kohane IS, Beggs AH (2004) Expression profiling and identification of novel genes involved in myogenic differentiation. FASEB J 18:403–405

Tseng BS, Zhao P, Pattison JS, Gordon SE, Granchelli JA, Madsen RW, Folk LC, Hoffman EP, Booth FW (2002) Regenerated mdx mouse skeletal muscle shows differential mRNA expression. J Appl Physiol 93:537–545

Turk R, Sterrenburg E, de Meijer EJ, van Ommen GJ, den Dunnen JT, 't Hoen PA (2005) Muscle regeneration in dystrophin-deficient mdx mice studied by gene expression profiling. BMC Genomics 6:98

Winokur ST, Chen YW, Masny PS, Martin JH, Ehmsen JT, Tapscott SJ, van der Maarel SM, Hayashi Y, Flanigan KM (2003) Expression profiling of FSHD muscle supports a defect in specific stages of myogenic differentiation. Hum Mol Genet 12:2895–2907

Wright WE, Sassoon DA, Lin VK (1989) Myogenin, a factor regulating myogenesis, has a domain homologous to MyoD. Cell 56:607–617

Yan Z, Choi S, Liu X, Zhang M, Schageman JJ, Lee SY, Hart R, Lin L, Thurmond FA, Williams RS (2003) Highly coordinated gene regulation in mouse skeletal muscle regeneration. J Biol Chem 278:8826–8836

Zhao P, Iezzi S, Carver E, Dressman D, Gridley T, Sartorelli V, Hoffman EP (2002) Slug is a novel downstream target of MyoD: temporal profiling in muscle regeneration. J Biol Chem 277: 30091–30101

Zhao P, Seo J, Wang Z, Wang Y, Shneiderman B, Hoffman EP (2003) In vivo filtering of in vitro expression data reveals MyoD targets. C R Biol 326:1049–1065

Zhao P, Hoffman EP (2004) Embryonic myogenesis pathways in muscle regeneration. Dev Dyn 229: 380–392

Zhao P, Caretti G, Mitchell S, McKeehan WL, Boskey AL, Pachman LM, Sartorelli V, Hoffman EP (2006) Fgfr4 is required for effective muscle regeneration in vivo: Delineation of a MyoD-Tead2-Fgfr4 transcriptional pathway. J Biol Chem 281:429–438

CHAPTER 6

THE INS AND OUTS OF SATELLITE CELL MYOGENESIS: THE ROLE OF THE RULING GROWTH FACTORS

GABI SHEFER[1] AND ZIPORA YABLONKA-REUVENI[2]

[1] *Department Cell and Developmental Biology, Sackler School of Medicine, Tel Aviv University, Tel Aviv, Israel*

[2] *Department of Biological Structure and Institute for Stem Cell and Regenerative Medicine, University of Washington School of Medicine, Seattle, WA, 98195, USA*

1. OVERVIEW: INTRA- AND EXTRA-CELLULAR ASPECTS OF SATELLITE CELL MYOGENESIS

1.1 The Focus of this Chapter

Extracellular cues such as growth factors and cytokines play critical roles in regulating myogensis during pre- and post-natal development (Buckingham, 2006; Grounds and Yablonka-Reuveni, 1993; Charge and Rudnicki, 2004; Hawke and Garry, 2001). Growth factors have been shown to regulate every phase of adult satellite cell myogenesis, including: (a) recruitment from their quiescent state (Nagata et al., 2006; Wozniak et al., 2005; Volonte et al., 2004); (b) proliferation (Kastner et al., 2000); (c) withdrawal from the cell cycle and terminal differentiation to myoblasts (Leshem et al., 2002); (d) migration (Bischoff, 1997) and (e) myoblast fusion to form growing myofibers (Horsley et al., 2003). Most studies on the impact of growth factors on myogenesis were performed with cell lines due to the ease in maintaining large quantities of these cells in culture. Such studies have contributed important insights to the field of myogenesis. However, cell lines, by their nature of immortality, may not truly represent bona fide satellite cells and their immediate progeny. In this chapter we discuss studies with satellite cells and their immediate progeny as well as studies with myogenic cell lines, with the goal of providing a comprehensive picture of this field of research.

We limit the discussion to several growth factors that are of direct relevance to our studies and that have clearly been shown to affect myogenesis of bona fide satellite

S. Schiaffino and T. Partridge (eds.), Skeletal Muscle Repair and Regeneration, 107–143.

cells. The three growth factor families discussed here are: Fibroblasts Growth Factor (FGF) family, Hepatocyte Growth Factor (HGF) family and Transforming Growth Factor beta (TGFβ) super family. The TGFβ super family is comprised of several families, here we mostly focus on the main TGFβ growth factors and myostatin. As discussed in the later part of this chapter, there is a general consensus that certain FGFs and HGF are positive regulators that promote satellite cell proliferation and in some instances can also delay differentiation of satellite cell progeny. Studies show that members of the TGFβ super family (myostatin included) may inhibit proliferation and are therefore considered negative regulators-whereas other studies suggest that TGFβ growth factors support proliferation and inhibit differentiation. A general overview of each family of growth factors is first introduced, followed by a discussion on specific studies related to myogenesis.

Insulin-like growth factors (IGFs) and platelet-derived growth factors (PDGFs) were also implicated to have a regulatory role in myoblast proliferation but their *direct* effect on bona fide satellite cells has not been demonstrated. IGFs presumably have a dual effect on myoblasts, supporting both proliferation during early stages of myogenesis and differentiation during later stages (Booth, 2006; Florini et al., 1996; Rosenthal and Cheng, 1995; Mourkioti and Rosenthal, 2005). Nevertheless, such a dual effect was observed only in cell lines and long-term proliferating myoblasts, thus, it remains unclear if IGFs indeed affect proliferation of satellite cells and their progeny or just play a central role in myofiber hypertrophy (Allen and Boxhorn, 1989; Bischoff, 1986). Platelet-derived growth factors (PDGFs) were originally shown to regulate myogenesis in long-term proliferating myoblasts and myogenic cell lines (Jin et al., 1991; Yablonka-Reuveni et al., 1990; Yablonka-Reuveni and Seifert, 1993; Yablonka-Reuveni and Rivera, 1997a; Jin et al., 1993; McFarland et al., 1997). However, unpublished results from our laboratory suggest that rodent satellite cells in isolated myofibers and their progeny do not respond to IGFs or PDGFs. The inconsistency between the two conclusions made with the different culture systems may reflect biological differences between myoblasts undergoing multiple passages in culture and bona fide satellite cells and their immediate progeny. It is however possible that difference in culture conditions-contribute to differing results and that all cell culture models provide physiologically relevant information. Caution should be taken when projecting conclusions from cell line studies onto bona fide satellite cell biology since long-term propagation of cell lines results in alternated expression of growth factor receptors and altered response to growth factors.

1.2 The Inside: Interplay Between Transcriptional Loops and Cell Cycle Regulators Govern Satellite Cell Transition from Quiescence to Differentiation

Adult skeletal muscle is composed of multinucleated myofibers (fibers) that are established during embryogenesis by fusion of myogenic cells (myoblasts). Typically, in a healthy muscle the myofiber nuclei (myonuclei) are mitotically inactive. Addition of new myonuclei or formation of new myofibers for supporting muscle growth and repair depends on satellite cells, myogenic stem cells located underneath the fiber basal lamina (Mauro, 1961; Collins et al., 2005; Shefer et al., 2006). During postnatal growth, activated satellite cells proliferate to form new myoblasts that fuse with the growing myofibers. In mature muscles, satellite cells are typically quiescent, but can be recruited to proliferate following subtle injuries (e.g., due to exercise) or massive muscle damage (e.g., due to trauma). Satellite cells support muscle integrity by giving rise to progeny that fuse with existing myofibers when minute repairs are needed, and by generating a large pool of progeny cells to form new myofibers upon massive damage (Grounds and Yablonka-Reuveni, 1993; Hawke and Garry, 2001). Since small myofiber injuries routinely occur during daily activity, the need for ongoing repair is essential for muscle maintenance. Therefore, a balance between satellite cell proliferation and differentiation must exist in order to maintain both functional fibers and the satellite cell reservoir.

At the molecular level, myogenesis of satellite cells is regulated in a highly orchestrated fashion to ensure that specific genes are turned on and off in a temporally organized manner according to genetic blueprints, cell cycle requirements, and environmental factors. The resulting pattern of gene expression yields the terminally differentiated myoblasts that are capable of adding myonuclei to existing myofibers in addition to fusing together to form new myofibers during muscle growth and repair. Both quiescent and proliferating satellite cells express the paired-homeobox transcription factor Pax7 (Collins et al., 2005; Shefer et al., 2006; Halevy et al., 2004; Seale et al., 2000) as well as Myf5, a member of the family of muscle specific transcription factors (MRFs, that include also MyoD, myogenin and MRF4; Ludolph and Konieczny, 1995). The expression of Myf5 in quiescent satellite cells has been demonstrated based on Myf5-lacZ reporter assays and endogenous Myf5 transcript expression (Beauchamp et al., 2000; Zammit et al., 2006; Day et al., 2007). Upregulation of MyoD in activated satellite cells marks the satellite cell's transition into a proliferative phase (Zammit et al., 2006; Yablonka-Reuveni and Rivera, 1994; Yablonka-Reuveni et al., 2007). The onset of myogenin expression marks a commitment of satellite cell progeny to differentiate (Yablonka-Reuveni and Rivera, 1994; Andres and Walsh, 1996). This differentiation commitment is also associated with a decline in Pax7 and Myf5 expression, withdrawal from the cell cycle and subsequent fusion of myoblasts into multinucleated myotubes (Shefer et al., 2006; Halevy et al., 2004; Yablonka-Reuveni and Rivera, 1997a; Zammit et al., 2006).

The role of Pax7 during satellite cell myogenesis has been under debate ever since its expression in these cells was first identified (Seale et al., 2000). Some studies suggest that Pax7 is required for satellite cell renewal while others concluded that Pax7 is actually required for satellite cell survival rather than renewal per se (Seale et al., 2000; Oustanina et al., 2004; Yablonka-Reuveni et al., 2007). Nevertheless, there solid evidence, indicating that satellite cells typically express Pax7 regardless of the type of parent fiber (i.e., fast versus slow) they are associated with, and that their self-renewed progeny also express Pax7 (and not MyoD), similar to their ancestors (Collins et al., 2005; Shefer et al., 2006; Day et al., 2007; Zammit et al., 2004, Yablonka-Reuveni et al., 2007).

The role of MRFs as myogenic determination factors during myogenic lineage establishment in early developed and as regulators of myogenic differentiation is well established (Ludolph and Konieczny, 1995; Kassar-Duchossoy et al., 2004). However, the role of MRFs during the life cycle of satellite cells is less clear. It is commonly held that MyoD serves as a master transcription factor that directs activation of differentiation-linked genes (Tapscott, 2005). Continuous MyoD expression in differentiated progeny of satellite cells seems to depend on the extra-cellular environment. In a serum-replacement based medium satellite cells undergo as little as one or two rounds of proliferation before rapidly entering differentiation, after which their progeny will express myogenin but not MyoD (Yablonka-Reuveni and Rivera, 1997b; Yablonka-Reuveni et al., 1999a; Yablonka-Reuveni et al., 1999b). Although MyoD is expressed in proliferating progeny of satellite cells (Shefer et al., 2006), the actual function of MyoD during myoblast proliferation remains to be determined (see Wyzykowski et al., 2002 for a proposed role). The findings that Myf5 expression declines when myoblasts enter differentiation, whereas MyoD expression persists well into the differentiation stage, suggest that these two MRFs have different roles during myogenesis of satellite cells (Zammit et al., 2006). Myogenin expression is critical for muscle formation during embryogenesis. However, conditional impairment of myogenin in the adult muscle does not interfere with myogenesis, raising further questions about the actual role of myogenin in adult life (Knapp et al., 2006). Lastly, the role of MRF4 during myogenesis of satellite cells is also unclear, as in different studies its expression was detected before, after or concurrently with myogenin expression (Smith et al., 1993; Smith et al., 1994).

Members of the myocyte enhancer factor 2 (MEF2) transcription factor family are also involved in myogenesis regulation (Molkentin et al., 1995; Black and Olson, 1998). MRFs and MEF2s function in concert to support the timely expression of muscle specific structural proteins following differentiation commitment. We showed that the transition into the MEF2A-expressing state occurs together with, or shortly after, the onset of myogenin expression in differentiating satellite cells (Kastner et al., 2000; Yablonka-Reuveni and Rivera, 1997a). The dual expression of myogenin and MEF2A is soon followed by the expression of sarcomeric myosin (Yablonka-Reuveni and Rivera, 1997a). The initial stage of myogenin expression marks myoblast commitment to differentiate. Terminal differentiation of such myogenin-expressing cells involves withdrawal from the cell cycle (Andres and Walsh, 1996; Wang and Walsh, 1996). Cell cycle regulators that are essential for this terminal differentiation include the cyclin dependent kinase cyclin D3, the cyclin-dependent kinase inhibitors p21 and pRb (Andres and Walsh, 1996; Halevy et al., 1995; Kiess et al., 1995; Cenciarelli et al., 1999). pRb is involved in the regulation of both cell cycle withdrawal and the expression of differentiation-linked structural genes (Halevy et al., 1995; Novitch et al., 1999; De Falco et al., 2006). The upregulation of p21, pRb and cyclin D3 in myogenic cells is thought to be governed by MyoD (Halevy et al., 1995; Cenciarelli et al., 1999). Hence, MyoD must be kept in a transcriptionaly inactive form in proliferating cells until appropriate signals for inducing differentiation are conveyed (Novitch et al., 1999; Song et al., 1998; Kitzmann et al., 1999; Puri et al., 2001; Perry et al., 2001).

An interplay between transcriptional loops and cell cycle regulators during myogenesis is typically investigated in cultures of myogenic cell lines and in some instances using long-term passaged progeny of satellite cells. In such models, myoblasts remain proliferative for a longer time when placed in a serum-rich environment and rapidly differentiate when placed in a serum-poor environment (Yablonka-Reuveni et al., 1990; Yablonka-Reuveni and Rivera, 1997a; Clegg et al., 1987; Yaffe and Saxel, 1977; Yaffe, 1969; Rando and Blau, 1994). It is however important to recognize that immediate progeny of satellite cells from adult skeletal muscle typically cannot be stopped from entering differentiation, regardless of medium composition or cell density (Shefer et al., 2006; Yablonka-Reuveni et al., 1987; Yablonka-Reuveni, 2004). Thus, the ability to stop the differentiation of long-term passaged myoblasts may reflect only a subpopulation of satellite cell progeny (especially when derived from individual myogenic clones). Moreover, cells that undergo long term passaging often transform and become immortal, introducing major variations in regulatory loops compared to founding ancestor cells.

1.3 The Outside: Extracellular Cues Regulate The "Built-In" Myogenic Program

1.3.1 Defining growth factors and their mode of action via transmembrane receptors

Growth factors are proteins capable of stimulating cellular proliferation and differentiation. Growth factors stimulate intracellular activities by binding to their transmembrane receptors. Chemokines, cytokines and growth factors are all peptide signaling molecules. Typically, the other groups of signaling proteins are categorized according to the following guidelines: (i) Chemokines – are small protein factors (8–10 kDa) that are released from a variety of cells in response to bacterial infection, viruses and agents that cause physical damage. (ii) Cytokines – are small water-soluble proteins and glycoproteins (8–30 kDa) that are produced by a wide variety of cell types and affect nearby as well as distant cells. The term growth factor is sometimes used interchangeably with the term cytokine. Historically, cytokines were associated with hematopoietic cells and immune system cells. However, some of the signaling proteins of the hematopoietic and immune systems are known today to be common to other cells and tissues as well. (iii) Hormones (which inlcude steroids in addition to peptide molecules) are released from an organ (usually an endocrine gland) directly into the blood stream and affect nearby (paracrine) or distant target cells as they are distributed throughout the body through the blood system (endocrine). In general, hormones that are secreted into the circulation are received by appropriate organs where they produce a specific effect on metabolism.

Growth factors typically do not commute via body fluids to their target cells, but rather are produced locally.

The actions of growth factors are mediated by their receptor specific binding. Growth factor receptors are classified into three major families: (i) tyrosine kinases; (ii) small G-protein-associated receptors; and (iii) serine/threonine kinases. Tyrosine phosphorylation is considered the most characteristic feature of growth factor receptors (Eswarakumar et al., 2005; DiMario, 2002). Ligand binding to its tyrosine kinase receptor causes receptor dimerization, which leads to autophosphorylation of conserved residues in its intracellular domain. Once activated, the receptor functions as a tyrosine kinase inside the cell. A cascade of downstream signaling enzymes carries the signal from the receptor tyrosine kinase domain through cytoplasmic target kinases and into the nucleus. The end targets of the cascade are transcription factors that, once phosphorylated, form multi-protein complexes with accessory proteins and bind specific promoter and enhancer sequences of target genes (Naar et al., 2001; Tartakoff, 1994). Nearly all tyrosine kinase receptors described thus far are composed of an extracellular ligand-binding domain, a single transmembrane domain, a region containing the tyrosine kinase activity, and a carboxy terminus extending into the cytoplasm (Perona, 2006). Various studies indicated that at times the receptor may also serve as a vehicle to shuttle its respective growth factor into the cell or nucleus and does not necessarily function to transduce a signaling cascade directly from its intracellular domain (see for example Haugsten et al., 2005 for the FGF receptor system).

Growth factors that are produced within the same cells that respond to these factors are considered to have an autocrine mode of action. Growth factors that act in a paracrine manner are produced in other sites within the tissue and reach target cells by diffusion for example. Often, one set of cells produces the ligand (e.g., growth factor) while the appropriate receptor is expressed on a separate cell type. For example, within the context of skeletal muscle, we demonstrated that satellite cell progeny express both PDGF-A and PDGF-B but only the surrounding connective tissue cells are able to proliferate in response to PDGF (Kastner et al., 2000 and unpublished results).

1.3.2 Extracellular matrix (ECM) and cell surface heparans facilitate growth factor functions

The complex set of signals conveyed to satellite cells by growth factors is often associated with components of the surrounding ECM, which is adjacent to the basal lamina of the myofiber. The ECM of the muscle tissue is composed of fibroblasts and a complex mesh of several types of collagen, glycoproteins, and proteoglycans. Blood vessels, especially the elaborated network of capillaries, also belong to the ECM constituents and affect myogenesis. In addition to serving as a structural scaffold, the ECM, especially the proteoglycan component, regulates cell behavior by interacting with growth factors and by activating cellular signal transduction pathways (Jenniskens et al., 2006; Velleman et al., 2006; Velleman, 2000).

Cell surface heparan sulphate proteoglycans often bind to transmembrane receptors (HSPGs) acting as co-receptors for enhanced binding to the respective growth factors. HSPGs can be found anchored to the outer membrane surface or in the ECM (Bernfield et al., 1999). HSPGs are able to recognize and bind soluble ligands, and this binding yields high local ligand concentration at the cell membrane proximity that is sufficient to activate signaling receptors (Bernfield et al., 1999; Carrino, 1998). HSPGs are present ubiquitously on cell surfaces and in the ECM of most mammalian cells. Cell surface heparan sulfate (HS) is found mainly attached to two families of proteoglycans: glypicans and syndecans. HS chains found in the extracellular matrix mainly attach to perlecans and agrins (Bernfield et al., 1999). Studies with primary myoblasts cultured on gelatin or Matrigel (Yablonka-Reuveni, 2004; Hartley and Yablonka-Reuveni, 1990) as well as extensive studies with myogenic cell lines demonstrated that the ECM is essential for normal myogenesis, both through direct interactions between ECM molecules with plasma membrane receptors and through modulation of growth factor activities, such as described above (Yablonka-Reuveni et al., 2007; Osses and Brandan, 2002; Melo et al., 1996).

Members of the FGF, HGF and TGFβ families are heparin binding growth factors and their function during myogenesis is most likely facilitated by their own or by their corresponding receptors' interactions with HS and HSPGs. Such interactions were suggested to influence various processes including: stabilization of the receptor-ligand complex; protection of the ligand from denaturation; enhancement or reduction of the activity of some members of a growth factor family (or of their alternative splice forms); generating specificity during development, growth and repair; and generating micro-niches with increased concentrations of the growth factors (Bernfield et al., 1999; Roghani et al., 1994; Aikawa and Esko, 1999; Lietha et al., 2001; Ornitz, 2000). Several studies with the mouse myogenic cell line C2C12 revealed that the expression of some HSPGs is differentially regulated during differentiation. For example, synthesis of the protoglycans decorin and glypican is increased whereas the synthesis of perlecan and syndecan-1 is decreased during differentiation (Larrain et al., 1997a; Larrain et al., 1997b; Olwin and Hall, 1985; Brandan et al., 1991; Brandan et al., 1996). Inhibition of proteoglycan sulfation by chlorate treatment of C2C12 cultures (Osses and Brandan, 2002; Melo et al., 1996), MM14 mouse myoblasts (Olwin and Rapraeger, 1992), or isolated myofibers (Cornelison et al., 2001) affects in vitro myogenesis. Moreover, in vivo administration of synthetic polymers that mimic HSPGs accelerates both regeneration and re-innervation of skeletal muscles (Desgranges et al., 1999; Meddahi et al., 2002). A recent study established the essential functions of Sulfs during satellite cell myogenesis and muscle regeneration; Sulfs are regulatory HS-modifying enzymes and regulate HS 6-O-desulfation of activated satellite cells (Langsdorf et al., in press; note added in proof).

In recent years much interest has been given to the role of decorin in satellite cell myogenesis and muscle regeneration, in view of its ability to modulate myoblast responsiveness to members of the TGFβ family which in turn affects fibrosis and muscle regeneration (Riquelme et al., 2001; Miura et al., 2006; Fadic et al., 2006; Sato et al., 2003; McCroskery et al., 2005; McFarland et al., 2006). Similarly, there is a growing interest in the role of syndecans in view of the finding that satellite cells express syndecans and that muscle regeneration is impaired in mice lacking certain syndecans (Cornelison et al., 2001; Cornelison et al., 2004).

1.4 Models for Analyzing the Role of Growth Factors in Satellite Cell Myogenesis

In vivo studies involving administration of growth factors into the muscle tissue and injury of mice lacking specific growth factors or their receptors were often used as means to assess satellite cell response to specific growth factors (Tatsumi et al., 1998; Miller et al., 2000; Pavlath et al., 1998; Floss et al., 1997). Such models are powerful tools for evaluating the end point in the physiological response of the skeletal muscle as well as for developing and testing new treatment strategies. For example, injection of molecules that inhibit myostatin improved muscle quality in certain pathological cases (Patel and Amthor, 2005) and reduced fibrosis during muscle regeneration (McCroskery et al., 2005). Yet, the above in vivo models fail to dissect out the effect of a growth factor on satellite cells from the effect on other cells in the muscle. Thus, conclusions about the biology of satellite cells made based on such models are largely indirect. For instance, the poor regeneration of the masseter muscle in comparison to that of limb muscles may reflect either progenitor heterogeneity or differences in available growth factors (Pavlath et al., 1998). Likewise, impaired muscle regeneration in old age, could be due to changes in satellite cells or, alternatively, in their environment (e.g., reduced availability of growth factors, reduced innervation and vascularization) (Shefer et al., 2006; Carlson et al., 2001; Conboy and Rando, 2005; Grounds, 2002).

Studies on the role of growth factors during myogenesis of bona fide satellite cells are scarce. Such studies include cultures of freshly isolated satellite cells dissociated from the muscle tissue by enzymatic digestion and harsh trituration, which disengages the cells from the muscle bulk. The satellite cells are subjected to massive trauma during this process and some recovery time is required before cells adhere to the matrix and analysis of growth factor effects can begin (Yablonka-Reuveni, 2004). As discussed in Section 1.2., many studies with myogenic cell lines also provide important insight to satellite cell research with the caveat that not all models and findings are applicable to the biology of bona fide satellite cells.

Isolated myofiber cultures provide the ideal milieu to investigate growth factor effects on satellite cells as the cells remain intact in their native position underneath the basement membrane of their parent myofiber during the time of culturing (Bischoff, 1986; Shefer et al., 2006; Yablonka-Reuveni and Rivera, 1994). Compared to the use of harsh conditions to release satellite cells from whole muscle (i.e., mechanical trituration to disengage the cells), the isolation of whole myofibers causes minimal stress to the satellite cells during the isolation procedure. This feature enables the investigation of the growth factors effect even within the early hours in culture (Kastner et al., 2000; Bischoff, 1986; Yablonka-Reuveni and Rivera, 1994; Yablonka-Reuveni and Rivera, 1997b; Bischoff, 1989). Progression through the myogenic program can by analyzed in myofiber cultures by immunocytochemistry (to trace MRF expression) (Yablonka-Reuveni and Rivera, 1994; Yablonka-Reuveni et al., 1999a; Yablonka-Reuveni et al., 1999b). Clearly, this model has its own limitations when it comes to biochemical analysis of intracellular signaling cascades; using Western blotting typically will not distinguish between the contribution of satellite cells and the surrounding cells, hence the results will mostly reflect events within the myofiber and not necessarily the satellite cells.

Notably, studies with primary myoblasts derived from satellite cells or even with isolated myofibers have their own limitations. To evaluate the effect of growth factors specifically on satellite cells, cultures need to be free of any possible contaminating interstitial cells that are often co-isolated with satellite cells. Such cells may respond to the same factors as satellite cells and in turn produce their own growth factors that can enhance or impair myogenesis of satellite cells. Ideally, investigators must use specific markers of satellite cells and myoblasts that are absent in other possible cell types, to verify the purity of the cultured cells before initiating detailed studies.

2. THE FGF SYSTEM AND ITS ROLE IN MYOGENESIS OF SATELLITE CELLS

2.1 The FGFs and Their Receptors: Overview

Fibroblast growth factors (FGFs) constitute a large family of related polypeptide growth factors, which are found in a variety of multicellular organisms, including invertebrates (Popovici et al., 2005). FGFs play regulatory roles in tissue- and organogenesis during development, postnatal life and in pathological processes. FGFs are known to act as mitogens, differentiation agents, and regulators of programmed cell death (Szebenyi and Fallon, 1999; Xu et al., 1999; Ornitz, 2005; Ornitz and Itoh, 2001). The nomenclature of the first identified FGFs (i.e., acidic and basic FGF) was based on their ability to stimulate proliferation of mouse 3T3 fibroblasts (Armelin, 1973; Gospodarowicz, 1974). However not all FGFs have fibroblast stimulating activities. With the subsequent identification of many more FGF proteins (over 20 to date), the nomenclature had changed with acidic FGF (aFGF) becoming-FGF1 and basic FGF (bFGF) becoming FGF2 (Eswarakumar et al., 2005; Szebenyi and Fallon, 1999; Goldfarb, 1996; Coulier et al., 1997; Botta et al., 2000). There are two main defining features of the FGF family: (i) strong affinity for heparin sulfate protoglycans (HSPGs; Burgess and Maciag, 1989; Pellegrini et al., 2000); and (ii) a central core of 140 amino acids that is highly homologous in all FGFs and serves for interactions with the FGF-receptor (Ornitz and Itoh, 2001; Plotnikov et al., 2000; Stauber et al., 2000). FGFs use a dual receptor system to activate downstream signal transduction pathways. The primary component of this system is a family of tyrosine kinase transmembrane FGF receptors (FGFR1 through FGFR4).

The prototype receptor contains an extracellular ligand-binding domain with three immunoglobulin-like domains (Ig-I, II, III), and an intracellular split tyrosine kinase domain. There are four independent FGFR genes, and numerous alternative splice forms, which contribute to FGF-FGFR ligand specificity (Eswarakumar et al., 2005; Szebenyi and Fallon, 1999; Xu et al., 1999; Johnson and Williams, 1993). FGFR1-4 genes encode polypeptides that are 55%–72% identical in their amino acid sequence (Johnson and Williams, 1993). A distant relative, FGFR5, was identified in human and mouse, showing an approximate 30% amino acid identity to other FGFR proteins (Sleeman et al., 2001; Kim et al., 2001b); the relevance of FGFR5 to the FGF system has remained unclear.

The second component of the FGF system consists of HSPGs that are required for FGF binding and activation of the FGFRs (Ornitz et al., 1992; Yayon et al., 1991; Rapraeger et al., 1991). Variations in HSPGs can generate tissue and age specific interactions between the FGFs and their receptors (Ornitz, 2000; Steinfeld et al., 1996; Kan et al., 1999). The binding of FGF to its receptor induces receptor dimerization and tyrosine phosphorylation of both the receptor itself and of intracellular target proteins (Eswarakumar et al., 2005; Hadari et al., 2001; Ong et al., 2001). FGFR activation can potentially be induced through FGF-independent mechanisms as had been shown for FGFR4 (Gao and Goldfarb, 1995; Cavallaro et al., 2001). FGFRs are generally thought to carry out roles in signal transduction at the cell surface, in addition evidence points to FGFR intracellular trafficking. In some contexts, FGF ligands and receptors were detected in the cytoplasm and in the nucleus (Feng et al., 1996; Maher, 1996; Reilly and Maher, 2001; Citores et al., 1999). FGF internalization requires receptor activation (Sorokin et al., 1994), but other studies point to receptor internalization by endocytosis (Citores et al., 1999; Klingenberg et al., 2000; Citores et al., 2001). An intracellular, cysteine-rich, FGF-binding protein was also identified and is thought to act in regulating the intracellular trafficking of FGFs (Zuber et al., 1997).

Special and temporal expressions of specific FGFs, FGFRs and HSPG modulate FGF signaling (Ornitz, 2000). Conserved FGF target genes that encode negative and positive feedback regulators of the FGF signaling itself were identified as well. Members of the Sprouty (Minowada et al., 1999; Klein et al., 2006; Shim et al., 2005; Reich et al., 1999) Sef (Kovalenko et al., 2006; Ziv et al., 2006), and mitogen-activated protein kinase phosphatase families are negative modulators of FGF signaling; positive factors that promote FGF signaling include the ETS transcription factors ERM and PEA3 and the transmembrane protein XFLRT3 (Tsang and Dawid, 2004). These molecules affect the FGF signaling cascade at different levels to regulate the final output of the pathway. Such multilayered regulation suggests that precise fine-tuning of FGF signaling is critical. The inhibitory effect of Sprouty proteins is not limited FGFR signaling; they inhibit other receptor tyrosine kinase signaling, including that of HGF (Lee et al., 2004). Overepxression of Sprouty in C2C12 cells abolished FGF-mediated delay of myogenic differentiation, suggesting a role for Sprouty in myogenesis (de Alvaro et al., 2005). However, more detailed studies are required to establish the relevance of Sprouty and Sef to myogenesis in general and specifically to satellite cell biology.

2.1.1 *Effect of FGFs on proliferation and differentiation of satellite cells*

Selected members of the FGF family are expressed in vivo in adult muscle tissue as well as in vitro in cultured myogenic cells. Satellite cells undergoing myogenesis were shown to also express FGFRs (mainly FGFR1 and FGFR4) (Kastner et al., 2000, Floss et al., 1997; Garrett and Anderson, 1995; Groux-Muscatelli et al., 1990; Lefaucheur and Sebille, 1995; Hannon et al., 1996). Correlative expression studies suggest that FGFs affect myogenesis in adult muscle via paracrine and/or autocrine mechanisms. Studies with primary satellite cell cultures and myogenic cell lines show that only selective FGFs support proliferation of myogenic cells (de Alvaro et al., 2005; Hannon et al., 1996; Johnson and Allen, 1995; Scata et al., 1999; Wilkie et al., 1995; McFarland et al., 2003). In certain myogenic cell line models and in cultures of long-term passaged myoblasts, FGFs were also shown to suppress myoblast differentiation (Clegg et al., 1987; Rando and Blau, 1994; Olwin and Rapraeger, 1992; Pizette et al., 1996; Kontaridis et al., 2002). Such findings with immortal cells have been considered as an indication that the FGF signaling system plays a role in enhancing proliferation and delaying differentiation of bona fide satellite cells. However, in cultures of isolated myofibers, certain FGFs (FGF1, 2, 4 and 6, but not 5 or 7) were shown to support proliferating satellite cells without an obvious effect on delaying subsequent differentiation (Kastner et al., 2000, Shefer et al., 2006; Yablonka-Reuveni and Rivera, 1994; Yablonka-Reuveni and Rivera, 1997b; Yablonka-Reuveni et al., 1999b; Yablonka-Reuveni and Anderson, 2006).

It is conceivable, that the effect of FGF on delaying myogenic differentiation, as seen with long-term passaged cells, is in fact dependant on various cofactors that were not necessarily present in serum-replacement medium used in our single fiber culture studies. Such cofactors could consist of factors present in the supplemented serum or specific HSPGs that were previously shown to be required for initiating the suppressing effect of FGF on mouse MM14 myogenic cell differentiation (Olwin and Rapraeger, 1992; Olwin et al., 1994). FGF1 supplementation or overexpression in rat myoblasts were shown to suppress myogenic differentiation. Conversely, endogenously expressed FGF1 did not suppress differentiation of rat myogenic cell lines. These findings suggest that FGF may initiate several different signaling mechanisms in myogenic cells (Uruno et al., 1999). Nevertheless, it is our view that the effect of FGF on delaying morphological and biochemical differentiation (i.e., delaying myotube formation and suppressing expression of MyoD and myogenin mRNA) as seen in the mouse myogenic cell line C2C12 (Kontaridis et al., 2002; Tortorella et al., 2001; Yoshida et al., 1996) does not necessarily reflect the biology of immediate progeny of satellite cells. Primary myogenic progeny maintained in basal medium respond to FGF by continuing to proliferate for 1–2 more rounds while maintaining MyoD$^+$ phenotype before upregulating myogenin rather than by suppressing MyoD expression as was seen in C2C12 cells (Z. Yablonka-Reuveni, unpublished studies). In contrast to FGF, lysophosphatidic acid (LPA), a bioactive phospholipid present in the serum, was shown to stimulate proliferation of C2C12 myoblasts while allowing the continued expression of MyoD (Yoshida et al., 1996), mimicking the effect seen with FGF in our primary myogenic cultures. A detailed immunostaining analysis of C2 cells (or of their subclone C2C12 cells) grown either as standard or in clonal cultures indicate that not all cells express MyoD at any given time (Yablonka-Reuveni and Rivera, 1997a). Conversely, proliferating progeny of satellite cells in primary cultures and clonal cultures all express MyoD (Shefer et al., 2004). Thus, in the previously discussed myogenic cell line studies, FGF may in fact support proliferation of cells before they enter the MyoD-expressing compartment rather than promoting proliferation of cells that already express MyoD. This can lead to the apparent suppression of MyoD expression whereas the MyoD expressing cells may simply be "diluted" by the expansion of cells that have not yet up-regulated MyoD. Different from primary myogenic cultures, the rat myogenic cell lines L6 and L8 express very little, if any MyoD (Graves and Yablonka-Reuveni, 2000). These differences in MyoD expression between extensively used myogenic cell lines and primary myogenic cultures should not be overlooked when inferring results from studies of long-term passaged cells to the biology of bona fide satellite cells and their progeny.

2.1.2 Does FGF6 play a unique role during myogenesis of satellite cells and muscle regeneration?

Studies suggest that FGF6 is an important ligand for satellite cell proliferation and for muscle maintenance. During embryogenesis, FGF6 is expressed at relatively high levels in the developing muscle while the expression of FGF2 is more wide spread (Coulier et al., 1994; deLapeyriere et al., 1993; Han and Martin, 1993; Gonzalez et al., 1990). FGF6 in both undamaged and regenerating adult skeletal muscles exhibits a restricted expression profile specific to myofibers rather than to the surrounding cells (Kastner et al., 2000, Floss et al., 1997; Armand et al., 2003 Armand et al., 2006). Moreover, a prolonged expression of the FGF6 gene accompanies the extended period of hyperplastic muscle growth in postembryonic fishes (Rescan, 1998). FGF6 was shown to have a dual role during myogenesis in culture, depending on dose and particular microenvironments. It was found to up-regulate the expression of several differentiation markers when given at a low dose to cultures of mouse-derived C2 cells, while a higher concentration of FGF6 enhanced proliferation (Pizette et al., 1996). FGF6 was also reported to expand a subset of cells with SP phenotype in the C2C12 myogenic cell line (Israeli et al; 2004). However, in our studies of satellite cells in single fibers and in primary cultures derived from satellite cells, FGF6 effect was indistinguishable from that of FGF2, with the caveat that addition of heparin to the medium was necessary for optimal FGF6 effect while FGF2 did not require addition of the heparin cofactor (Kastner et al., 2000; and additional unpublished studies with primary myogenic cultures).

FGF6 knockout mice yielded conflicting results regarding its role in skeletal muscle regeneration. Depending on the genetic background, mice lacking FGF6 displayed either reduced or unaltered muscle regeneration (Floss et al., 1997; Fiore et al., 2000). Nevertheless, more recent studies on this mouse model of Fiore and colleagues did demonstrate specific effects of FGF6 on regeneration and hypertrophy of the soleus muscle (Armand et al., 2003; Armand et al., 2005; Armand et al., 2006). In order to establish the role of FGF6, more experiments must be conducted, as the above mentioned experimental models do not provide adequate insight regarding the effect of FGF6 in satellite cells and their immediate progeny. We were unable to demonstrate any postnatal growth defect in various examined hindlimb muscles from growing and adult FGF6^{-}/$^{-}$ mice (described by Floss et al., 1997). We also did not identify any evidence for impaired myogenesis in cultures from this mouse model (Z. Yablonka-Reuveni, unpublished work). It is conceivable that FGF6 is important, but not essential, for myogenesis because of its overlapping roles with other FGF ligands. Indeed, genetic elimination of both FGF2 and FGF6 in mdx mice (a murine model of Duchenne muscular dystrophy) aggravates muscle impairment, suggesting that these FGFs may be able to compensate for each other during myogenesis in vivo (Neuhaus et al., 2003).

2.1.3 FGFR1 and FGFR4 may play a different role during myogenesis

Developing skeletal muscles show selective expression of FGFR4, while FGFR1 is expressed by both skeletal muscles and the surrounding tissues (Stark et al., 1991; Korhonen et al., 1992; Gonzalez et al., 1996). Muscle development is differentially affected when the function of either FGFR1 or FGFR4 is abrogated (Marics et al., 2002). Moreover, regenerating skeletal muscles exhibit expression of FGFR4 in a specific temporal pattern, peaking at the time of myofiber formation (Zhao and Hoffman, 2004). We demonstrated that progeny of satellite cells, but not surrounding interstitial cells, express FGFR4 in a pattern similar to FGF6. FGFR1 was shown to be expressed by both myogenic and non-myogenic cells from adult skeletal muscle (Kastner et al., 2000). Unpublished studies from our laboratory further demonstrate that FGFR4 expression correlates with differentiation while FGFR1 might be of relevance to cell proliferation but without any specific impact on differentiation. This was concluded based on studies with myogenic models where differentiation was delayed (i.e., satellite cell cultures from MyoD^{-}/$^{-}$ mice) as well as on studies with inducible cell lines where myogenic differentiation was regulated by synthetic ligands. In all, biochemical analysis of the interactions between over-expressed FGFRs and the various FGF ligands, together with the published coexpression patterns of FGF6 and FGFR4 during myogenic development, suggested that a receptor-ligand FGFR4-FGF6 complex exists in these cells and may have a specific role in regulating myogenic differentiation (Kastner et al., 2000, Coulier et al., 1994; Ornitz et al., 1996). Notably, studies with the MM14 mouse myogenic cell line indicate that these cells only express FGFR1 and not FGFR4 (Templeton and Hauschka, 1992; Kudla et al., 1998), while studies with the C2 mouse myogenic cell line express both FGFRs (Pizette et al., 1996;

B. Kwiatkowski ans Z. Yablonka-Reuveni, submitted). These differences in FGFR expression should be taken in account when considering the use of these models for studies on the effect of FGF during myogenesis.

The possibility that FGFR1 and FGFR4 may operate differently is supported by earlier FGFR transfection studies, where the effects of different FGFRs on the activation of intracellular targets were compared (Vainikka et al., 1994; Vainikka et al., 1996; Wang et al., 1994; Shaoul et al., 1995; Johnston et al., 1995). The activation of an 85-kDa-serine kinase (Vainikka et al., 1994) and the induction of cell membrane ruffling (Johnston et al., 1995) were associated only with FGFR4 but not with other FGFRs. In most such transfection models FGFR4 elicited a poor mitogenic response compared to FGFR1. These findings suggest that FGFR4 is not necessarily involved in mitogenesis regulation. Notably, a certain comprehensive study on FGF-FGFR interactions is frequently cited as evidence for the ability of various FGFs to induce mitogenesis via FGFR4 (Ornitz et al., 1996). In this particular study a chimeric FGFR4 receptor was used. The receptor had only the external portion of FGFR4 while the cytoplasmic domain was that of FGFR1. Thus, the latter study could be used as evidence for FGFR4 binding to certain FGF ligands via its external domain, but not referred to as evidence for the capacity of FGFR4 to induce cell proliferation. Recent studies from our laboratory indicate a possible heterologous interaction between FGFR4 and FGFR1 as a means that contributes to FGFR4 tyrosine phosphorylation. (B. Kwiatkowski and Z. Yablonka-Reuveni, submitted).

Collectively, our studies put forth the hypothesis that FGFR4 is involved in the transition of satellite cells from proliferation to differentiation while FGFR1 is involved in maintaining the proliferative state of myoblasts. To further investigate this hypothesis, we performed extensive analysis of satellite cells from mice lacking FGFR4. These FGFR4^{-}/$^{-}$ mice (Weinstein et al., 1998) did not demonstrate any apparent skeletal muscle abnormality during postnatal growth or in the mature muscle. However, crossing FGFR4 null mice with mice lacking FGFR3 produced double mutant mice whose postnatal muscle growth and maturation were severely retarded (Z. Yablonka-Reuveni, unpublished). Only a subtle delay in the transition from proliferation to differentiation occurred when myogenesis of satellite cells was examined in single myofibers and in primary cultures isolated from these FGFR4 null mice (ongoing studies). Thus, the report that muscle regeneration was impaired in FGFR4^{-}/$^{-}$ mice compared to control mice (Zhao et al., 2006), was not confirmed by us when control mice of the same genetic background as the mutant mice were used. Recent studies suggested FGF19 as the specific ligand for FGFR4 in human (Xie et al., 1999; Harmer et al., 2004) with FGF15 being its mouse ortholog (Wright et al., 2004). Nevertheless, to the best of our knowledge, it has not been determined whether FGF15 is indeed expressed in skeletal muscles. Furthermore, it remains to be proven that FGF19 functions strictly through FGFR4, since this ligand did exert a metabolic influence when overexpressed in FGFR4^{-}/$^{-}$ mice (Strack and Myers, 2004). Altogether, final conclusions about the role of FGFR4 and its ligand(s) during myogenesis await future studies.

2.1.4 Does FGF promote satellite cell activation?

There has been a continuous debate as to whether FGF is involved in activation of satellite cells or, alternatively, only supports satellite cell proliferation once activation had already been initiated. We previously demonstrated that both FGF and HGF are able to initiate proliferation of quiescent satellite cells, including in myofibers from old rodents (Kastner et al., 2000, Shefer et al., 2006; Yablonka-Reuveni et al., 1999b). These findings disagree with other studies suggesting that only HGF is capable of supporting satellite cell recruitment and that the response to FGF is secondary due to HGF-mediated upregulation of FGF receptors (Johnson and Allen, 1995; Allen et al., 1995). The latter studies are based on cells isolated from whole muscle tissue following enzymatic digestion; our conclusion that FGF and HGF have similar effects on entry of satellite cells into the proliferative phase is based on isolated myofibers. This discrepancy can be due to the use of different cell culture models. Both FGF and HGF may regulate satellite cell recruitment, however, the FGF receptor system may be more sensitive to the physical and enzymatic digestion steps involved in satellite cell isolation from the muscle tissue. Conversely, the gentle procedure of myofiber isolation perhaps better preserve the FGF receptors, permitting immediate response of satellite cells to selective FGFs. Therefore the effect of FGF system is less robust when observations are made with primary myogenic cultures compared to observations with myofiber cultures, where the action if the FGF system is preserved. Studies by Jones et al. (Jones et al., 2005) with mouse satellite cells and the MM14 myogenic cell-line suggest a role for FGF in activating satellite cells via p38α/β MAPKs. These studies suggest that p38α/β MAPKs function as a molecular switch to activate quiescent satellite cells. However, these MAPKs were not unique to the activation stage and were also shown to be involved in satellite cell proliferation and differentiation. Further studies with genetic models in which satellite cells can be specifically traced are needed to conclude on the regulatory role of the FGF system during satellite cell activation.

3. THE HGF SYSTEM AND ITS ROLE IN MYOGENESIS OF SATELLITE CELLS

3.1 HGF and its c-met Receptor: Overview

Hepatocyte growth factor (HGF), also known historically as scatter factor (SF), was first identified and purified as a potent mitogen of primary cultured hepatocytes (Nakamura et al., 1984; Nakamura et al., 1986) and independently, as a human embryonic fibroblast-derived factor that stimulates the motility of epithelial cells in vitro (Stoker et al., 1987). These two activities were subsequently ascribed to the same growth factor by both physiological effects and molecular cloning (Weidner et al., 1991). Another factor now established as identical to HGF is the tumor cytotoxic factor (TCF) derived from fibroblasts (Higashio et al., 1990). HGF participates in multiple physiological activities, including tissue development, regeneration and wound healing (Birchmeier and Gherardi, 1998; Matsumoto and Nakamura, 1997). In general, cells that produce or respond to HGF are located in close proximity to each other, reflecting the reduced biochemical properties of the factor and its limited capacity to diffuse in vivo (Birchmeier and Gherardi, 1998). In vivo, HGF exists in two forms, a biologically inactive monomeric molecule (pro-HGF) and a biologically active heterodimer. Pro-HGF is synthesized and secreted as a single chain, it binds to ECM molecules until it is cleaved by a serine protease (HGF activator) and converted to the mature, physiologically active form (Gohda et al., 1998; Naka et al., 1992). Mature HGF consists of a light chain (β-chain) of 34-kDa and a heavy chain (α-chain) of 69-kDa, the α and β chains are linked together by a disulfide bond. The α-chain contains an N-terminal hairpin domain and subsequent four-kringle domains, and the β-chain contains a serine protease-like domain with no enzymatic activity (Nakamura et al., 1989; Tashiro et al., 1990; Funakoshi and Nakamura, 2003).

For its function, HGF needs to bind to its cell surface receptor, c-met (Naldini et al., 1991; Bottaro et al., 1991). C-met is a heterodimeric receptor tyrosine kinase that was initially discovered as a transforming gene from chemically treated osteosarcoma cells (Cooper et al., 1984). HGF also binds the glycosaminoglycan (GAG) chains of heparan sulfate (HS) and dermatan sulfate (DS) proteoglycans (Lyon et al., 1994; Lyon et al., 1998), although with lower affinity than to the c-met receptor. Evidence suggests that an active ternary complex forms between HGF, c-met and appropriate proteoglycans (Lyon et al., 2002). The primary c-met transcript is translated into a 150-kDa polypeptide that is further glycosylated to give a 195-kDa precursor protein. This precursor is then cleaved into a 50-kDa α-chain and a 145-kDa β-chain, which are linked via disulfide bonds (Comoglio, 1993). The mature c-met heterodimer consists of a highly glycosylated extracellular α-subunit, a β-subunit with a large extracellular region, a membrane spanning segment, and an intracellular tyrosine kinase domain. Upon HGF binding, c-met undergoes autophosphorylation of specific tyrosine residues within the intracellular region of the β chain and ignites downstream signaling (Leshem et al., 2002; Ponzetto et al., 1994; Schaeper et al., 2000; Sachs et al., 2000).

3.2 Effect of HGF on Activation, Proliferation and Differentiation of Satellite Cells

HGF is expressed in intact and regenerating muscle (Kastner et al., 2000, Tatsumi et al., 1998; Jennische et al., 1993; Hayashi et al., 2000). Transcripts and protein levels of HGF are increased during the early phase of muscle regeneration, and this increase is proportional to the degree of injury (Suzuki et al., 2002; Tatsumi et al., 2001). Studies demonstrated that HGF is produced by muscle cells in vitro and in vivo and is secreted to the extracellular environment where it is stored in its heterodimeric form. The c-met receptor is expressed by satellite cells and proliferating myoblasts and exogenous HGF promotes satellite cell activation and myoblast proliferation, indicating a direct role of the HGF system in satellite cell myogenesis (Kastner et al., 2000; Yablonka-Reuveni et al., 1999a; Tatsumi et al., 1998; Gal-Levi et al., 1998; Cornelison and Wold, 1997).

HGF was also suggested to play a role in preventing proliferating satellite cells from differentiating; this inhibitory effect may occur via the involvement of the basic helix loop helix protein Twist and the cyclin-dependent kinase inhibitor p27 (Leshem et al., 2002; Tatsumi et al., 1998; Anastasi et al., 1997; Zeng et al., 2002). Nonetheless, the pattern of satellite cell proliferation on isolated rat myofibers did not support the notion that HGF delays satellite cells differentiation (Kastner et al., 2000, Yablonka-Reuveni et al., 1999b). In-vivo administration of HGF to injured mice indeed revealed enhancement of satellite cell proliferation and delayed differentiation. However, such sustained HGF administration resulted in impaired regeneration (Miller et al., 2000). The latter study further exemplified the difficulties associated with controlling muscle regeneration by supplementation of growth factors. The interplay between myoblast proliferation and differentiation is a complex process that requires an optimal spatial and temporal milieu of multiple growth factors, each present in the right amount at the right time.

In vitro and in vivo data demonstrate that release of nitric oxide synthase from the basal lamina, in response to myofiber stretch or damage, leads to the production of nitric oxide. Nitric oxide then activates matrix metalloproteinases, which in turn can cause release of HGF from its association to HSPGs, making HGF available for binding to the c-met receptor and to activate satellite cells (Anderson, 2000; Tatsumi et al., 2002; Tatsumi et al., 2006; Yamada et al., 2006). In addition to this autocrine/paracrine mechanism that provides HGF from cellular sources near to satellite cells, an endocrine delivery of HGF to the injured muscle was also suggested based on the rapid upregulation of HGF in the spleen following muscle injury (Suzuki et al., 2002).

Aside from the effect on proliferation and differentiation, HGF is also involved in promoting satellite cell migration to the site of injury via activation of the Ras-Ral pathway, as demonstrated by the in vitro chemotactic activity of this factor in primary myogenic cultures and the C2C12 cell line (Bischoff, 1997; Suzuki et al., 2000). Taken together, these data demonstrate the pleiotropic role that HGF probably plays during muscle regeneration by boosting the proliferating myoblast population due to its mitogenic and chemotactic activities. These may be important for accomplishing a threshold myoblast density needed to start the fusion phase.

4. THE TGFβ SYSTEM AND ITS ROLE IN MYOGENESIS OF SATELLITE CELLS

4.1 The TGFβ Superfamily and its Receptors: Overview

The TGFβ superfamily consists of more than 40 members, such as TGFβs, bone morphogenetic proteins (BMPs) and growth differentiation factors (GDFs) (Shi and Massague, 2003). One of the more recently discovered members of the GDF family, named GDF8 or myostatin, is a negative regulator of embryonic and postnatal skeletal muscle growth that functions to maintain proper muscle mass (Dominique and Gerard, 2006). The diverse TGFβ ligands share a common sequence and some structural features but elicit different cellular responses during pre- and postnatal development as well as in disease. Members of the TGFβ superfamily are known to participate in the regulation of various biological processes, such as tissue homeostasis, cell-cycle progression, differentiation, reproductive function, motility, adhesion, neuronal growth, bone morphogenesis, wound healing, and immune surveillance (Attisano and Wrana, 2002; Chang et al., 2002; Massague, 2000; Massague et al., 2000).

TGFβs, the prototype members of this superfamily, are released from cells as an inactive complex where their active domain is masked by a propeptide termed Latency Associated Peptide (LAP). They have little or no biological activity until LAP is cleaved by a furin-like endoproteinase (Dubois et al., 1995). Members of the TGFβ superfamily signal through transmembrane receptors that have a cytoplasmic serine/threonine kinase domain. The TGFβ receptors are divided into two subfamilies, type I and type II, which interact to initiate TGFβ signaling. Both receptor types are glycoproteins of approximately 55 kDa and 70–85 kDa, respectively. Their extracellular region contains about 150 amino acids, including 10 or more cysteines that determine the folding of this region. A unique feature of type I TGFβ receptors is a highly conserved intracellular region, composed of 30 amino acids, located upstream to the cytoplasmatic kinase domain named GS domain for its SGSGSG sequence (Wrana et al., 1994). Binding of a TGFβ ligand induces the type II receptor kinase to phosphorylate multiple serine and threonine residues in the TTSGSGSG sequence of the cytoplasmic GS region of the type I receptor, leading to its activation (Wrana et al., 1994; Souchelnytskyi et al., 1996). Binding of the ligand triggers the assembly of a receptor complex and thus initials the phosphorylation of signaling transducers of the SMAD protein family; once phosphorylatied, SMADs migrate into the nucleus, where they assemble to form protein DNA binding complexes that control gene expression (Massague, 2000). A TGFβ type III receptor was also described (Cheifetz et al., 1988; Wickert et al., 2004). It is a large (250–350 kDa) transmembrane proteoglycan with a large extracellular domain and a 43 amino acid residue cytoplasmic domain. The cytoplasmic domain of the Type III receptor lacks an obvious signaling motif and the receptor may not be involved directly in signal transduction. The Type III receptor binds TGFβ2 with the highest affinity. Other TGFβ isoforms also bind the Type III receptor, but with lower affinities. Cellular responsiveness to TGFβ2 appears to be dependent on the presence of the Type III receptor, which can interact with the signaling receptor complex. In addition to the transmembrane Type III receptor, a soluble form of the receptor is secreted by some cell types (Venkatesha et al., 2006). The physiological role of this soluble receptor remains to be determined.

4.2 Effects of the TGFβ Family on Proliferation and Differentiation of Satellite Cells

The TGFβ family comprises of three typical members (1, 2, and 3). Two additional members, TGFβ4 and TGFβ5, were found in chicken and xenopus, respectively.

Selective TGFβs were shown to regulate myogenesis of adult-derived myobasts. It is generally accepted that TGFβs suppress myogenic differentiation. Nevertheless, some studies indicate a positive effect of TGFβ on mammalian myoblast proliferation while in other instances they were shown to suppress proliferation (Allen and Boxhorn, 1989; Cook et al., 1993; Hathaway et al., 1991; Hathaway et al., 1994; Quinn et al., 1994). Importantly, addition of TGFβ1 to isolated myofiber cultures resulted in a drastic reduction in the number of proliferating satellite cells both in the absence or presence of FGF2 (Yablonka-Reuveni and Rivera, 1997b; Bischoff, 1990). These findings clearly demonstrate that TGFβ1 suppresses proliferation of bona fide satellite cells. On another note, administration of TGFβ1 to C2C12 myogenic cell line or to the muscle tissue, initiated fibrosis (Li et al., 2004). Taken together, the latter studies indicate that TGFβ might directly affect satellite cell myogenesis within their niche, but when its physiological levels are increased it may contribute to muscle pathology. The correlation between elevated expression levels of TGFβ in the mdx mouse, model of human Duchenne muscular dystrophy (Zhou et al., 2006), further supports involvement of TGFβ in this muscle pathology. In contrast, there is no evidence for increased TGFβ expression in the laminin alpha 2 (merosin)-deficient dy mouse, which shows progressive muscle fiber necrosis and ineffective muscle regeneration (Sakuma et al., 2000).

Clearly, there is a need for more studies on the role of TGFβ during myogenesis of satellite cells. We demonstrated that freshly isolated myofibers express high levels of TGFβ1 transcripts and it is conceivable that age-associated changes in this factor within the context of the myofiber could be involved in reduced performance of satellite cells in old age (S. Kastner and Z. Yablonka-Reuveni, unpublished studies). A gene array study of myogenic cells propagated for long term in culture demonstrated alterations in the expression level of many genes directly or indirectly involved with the TGFβ signaling pathway (Beggs et al., 2004). This study suggested that with age, myogenic progenitors acquire the paradoxical phenotype of being both TGFβ-activated based on overexpression of TGFβ-inducible genes, but resistant to the differentiation-inhibiting effects of exogenous TGFβ. Additionally, over expression of TGFβ-regulated genes, such as connective tissue growth factor, was proposed to play a role in increasing fibrosis in aging muscle (Beggs et al., 2004). The caveat that comes with this study is that the cells were first passaged to generate sufficient cells for the analyses and were not necessarily free of muscle connective tissue cells. Thus, results can be affected by the contribution of genes expressed by non-myogenic cells. If such contribution is higher in preparations from aged animals, it can lead to the conclusion that that myoblasts from old age mice undergo alterations with regard to their TGFbeta signaling system. In all, it is clear that studies on the role of the TGFβ system of the satellite cells are greatly needed. However, the research effort on the role of TGFβ during myogenesis has been shifted toward myostatin upon the discovery of this presumably muscle-specific member of the TGFβ superfamily.

4.3 The Role of Myostatin (GDF8)

4.3.1 Myostatin regulates muscle mass

Myostatin, a negative regulator of embryonic and postnatal skeletal muscle growth, functions to maintain a proper muscle mass during development and in adult life (Dominique and Gerard, 2006; McPherron and Lee, 1997; Kambadur et al., 1997; McPherron et al., 1997; Schuelke et al., 2004; Carnac et al., 2006). A myostatin deficiency results in an enhanced muscular phenotype that is maintained throughout life, resulting in reduced age-linked muscle atrophy (Wagner, 2005; Siriett et al., 2006; Yablonka-Reuveni, 2007). The production and effect of myostatin is generally held to be skeletal muscle-specific during pre and postnatal growth (McPherron and Lee, 1997), however, myostatin mRNA or protein have been detected in other tissues of vertebrates, including mammary gland (Ji et al., 1998), adipose tissue (McPherron and Lee, 1997) and in plasma (Gonzalez-Cadavid et al., 1998). Myostatin was found in several tissues in the fish (Ostbye et al., 2001). Loss of myostatin activity in cattle, mice, and humans leads to a profound phenotype of muscle overgrowth, associated with increased fiber numbers and size (McPherron and Lee, 1997; Schuelke et al., 2004; Grobet et al., 1997; Nishi et al., 2002). Myostatin null animals and transgenic mice overexpressing signaling inhibitors of myostatin such as follistatin and myostatin propeptide, exhibit increased muscle mass that results both from increased number of muscle fibers, and/or larger than normal fibers (Lee and McPherron, 2001; Yang et al., 2001).

Injured muscles lacking functional myostatin, exhibit improved regeneration and reduced fibrosis, while over expression of myostatin leads to reduced muscle size and increased wasting (cachexia) (Wagner, 2005; Reisz-Porszasz et al., 2003; Jespersen et al., 2006). Animal models with constitutive over- or under-expression of myostatin do not permit direct evaluation of myostatin role in adult life, as the observed mass increase could be a consequence of events taking place during muscle histogenesis and prenatal development. Nevertheless, conditional gene targeting approach exploiting the cre-lox system, demonstrated that *postnatal* inactivation of the myostatin gene is sufficient to cause a generalized muscular hypertrophy of the same magnitude as that observed for constitutive myostatin knockout mice (Grobet et al., 2003). Additionally, the increased expression of myostatin associated with muscle atrophy after periods of muscle inactivity and upon the induction of cachexia in mice, by systemically administered myostatin, also provides evidence for a role of myostatin in adult muscle (Zimmers et al., 2002; Carlson et al., 1999; Wehling et al., 2000; Morley et al., 2006).

Ablation of myostatin function was also shown to ameliorate the dystrophic phenotype in certain myopathies. In the mdx mouse model of Duchenne muscular dystrophy, deletion of the myostatin gene or treatment with a myostatin dominant-negative polypeptide enhanced muscle mass and reduced disease severity (Wagner et al., 2002; Bogdanovich et al., 2002; Bogdanovich et al., 2005). In contrast, loss of myostatin activity in the dyW/dyW mouse model of laminin-deficient congenital muscular dystrophy, a more severe and lethal disease model, did not improve all aspects of muscle pathology (Li et al., 2005). Genetic manipulation or antibody-mediated inhibition of myostatin function in a model of limb-girdle muscular dystrophy (mice lacking delta-sarcoglycan), improved muscle mass, regeneration, and reduced fibrosis. However, this improvement was achieved only during the phase of postnatal growth but not in adults (Parsons et al., 2006). Altogether, the aforementioned in vivo studies suggest that myostatin inhibition may benefit muscle function in dystrophic and atrophic conditions.

4.3.2 Does myostatin regulate myogenesis of satellite cells?

Studies with the myogenic cell line C2C12, suggest that myoblast proliferation is negatively regulated by myostatin via the up-regulation of p21 and inactivation of Cdk (cyclin-dependent kinase), resulting in retinoblastoma (Rb) hypophosphorylation and myoblast cell cycle arrest (Rios et al., 2001; Thomas et al., 2000). The effect of myostatin on differentiation appears to occur through down-regulation of the myogenic differentiation factors MyoD, Myf5 and myogenin. In response to differentiation signals, MyoD becomes activated and induces downstream gene expression, including myogenin and p21, resulting in committed differentiated myoblasts that further fuse to form myotubes. Myostatin inhibits MyoD expression via Smad3 (resulting in the loss of myogenic gene expression and differentiation; Langley et al., 2002; Spiller et al., 2002; Joulia et al., 2003). Recent studies demonstrated that overexpression of Smad7 in C2C12 rescues the inhibitory effects of myostatin (or TGFβ) on myogenic differentiation. Additionally, regardless of myostatin signaling, Smad7 was suggested to directly interact with MyoD, to promote its activity and consequently enhance myogenic differentiation (Kollias et al., 2006).

Evidence from cell culture studies of immediate progeny of mouse satellite cells (McCroskery et al., 2003) and of long-term proliferating myoblasts isolated from poultry muscle (McFarland et al., 2006) also imply a direct inhibitory effect of myostatin on proliferation. Cell cycle analysis confirmed that myostatin up-regulated p21, a Cdk inhibitor, and decreased the levels and activity of Cdk2 protein in satellite cells (McCroskery et al., 2003). In contrast to the previously discussed C2C12 studies where p21 up-regulation was investigated within the context of myogenic differentiation, McCroskery and colleagues proposed that their findings indicate that myostatin negatively regulates the G_1 to S progression and proposed that myostatin maintains satellite cells in a quiescent state (McCroskery et al., 2003). This team also conducted in vivo BrdU-labeling of proliferating cells and concluded that the number of proliferating satellite cells was higher in the myostatin null mice compared to wildtype mice (McCroskery et al., 2003). While the labeling was done in vivo, the analysis of the cells was performed after they were isolated from whole muscle. Hence, it is unknown if there is a general increase in proliferating satellite cells throughout the muscle or alternatively, satellite cells in some muscles might be contributing to the apparent general increase in proliferative activity of satellite cells in myostatin null mice. The increased satellite cell activity in myostatin null mice can also be due to high demand for progenitors to sustained the increased muscle mass and not only because of the release from the negative proliferative effect of myostatin in the null mouse.

The inverse relationship between myostatin levels and satellite cell numbers was also suggested based on the correlation between fiber type and satellite cell numbers: fast type muscle fibers express higher myostatin levels and contain fewer satellite cells compared to slow twitch muscles (Wehling et al., 2000). Also, the increase in myostatin expression occurring during limb immobilization could be associated with suppression of satellite cell function in this physiological condition (Carlson et al., 1999; Wehling et al., 2000). The increased number of satellite cells may be reflective of overall increased muscularity (hyperplasia and hypertrophy) that has developed during embryogenesis and not necessarily an outcome of the lack of myostatin function in postnatal life. Therefore, the myostatin knockout mouse model does not permit direct conclusion regarding the mechanism involved in the increased number of satellite cells in myostatin-null myofibers.

Unpublished studies from our laboratory could not identify obvious differences between myostatin-knockout mice and their wildtype counterparts in gene expression patterns characterizing satellite cell activation, proliferation, differentiation and renewal in single myofiber cultures, primary cultures and clonal cultures from adult and aging mice (see Shefer and Yablonka-Reuveni, 2007 for similar studies with wildtype mice). While age-linked muscle atrophy is reduced in myostatin-null mice (Siriett et al., 2006), our preliminary studies indicate that the number of satellite cells per individual myofiber from the extensor digitorum longus muscle of myostatin-null mice *does* decline with age as observed by us in wildtype mice. While, for both young and old mice, the average number of satellite cells per myofiber is increased in myostatin null mice in comparison to wildtype mice, there is still a profound decline in the numbers of satellite cells between 6 and 24 month of age. Based on these data we suggest that the absence of myostatin leads to increased muscle mass but does not protect from the age-linked decline in satellite cells numbers.

4.3.3 Effect of myostatin on whole body fat, muscle adiposity and adipogenic differentiation

The expression of myostatin mRNA was found primarily in skeletal muscle, but was also detected in the adipose tissue of mammals (McPherron and Lee, 1997). Whether or not the latter expression reflects production of functional myostatin outside of skeletal muscle is as yet unknown. However, it is well established from both natural mutations in cattle and from targeted mutations in mice that animals lacking functional myostatin or overexpressing the inhibitory myostatin propeptide are leaner than wildtype counterparts, even following feeding with high fat diet (McPherron and Lee, 2002; Yang and Zhao, 2006; Zhao et al., 2005; Lin et al., 2002).

There are conflicting reports about the effect of myostatin on adipocyte differentiation as some studies report inhibition of adipogenic differentiation upon its addition to preadipocytes (Kim et al., 2001a; Hirai et al., 2007) whereas other studies report promoted adipogenesis (Artaza et al., 2005; Feldman et al., 2006). Hence more studies are needed to establish the effect of myostatin on adipogenic differentiation. Furthermore, the reduced whole body adiposity seen in vivo might result from a metabolic shift due to the increased muscularity and subsequent enhanced metabolic demands. We suggest that studies with primary cells, rather than C3H10T1/2 cells, are required for achieving further resolution on the role of myostatin in adipogenic differentiation. As in the case of myogenic cell lines, these C3H10T1/2 cells have provided important insights but their adipogenesis (or myogenesis) that is induced by chemical reagents does not necessarily fully mirror the biology of bona fide pre-adipocytes.

We identified adipogenic differentiation in single muscle fibers, suggesting that satellite cells either enter adipogenesis or that myofiber-associated mesenchymal-like progenitors may reside in the satellite cell niche and contribute to this adipogenic differentiation (Shefer and Yablonka-Reuveni, 2007; Shefer et al., 2004). Our preliminary results established that this adipogenic differentiation in isolated myofibers is maintained in myostatin null mice throughout life (although aging muscle from myostatin null mice does not demonstrate a build up of age-linked muscle adiposity as seen in aging wildtype mice). Comparing clones of myogenic and peradipogenic/adipogenic progenitors isolated from myofibers (Shefer et al., 2004), we concluded that only the myogenic clones express myostatin (unpublished). These observations on the absence of myostatin transcripts in adipogenic clones contrast with the recent report that chemical induction of C3H10T1/2 cells to enter adipogenesis also resulted in myostatin gene expression (Feldman et al., 2006). The origin of the cells undergoing adpogenesis in myofiber cultures is most likely different from that of C3H10T1/2 cells that represent a cell line propagated in culture over a long term and this difference may underlie the observed differences in absence or presence of myostatin transcripts. Clearly, more studies are needed to establish the role of myostatin during adipogenesis.

5. CONCLUDING REMARKS

Although there are numerous publications on the effect of growth factors during myogenesis, very little is really known regarding how growth factors permit continuous proliferation of myogenic cells, versus promoting exit form the cell cycle to the quiescent G_0 state, versus promoting the terminal withdrawal from the cell cycle toward differentiation. The regulation of satellite cell recruitment from quiescence is also poorly understood. Cell culture studies do not always precisely recapitulate events occurring in vivo but they permit some gain of insight into the regulation of satellite cell myogenesis. The amounts and forms at which growth factors are added in culture or in vivo can result in inhibitory or enhancing effects that exceed physiological effect of growth factors. Transcriptional, translational and posttranslational events can control the biological availability of growth factors discussed above. HSPG and receptor tyrosine kinase types and specificity can add yet another level of control. For example, members of the TGFβ family (i.e., TGFβ1-3) and myostatin may have very different functions based on their association with different type II TGFβ receptors. Another example is receptor heterogeneity in the FGF system with four different FGFR genes (1–4) and numerous splice forms of these gene products. In addition, regulation of various steps within the signaling machinery, down stream of the initial receptor activation, can enhance or prevent ligand effects. All these issues need to be taken into account when considering future therapies for enhancing muscle regeneration and counter-acting degenerative muscle diseases and age linked-muscle atrophy. Most likely, future availability of conditional knockout mice where growth factors and their receptors can be eliminated during adult life in a tissue specific manner will have a major impact on our understanding of the role of growth factors during the myogenesis of satellite cells and routine muscle maintenance. Such findings will in turn make important contributions to the skeletal muscle regenerative biology.

ACKNOWLEDGEMENTS

Z.Y.-R is presently supported by the National Institute on Aging (NIH, AG 021566 and AG 013798) and the USDA Cooperative State Research, Education and Extension Service (NRI, 99-35206-7934). G.S. is an Eshkol Fellow of the Israeli Ministry of Science and Technology and of the Israeli 'Budgeting Committee & Planning' (VATAT).

REFERENCES

Aikawa J, Esko JD (1999) Molecular cloning and expression of a third member of the heparan sulfate/heparin GlcNAc N-deacetylase/ N-sulfotransferase family. J Biol Chem 274(5):2690–2695

Allen RE, Boxhorn LK (1989) Regulation of skeletal muscle satellite cell proliferation and differentiation by transforming growth factor-beta, insulin-like growth factor I, and fibroblast growth factor. J Cell Physiol 138(2):311–315

Allen RE, Sheehan SM, Taylor RG, Kendall TL, Rice GM (1995) Hepatocyte growth factor activates quiescent skeletal muscle satellite cells in vitro. J Cell Physiol 165(2):307–312

Anastasi S, Giordano S, Sthandier O, Gambarotta G, Maione R, Comoglio PM et al (1997) A natural hepatocyte growth factor/scatter factor autocrine loop in myoblast cells and the effect of the constitutive met kinase activation on myogenic differentiation. J Cell Biol 137(5):1057–1068

Anderson JE (2000) A role for nitric oxide in muscle repair: nitric oxide-mediated activation of muscle satellite cells. Mol Cell Biol 11(5):1859–1874

Andres V, Walsh K (1996) Myogenin expression, cell cycle withdrawal, and phenotypic differentiation are temporally separable events that precede cell fusion upon myogenesis. J Cell Biol 132(4):657–666

Armand AS, Launay T, Pariset C, Della Gaspera B, Charbonnier F, Chanoine C (2003) Injection of FGF6 accelerates regeneration of the soleus muscle in adult mice. Biochim Biophys Acta 1642(1–2):97–105

Armand AS, Laziz I, Chanoine C (2006) FGF6 in myogenesis. Biochim Biophys Acta 1763(8):773–778

Armand AS, Pariset C, Laziz I, Launay T, Fiore F, Della Gaspera B et al (2005) FGF6 regulates muscle differentiation through a calcineurin-dependent pathway in regenerating soleus of adult mice. J Cell Physiol 204(1):297–308

Armelin HA (1973) Pituitary extracts and steroid hormones in the control of 3T3 cell growth. Proc Natl Acad Sci USA 70(9):2702–2706

Artaza JN, Bhasin S, Magee TR, Reisz-Porszasz S, Shen R, Groome NP et al (2005) Myostatin inhibits myogenesis and promotes adipogenesis in C3H 10T(1/2) mesenchymal multipotent cells. Endocrinology 146(8):3547–3557

Attisano L, Wrana JL (2002) Signal transduction by the TGF-beta superfamily. Science 296(5573):1646–1647

Beauchamp JR, Heslop L, Yu DSW, Tajbakhsh S, Kelly RG, Wernig A et al (2000) Expression of CD34 and Myf5 defines the majority of quiescent adult skeletal muscle satellite cells. J Cell Biol 151(6):1221–1234

Beggs ML, Nagarajan R, Taylor-Jones JM, Nolen G, MacNicol M, Peterson CA (2004) Alterations in the TGFbeta signaling pathway in myogenic progenitors with age. Aging Cell 3(6):353–361

Bernfield M, Gotte M, Park PW, Reizes O, Fitzgerald ML, Lincecum J et al., (1999) Functions of cell surface heparan sulfate proteoglycans. Annu Rev Biochem 68:729–777

Birchmeier C, Gherardi E (1998) Developmental roles of HGF/SF and its receptor, the c-Met tyrosine kinase. Trends Cell Biol 8(10):404–410

Bischoff R (1986) A satellite cell mitogen from crushed adult muscle. Dev Biol 115(1):140–147

Bischoff R (1989) Analysis of muscle regeneration using single myofibers in culture. Med Sci Sports Exerc 21(5 Suppl):S164–S172

Bischoff R (1990) Control of satellite cell proliferation. Adv Exp Med Biol 280:147–157

Bischoff R (1997) Chemotaxis of skeletal muscle satellite cells. Dev Dyn 208(4):505–515

Black BL, Olson EN (1998) Transcriptional control of muscle development by myocyte enhancer factor-2 (MEF2) proteins. Annu Rev Cell Dev Biol 14(1):167–196

Bogdanovich S, Krag TOB, Barton ER, Morris LD, Whittemore LA, Ahima RS et al (2002) Functional improvement of dystrophic muscle by myostatin blockade. Nature 420(6914):418–421

Bogdanovich S, Perkins KJ, Krag TOB, Whittemore LA, Khurana TS (2005) Myostatin propeptide-mediated amelioration of dystrophic pathophysiology. FASEB J 19(6):543–549

Booth FW (2006) The many flavors of IGF-I. J Appl Physiol 100(6):1755–1756

Botta M, Manetti F, Corelli F (2000) Fibroblast growth factors and their inhibitors. Curr Pharm Des 6:1897–1924

Bottaro DP, Rubin JS, Faletto DL, Chan AM, Kmiecik TE, Vande Woude GF et al (1991) Identification of the hepatocyte growth factor receptor as the c-met proto-oncogene product. Science 251(4995):802–804

Brandan E, Carey DJ, Larrain J, Melo F, Campos A (1996) Synthesis and processing of glypican during differentiation of skeletal muscle cells. Eur J Cell Biol 71(2):170–176

Brandan E, Fuentes ME, Andrade W (1991) The proteoglycan decorin is synthesized and secreted by differentiated myotubes. Eur J Cell Biol 55(2):209–216

Buckingham M (2006) Myogenic progenitor cells and skeletal myogenesis in vertebrates. Curr Opin Genet Dev 16(5):525–532

Burgess WH, Maciag T (1989) The heparin-binding (fibroblast) growth factor family of proteins. Annu Rev Biochem 58(1):575–602

Carlson BM, Dedkov EI, Borisov AB, Faulkner JA (2001) Skeletal muscle regeneration in very old rats. Journal of Gerontology. Series A, J Gerontol A Biol Sci Med Sci 56(5):B224–B233

Carlson CJ, Booth FW, Gordon SE (1999) Skeletal muscle myostatin mRNA expression is fiber-type specific and increases during hindlimb unloading. Am J Physiol 277(2, Pt 2):R601-R606

Carnac G, Ricaud S, Vernus B, Bonnieu A (2006) Myostatin: Biology and clinical relevance. Mini Rev Med Chem 6:765–770

Carrino DA (1998) Dynamic expression of proteoglycans during skeletal muscle development. Basic Appl Myology 8(2):95–106

Casar JC, Cabello-Verrugio C, Olguin H, Aldunate R, Inestrosa NC, Brandan E (2004) Heparan sulfate proteoglycans are increased during skeletal muscle regeneration: requirement of syndecan-3 for successful fiber formation. J Cell Sci 117(1):73–84

Cavallaro U, Niedermeyer J, Fuxa M, Christofori G (2001) N-CAM modulates tumour-cell adhesion to matrix by inducing FGF-receptor signalling. Nat Cell Biol 3(7):650–657

Cenciarelli C, De Santa F, Puri PL, Mattei E, Ricci L, Bucci F et al (1999) Critical role played by Cyclin D3 in the MyoD-mediated arrest of cell cycle during myoblast differentiation. Mol Cell Biol 19(7):5203–5217

Chang H, Brown CW, Matzuk MM (2002) Genetic analysis of the mammalian transforming growth factor-beta superfamily. Endocr Rev 23(6):787–823

Charge SB, Rudnicki MA (2004) Cellular and molecular regulation of muscle regeneration. Physiol Rev 84(1):209–238

Cheifetz S, Andres JL, Massague J (1988) The transforming growth factor-beta receptor type III is a membrane proteoglycan. Domain structure of the receptor. J Biol Chem 263(32):16984–16991

Citores L, Khnykin D, Sorensen V, Wesche J, Klingenberg O, Wiedlocha A et al (2001) Modulation of intracellular transport of acidic fibroblast growth factor by mutations in the cytoplasmic receptor domain. J Cell Sci 114(9):1677–1689

Citores L, Wesche J, Kolpakova E, Olsnes S (1999) Uptake and Intracellular transport of acidic fibroblast growth factor: evidence for free and cytoskeleton-anchored fibroblast growth factor Receptors. Mol Biol Cell 10(11):3835–3848

Clegg CH, Linkhart TA, Olwin BB, Hauschka SD (1987) Growth factor control of skeletal muscle differentiation: commitment to terminal differentiation occurs in G1 phase and is repressed by fibroblast growth factor. J Cell Biol 105(2):949–956

Collins CA, Olsen I, Zammit PS, Heslop L, Petrie A, Partridge TA et al (2005) Stem cell function, self-renewal, and behavioral heterogeneity of cells from the adult muscle satellite cell niche. Cell 122(2):289–301

Comoglio PM (1993) Structure, biosynthesis and biochemical properties of the HGF receptor in normal and malignant cells. EXS 65:131–165

Conboy IM, Rando TA (2005) Aging, stem cells and tissue regeneration: lessons from muscle. Cell Cycle 4(3):407–410

Cook DR, Doumit ME, Merkel RA (1993) Transforming growth factor-beta, basic fibroblast growth factor, and platelet-derived growth factor-BB interact to affect proliferation of clonally derived porcine satellite cells. J Cell Physiol 157(2):307–312

Cooper CS, Park M, Blair DG, Tainsky MA, Huebner K, Croce CM et al (1984) Molecular cloning of a new transforming gene from a chemically transformed human cell line. Nature 311(5981):29–33

Cornelison DDW, Filla MS, Stanley HM, Rapraeger AC, Olwin BB (2001) Syndecan-3 and syndecan-4 specifically mark skeletal muscle satellite cells and are implicated in satellite cell maintenance and muscle regeneration. Dev Biol 239(1):79–94

Cornelison DDW, Wilcox-Adelman SA, Goetinck PF, Rauvala H, Rapraeger AC, Olwin BB (2004) Essential and separable roles for Syndecan-3 and Syndecan-4 in skeletal muscle development and regeneration. Genes Dev 18(18):2231–2236

Cornelison DDW, Wold BJ (1997) Single-cell analysis of regulatory gene expression in quiescent and activated mouse skeletal muscle satellite cells. Dev Biol 191(2):270–283

Coulier F, Pizette S, Ollendorff V, deLapeyriere O, Birnbaum D (1994) The human and mouse fibroblast growth factor 6 (FGF6) genes and their products: possible implication in muscle development. Prog Growth Factor Res 5(1):1–14

Coulier F, Pontarotti P, Roubin R, Hartung H, Goldfarb M, Birnbaum D (1997) Of worms and men: an evolutionary perspective on the fibroblast growth factor (FGF) and FGF receptor families. J Mol Evol 44(1):43–56

Day K, Shefer G, Richardson JB, Enikolopov G, Yablonka-Reuveni Z (2007) Nestin-GFP reporter expression defines the quiescent state of skeletal muscle satellite cells. Dev Biol 304(1):246–259

de Alvaro C, Martinez N, Rojas JM, Lorenzo M (2005) Sprouty-2 overexpression in C2C12 Cells confers myogenic differentiation properties in the presence of FGF2. Mol Biol Cell 16(9):4454–4461

deLapeyriere O, Ollendorff V, Planche J, Ott MO, Pizette S, Coulier F et al (1993) Expression of the Fgf6 gene is restricted to developing skeletal muscle in the mouse embryo. Development 118(2):601–611

De Falco G, Comes F, Simone C (2006) pRb: master of differentiation. Coupling irreversible cell cycle withdrawal with induction of muscle-specific transcription. Oncogene 25(38):5244–5249

Desgranges P, Barbaud C, Caruelle JP, Barritault D, Gautron J (1999) A substituted dextran enhances muscle fiber survival and regeneration in ischemic and denervated rat EDL muscle. FASEB J 13(6):761–766

DiMario JX (2002) Activation and repression of growth factor receptor gene transcription (Review). Int J Mol Med 10(1):65–71

Dominique JE, Gerard C (2006) Myostatin regulation of muscle development: Molecular basis, natural mutations, physiopathological aspects. Exp Cell Res 312(13):2401–2414

Dubois CM, Laprise MH, Blanchette F, Gentry LE, Leduc R (1995) Processing of transforming growth factor 1 precursor by human furin convertase. J Biol Chem 270(18):10618–10624

Eswarakumar VP, Lax I, Schlessinger J (2005) Cellular signaling by fibroblast growth factor receptors. Cytokine Growth Factor Rev 16(2):139–149

Fadic R, Mezzano V, Alvarez K, Cabrera D, Holmgren J, Brandan E (2006) Increase in decorin and biglycan in Duchenne Muscular Dystrophy: role of fibroblasts as cell source of these proteoglycans in the disease J Cell Mol Med 3:758–769

Feldman BJ, Streeper RS, Farese RVJ, Yamamoto KR (2006) Myostatin modulates adipogenesis to generate adipocytes with favorable metabolic effects. Proceedings of the National Academy of Sciences of the United States of America 103(42):15675–15680

Feng S, Xu J, Wang F, Kan M, McKeehan WL (1996) Nuclear localization of a complex of fibroblast growth factor(FGF)-1 and an NH2-terminal fragment of FGF receptor isoforms R4 and R1alpha in human liver cells. Biochimica et Biophysica Acta 1310(1):67–73

Fiore F, Sebille A, Birnbaum D (2000) Skeletal muscle regeneration is not impaired in Fgf6 $^{-/-}$ mutant mice. Biochem Biophys Res Commun 272(1):138–143

Florini JR, Ewton DZ, Coolican SA (1996) Growth hormone and the insulin-like growth factor system in myogenesis. Endocr Rev 17(5):481–517

Floss T, Arnold HH, Braun T (1997) A role for FGF-6 in skeletal muscle regeneration. Genes Dev 11(16):2040–2051

Funakoshi H, Nakamura T (2003) Hepatocyte growth factor: from diagnosis to clinical applications. Clin Chim Acta 327(1–2):1–23

Gal-Levi R, Leshem Y, Aoki S, Nakamura T, Halevy O (1998) Hepatocyte growth factor plays a dual role in regulating skeletal muscle satellite cell proliferation and differentiation. Biochim Biophys Acta 1402(1):39–51

Gao G, Goldfarb M (1995) Heparin can activate a receptor tyrosine kinase. EMBO J 14(10):2183–2190

Garrett KL, Anderson JE (1995) Colocalization of bFGF and the myogenic regulatory gene myogenin in dystrophic mdx muscle precursors and young myotubes in vivo. Dev Biol 169(2):596–608

Gohda E, Tsubouchi H, Nakayama H, Hirono S, Sakiyama O, Takahashi K et al (1998) Purification and partial characterization of hepatocyte growth factor from plasma of a patient with fulminant hepatic failure. J Clin Invest 81(2):414–419

Goldfarb M (1996) Functions of fibroblast growth factors in vertebrate development. Cytokine Growth Factor Rev 7(4):311–325

Gonzalez AM, Buscaglia M, Ong M, Baird A (1990) Distribution of basic fibroblast growth factor in the 18-day rat fetus: localization in the basement membranes of diverse tissues. J Cell Biol 110(3):753–765

Gonzalez AM, Hill DJ, Logan A, Maher PA, Baird A (1996) Distribution of fibroblast growth factor (FGF)-2 and FGF receptor-1 messenger RNA expression and protein presence in the mid-trimester human fetus. Pediatr Res 39(3):375–385

Gonzalez-Cadavid NF, Taylor WE, Yarasheski K, Sinha-Hikim I, Ma K, Ezzat S et al (1998) Organization of the human myostatin gene and expression in healthy men and HIV-infected men with muscle wasting. Proc Natl Acad Sci USA 95(25):14938–14943

Gospodarowicz D (1974) Localisation of a fibroblast growth factor and its effect alone and with hydrocortisone on 3T3 cell growth. Nature 249(453):123–127

Graves DC, Yablonka-Reuveni Z (2000) Vascular smooth muscle cells spontaneously adopt a skeletal muscle phenotype: a unique Myf5−/MyoD+ myogenic program. J Histochem Cytochem 48(9):1173–1194

Grobet L, Pirottin D, Farnir F, Poncelet D, Royo LJ, Brouwers B et al (2003) Modulating skeletal muscle mass by postnatal, muscle-specific inactivation of the myostatin gene. Genesis 35(4):227–238
Grobet L, Royo Martin LJ, Poncelet D, Pirottin D, Brouwers B, Riquet J et al (1997) A deletion in the bovine myostatin gene causes the double-muscled phenotype in cattle. Nature Genetics 17(1):71–74
Grounds MD (2002) Reasons for the degeneration of ageing skeletal muscle: a central role for IGF-1 signalling. Biogerontology 3(1–2):19–24
Grounds MD, Yablonka-Reuveni Z (1993) Molecular and cell biology of skeletal muscle regeneration. Mol Cell Biol Hum Dis Ser 3:210–256
Groux-Muscatelli B, Bassaglia Y, Barritault D, Caruelle JP, Gautron J (1990) Proliferating satellite cells express acidic fibroblast growth factor during in vitro myogenesis. Dev Biol 142(2):380–385
Hadari YR, Gotoh N, Kouhara H, Lax I, Schlessinger J (2001) Critical role for the docking-protein FRS2alpha in FGF receptor-mediated signal transduction pathways. Proc Natl Acad Sci USA 98(15):8578–8583
Halevy O, Novitch BG, Spicer DB, Skapek SX, Rhee J, Hannon GJ et al (1995) Correlation of terminal cell cycle arrest of skeletal muscle with induction of p21 by MyoD. Science 267(5200):1018–1021
Halevy O, Piestun Y, Allouh MZ, Rosser BW, Rinkevich Y, Reshef R et al (2004) Pattern of Pax7 expression during myogenesis in the posthatch chicken establishes a model for satellite cell differentiation and renewal. Dev Dyn 231(3):489–502
Han JK, Martin GR (1993) Embryonic expression of Fgf-6 is restricted to the skeletal muscle lineage. Dev Biol 158(2):549–554
Hannon K, Kudla A, J., McAvoy MJ, Clase KL, Olwin BB (1996) Differentially expressed fibroblast growth factors regulate skeletal muscle development through autocrine and paracrine mechanisms. J Cell Biol 132(6):1151–1159
Harmer NJ, Pellegrini L, Chirgadze D, Fernandez-Recio J, Blundell TL (2004) The crystal structure of fibroblast growth factor (FGF) 19 reveals novel features of the FGF Family and offers a structural basis for its unusual receptor affinity. Biochemistry 43(3):629–640
Hartley RS, Yablonka-Reuveni Z (1990) Long-term maintenance of primary myogenic cultures on a reconstituted basement membrane. In Vitro Cell Dev Biol 26(10):955–961
Hathaway MR, Hembree JR, Pampusch MS, Dayton WR (1991) Effect of transforming growth factor beta-1 on ovine satellite cell proliferation and fusion. J Cell Physiol 146(3):435–441
Hathaway MR, Pampusch MS, Hembree JR, Dayton WR (1994) Transforming growth factor beta-1 facilitates establishing clonal populations of ovine muscle satellite cells. J Anim Sci 72(8):2001–2007
Haugsten EM, Sorensen V, Brech A, Olsnes S, Wesche J (2005) Different intracellular trafficking of FGF1 endocytosed by the four homologous FGF receptors. J Cell Sci 118(17):3869–3881
Hawke TJ, Garry DJ (2001) Myogenic satellite cells: physiology to molecular biology. J Appl Physiol 91:534–551
Hayashi S, Aso H, Watanabe K, Nara H, Rose MT, Ohwada S et al (2000) Sequence of IGF-I, IGF-II, and HGF expression in regenerating skeletal muscle. Histochem Cell Biol 122(5):427–434
Higashio K, Shima N, Goto M, Itagaki Y, Nagao M, Yasuda H et al (1990) Identity of a tumor cytotoxic factor from human fibroblasts and hepatocyte growth factor. Biochem Biophys Res Commun 170(1):397–404
Hirai S, Matsumoto H, Hino N, Kawachi H, Matsui T, Yano H (2007) Myostatin inhibits differentiation of bovine preadipocyte. Domest Anim Endocrinol 32(1):1–14
Horsley V, Jansen KM, Mills ST, Pavlath GK (2003) IL-4 acts as a myoblast recruitment factor during mammalian muscle growth. Cell 113(4):483–494
Israeli D, Benchaouir R, Ziaei S, Rameau P, Gruszczynski C, Peltekian E et al (2004) FGF6 mediated expansion of a resident subset of cells with SP phenotype in the C2C12 myogenic line. J Cell Physiol 201(3):409–419
Jennische E, Ekberg S, Matejka GL (1993) Expression of hepatocyte growth factor in growing and regenerating rat skeletal muscle. Am J Physiol Cell Physiol 265(1):C122–C128
Jenniskens GJ, Veerkamp JH, van Kuppevelt TH (2006) Heparan sulfates in skeletal muscle development and physiology. J Cell Physiol 206(2):283–294

Jespersen J, Kjaer M, Schjerling P (2006) The possible role of myostatin in skeletal muscle atrophy and cachexia. Scand J Med Sci Sports 16(2):74–82

Ji S, Losinski RL, Cornelius SG, Frank GR, Willis GM, Gerrard DE et al (1998) Myostatin expression in porcine tissues: tissue specificity and developmental and postnatal regulation. Am J Physiol 275(4, Pt 2):R1265–R1273

Jin P, Farmer K, Ringertz NR, Sejersen T (1993) Proliferation and differentiation of human fetal myoblasts is regulated by PDGF-BB. Differentiation 54(1):47–54

Jin P, Sejersen T, Ringertz NR (1991) Recombinant platelet-derived growth factor-BB stimulates growth and inhibits differentiation of rat L6 myoblasts. J Biol Chem 266(2):1245–1249

Johnson DE, Williams LT (1993) Structural and functional diversity in the FGF receptor multigene family. Adv Cancer Res 60:1–41

Johnson SE, Allen RE (1995) Activation of skeletal muscle satellite cells and the role of fibroblast growth factor receptors. Exp Cell Res 219(2):449–453

Johnston CL, Cox HC, Gomm JJ, Coombes RC (1995) bFGF and aFGF induce membrane ruffling in breast cancer cells but not in normal breast epithelial cells: FGFR-4 involvement. Biochem J 306(Pt 2):609–616

Jones NC, Tyner KJ, Nibarger L, Stanley HM, Cornelison DDW, Fedorov YV et al (2005) The p38{alpha}/{beta} MAPK functions as a molecular switch to activate the quiescent satellite cell. J Cell Biol 169(1):105–116

Joulia D, Bernardi H, Garandel V, Rabenoelina F, Vernus B, Cabello G (2003) Mechanisms involved in the inhibition of myoblast proliferation and differentiation by myostatin. Exp Cell Res 286(2):263–275

Kambadur R, Sharma M, Smith TPL, Bass JJ (1997) Mutations in myostatin (GDF8) in double-muscled belgian blue and piedmontese cattle. Genome Res 7(9):910–915

Kan M, Wu X, Wang F, McKeehan WL (1999) Specificity for fibroblast growth factors determined by heparan sulfate in a binary complex with the receptor kinase. J Biol Chem 1999;274(22):15947–15952

Kassar-Duchossoy L, Gayraud-Morel B, Gomes D, Rocancourt D, Buckingham M, Shinin V et al (2004) Mrf4 determines skeletal muscle identity in Myf5:Myod double-mutant mice. Nature 431(7007):466–471

Kastner S, Elias MC, Rivera AJ, Yablonka-Reuveni Z (2000) Gene expression patterns of the fibroblast growth factors and their receptors during myogenesis of rat satellite cells. J Histochem Cytochem 48(8):1079–1096

Kiess M, Gill RM, Hamel PA (1995) Expression of the positive regulator of cell cycle progression, cyclin D3, is induced during differentiation of myoblasts into quiescent myotubes. Oncogene 10(1):159–166

Kim HS, Liang L, Dean RG, Hausman DB, Hartzell DL, Baile CA (2001a) Inhibition of preadipocyte differentiation by myostatin treatment in 3T3-L1 cultures. Biochem Biophys Res Commun 281(4):902–906

Kim I, Moon S, Yu K, Kim U, Koh GY (2001b) A novel fibroblast growth factor receptor-5 preferentially expressed in the pancreas. Biochim Biophys Acta 1518(1–2):152–156

Kitzmann M, Vandromme M, Schaeffer V, Carnac G, Labbe JC, Lamb N et al (1999) cdk1- and cdk2-mediated phosphorylation of MyoD Ser200 in growing C2 myoblasts: Role in modulating MyoD half-life and myogenic activity. Mol Cell Biol 19(4):3167–3176

Klein OD, Minowada G, Peterkova R, Kangas A, Yu BD, Lesot H et al (2006) Sprouty genes control diastema tooth development via bidirectional antagonism of epithelial-mesenchymal FGF signaling. Dev Cell 11(2):181–190

Klingenberg O, Wiedlocha A, Rapak A, Khnykin D, Citores L, Olsnes S (2000) Requirement for C-terminal end of fibroblast growth factor receptor 4 in translocation of acidic fibroblast growth factor to cytosol and nucleus. J Cell Sci 113(10):1827–1838

Knapp JR, Davie JK, Myer A, Meadows E, Olson EN, Klein WH (2006) Loss of myogenin in postnatal life leads to normal skeletal muscle but reduced body size. Development 133(4):601–610

Kollias HD, Perry RLS, Miyake T, Aziz A, McDermott JC (2006) Smad7 promotes and enhances skeletal muscle differentiation. Mol Cell Biol 26(16):6248–6260

Kontaridis MI, Liu X, Zhang L, Bennett AM (2002) Role of SHP-2 infibroblast growth factor receptor-mediated suppression of myogenesis in C2C12 myoblasts. Mol Cell Biol (11):3875–3891

Korhonen J, Partanen J, Eerola E, Vainikka S, Alitalo R, Makela TP et al (1992) Five FGF receptors with distinct expression patterns. EXS 61:91–100

Kovalenko D, Yang X, Chen PY, Nadeau RJ, Zubanova O, Pigeon K et al (2006) A role for extracellular and transmembrane domains of Sef in Sef-mediated inhibition of FGF signaling. Cell Signal 18(11):1958–1966

Kudla AJ, Jones NC, Rosenthal RS, Arthur K, Clase KL, Olwin BB (1998) The FGF Receptor-1 tyrosine kinase domain regulates myogenesis but is not sufficient to stimulate proliferation. J Cell Biol 142(1):241–250

Langley B, Thomas M, Bishop A, Sharma M, Gilmour S, Kambadur R (2002) Myostatin inhibits myoblast differentiation by down-regulating MyoD expression. J Biol Chem 277(51):49831–49840

Langsdorf A, Do AT, Kusche-Gullberg M, Emerson CP Jr, Ai X (2007) Sulfs are regulators of growth factor signaling for satellite cell differentiation and muscle regeneration. Dev Biol in press

Larrain J, Alvarez J, Hassell JR, Brandan E (1997a) Expression of perlecan, a proteoglycan that binds myogenic inhibitory basic fibroblast growth factor, is down regulated during skeletal muscle differentiation. Exp Cell Res 234(2):405–412

Larrain J, Cizmeci-Smith G, Troncoso V, Stahl RC, Carey DJ, Brandan E (1997b) Syndecan-1 expression is down-regulated during myoblast terminal differentiation. Modulation by growth factors and retinoic acid. J Biol Chem 272(29):18418–18424

Lee SJ, McPherron AC (2001) Regulation of myostatin activity and muscle growth. Proc Natl Acad Sci USA 98(16):9306–9311

Lee CC, Putnam AJ, Miranti CK, Gustafson M, Wang LM, Vande Woude GF et al (2004) Overexpression of sprouty 2 inhibits HGF/SF-mediated cell growth, invasion, migration, and cytokinesis. Oncogene 23(30):5193–5202

Lefaucheur JP, Sebille A (1995) Muscle regeneration following injury can be modified in vivo by immune neutralization of basic fibroblast growth factor, transforming growth factor beta 1 or insulin-like growth factor I. J Neuroimmunol 57(1–2):85–91

Leshem Y, Gitelman I, Ponzetto C, Halevy O (2002) Preferential binding of grb2 or phosphatidylinositol 3-kinase to the met receptor has opposite effects on hgf-induced myoblast proliferation. Exp Cell Res 274(2):288–298

Li Y, Foster W, Deasy BM, Chan Y, Prisk V, Tang Y et al (2004) Transforming growth factor-beta 1 induces the differentiation of myogenic cells into fibrotic cells in injured skeletal muscle: A key event in muscle fibrogenesis. Am J Pathol 164(3):1007–1019

Li ZF, Shelton GD, Engvall E (2005) Elimination of myostatin does not combat muscular dystrophy in dy mice but increases postnatal lethality. Am J Pathol 166(2):491–497

Lietha D, Chirgadze DY, Mulloy B, Blundell TL, Gherardi E (2001) Crystal structures of NK1-heparin complexes reveal the basis for NK1 activity and enable engineering of potent agonists of the MET receptor. EMBO J 20(20):5543–5555

Lin J, Arnold HB, Della-Fera MA, Azain MJ, Hartzell DL, Baile CA (2002) Myostatin knockout in mice increases myogenesis and decreases adipogenesis. Biochem Biophys Res Commun 291(3): 701–706

Ludolph DC, Konieczny SF (1995) Transcription factor families: muscling in on the myogenic program. FASEB J 9(15):1595–1604

Lyon M, Deakin JA, Gallagher JT (2002) The mode of action of heparan and dermatan sulfates in the regulation of hepatocyte growth factor/scatter factor. J Biol Chem 277(2):1040–1046

Lyon M, Deakin JA, Mizuno K, Nakamura T, Gallagher JT (1994) Interaction of hepatocyte growth factor with heparan sulfate. Elucidation of the major heparan sulfate structural determinants. J Biol Chem 269(15):11216–11223

Lyon M, Deakin JA, Rahmoune H, Fernig DG, Nakamura T, Gallagher JT (1998) Hepatocyte growth factor/scatter factor binds with high affinity to dermatan sulfate. J Biol Chem 273(1):271–278

Maher PA (1996) Nuclear translocation of fibroblast growth factor (FGF) receptors in response to FGF-2. J Cell Biol 134(2):529–536

Marics I, Padilla F, Guillemot JF, Scaal M, Marcelle C (2002) FGFR4 signaling is a necessary step in limb muscle differentiation. Development 129(19):4559–4569

Massague J (2000) How cells read TGF-beta signals. Nat Rev Mol Cell Biol 1(3):169–178
Massague J, Blain SW, Lo RS (2000) TGF beta signaling in growth control, cancer, and heritable disorders. Cell 103(2):295–309
Matsumoto K, Nakamura T (1997) Hepatocyte growth factor (HGF) as a tissue organizer for organogenesis and regeneration. Biochem Biophys Res Commun 239(3):639–644
Mauro A (1961) Satellite cells of skeletal muscle fibres. J Biophys Biochem 9:493–495
McCroskery S, Thomas M, Maxwell L, Sharma M, Kambadur R (2003) Myostatin negatively regulates satellite cell activation and self-renewal. J Cell Biol 162(6):1135–1147
McCroskery S, Thomas M, Platt L, Hennebry A, Nishimura T, McLeay L et al (2005) Improved muscle healing through enhanced regeneration and reduced fibrosis in myostatin-null mice. J Cell Sci 118(15):3531–3541
McFarland DC, Gilkerson KK, Pesall JE, Wellenreiter RH, Ferrin NH, Ye WV et al (1997) Comparison of growth factor receptors and metabolic characteristics of satellite cells derived from the biceps femoris and pectoralis major muscles of the turkey. Gen Comp Endocrinol 105(1):114–120
McFarland DC, Liu X, Velleman SG, Zeng C, Coy CS, Pesall JE (2003) Variation in fibroblast growth factor response and heparan sulfate proteoglycan production in satellite cell populations. Comparative Biochemistry and Physiology Part C: Toxicol Pharmacol 134(3):341–351
McFarland DC, Velleman SG, Pesall JE, Liu C (2006) Effect of myostatin on turkey myogenic satellite cells and embryonic myoblasts. Comp Biochem Physiol A Mol Integr Physiol 144(4):501–508
McPherron AC, Lawler AM, Lee SJ (1997) Regulation of skeletal muscle mass in mice by a new TGF-beta superfamily member. Nature 387(6628):83–90
McPherron AC, Lee SJ (1997) Double muscling in cattle due to mutations in the myostatin gene. Proc Natl Acad Sci USA 94(23):12457–12461
McPherron AC, Lee SJ (2002) Suppression of body fat accumulation in myostatin-deficient mice. J Clin Invest;109(5):595–601
Meddahi A, Bree F, Papy-Garcia D, Gautron J, Barritault D, Caruelle JP (2002) Pharmacological studies of RGTA(11), a heparan sulfate mimetic polymer, efficient on muscle regeneration. J Biomed Mater Res 62(4):525–531
Melo F, Carey DJ, Brandan E (1996) Extracellular matrix is required for skeletal muscle differentiation but not myogenin expression. J Cell Biochem 1996;62(2):227–239
Miller KJ, Thaloor D, Matteson S, Pavlath GK (2000) Hepatocyte growth factor affects satellite cell activation and differentiation in regenerating skeletal muscle. Am J Physiol Cell Physiol 278(1):C174–C181
Minowada G, Jarvis LA, Chi CL, Neubuser A, Sun X, Hacohen N et al (1999) Vertebrate Sprouty genes are induced by FGF signaling and can cause chondrodysplasia when overexpressed. Development 126(20):4465–4475
Miura T, Kishioka Y, Wakamatsu J, Hattori A, Hennebry A, Berry CJ et al (2006) Decorin binds myostatin and modulates its activity to muscle cells. Biochem Biophys Res Commun 340(2):675–680
Molkentin JD, Black BL, Martin JF, Olson EN (1995) Cooperative activation of muscle gene expression by MEF2 and myogenic bHLH proteins. Cell 83(7):1125–1136
Morley JE, Thomas DR, Wilson MMG (2006) Cachexia: pathophysiology and clinical relevance. Am J Clin Nutr 83(4):735–743
Mourkioti F, Rosenthal N (2005) IGF-1, inflammation and stem cells: interactions during muscle regeneration. Trends Immunol 26(10):535–542
Naar AM, Lemon BD, Tjian R (2001) Transcriptional coactivator complexes. Annu Rev Biochem 70:475–501
Nagata Y, Partridge TA, Matsuda R, Zammit PS (2006) Entry of muscle satellite cells into the cell cycle requires sphingolipid signaling. J Cell Biol 174(2):245–253
Naka D, Ishii T, Yoshiyama Y, Miyazawa K, Hara H, Hishida T et al (1992) Activation of hepatocyte growth factor by proteolytic conversion of a single chain form to a heterodimer. J Biol Chem 267(28):20114–20119
Nakamura T, Nawa K, Ichihara A (1984) Partial purification and characterization of hepatocyte growth factor from serum of hepatectomized rats. Biochem Biophys Res Commun 122(3):1450–1459

Nakamura T, Nishizawa T, Hagiya M, Seki T, Shimonishi M, Sugimura A et al (1989) Molecular cloning and expression of human hepatocyte growth factor. Nature 342(6248):440–443

Nakamura T, Teramoto H, Ichihara A (1986) Purification and characterization of a growth factor from rat platelets for mature parenchymal hepatocytes in primary cultures. Proc Natl Acad Sci USA 83(17):6489–6493

Naldini L, Vigna E, Narsimhan RP, Gaudino G, Zarnegar R, Michalopoulos GK et al (1991) Hepatocyte growth factor (HGF) stimulates the tyrosine kinase activity of the receptor encoded by the proto-oncogene c-MET. Oncogene 6(4):501–504

Neuhaus P, Oustanina S, Loch T, Kruger M, Bober E, Dono R et al (2003) Reduced mobility of fibroblast growth factor (FGF)-deficient myoblasts might contribute to dystrophic changes in the musculature of FGF2/FGF6/mdx triple-mutant mice. Mol Cell Biochem 23(17):6037–6048

Nishi M, Yasue A, Nishimatu S, Nohno T, Yamaoka T, Itakura M et al (2002) A missense mutant myostatin causes hyperplasia without hypertrophy in the mouse muscle. Biochem Biophys Res Commun 293(1):247–251

Novitch BG, Spicer DB, Kim PS, Cheung WL, Lassar AB (1999) pRb is required for MEF2-dependent gene expression as well as cell-cycle arrest during skeletal muscle differentiation. Curr Biol 9(9):449–459

Olwin BB, Arthur K, Hannon K, Hein P, McFall A, Riley B et al (1994) Role of FGFs in skeletal muscle and limb development. Mol Reprod Dev 39(1):90–100

Olwin BB, Hall ZW (1985) Developmental regulation of laminin accumulation in the extracellular matrix of a mouse muscle cell line. Dev Biol 112(2):359–367

Olwin BB, Rapraeger A (1992) Repression of myogenic differentiation by aFGF, bFGF, and K-FGF is dependent on cellular heparan sulfate. J Cell Biol 118(3):631–639

Ong SH, Hadari YR, Gotoh N, Guy GR, Schlessinger J, Lax I (2001) Stimulation of phosphatidylinositol 3-kinase by fibroblast growth factor receptors is mediated by coordinated recruitment of multiple docking proteins. Proceedings of the National Academy of Sciences of the United States of America 98(11):6074–6079

Ornitz DM (2000) FGFs, heparan sulfate and FGFRs: complex interactions essential for development. BioEssays 22(2):108–112

Ornitz DM (2005) FGF signaling in the developing endochondral skeleton. Cytokine Growth Factor Rev 16(2):205–213

Ornitz DM, Itoh N (2001) Fibroblast growth factors. Genome Biology 2(3):reviews3005.1–3005.12

Ornitz DM, Xu J, Colvin JS, McEwen DG, MacArthur CA, Coulier F et al (1996) Receptor specificity of the fibroblast growth factor family. J Biol Chem 271(25):15292–15297

Ornitz DM, Yayon A, Flanagan JG, Svahn CM, Levi E, Leder P (1992) Heparin is required for cell-free binding of basic fibroblast growth factor to a soluble receptor and for mitogenesis in whole cells. Mol Cell Biol 12(1):240–247

Osses N, Brandan E (2002) ECM is required for skeletal muscle differentiation independently of muscle regulatory factor expression. Am J Physiol Cell Physiol 282(2):C383–394

Ostbye TK, Galloway TF, Nielsen C, Gabestad I, Bardal T, Andersen O (2001) The two myostatin genes of Atlantic salmon (Salmo salar) are expressed in a variety of tissues. Eur J Biochem 268(20):5249–5257

Oustanina S, Hause G, Braun T (2004) Pax7 directs postnatal renewal and propagation of myogenic satellite cells but not their specification. EMBO J 2004;23(16):3430–3439

Parsons SA, Millay DP, Sargent MA, McNally EM, Molkentin JD (2006) Age-dependent effect of myostatin blockade on disease severity in a murine model of limb-girdle muscular dystrophy. Am J Pathol 168(6):1975–1985

Patel K, Amthor H (2005) The function of Myostatin and strategies of Myostatin blockade–new hope for therapies aimed at promoting growth of skeletal muscle. Neuromuscul Disord 15(2):117–126

Pavlath GK, Thaloor D, Rando TA, Cheong M, English AW, Zheng B (1998) Heterogeneity among muscle precursor cells in adult skeletal muscles with differing regenerative capacities. Dev Dyn 212(4):495–508

Pellegrini L, Burke DF, von Delft F, Mulloy B, Blundell TL (2000) Crystal structure of fibroblast growth factor receptor ectodomain bound to ligand and heparin. Nature 407(6807):1029–1034
Perona R (2006) Cell signalling: growth factors and tyrosine kinase receptors. Clin Translational Oncol 8(2):77–82
Perry RLS, Parker MH, Rudnicki MA (2001) Activated MEK1 binds the nuclear MyoD transcriptional complex to repress transactivation. Mol Cell 8(2):291–301
Pizette S, Coulier F, Birnbaum D, deLapeyriere O (1996) FGF6 modulates the expression of fibroblast growth factor receptors and myogenic genes in muscle cells. Exp Cell Res 224(1):143–151
Plotnikov AN, Hubbard SR, Schlessinger J, Mohammadi M (2000) Crystal structures of two FGF-FGFR complexes reveal the determinants of ligand-receptor specificity. Cell 101(4):413–424
Ponzetto C, Bardelli A, Zhen Z, Maina F, dalla Zonca P, Giordano S et al (1994) A multifunctional docking site mediates signaling and transformation by the hepatocyte growth factor/scatter factor receptor family. Cell 77(2):261–271
Popovici C, Roubin R, Coulier F, Birnbaum D (2005) An evolutionary history of the FGF superfamily. BioEssays 27(8):849–857
Puri PL, Iezzi S, Stiegler P, Chen TT, Schiltz RL, Muscat GEO et al (2001) Class I histone deacetylases sequentially interact with MyoD and pRb during skeletal myogenesis. Mol Cell 8(4):885–897
Quinn LS, Steinmetz B, Maas A, Ong L, Kaleko M (1994) Type-1 insulin-like growth factor receptor overexpression produces dual effects on myoblast proliferation and differentiation. J Cell Physiol 159(3):387–398
Rando TA, Blau HM (1994) Primary mouse myoblast purification, characterization, and transplantation for cell-mediated gene therapy. J Cell Biol 125(6):1275–1287
Rapraeger AC, Krufka A, Olwin BB (1991) Requirement of heparan sulfate for bFGF-mediated fibroblast growth and myoblast differentiation. Science 252(5013):1705–1708
Reich A, Sapir A, Shilo B (1999) Sprouty is a general inhibitor of receptor tyrosine kinase signaling. Development 126(18):4139–4147
Reilly JF, Maher PA (2001) Importin {beta}-mediated nuclear import of fibroblast growth factor receptor: role in cell proliferation. J Cell Biol 152(6):1307–1312
Reisz-Porszasz S, Bhasin S, Artaza JN, Shen R, Sinha-Hikim I, Hogue A et al (2003) Lower skeletal muscle mass in male transgenic mice with muscle-specific overexpression of myostatin. Am J Physiol Endocrinol Metab 285(4):E876–E888
Rescan PY (1998) Identification of a fibroblast growth factor 6 (FGF6) gene in a non-mammalian vertebrate: continuous expression of FGF6 accompanies muscle fiber hyperplasia. Biochim Biophys Acta 1443(3):305–314
Rios R, Carneiro I, Arce VM, Devesa J (2001) Myostatin regulates cell survival during C2C12 myogenesis. Biochem Biophys Res Commun 280(2):561–566
Riquelme C, Larrain J, Schonherr E, Henriquez JP, Kresse H, Brandan E (2001) Antisense inhibition of decorin expression in myoblasts decreases cell responsiveness to transforming growth factor beta and accelerates skeletal muscle differentiation. J Biol Chem 276(5):3589–3596
Roghani M, Mansukhani A, Dell'Era P, Bellosta P, Basilico C, Rifkin DB et al (1994) Heparin increases the affinity of basic fibroblast growth factor for its receptor but is not required for binding. J BiolChem 269(6):3976–3984
Rosenthal SM, Cheng Z (1995) Opposing early and late effects of insulin-like growth factor I on differentiation and the cell cycle regulatory retinoblastoma protein in skeletal myoblasts. Proc Natl Acad Sci USA 92(22):10307–10311
Sachs M, Brohmann H, Zechner D, Muller TS, Hulsken J, Walther I et al (2000) Essential role of Gab1 for signaling by the c-Met receptor in vivo. J Cell Biol 150(6):1375–1384
Sakuma K, Watanabe K, Sano M, Uramoto I, Totsuka T (2000) Postnatal profiles of myogenic regulatory factors and the receptors of TGF-beta 2, LIF and IGF-I in the gastrocnemius and rectus femoris muscles of dy mouse. Acta Neuropathol 99(2):169–176
Sato K, Li Y, Foster W, Fukushima K, Badlani N, Adachi N et al (2003) Improvement of muscle healing through enhancement of muscle regeneration and prevention of fibrosis. Muscle Nerve 28(3):365–372

Scata KA, Bernard DW, Fox J, Swain JL (1999) FGF receptor availability regulates skeletal myogenesis. Exp Cell Res 250(1):10–21

Schaeper U, Gehring NH, Fuchs KP, Sachs M, Kempkes B, Birchmeier W (2000) Coupling of Gab1 to c-Met, Grb2, and Shp2 mediates biological responses. J Cell Biol 149(7):1419–1432

Schuelke M, Wagner KR, Stolz LE, Hubner C, Riebel T, Komen W et al (2004) Myostatin mutation associated with gross muscle hypertrophy in a child. New Engl J Med 350(26):2682–2688

Seale P, Sabourin LA, Girgis-Gabardo A, Mansouri A, Gruss P, Rudnicki MA (2000) Pax7 is required for the specification of myogenic satellite cells. Cell 102(6):777–786

Shaoul E, Reich-Slotky R, Berman B, Ron D (1995) Fibroblast growth factor receptors display both common and distinct signaling pathways. Oncogene 10(8):1553–1561

Shefer G, Van de Mark DP, Richardson JB, Yablonka-Reuveni Z (2006) Satellite-cell pool size does matter: Defining the myogenic potency of aging skeletal muscle. Dev Biol 294(1):50–66

Shefer G, Wleklinski-Lee M, Yablonka-Reuveni Z (2004) Skeletal muscle satellite cells can spontaneously enter an alternative mesenchymal pathway. J Cell Sci 117:5393–5404

Shefer G, Yablonka-Reuveni Z (2007) Reflections on lineage potential of skeletal muscle satellite cells: Do they sometimes go MAD? Crit Rev Eukaryot Gene Expr. 17(1):13–29

Shi Y, Massague J (2003) Mechanisms of TGF-beta signaling from cell membrane to the nucleus. Cell 113(6):685–700

Shim K, Minowada G, Coling DE, Martin GR (2005) Sprouty2, a mouse deafness gene, regulates cell fate decisions in the auditory sensory epithelium by antagonizing FGF signaling. Dev Cell 8(4):553–564

Siriett V, Platt L, Salerno MS, Ling N, Kambadur R, Sharma M (2006) Prolonged absence of myostatin reduces sarcopenia. J Cell Physiol 209(3):866–873

Sleeman M, Fraser J, McDonald M, Yuan S, White D, Grandison P et al (2001) Identification of a new fibroblast growth factor receptor, FGFR5. Gene 2001;271(2):171–182

Smith TH, Block NE, Rhodes SJ, Konieczny SF, Miller JB (1993) A unique pattern of expression of the four muscle regulatory factor proteins distinguishes somitic from embryonic, fetal and newborn mouse myogenic cells. Development 117(3):1125–1133

Smith TH, Kachinsky AM, Miller JB (1994) Somite subdomains, muscle cell origins, and the four muscle regulatory factor proteins. J Cell Biol 127(1):95–105

Song A, Wang Q, Goebl MG, Harrington MA (1998) Phosphorylation of nuclear MyoD is required for its rapid degradation. Mol Cell Biol 18(9):4994–4999

Sorokin A, Mohammadi M, Huang J, Schlessinger J (1994) Internalization of fibroblast growth factor receptor is inhibited by a point mutation at tyrosine 766. J Biol Chem 269(25):17056–17061

Souchelnytskyi S, ten Dijke P, Miyazono K, Heldin CH (1996) Phosphorylation of Ser165 in TGF-beta type I receptor modulates TGF-beta1-induced cellular responses. EMBO J 15(22):6231–6240

Spiller MP, Kambadur R, Jeanplong F, Thomas M, Martyn JK, Bass JJ et al (2002) The myostatin gene is a downstream target gene of basic Helix-Loop-Helix transcription factor MyoD. Mol Cell Biochem 22(20):7066–7082

Stark KL, McMahon JA, McMahon AP (1991) FGFR-4, a new member of the fibroblast growth factor receptor family, expressed in the definitive endoderm and skeletal muscle lineages of the mouse. Development 113(2):641–651

Stauber DJ, DiGabriele AD, Hendrickson WA (2000) Structural interactions of fibroblast growth factor receptor with its ligands. Proc Natl Acad Sci USA 97(1):49–54

Steinfeld R, Van Den Berghe H, David G (1996) Stimulation of fibroblast growth factor receptor-1 occupancy and signaling by cell surface-associated syndecans and glypican. J Cell Biol 133(2):405–416

Stoker M, Gherardi E, Perryman M, Gray J (1987) Scatter factor is a fibroblast-derived modulator of epithelial cell mobility. Nature 327(6119):239–242

Strack AM, Myers RW (2004) Modulation of metabolic syndrome by fibroblast growth factor 19 (FGF19)? Endocrinology 145(6):2591–2593

Suzuki J, Yamazaki Y, Li G, Kaziro Y, Koide H (2000) Involvement of Ras and Ral in chemotactic migration of skeletal myoblasts. Mol Cell Biol 20(13):4658–4665

Suzuki S, Yamanouchi K, Soeta C, Katakai Y, Harada R, Naito K et al (2002) Skeletal muscle injury induces hepatocyte growth factor expression in spleen. Biochem Biophys Res Commun 292(3):709–714

Szebenyi G, Fallon JF (1999) Fibroblast growth factors as multifunctional signaling factors. Int Rev Cytology 185:45–106

Tapscott SJ (2005) The circuitry of a master switch: Myod and the regulation of skeletal muscle gene transcription. Development 132(12):2685–2695

Tartakoff AM (1994) Signal transduction through growth factor receptors. Nagoya: Elsevier Science and Technology

Tashiro K, Hagiya M, Nishizawa T, Seki T, Shimonishi M, Shimizu S et al (1990) Deduced primary structure of rat hepatocyte growth factor and expression of the mRNA in rat tissues. Proc Natl Acad Sci USA 87(8):3200–3204

Tatsumi R, Anderson JE, Nevoret CJ, Halevy O, Allen RE (1998) HGF/SF is present in normal adult skeletal muscle and is capable of activating satellite cells. Dev Biol 194(1):114–128

Tatsumi R, Hattori A, Ikeuchi Y, Anderson JE, Allen RE (2002) Release of hepatocyte growth factor from mechanically stretched skeletal muscle satellite cells and role of ph and nitric oxide. Mol Biol Cell 13(8):2909–2918

Tatsumi R, Liu X, Pulido A, Morales M, Sakata T, Dial S et al (2006) Satellite cell activation in stretched skeletal muscle and the role of nitric oxide and hepatocyte growth factor. Am J Physiol Cell Physiol 290(6):C1487–1494

Tatsumi R, Sheehan SM, Iwasaki H, Hattori A, Allen RE (2001) Mechanical stretch induces activation of skeletal muscle satellite cells in vitro. Exp Cell Res 267(1):107–114

Templeton TJ, Hauschka SD (1992) FGF-mediated aspects of skeletal muscle growth and differentiation are controlled by a high affinity receptor, FGFR1. Dev Biol 154(1):169–181

Thomas M, Langley B, Berry C, Sharma M, Kirk S, Bass JJ et al (2000) Myostatin, a negative regulator of muscle growth, functions by inhibiting myoblast proliferation. J Biol Chem 275(51):40235–40243

Tortorella LL, Milasincic DJ, Pilch PF (2001) Critical proliferation-independent window for basic fibroblast growth factor repression of myogenesis via the p42/p44 MAPK signaling pathway. J Biol Chem 276(17):13709–13717

Tsang M, Dawid IB (2004) Promotion and attenuation of FGF signaling through the Ras-MAPK pathway. Science STKE 2004(228):pe17

Uruno T, Oki J, Ozawa K, Miyakawa K, Ueno H, Imamura T (1999) Distinct regulation of myoblast differentiation by intracellular and extracellular fibroblast growth factor-1. Growth Factors 17(2):93–113

Vainikka S, Joukov V, Klint P, Alitalo K (1996) Association of a 85-kDa Serine Kinase with Activated Fibroblast Growth Factor Receptor-4. J Biol Chem 271(3):1270–1273

Vainikka S, Joukov V, Wennstrom S, Bergman M, Pelicci PG, Alitalo K (1994) Signal transduction by fibroblast growth factor receptor-4 (FGFR-4). Comparison with FGFR-1. J Biol Chem 269(28):18320–18326

Velleman SG (2000) The role of the extracellular matrix in skeletal development. Poult Sci 79(7):985–989

Velleman SG, Liu C, Coy CS, McFarland DC (2006) Effects of glypican-1 on turkey skeletal muscle cell proliferation, differentiation and fibroblast growth factor 2 responsiveness. Dev Growth Differ 48(4):271–276

Venkatesha S, Toporsian M, Lam C, Hanai J, Mammoto T, Kim YM et al (2006) Soluble endoglin contributes to the pathogenesis of preeclampsia. Nature Med 12(6):642–649

Volonte D, Liu Y, Galbiati F (2004) The modulation of caveolin-1 expression controls satellite cell activation during muscle repair. FASEB J 19(2):237–239

Wagner KR, Liu X, Chang X, Allen RE (2005) Muscle regeneration in the prolonged absence of myostatin. Proc Natl Acad Sci USA 102(7):2519–2524

Wagner KR, McPherron AC, Winik N, Lee SJ (2002) Loss of myostatin attenuates severity of muscular dystrophy in mdx mice. Ann of Neurol 52(6):832–836

Wagner KR (2005) Muscle regeneration through myostatin inhibition. Curr Opin Rheumatol 17(6):720–724

Wang J, Walsh K (1996) Resistance to apoptosis conferred by Cdk inhibitors during myocyte differentiation. Science 273(5273):359–361

Wang JK, Gao G, Goldfarb M (1994) Fibroblast growth factor receptors have different signaling and mitogenic potentials. Mol Cell Biol 14(1):181–188

Wehling M, Cai B, Tidball JG (2000) Modulation of myostatin expression during modified muscle use. FASEB J 14(1):103–110

Weidner KM, Arakaki N, Hartmann G, Vandekerckhove J, Weingart S, Rieder H et al (1991) Evidence for the identity of human scatter factor and human hepatocyte growth factor. Proc Natl Acad Sci USA 88(16):7001–7005

Weinstein M, Xu X, Ohyama K, Deng CX (1998) FGFR-3 and FGFR-4 function cooperatively to direct alveogenesis in the murine lung. Development 125(18):3615–3623

Wickert L, Abiaka M, Bolkenius U, Gressner AM (2004) Corticosteroids stimulate selectively transforming growth factor (TGF)-[beta] receptor type III expression in transdifferentiating hepatic stellate cells. J Hepatol 40(1):69–76

Wilkie RS, O'Neill IE, Butterwith SC, Duclos MJ, Goddard C (1995) Regulation of chick muscle satellite cells by fibroblast growth factors: interaction with insulin-like growth factor-I and heparin. Growth Regul 5(1):18–27

Wozniak AC, Kong J, Bock E, Pilipowicz O, Anderson JE (2005) Signaling satellite-cell activation in skeletal muscle: Markers, models, stretch, and potential alternate pathways. Muscle Nerve 31(3):283–300

Wrana JL, Attisano L, Wieser R, Ventura F, Massague J (1994) Mechanism of activation of the TGF-beta receptor. Nature 370(6488):341–347

Wright TJ, Ladher R, McWhirter J, Murre C, Schoenwolf GC, Mansour SL (2004) Mouse FGF15 is the ortholog of human and chick FGF19, but is not uniquely required for otic induction. Dev Biol 269(1):264–275

Wyzykowski JC, Winata TI, Mitin N, Taparowsky EJ, Konieczny SF (2002) Identification of novel MyoD gene targets in proliferating myogenic stem cells. Mol Cell Biol 22(17):6199–6208

Xie MH, Holcomb I, Deuel B, Dowd P, Huang A, Vagts A et al (1999) FGF-19, a novel fibroblast growth factor with unique specificity for fgfr4. Cytokine 11(10):729–735

Xu X, Weinstein M, Li C, Deng C-X (1999) Fibroblast growth factor receptors (FGFRs) and their roles in limb development. Cell Tissue Res 296(1):33–43

Yablonka-Reuveni Z (2007) Myostatin blockade: a new way to enhance skeletal muscle repair in old age? Mol Ther 15(8):1407–1409

Yablonka-Reuveni Z, Anderson JE (2006) Satellite cells from dystrophic (Mdx) mice display accelerated differentiation in primary cultures and in isolated myofibers. Dev Dyn 235(1):203–212

Yablonka-Reuveni Z, Balestreri TM, Bowen-Pope DF (1990) Regulation of proliferation and differentiation of myoblasts derived from adult mouse skeletal muscle by specific isoforms of PDGF. J Cell Biol 111(4):1623–1629

Yablonka-Reuveni Z, Day K, Vine A, Shefer G (2007) Defining the transcriptional signature of skeletal muscle stem cells. J Anim Sci [Epub ahead of print; Sept 18, 2007]

Yablonka-Reuveni Z, Quinn LS, Nameroff M (1987) Isolation and clonal analysis of satellite cells from chicken pectoralis muscle. Dev Biol 119(1):252–259

Yablonka-Reuveni Z, Rivera AJ (1994) Temporal expression of regulatory and structural muscle proteins during myogenesis of satellite cells on isolated adult rat fibers. Dev Biol 164(2):588–603

Yablonka-Reuveni Z, Rivera AJ (1997a) Influence of PDGF-BB on proliferation and transition through the MyoD-myogenin-MEF2A expression program during myogenesis in mouse C2 myoblasts. Growth Factors 15(1):1–27

Yablonka-Reuveni Z, Rivera AJ (1997b) Proliferative dynamics and the role of FGF2 during myogenesis of rat satellite cells on isolated fibers. Basic Appl Myology 7(3&4):189–202

Yablonka-Reuveni Z, Rudnicki MA, Rivera AJ, Primig M, Anderson JE, Natanson P (1999a) The transition from proliferation to differentiation is delayed in satellite cells from mice lacking MyoD. Dev Biol 210(2):440–455

Yablonka-Reuveni Z, Seger R, Rivera AJ (1999b) Fibroblast growth factor promotes recruitment of skeletal muscle satellite cells in young and old rats. J Histochem Cytochem 47(1):23–42

Yablonka-Reuveni Z, Seifert RA (1993) Proliferation of chicken myoblasts is regulated by specific isoforms of platelet-derived growth factor: Evidence for differences between myoblasts from mid and late stages of embryogenesis. Dev Biol 156(2):307–318

Yablonka-Reuveni Z (2004) Isolation and culture of myogenic stem cells. In: Lanza R, Blau H, Melton D, Moore M, Thomas ED, Verfaillie C, et al., (eds) Handbook of Stem Cells – Vol 2: Adult and Fetal Stem Cells. San Diego: Elsevier – Academic Press, pp. 571–580

Yaffe D, Saxel O (1977) A myogenic cell line with altered serum requirements for differentiation. Differentiation 7(3):159–166

Yaffe D (1969) Cellular aspects of muscle differentiation in vitro. Curr Top Dev Biol 4:37–77

Yamada M, Tatsumi R, Kikuiri T, Okamoto S, Nonoshita S, Mizunoya W et al al (2006) Matrix metalloproteinases are involved in mechanical stretch-induced activation of skeletal muscle satellite cells. Muscle & Nerve; Epub ahead of print:NA

Yang J, Ratovitski T, Brady JP, Solomon MB, Wells KD, Wall RJ (2001) Expression of myostatin pro domain results in muscular transgenic mice. Mol Reprod Dev 60(3):351–361

Yang J, Zhao B (2006) Postnatal expression of myostatin propeptide cDNA maintained high muscle growth and normal adipose tissue mass in transgenic mice fed a high-fat diet. Mol Reprod Dev 73(4):462–469

Yayon A, Klagsbrun M, Esko JD, Leder P, Ornitz DM (1991) Cell surface, heparin-like molecules are required for binding of basic fibroblast growth factor to its high affinity receptor. Cell 64(4):841–848

Yoshida S, Fujisawa-Sehara A, Taki T, Arai K, Nabeshima Y (1996) Lysophosphatidic acid and bFGF control different modes in proliferating myoblasts. J Cell Biol 132(1):181–193

Zammit PS, Golding JP, Nagata Y, Hudon V, Partridge TA, Beauchamp JR (2004) Muscle satellite cells adopt divergent fates: A mechanism for self-renewal? J Cell Biol 166(3):347–357

Zammit PS, Partridge TA, Yablonka-Reuveni Z (2006) The skeletal muscle satellite cell: A stem cell that came in from the cold. J Histochem Cytochem 54(11):1177–1191

Zeng C, Pesall JE, Gilkerson KK, McFarland DC (2002) The effect of hepatocyte growth factor on turkey satellite cell proliferation and differentiation. Poult Sci 81(8):1191–1198

Zhao B, Wall RJ, Yang J (2005) Transgenic expression of myostatin propeptide prevents diet-induced obesity and insulin resistance. Biochem Biophys Res Commun 337(1):248–255

Zhao P, Caretti G, Mitchell S, McKeehan WL, Boskey AL, Pachman LM et al (2006) Fgfr4 Is Required for Effective Muscle Regeneration in vivo: Delineation of a MyoD-Tead2-Fgfr4 Transcriptional Pathway. J Biol Chem 281(1):429–438

Zhao P, Hoffman EP (2004) Embryonic myogenesis pathways in muscle regeneration. Dev Dyn 229(2):380–392

Zhou L, Porter JD, Cheng G, Gong B, Hatala DA, Merriam AP et al (2006) Temporal and spatial mRNA expression patterns of TGF-beta1, 2, 3 and TbetaRI, II, III in skeletal muscles of mdx mice. Neuromuscul Disord 16(1):32–38

Zimmers TA, Davies MV, Koniaris LG, Haynes P, Esquela AF, Tomkinson KN et al (2002) Induction of cachexia in mice by systemically administered myostatin. Science 296(5572):1486–1488

Ziv I, Fuchs Y, Preger E, Shabtay A, Harduf H, Zilpa T et al (2006) The human Sef-a isoform utilizes different mechanisms to regulate FGFR signaling pathways and subsequent cell fate. J Biol Chem; Epub ahead of print:M607327200

Zuber ME, Zhou Z, Burrus LW, Olwin BB (1997) Cysteine-rich FGF receptor regulates intracellular FGF-1 and FGF-2 levels. J Cell Physiol 170(3):217–227

CHAPTER 7

RELAYING THE SIGNAL DURING MYOGENESIS: INTRACELLULAR MEDIATORS AND TARGETS

RODDY S. O'CONNOR[1,2] AND GRACE K. PAVLATH[2]

[1] *Graduate Program in Molecular and Systems Pharmacology*

[2] *Department of Pharmacology, Emory University, Atlanta, GA 30322*

1. INTRODUCTION

Myofibers, the cellular units of skeletal muscle, contain multiple nuclei within a continuous cytoplasm. As myonuclei are terminally differentiated, skeletal muscle repair following injury is dependent upon a local pool of satellite cells. Satellite cells are myogenic stem cells situated beneath the basal lamina that surrounds each myofiber (Charge and Rudnicki, 2004). In response to injury, satellite cells proliferate and their progeny myoblasts migrate to sites of damage. Eventually myoblasts withdraw from the cell cycle and undergo differentiation leading to their fusion either with each other to form nascent myofibers or with existing myofibers, and muscle architecture is restored.

Signal transduction regulates all aspects of satellite cell behavior. During regeneration, extracellular stimuli are transmitted intracellularly through diverse classes of receptors. The generation and/or release of second messenger molecules relays the signal through molecules such as downstream kinases, phosphatases and ultimately transcription factors. Additionally, chromatin remodeling enzymes promote posttranslational modification of the transcriptional machinery regulating gene transcription both positively and negatively during regeneration.

Much of the research on signaling pathways in myogenesis has been performed using permanent cell lines. As cell lines do not always recapitulate events occurring in primary muscle cells, we have mainly limited our review of the literature to those papers that deal with primary myoblasts in vitro or satellite cells in vivo. Unfortunately, defining the role of specific signaling pathways in vivo has been difficult. Developmental complications in generating loss of function knock-out

S. Schiaffino and T. Partridge (eds.), Skeletal Muscle Repair and Regeneration, 145–162.

mice combined with a lack of conditional transgenics has limited our understanding of signaling during regeneration. However, the use of pharmacologic inhibitors and plasmid based dominant-negative and constitutively active constructs has provided important insight. Additionally, immunocytochemical analyses have demonstrated the activation of key signaling molecules during various phases of regeneration.

In this review, we focus on signaling pathways where the literature is reasonably extensive at the in vitro and in vivo levels. We discuss the role of G proteins in myogenesis, including prostaglandin-dependent activation of G protein-coupled receptors. As calcium is a major second messenger in myogenesis, we review the roles of multiple calcium-dependent pathways that include kinases, phosphatases, transcription factors and proteolytic enzymes. The role of p38 MAPK during myogenesis is also discussed. Finally, the role of microRNA in the regulation of gene expression during myogenesis is reviewed with particular emphasis on future work in this exciting field.

2. GPCR-DEPENDENT SIGNALING

G protein-coupled receptors (GPCR) are a large group of cell surface receptors that respond to hormones, neurotransmitters and cytokines promoting diverse cellular responses (Hubbard and Hepler, 2006). Agonist-induced activation of GPCRs promotes an interaction with heterotrimeric G proteins that are guanine nucleotide binding proteins containing α, β and γ subunits. Guanine nucleotide exchange factors (GEFs) exchange GDP for GTP on Gα leading to subunit dissociation and interactions with effector molecules.

GPCRs couple to multiple families of G proteins defined by their Gα subunit. Gsα and Giα activate and inhibit adenylyl cyclases, respectively. G12/13α activates GDP/GTP exchange factors called Rho family guanine nucleotide exchange factors (RhoGEFs). Direct effectors for Goα are unknown. The Gqα family is of particular interest during myogenesis. Upon activation by Gqα, phospholipase C-β catalyzes the hydrolysis of membrane phospholipids that increase intracellular calcium in an inositol 1,4,5 triphosphate (IP3)-dependent manner. Calcium regulates multiple phases of myogenesis including differentiation and fusion as will be discussed later.

The intensity and duration of G protein signaling is GTP-dependent and regulated by the intrinsic GTPase activity of the Gα subunit and effector molecules. Hydrolysis of GTP results in reassociation of the heterotrimeric G protein and cessation of signaling. RGS proteins are a family of proteins containing a conserved 120 amino acid "RGS domain" that modulate G protein signaling. RGS proteins bind to Gα subunits and accelerate the rate of GTP hydrolysis thus regulating the duration of G protein signaling (Hollinger and Hepler, 2002). G12/13α proteins activate RhoGEFs that stimulate RhoA, a monomeric GTPase that regulates both myoblast differentiation (Castellani et al., 2006) and fusion (Charrasse et al., 2006). RhoGEFs contain the canonical RGS domain and may therefore act as both modulators (accelerate the GTPase activity of G12/13α) and effectors (targets of G12/13α stimulating RhoA activity) of G protein-dependent signaling. The Rho family guanine exchange factor, GEFT, is highly expressed in skeletal muscle and upregulated during regeneration in vivo (Bryan et al., 2005). Adenoviral expression of GEFT in regenerating skeletal muscle enhanced myofiber formation and myofiber size measured 15 days after injury (Bryan et al., 2005).

Several studies have demonstrated roles for G protein-dependent signaling in myogenesis (Fedorov et al., 1998; Vandenburgh et al., 1995). Pertussis toxin is an exotoxin that selectively inhibits Gi/oα protein-dependent signaling. Pertussis toxin induces ADP ribosylation of heteromeric Gi/o proteins preventing guanine nucleotide exchange and subunit dissociation. Myofiber growth following mechanical stretch was sensitive to pertussis toxin (Vandenburgh et al., 1995). Moreover, stretch-induced prostaglandin ($PGF_{2\alpha}$) synthesis was decreased following pertussis toxin treatment suggesting Gi/oα-dependent signaling regulates cyclooxygenase (COX) activity during muscle growth.

Growth factors promote myoblast proliferation through activation of the MAP kinase (MAPK) ERK1/2 (Bennett and Tonks, 1997). Pertussis toxin induced a dose-dependent inhibition of myoblast proliferation suggesting cross talk between G protein and growth factor receptor signaling during myoblast proliferation (Fedorov et al., 1998). In addition, pertussis toxin promoted differentiation of myoblasts cultured in high serum conditions containing FGF-2. Ectopic expression of Gβγ subunits prevented differentiation of myoblasts cultured in high serum, FGF-2 and pertussis toxin. In complementary assays, a peptide inhibitor of Gβγ subunits induced myoblast differentiation in the presence of FGF. Taken together, Gβγ signaling activates MAPK illustrating 1) crosstalk/convergence between growth factor (MAPK) and G protein signaling (at the level of Gβγ) during myoblast proliferation and 2) all G protein subunits participate in signal transduction during myogenesis.

2.1 Prostaglandins

Prostaglandins are membrane-derived lipid signaling molecules synthesized by regenerating muscle (McArdle et al., 1994; Palmer et al., 1983). Prostaglandins regulate multiple cellular responses through selective activation of G protein-coupled receptors (Funk, 2001; Narumiya et al., 1999). Prostaglandins regulate myoblast proliferation (Otis et al., 2005; Zalin, 1977; Zalin, 1987), differentiation (Schutzle et al., 1984; Zalin, 1987) and fusion (Entwistle et al., 1986; Horsley and Pavlath, 2003; Shen et al., 2006; Zalin, 1977). Muscle cells synthesize both $PGF_{2\alpha}$ and PGE_2 (Otis et al., 2005; Shen et al., 2005). $PGF_{2\alpha}$ and PGE_2 bind with high affinity (Narumiya et al., 1999) to the FP and EP1-4 receptors, respectively. Both FP and EP1 receptors signal through Gqα proteins. $PGF_{2\alpha}$ activated the calcium-sensitive phosphatase calcineurin and downstream transcription factor NFATc2 in nascent myotubes leading to the fusion of myoblasts with newly forming myotubes (Horsley and Pavlath, 2003). Fluoprostenol, a specific agonist of the FP receptor (Griffin et al., 1999; Griffin et al., 1998), elicited similar results demonstrating selective activation of the FP receptor during myoblast fusion. Additionally, AL-8810, a competitive antagonist the FP receptor, decreased myotube fusion. These data demonstrate a role for endogenous prostaglandins in myoblast fusion in an FP receptor-dependent manner.

Cyclooxygenase (COX) enzymes catalyze the production of prostaglandins from arachidonic acid. In contrast to the ubiquitous isoform COX-1, COX-2 was upregulated during muscle regeneration suggesting a role in regulating myogenesis (Bondesen et al., 2004). Moreover, pharmacologic inhibition of COX-2 impaired muscle regeneration (Bondesen et al., 2004). Selective inhibition of COX-2 with the compound SC-236 attenuated the growth of regenerating myofibers 7 and 14 days following injury. Both MyoD expression and myoblast number were decreased in mice treated with SC-236 suggesting COX-2 derived prostaglandins are required for myoblast proliferation and/or survival during regeneration. Complementary studies in vivo demonstrated that pharmacologic inhibition of COX-2 with NS-398 also decreased myofiber formation and size during regeneration (Shen et al., 2005). Similarily, myofiber formation (Shen et al., 2006) and myofiber growth (Bondesen et al., 2004; Shen et al., 2006) were decreased after injury in regenerating muscles of COX-$2^{-/-}$ mice. Additional studies demonstrated inhibition of myoblast fusion following treatment with the non-selective pharmacologic antagonist of COX, indomethacin in vitro (Entwistle et al., 1986; Zalin, 1977). The addition of PGE1 returned myoblast fusion to normal values suggesting prostanoid signaling through Gsα also regulates myoblast fusion. Taken together, these data suggest roles for both prostaglandins and their G protein-coupled receptors in myoblast proliferation, differentiation and fusion during regeneration.

3. CALCIUM

3.1 Requirement for Calcium in Myogenesis

Receptor coupling leads to the release of second messengers that propagate and amplify the signal through the cell. Calcium is a highly regulated second messenger involved in multiple phases of myogenesis. Myogenesis is dependent on both extracellular (Przybylski et al., 1994; Salzberg et al., 1995; Shainberg et al., 1969) and intracellular calcium (Constantin et al., 1996; David et al., 1981; Przybylski et al., 1989). The requirement for extracellular calcium may be related to extracellular proteins that require calcium for normal activity during either myoblast adhesion or fusion (Knudsen, 1985; Knudsen et al., 1990). In addition, extracellular calcium may be important in the fusion of lipid bilayers (Papahadjopoulos et al., 1990) that occurs during myogenic cell fusion. Increases in intracellular calcium occur during myogenesis and activate various signaling pathways as discussed below. Increases in intracellular calcium were observed as early as 20 hours in differentiation medium (Constantin et al., 1996) and also occured prior to cell fusion (David et al., 1981; Liu et al., 2003).

3.2 Mechanisms Leading to Increased Intracellular Calcium

Increases in intracellular calcium during myogenesis could arise from either influxes from the extracellular environment or release from intracellular stores (e.g. sarcoplasmic reticulum). Influx from extracellular sources occurs mainly through voltage-gated channels, ligand-gated channels and store-operated channels. Voltage-gated ion channels are transmembrane protein complexes that regulate calcium permeability in response to changes in membrane potential. In contrast, ligand-gated channels increase calcium permeability in an agonist-dependent manner. Store-operated channels are calcium permeable integral membrane proteins that are activated following depletion of intracellular calcium stores. Voltage-gated channels have been extensively studied in myoblast differentiation. Calcium influx through T-type calcium channels was required for the earliest steps in differentiation of human myoblasts (Bijlenga et al., 2000). This calcium influx arose from membrane hyperpolarization due to Kir2.1 potassium channel activity (Fischer-Lougheed et al., 2001; Liu et al., 1998). Blocking this hyperpolarization step inhibited myoblast differentiation. T-type channels are not involved in differentiation of primary mouse myoblasts suggesting that different mechanisms may be responsible for increases in intracellular calcium in different species (Bidaud et al., 2006). Other sources include L-type voltage-gated calcium channels (David et al., 1981; Entwistle et al., 1988a), stretch activated cation channels (Shin et al., 1996) and ligand-gated channels such as nicotinic acetylcholine receptors (Constantin et al., 1996; Entwistle et al., 1988b) and P2X ionotropic purinergic receptors (Cseri et al., 2002; Szigeti et al., 2006).

Release of calcium from intracellular stores also occurs during myogenesis through activation of IP3 receptors (IP3R) and ryanodine receptors (RyR). Both are ligand-gated ion channels located on the surface of the sarcoplasmic reticulum. A major signal for release of calcium from intracellular stores is IP3. Activation of various GPCR receptors such as prostaglandin receptors leads to generation of IP3 (David et al., 1981; Entwistle et al., 1988a). The cellular IP3 concentration and the activity of phospholipase C, the enzyme responsible for IP3 production, increase during differentiation (Carrasco et al., 1997). RyR do not appear to be involved in either human or fetal mouse differentiation (Pisaniello et al., 2003).

Recent evidence suggests heterogeneity among human myoblasts as to the mechanism by which intracellular calcium is increased within individual cells (Arnaudeau et al., 2006). Three different mechanisms of increasing intracellular calcium at the onset of differentiation were observed among clones of human myoblasts: T-type calcium channels, store operated calcium channels and release from intracellular calcium stores through IP3R. Furthermore, myoblast clones could switch between different mechanisms indicating plasticity in the choice of calcium source. Such plasticity may ensure that the critical increase in cytoplasmic calcium necessary for differentiation occurs regardless of environmental conditions.

3.3 Intracellular Signaling Pathways Regulated by Calcium

3.3.1 Protein kinase C

Protein kinase C (PKC) enzymes are a family of serine/threonine kinases involved in the control of cell proliferation and differentiation of many cell types. The family consists of three subgroups (conventional, novel and atypical) that differ in their cellular localization and requirements for the cofactors calcium and diacylglycerol (DAG). The conventional subfamily contains the calcium-sensitive PKCs (cPKC): α, βI, βII and γ. Activation of conventional PKC isozymes is regulated by receptor-mediated production of DAG through phospholipase C or D pathways. Activation of PKC involves a partial translocation from the cytoplasm to membrane fractions.

PKC has been implicated in several steps during myogenesis. Total PKC activity was high in proliferating myoblasts and decreased in differentiating cultures (Capiati et al., 1999) (Adamo et al., 1989). However, total PKC activity reflects the activity of all PKC isoforms present not just the calcium-sensitive ones. Isoforms of all three PKC subgroups are expressed in myoblasts (Boczan et al., 2000; Capiati et al., 1999; Disatnik et al., 2002). Little work has been done on the role of individual PKC isoforms in myogenesis thus the relative importance of each isoform is unknown. However, PKCα is expressed in myoblasts (Boczan et al., 2000; Capiati et al., 1999). Downregulation of PKCα either by treatment with the phorbol ester TPA (Capiati et al., 1999) or with antisense oligonucleotides (Capiati et al., 2000) inhibited myoblast proliferation. In contrast, the expression of PKC β (Capiati et al., 1999) and γ (Boczan et al., 2000) increases as myoblasts differentiate but the specific role of these PKC isoforms is unknown. Additional studies using isoform-specific inhibitors suggest a role for PKCα in myoblast spreading after attachment to fibronectin (Disatnik et al., 2002).

Only one study to date has analyzed PKC during muscle regeneration in vivo (Moraczewski et al., 2002). After crush injury of rat EDL and soleus muscles, PKC activity initially decreased during the first 3 days but subsequently increased during regeneration reaching levels that were 40% higher after 2 weeks of regeneration. This increase correlated with the increase in size of regenerating myofibers. Multiple PKC isoforms were sequentially expressed in the regenerating muscles with some isoforms being differentially expressed between the regenerating soleus and EDL muscles. The role of specific PKC isoforms during muscle regeneration is unknown.

Several PKC substrates in muscle cells are known that could account for the effects of PKC on myogenesis. The bHLH motif of myogenic regulatory factors contains a consensus phosphorylation sequence for PKC. Indeed, both MyoD (Baudier et al., 1995) and myogenin (Li et al., 1992) are targets of PKC signaling although the specific isoform responsible for this activity is unknown. Phosphorylation of myogenin by PKC resulted in a loss of its DNA binding ability and hence repression of myogenic differentiation (Li et al., 1992). Myristoylated alanine-rich C kinase substrate (MARCKS), which regulates actin dynamics, is another target of PKC signaling in muscle (Disatnik et al., 2002; Kim et al., 2000; Poussard et al., 2001). The actin cross-linking activity of MARCKS was preventing by phosphorylation with PKC (Hartwig et al., 1992). MARCKS appears to play a role in both myoblast attachment and spreading (Disatnik et al., 2002) as well as myoblast fusion (Kim et al., 2000). Phosphorylation levels of MARCKS change during myogenesis (Kim et al., 2002). MARCKS proteolysis increased with cell fusion and its proteolysis was dependent on its phosphorylation by PKCα (Dulong et al., 2004).

3.3.2 Calcineurin and NFAT

Numerous studies have examined the role of calcineurin both in muscle differentiation in vitro as well as in muscle regeneration. Calcineurin activity increases during differentiation and is required for the initiation of differentiation (Friday et al., 2000). Expression of a constitutively active form of calcineurin stimulated differentiation even in the presence of high serum, whereas a dominant-negative form of calcineurin inhibited myogenesis. Activation of calcineurin led to downstream activation of the myogenic regulatory factors myogenin and MEF2 (Friday et al., 2003; Konig et al., 2006; Konig et al., 2004; Xu et al., 2002).

Calcineurin expression and activity increase early in regeneration and remain elevated for approximately 2 weeks (Armand et al., 2005; Sakuma et al., 2003). Muscle regeneration was delayed when rodents were treated with a calcineurin inhibitor starting at the time of injury (Abbott et al., 1998; Miyabara et al., 2005; Sakuma et al., 2005) but not 3 days later (Serrano et al., 2001). These studies support the in vitro requirement for calcineurin activity in the initiation of muscle differentiation but do not rule out additional requirements for calcineurin in the inflammatory response, which is critical for muscle regeneration. Calcineurin may also be required prior to differentiation as calcineurin increased expression of Rad, an inhibitor of L type calcium channels, expressed in myogenic progenitor cells during early phases of regeneration (Hawke et al., 2006). Inhibition of L- type calcium channels in the C2 muscle cell line inhibited muscle differentiation (Porter et al., 2002) suggesting that calcineurin signaling may be necessary to help promote myoblast proliferation early in regeneration.

A well-studied calcineurin substrate is the nuclear factor of activated T cells (NFAT) family of transcription factors that consists of 4 calcium sensitive members, NFATc1-c4. Under basal conditions, NFAT proteins are phosphorylated and localized to the cytoplasm. During periods of sustained elevations in intracellular calcium, the phosphatase calcineurin becomes activated and dephosphorylates NFAT proteins, thereby allowing their nuclear translocation. Once in the nucleus, NFAT proteins, in association with other transcription factors, bind to a consensus DNA sequence and activate gene transcription. Rephosphorylation of NFAT proteins by several kinases results in nuclear export.

NFAT5 is the newest member of the NFAT family of transcription factors based upon homology of its DNA binding domain. The DNA binding domain of NFAT5 exhibits approximately 40% homology to the NFATc1-c4 family members (Lopez-Rodriguez et al., 1999). A number of key differences exist between NFAT5 and the NFATc1-c4 isoforms described above in terms of their activation and DNA binding.

NFAT5 is constitutively nuclear in many cell types (Lopez-Rodriguez et al., 1999) regardless of whether calcineurin is active. NFAT5 binds to DNA as a dimer (Lee et al., 2002; Lopez-Rodriguez et al., 2001; Lopez-Rodriguez et al., 1999). The DNA binding sequences recognized by NFATc1-c4 vs. NFAT5, although similar in a core GGAA recognition sequence, otherwise differ in the complete consensus sequence necessary for optimal binding (Lopez-Rodriguez et al., 1999). NFAT5 is activated in response to hypertonicity. It is expressed in virtually all tissues, including many never normally exposed to extremes of hypertonicity (Dalski et al., 2000; Lopez-Rodriguez et al., 1999; Miyakawa et al., 1999; Trama et al., 2000; Zhang et al., 2003). In isotonic environments NFAT5 can be activated by amino acid depletion (Franchi-Gazzola et al., 2001; Trama et al., 2002) as well as engagement of extracellular receptors such as the T cell receptor (Trama et al., 2000) and $\alpha 6\beta 4$ integrin (Jauliac et al., 2002). In contrast, NFATc1-c4 are not known to be activated by hypertonicity. NFAT5 activity is regulated at multiple stages: nuclear translocation (Cha et al., 2001; Dahl et al., 2001; Miyakawa et al., 1999; Woo et al., 2000) increase of its transcriptional activity (Ferraris et al., 2002) and increased expression (Jauliac et al., 2002; Miyakawa et al., 1999; Trama et al., 2000) Calcineurin activity is required for increases in NFAT5 expression (Trama et al., 2000).

Primary skeletal muscle cells express four isoforms of NFAT: NFATc1-c3 (Abbott et al., 1998) and NFAT5 (O'Connor et al., 2007). The expression levels of NFATc1-c3 are consistent throughout all stages of myogenesis (Abbott et al., 1998). However, specific NFAT isoforms translocated to the nucleus in response to a calcium signal at specific stages of myogenesis. The complexity of NFAT signaling is further highlighted by the fact that mobilization of the IP3R calcium pool caused nuclear entry of NFATc1 in myoblasts but nuclear exit in myotubes (Stiber et al., 2005). Furthermore, calcium transients evoked by activation of RyR led to nuclear entry of NFATc1 in myotubes. Thus, skeletal muscle cells can use discrete sources of intracellular calcium to activate NFAT signaling. In contrast, NFAT5 is expressed at higher levels in myotubes and is constitutively nuclear at all stages of myogenesis (O'Connor et al., 2007). Together, these data suggest different NFAT isoforms have non-redundant roles in skeletal muscle.

Analyses of myogenesis in various NFAT null mice showed distinct roles for each NFAT isoform. Muscle regeneration in NFATC2$^{-/-}$ mice was characterized by normal formation of regenerating myofibers but these myofibers were unable to grow at the same rate as wild type mice. NFATc2$^{-/-}$ myoblasts formed small multinucleated muscle cells in vitro due to a defect in the recruitment and/or fusion of myogenic cells with nascent multinucleated muscle cells (Horsley et al., 2001). Subsequent experiments demonstrated NFATc2 was required for the production of IL4 in muscle cells during the fusion of myoblasts with myotubes. IL4$^{-/-}$ myoblasts were also defective in the recruitment of myogenic cells with nascent myotubes (Horsley, 2003). Furthermore, IL4 was required for the fusion of mesenchymal stem cells with muscle (Schulze et al., 2005). NFATc2 was required for the increase in myonuclear number due to prostaglandin $F_{2\alpha}$ (Horsley and Pavlath, 2003) as well as growth hormone (Sotiropoulos et al., 2006). Recent evidence suggests

NFATc2 may be downstream of a c-src pathway in muscle (Fornaro et al., 2006). Fornaro et al demonstrated that the phosphatase SHP-2 stimulates c-src leading to activation of NFAT and subsequent fusion of myoblasts with myotubes. IL-4 was decreased in SHP-2 null muscle cells. These results support the idea that NFATc2 is a target for positive regulation by SHP-2 in skeletal muscle. However, whether the phenotype of MCK-SHP-2 null mice is a consequence of disrupting NFATc2 solely as opposed to other NFAT family members needs to be formally determined. NFAT5$^{+/-}$ mice displayed a defect in muscle regeneration with fewer myofibers formed at early times after injury but these myofibers were similar in size to wild type (O'Connor et al., 2007). NFAT5 likely has a muscle-intrinsic function as inhibition of NFAT5 transcriptional activity caused both a migratory and differentiation defect in cultured myoblasts. The secreted cysteine-rich CCN (connective tissue growth factor) matrix protein Cyr61 is a target gene of NFAT5 signaling that regulates myoblast migration.

In contrast, less is known about the roles of NFATc3 and NFATc1 in muscle. NFATc3$^{-/-}$ mice displayed embryonic defects in the formation of primary myofibers, the first multinucleated muscle cell (Kegley et al., 2001), but no difference in the size of regenerating myofibers after injury. The mechanism by which defects in formation of primary myofibers occurred is unknown. NFATc1 may regulate a subpopulation of myogenic cells called reserve cells (Yoshida et al., 1998). These cells remain unfused in cultures of multinucleated muscle cells and express the muscle regulatory protein, myf-5. NFATc1 enhanced the expression of myf-5 in these cells, suggesting that NFAT may regulate properties of reserve cells (Friday and Pavlath, 2001). In vivo, the levels of activated NFATc1 increased post injury (Sakuma et al., 2003). Loss of function studies have not been performed with NFATc1as NFATc1$^{-/-}$ mice are embryonic lethal (de la Pompa et al., 1998). However, recent transgenic models in which the embryonic lethality of NFATc1 was rescued (Winslow et al., 2006) will allow further studies on the role of this NFAT isoform in myogenesis.

3.3.3 Calpain

Calpains are calcium-activated intracellular cysteine proteases (Huang and Wang, 2001). Calpain activity is activated by a variety of factors, including calcium, phospholipids and phosphorylation by the MAP kinase ERK (Glading et al., 2004). Conversely, calpastatin is a specific intracellular inhibitor of calpain. The calpain family is composed of two ubiquitously expressed members (μ-calpain and m-calpain). Skeletal muscle also expresses a muscle-specific calpain (calpain 3), which is the major isoform expressed in adult skeletal muscles. Alterations in calpain 3 cause limb-girdle muscular dystrophy type 2A (Richard et al., 1995).

Calpain activity is modulated during muscle regeneration. Calpain activity was elevated during the time of extensive myofiber formation (Duguez et al., 2003; Zimowska et al., 2001). M-calpain immunostaining was absent in quiescent satellite cells but was present in proliferating myoblasts after injury (Raynaud et al., 2004). Both μ and m-calpain were present in regenerating myofibers by immunostaining

(Zimowska et al., 2001). Similarly during in vitro muscle differentiation the activity of calpains was increased as myotubes are formed (Elamrani et al., 1995; Joffroy et al., 2000; Stockholm et al., 1999). Calpain activity is important for myoblast fusion. Addition of a calpain inhibitor decreased fusion in myoblasts (Elamrani et al., 1995; Joffroy et al., 2000) whereas preventing synthesis of calpastatin, the endogenous calpain inhibitor, enhanced fusion (Dourdin et al., 1999). M-calpain appears to be particularly important for the fusion process as either antibodies (Dourdin et al., 1997) or antisense oligodeoxyribonucleotides (Balcerzak et al., 1995) to m-calpain decreased fusion. Calpain activity may be required for alteration of matrix proteins (Dourdin et al., 1999; Dourdin et al., 1997), reorganization of the cytoskeleton (Dourdin et al., 1999; Elamrani et al., 1995) and changes in intracellular proteins. Indeed, calpain 3 acts during myogenesis to control the levels of membrane associated μ-catenin and M-cadherin during myogenesis (Kramerova et al., 2006). Furthermore, calpain 3 knockout myotubes displayed an increased number of myonuclei per myotube (Kramerova et al., 2004).

4. p38 MAPK

MAP kinases (p38, ERK and JNK) are serine/threonine kinases that transduce intracellular signals following activation of either growth factor receptors or GPCRs (Hubbard and Hepler, 2006; Larsen et al., 1997; Williams et al., 1998; Yamauchi et al., 1997). p38 is the most extensively studied MAP kinase in myogenesis. To date, all work has been performed in vitro as the absence of p38 during development leads to embryonic lethality (Allen et al., 2000). p38 is activated by phosphorylation on a canonical TxY motif by dual specificity kinases called MAP kinase kinases (MKK6 and MKK3). The upstream activating kinase MKK6 is most abundant in skeletal muscle (Han et al., 1996) and displays minimal substrate selectively among all p38 isoforms (α, β, γ and δ). Conversely, MKK3 is less promiscuous, preferentially activating p38α, γ and δ isoforms. Multiple splice variants of these upstream kinases with differential kinetics and amplitude of activation have been demonstrated in other cell types (Han et al., 1997).

p38 MAPK displays a dual role in myogenesis, regulating both satellite cell activation and myoblast differentiation. Both p38 MAPK and activated p-p38 MAPK were present in a subpopulation of satellite cells in single myofiber preparations (Jones et al., 2005). Selective pharmacologic inhibition of p38α and β using SB203580 for 24 hrs had no effect on satellite cell number but significantly decreased the numbers of cells expressing MyoD. Further growth of these cells was retarded as they failed to incorporate BrdU by 48 hrs in culture suggesting roles for both p38α and β in satellite cell activation and proliferation.

Multiple studies point to a role for p38 MAPK in regulation of myoblast differentiation. The transmembrane protein Cdo is an integral component of cell-cell adhesion complexes including M-cadherin, Boc, neogenin and netrin-3 that transduce signals to activate myogenic gene transcription during differentiation

(Krauss et al., 2005). Cdo regulates p38 MAPK activity during myoblast differentiation (Takaesu et al., 2006). Primary myoblasts derived from $Cdo^{-/-}$ mice lacked p38 MAPK activity and failed to differentiate. Ectopic expression of MKK6, an upstream activator of p38 MAPK, rescued myoblast differentiation. Furthermore, P38 MAPK was a member of a multiprotein complex with Cdo that transduced adhesion-dependent signals to activate myoblast differentiation.

p38 MAPK also promotes myoblast differentiation following ligand binding to receptors. Stimulation of primary myoblasts with extracellular ATP promoted differentiation in a p38 MAPK-dependent manner (Ryten et al., 2002). Similarily, the expression of biochemical differentiation markers by ATP was inhibited by the p38 inhibitor SB203580 suggesting a selective activation of p38 α and β during myoblast differentiation. Indeed, selective immunodepletion of p38 isoforms α, β, γ and δ during myogenesis demonstrated a rank order of potency of $\alpha > \beta/\delta > \gamma$ during myoblast differentiation (Zetser et al., 1999). MEF2 transcription factors heterodimerize with basic helix-loop helix transcription factors on the promoters of muscle-specific genes driving myoblast differentiation. Insulin-induced activation of MEF2 in primary human myotubes was sensitive to p38 MAPK inhibition (Al-Khalili et al., 2004). Similarily, MEF2 activation following contractile activity was decreased following pharmacological inhibiton of p38 MAPK suggesting p38 signaling contributes to myoblast differentiation during muscle adaptation and growth.

Taken together, these results suggest p38 MAPK functions in both proliferating and differentiating cells eliciting distinct cellular responses depending upon the stage of myogenesis. Differential expression of upstream kinases, phosphatases and downstream targets in proliferating and differentiating cells may elicit such heterogeneous effects.

5. MicroRNA

Small noncoding microRNAs (miR) are a rapidly emerging area of research. MiR regulate gene expression by posttranscriptional mechanisms through mRNA degradation, translational repression and/or chromatin inactivation (Bartel, 2004). MiR-1, 133 and 206 are specifically expressed in muscle and inhibit expression of multiple genes with important roles in myoblast proliferation and differentiation (Chen et al., 2006; Kim et al., 2006; Kwon et al., 2005). Muscle-nonspecific miR are also important for myogenesis. For example, miR-181 was upregulated in nascent myotubes during muscle regeneration suggesting roles in muscle differentiation (Naguibneva et al., 2006). Moreover, miR-181 decreased expression of MyoD, myogenin and HoxA11 in differentiating myoblasts (Naguibneva et al., 2006).

Additional research has focused on miR induction by basic helix-loop-helix (bHLH) muscle regulatory factors. MyoD-mediated transcription of miR-206 resulted in inhibition of follistatin-like 1 (Fstl1) and utrophin expression (Rosenberg et al., 2006). While utrophin repression is important for muscle differentiation, the role of Fstl1 in myogenesis is unknown. Similarily, myogenin and MyoD bound to the promoters of miR-1, 133 and 206 in differentiating myoblasts, presumably regulating transcription of these miR during myogenesis (Rao et al., 2006). During *Drosophila* muscle development, expression of miR-1 was directly regulated by twist (Sokol and Ambros, 2005), serum response factor, MyoD and MEF2 (Zhao et al., 2005). Additional targets and roles of miR during myogenesis are likely to be elucidated.

6. FUTURE DIRECTIONS

Muscle regeneration is a complex multi-step process involving satellite cell activation and proliferation followed by differentiation and fusion to form new myofibers. Myogenesis is often described as being regulated by linear signal cascades but in reality results from the integration of multiple signals following receptor activation and cross talk between a host of signaling molecules such as second messengers, kinases, phosphatases and transcription factors. This results in complex signal circuits that we are only beginning to understand. Future studies are needed to identify the signaling circuits that regulate myogenic cells at each step of myogenesis. Non-muscle cells also contribute to the effectiveness of muscle regeneration introducing further complexity into how multiple cell types coordinate their signaling.

The role of microRNAs in regulating gene expression in skeletal muscle is likely to greatly expand in the future. The regulatory pathways from receptors to the nucleus involved in miR transcription in skeletal muscle are poorly understood. Additional research may focus on miR induction by muscle specific and nonspecific transcription factors during myogenesis. The functional role of miR during myogenesis in vitro and regeneration in vivo can be analyzed using ectopic expression of miR mimetics and inhibition of miR activity by antisense oligonucleotides to fully understand the multitude of processes likely regulated by microRNAs during myogenesis.

REFERENCES

Abbott, KL, Friday BB, Thaloor D, TJ Murphy, Pavlath GK (1998) Activation and cellular localization of the cyclosporine A-sensitive transcription factor NF-AT in skeletal muscle cells. Mol Biol Cell 9:2905–2916

Adamo S, Caporale C, Nervi C, Ceci R, Molinaro M (1989) Activity and regulation of calcium-phospholipid-dependent protein kinase in differentiating chick myogenic cells. J Cell Biol 108: 153–158

Al-Khalili L, Chibalin AV, Yu M, Sjodin B, Nylen C, Zierath JR, Krook A (2004) MEF2 activation in differentiated primary human skeletal muscle cultures requires coordinated involvement of parallel pathways. Am J Physiol Cell Physiol 286:C1410–C1416

Allen, M, L Svensson, M Roach, J Hambor, J McNeish, CA Gabel. (2000) Deficiency of the stress kinase p38alpha results in embryonic lethality: characterization of the kinase dependence of stress responses of enzyme-deficient embryonic stem cells. J Exp Med 191:859–870

Armand AS, Pariset C, Laziz I, Launay T, Fiore F, Della B, Gaspera, Birnbaum D, Charbonnier F, Chanoine C (2005) FGF6 regulates muscle differentiation through a calcineurin-dependent pathway in regenerating soleus of adult mice. J Cell Physiol 204:297–308

Arnaudeau S, Holzer N, Konig S, Bader CR, Bernheim L (2006) Calcium sources used by post-natal human myoblasts during initial differentiation. J Cell Physiol 208:435–445

Balcerzak D, Poussard S, Brustis JJ, Elamrani N, Soriano M, Cottin P, Ducastaing A (1995) An antisense oligodeoxyribonucleotide to m-calpain mRNA inhibits myoblast fusion. J Cell Sci 108 (Pt 5):2077–2082

Bartel DP (2004) MicroRNAs: genomics, biogenesis, mechanism and function. Cell 116:281–297

Baudier J, Bergeret E, Bertacchi N, Weintraub H, Gagnon J, Garin J (1995) Interactions of myogenic bHLH transcription factors with calcium-binding calmodulin and S100a (alpha) proteins. Biochemistry 34:7834–7846

Bennett AM, Tonks NK (1997) Regulation of distinct stages of skeletal muscle differentiation by mitogen-activated protein kinases. Science 278:1288–1291

Bidaud, I, Monteil A, Nargeot J, Lory P (2006) Properties and role of voltage-dependent calcium channels during mouse skeletal muscle differentiation. J Muscle Res Cell Motil 27:75–81

Bijlenga P, Liu JH, Espinos E, Haenggeli CA, Fischer-Lougheed J, Bader CR, Bernheim L (2000) T-type alpha 1H Ca2+ channels are involved in Ca2+ signaling during terminal differentiation (fusion) of human myoblasts. Proc Natl Acad Sci U S A 97:7627–7632

Boczan J, Boros S, Mechler F, Kovacs L, Biro T (2000) Differential expressions of protein kinase C isozymes during proliferation and differentiation of human skeletal muscle cells in vitro. Acta Neuropathol (Berl) 99:96–104

Bondesen BA, Mills ST, Kegley KM, Pavlath GK (2004) The COX-2 pathway is essential during early stages of skeletal muscle regeneration. Am J Physiol Cell Physiol 287:C475–C483

Bryan BA, Mitchell DC, Zhao L, Ma W, Stafford LJ, Teng BB, Liu M (2005) Modulation of muscle regeneration, myogenesis, adipogenesis by the Rho family guanine nucleotide exchange factor GEFT Mol Cell Biol 25:11089–11101

Capiati DA, Limbozzi F, Tellez-Inon MT, Boland RL (1999) Evidence on the participation of protein kinase C alpha in the proliferation of cultured myoblasts. J Cell Biochem 74:292–300

Capiati, DA, Vazquez G, Tellez Inon MT, Boland RL (2000) Antisense oligonucleotides targeted against protein kinase c alpha inhibit proliferation of cultured avian myoblasts. Cell Prolif 33:307–315

Carrasco, MA, Marambio P, Jaimovich E (1997) Changes in IP3 metabolism during skeletal muscle development in vivo and in vitro. Comp Biochem Physiol B Biochem Mol Biol 116:173–181

Castellani L, Salvati E, Alema S, Falcone G (2006) Fine regulation of RhoA and Rock is required for skeletal muscle differentiation. J Biol Chem 281:15249–15257

Cha JH, Woo SK, Han KH, Kim YH, Handler JS, Kim J, Kwon HM (2001) Hydration status affects nuclear distribution of transcription factor tonicity responsive enhancer binding protein in rat kidney. J Am Soc Nephrol 12:2221–2230

Charge SB, Rudnicki MA (2004) Cellular and molecular regulation of muscle regeneration. Physiol Rev 84:209–238

Charrasse S, Comunale F, Grumbach Y, Poulat F, Blangy A, Gauthier-Rouviere C (2006) RhoA GTPase regulates M-cadherin activity and myoblast fusion. Mol Biol Cell 17:749–759

Chen JF, Mandel EM, Thomson JM, Wu Q, Callis TE, Hammond SM, Conlon FL, Wang DZ (2006) The role of microRNA-1 and microRNA-133 in skeletal muscle proliferation and differentiation. Nat Genet 38:228–233

Constantin B, Cognard C, Raymond G (1996) Myoblast fusion requires cytosolic calcium elevation but not activation of voltage-dependent calcium channels. Cell Calcium 19:365–374

Cseri, J, H Szappanos, Szigeti GP, Csernatony Z, Kovacs L, Csernoch L (2002) A purinergic signal transduction pathway in mammalian skeletal muscle cells in culture. Pflugers Arch 443:731–738

Dahl SC, Handler JS, Kwon HM (2001) Hypertonicity-induced phosphorylation and nuclear localization of the transcription factor TonEBP Am J Physiol Cell Physiol 280:C248–C253

Dalski A, Wagner HJ, Schwinger E, Zuhlke C (2000) Quantitative PCR analysis of different splice forms of NFAT5 revealed specific gene expression in fetal and adult brain. Brain Res Mol Brain Res 83:125–127

David JD, Higginbotham CA (1981) Fusion of chick embryo skeletal myoblasts: interactions of prostaglandin E1, adenosine 3′ : 5′ monophosphate, and calcium influx. Dev Biol 82:308–316

David JD, See WM, Higginbotham CA (1981) Fusion of chick embryo skeletal myoblasts: role of calcium influx preceding membrane union. Dev Biol 82:297–307

de la Pompa, JL, Timmerman LA, Takimoto H, Yoshida H, Elia AJ, Samper E, Potter J, Wakeham A, Marengere L, Langille BL, Crabtree GR, Mak TW (1998) Role of the NF-ATc transcription factor in morphogenesis of cardiac valves and septum. Nature 392:182–186

Disatnik MH, Boutet SC, Lee CH, Mochly-Rosen D, Rando TA (2002) Sequential activation of individual PKC isozymes in integrin-mediated muscle cell spreading: a role for MARCKS in an integrin signaling pathway. J Cell Sci 115:2151–2163

Dourdin N, Balcerzak D, Brustis JJ, Poussard S, Cottin P, Ducastaing A (1999) Potential m-calpain substrates during myoblast fusion. Exp Cell Res 246:433–442

Dourdin N, Brustis JJ, Balcerzak D, Elamrani N, Poussard S, Cottin P, Ducastaing A (1997) Myoblast fusion requires fibronectin degradation by exteriorized m-calpain. Exp Cell Res 235: 385–394

Duguez, S, Bihan MC, Gouttefangeas D, Feasson L, Freyssenet D (2003) Myogenic and nonmyogenic cells differentially express proteinases, Hsc/Hsp70, and BAG-1 during skeletal muscle regeneration. Am J Physiol Endocrinol Metab 285:E206–E215

Dulong S, Goudenege S, Vuillier-Devillers K, Manenti S, Poussard S, Cottin P (2004) Myristoylated alanine-rich C kinase substrate (MARCKS) is involved in myoblast fusion through its regulation by protein kinase Calpha and calpain proteolytic cleavage. Biochem J 382:1015–1023

Elamrani N, Brustis JJ, Dourdin N, Balcerzak D, Poussard S, Cottin P, Ducastaing A (1995) Desmin degradation and Ca(2+)-dependent proteolysis during myoblast fusion. Biol Cell 85: 177–183

Entwistle A, Curtis DH, Zalin RJ (1986) Myoblast fusion is regulated by a prostanoid of the one series independently of a rise in cyclic AMP J Cell Biol 103:857–866

Entwistle A, Zalin RJ, Bevan S, Warner AE (1988a) The control of chick myoblast fusion by ion channels operated by prostaglandins and acetylcholine. J Cell Biol 106:1693–1702

Entwistle A, Zalin RJ, Warner AE, Bevan S (1988b) A role for acetylcholine receptors in the fusion of chick myoblasts. J Cell Biol 106:1703–1712

Fedorov YV, Jones NC, Olwin BB (1998) Regulation of myogenesis by fibroblast growth factors requires beta-gamma subunits of pertussis toxin-sensitive G proteins. Mol Cell Biol 18:5780–5787

Ferraris JD, Williams CK, Persaud P, Zhang Z, Chen Y, Burg MB (2002) Activity of the TonEBP/OREBP transactivation domain varies directly with extracellular NaCl concentration. Proc Natl Acad Sci U S A 99:739–744

Fischer-Lougheed J, Liu JH, Espinos E, Mordasini D, Bader CR, Belin D, Bernheim L (2001) Human myoblast fusion requires expression of functional inward rectifier Kir2.1 channels. J Cell Biol 153: 677–686

Fornaro M, Burch PM, Yang W, Zhang L, Hamilton CE, Kim JH, Neel BG, Bennett AM (2006) SHP-2 activates signaling of the nuclear factor of activated T cells to promote skeletal muscle growth. J Cell Biol 175:87–97

Franchi-Gazzola R, Visigalli R, Dall'Asta V, Sala R, Woo SK, Kwon HM, Gazzola GC, Bussolati O (2001) Amino acid depletion activates TonEBP and sodium-coupled inositol transport. Am J Physiol Cell Physiol 280:C1465–C1474

Friday BB, Horsley V, Pavlath GK (2000) Calcineurin activity is required for the initiation of skeletal muscle differentiation. J Cell Biol 149:657–666

Friday BB, Mitchell PO, Kegley KM, Pavlath GK (2003) Calcineurin initiates skeletal muscle differentiation by activating MEF2 and MyoD Differentiation 71:217–227

Friday BB, Pavlath GK (2001) A calcineurin- and NFAT-dependent pathway regulates Myf5 gene expression in skeletal muscle reserve cells. J Cell Sci 114:303–310

Funk CD (2001) Prostaglandins and leukotrienes: advances in eicosanoid biology. Science 294: 1871–1875

Glading A, Bodnar RJ, Reynolds IJ, Shiraha H, Satish L, Potter DA, Blair HC, Wells A (2004) Epidermal growth factor activates m-calpain (calpain II), at least in part, by extracellular signal-regulated kinase-mediated phosphorylation. Mol Cell Biol 24:2499–2512

Griffin BW, Klimko P, Crider JY, Sharif NA (1999) AL-8810: a novel prostaglandin F2 alpha analog with selective antagonist effects at the prostaglandin F2 alpha (FP) receptor. J Pharmacol Exp Ther 290:1278–1284

Griffin BW, Magnino PE, Pang IH, Sharif NA (1998) Pharmacological characterization of an FP prostaglandin receptor on rat vascular smooth muscle cells (A7r5) coupled to phosphoinositide turnover and intracellular calcium mobilization. J Pharmacol Exp Ther 286:411–418

Han J, Jiang Y, Li Z, Kravchenko VV, Ulevitch RJ (1997) Activation of the transcription factor MEF2C by the MAP kinase p38 in inflammation. Nature 386:296–299

Han J, Lee JD, Jiang Y, Li Z, Feng L, Ulevitch RJ (1996) Characterization of the structure and function of a novel MAP kinase (MKK6). J Biol Chem 271:2886–2891

Hartwig JH, Thelen M, Rosen A, Janmey PA, Nairn AC, Aderem A (1992) MARCKS is an actin filament crosslinking protein regulated by protein kinase C and calcium-calmodulin. Nature 356:618–622

Hawke TJ, Kanatous SB, Martin CM, Goetsch SC, Garry DJ (2006) Rad is temporally regulated within myogenic progenitor cells during skeletal muscle regeneration. Am J Physiol Cell Physiol 290: C379–C387

Hollinger S, Hepler JR (2002) Cellular regulation of RGS proteins: modulators and integrators of G protein signaling. Pharmacol Rev 54:527–559

Horsley V, Friday BB, Matteson S, Kegley KM, Gephart J, Pavlath GK (2001) Regulation of the growth of multinucleated muscle cells by an NFATC2-dependent pathway. J Cell Biol 153:329–338

Horsley V, Jansen KM, Mills ST, Pavlath GK (2003) IL-4 acts as a myoblast recruitment factor during mammalian muscle growth. Cell 113:483–494

Horsley V, Pavlath GK (2003) Prostaglandin F2(alpha) stimulates growth of skeletal muscle cells via an NFATC2-dependent pathway. J Cell Biol 161:111–118

Huang Y, Wang KK (2001) The calpain family and human disease. Trends Mol Med 7:355–362

Hubbard KB, Hepler JR (2006) Cell signalling diversity of the Gqalpha family of heterotrimeric G proteins. Cell Signal 18:135–150

Jauliac S, Lopez-Rodriguez C, Shaw LM, Brown LF, Rao A, Toker A (2002) The role of NFAT transcription factors in integrin-mediated carcinoma invasion. Nat Cell Biol 4:540–544

Joffroy S, Dourdin N, Delage JP, Cottin P, Koenig J, Brustis JJ (2000) M-calpain levels increase during fusion of myoblasts in the mutant muscular dysgenesis (mdg) mouse. Int J Dev Biol 44:421–428

Jones NC, Tyner KJ, Nibarger L, Stanley HM, Cornelison DD, Fedorov YV, Olwin BB (2005) The p38alpha/beta MAPK functions as a molecular switch to activate the quiescent satellite cell. J Cell Biol 169:105–116

Kegley KM, Gephart J, Warren GL, Pavlath GK (2001) Altered primary myogenesis in NFATC3(-/-) mice leads to decreased muscle size in the adult. Dev Biol 232:115–126

Kim HK, Lee YS, Sivaprasad U, Malhotra A, Dutta A (2006) Muscle-specific microRNA miR-206 promotes muscle differentiation. J Cell Biol 174:677–687

Kim SS, Kim JH, Kim HS, Park DE, Chung CH (2000) Involvement of the theta-type protein kinase C in translocation of myristoylated alanine-rich C kinase substrate (MARCKS) during myogenesis of chick embryonic myoblasts. Biochem J 347(Pt 1):139–146

Kim SS, Kim JH, Lee SH, Chung SS, Bang OS, Park D, Chung CH (2002) Involvement of protein phosphatase-1-mediated MARCKS translocation in myogenic differentiation of embryonic muscle cells. J Cell Sci 115:2465–2473

Knudsen KA (1985) The calcium-dependent myoblast adhesion that precedes cell fusion is mediated by glycoproteins. J Cell Biol 101:891–897

Knudsen KA, Myers L, McElwee SA (1990) A role for the Ca2(+)-dependent adhesion molecule, N-cadherin, in myoblast interaction during myogenesis. Exp Cell Res 188:175–184

Konig S, Beguet A, Bader CR, Bernheim L (2006) The calcineurin pathway links hyperpolarization (Kir2.1)-induced Ca2+ signals to human myoblast differentiation and fusion. Development 133: 3107–3114

Konig S, Hinard V, Arnaudeau S, Holzer N, Potter G, Bader CR, Bernheim L (2004) Membrane hyperpolarization triggers myogenin and myocyte enhancer factor-2 expression during human myoblast differentiation. J Biol Chem 279:28187–28196

Kramerova I, Kudryashova E, Tidball JG, Spencer MJ (2004) Null mutation of calpain 3 (p94) in mice causes abnormal sarcomere formation in vivo and in vitro. Hum Mol Genet 13:1373–1388

Kramerova I, Kudryashova E, Wu B, Spencer MJ (2006) Regulation of M-cadherin-{beta}-catenin complex by calpain 3 during terminal stages of myogenic differentiation. Mol Cell Biol 26:8437–8447

Krauss RS, Cole F, Gaio U, Takaesu G, Zhang W, Kang JS (2005) Close encounters: regulation of vertebrate skeletal myogenesis by cell-cell contact. J Cell Sci 118:2355–2362

Kwon C, Han Z, Olson EN, Srivastava D (2005) MicroRNA1 influences cardiac differentiation in Drosophila and regulates Notch signaling. Proc Natl Acad Sci U S A 102:18986–18991

Larsen JK, Yamboliev IA, Weber LA, Gerthoffer WT (1997) Phosphorylation of the 27-kDa heat shock protein via p38 MAP kinase and MAPKAP kinase in smooth muscle. Am J Physiol 273:L930–L940

Lee SD, Woo SK, Kwon HM (2002) Dimerization is required for phosphorylation and DNA binding of TonEBP/NFAT5. Biochem Biophys Res Commun 294:968–975

Li L, Zhou J, James G, Heller-Harrison R, Czech MP, Olson EN (1992) FGF inactivates myogenic helix-loop-helix proteins through phosphorylation of a conserved protein kinase C site in their DNA-binding domains. Cell 71:1181–1194

Liu JH, Bijlenga P, Fischer-Lougheed J, Occhiodoro T, Kaelin A, Bader CR, Bernheim L (1998) Role of an inward rectifier K+ current and of hyperpolarization in human myoblast fusion. J Physiol 510(Pt 2):467–476

Liu JH, Konig S, Michel M, Arnaudeau S, Fischer-Lougheed J, Bader CR, Bernheim L (2003) Acceleration of human myoblast fusion by depolarization: graded Ca2+ signals involved. Development 130:3437–3446

Lopez-Rodriguez C, Aramburu J, Jin L, Rakeman AS, Michino M, Rao A (2001) Bridging the NFAT and NF-kappaB families: NFAT5 dimerization regulates cytokine gene transcription in response to osmotic stress. Immunity 15:47–58

Lopez-Rodriguez C, Aramburu J, Rakeman AS, Rao A (1999) NFAT5, a constitutively nuclear NFAT protein that does not cooperate with Fos and Jun. Proc Natl Acad Sci U S A 96:7214–7219

McArdle A, Edwards RH, Jackson MJ (1994) Release of creatine kinase and prostaglandin E2 from regenerating skeletal muscle fibers. J Appl Physiol 76:1274–1278

Miyabara EH, Aoki MS, Moriscot AS (2005) Cyclosporin A preferentially attenuates skeletal slow-twitch muscle regeneration. Braz J Med Biol Res 38:559–563

Miyakawa H, Woo SK, Dahl SC, Handler JS, Kwon HM (1999) Tonicity-responsive enhancer binding protein, a rel-like protein that stimulates transcription in response to hypertonicity. Proc Natl Acad Sci U S A 96:2538–2542

Moraczewski J, Nowotniak A, Wrobel E, Castagna M, Gautron J, Martelly I (2002) Differential changes in protein kinase C associated with regeneration of rat extensor digitorum longus and soleus muscles. Int J Biochem Cell Biol 34:938–949

Naguibneva, I, M Ameyar-Zazoua, A Polesskaya, S Ait-Si-Ali, R Groisman, M Souidi, S Cuvellier, A Harel-Bellan. (2006) The microRNA miR-181 targets the homeobox protein Hox-A11 during mammalian myoblast differentiation. Nat Cell Biol 8:278–284

Narumiya S, Sugimoto Y, Ushikubi F (1999) Prostanoid receptors: structures, properties, and functions. Physiol Rev 79:1193–1226

O'Connor RS, Mills ST, Jones KA, Ho SN, Pavlath GK (2007) A combinatorial role for NFAT5 in both myoblast migration and differentiation during skeletal muscle myogenesis. J Cell Sci 120:149–159

Otis JS, Burkholder TJ, Pavlath GK (2005) Stretch-induced myoblast proliferation is dependent on the COX2 pathway. Exp Cell Res 310:417–425

Palmer RM, Reeds PJ, Atkinson T, Smith RH (1983) The influence of changes in tension on protein synthesis and prostaglandin release in isolated rabbit muscles. Biochem J 214:1011–1014

Papahadjopoulos D, Nir S, Duzgunes N (1990) Molecular mechanisms of calcium-induced membrane fusion. J Bioenerg Biomembr 22:157–179

Pisaniello A, Serra C, Rossi D, Vivarelli E, Sorrentino V, Molinaro M, Bouche M (2003) The block of ryanodine receptors selectively inhibits fetal myoblast differentiation. J Cell Sci 116:1589–1597

Porter GA Jr, Makuck RF, Rivkees SA (2002) Reduction in intracellular calcium levels inhibits myoblast differentiation. J Biol Chem 277:28942–28947

Poussard S, Dulong S, Aragon B, Jacques Brustis J, Veschambre P, Ducastaing A, Cottin P (2001) Evidence for a MARCKS-PKCalpha complex in skeletal muscle. Int J Biochem Cell Biol 33:711–721
Przybylski RJ, MacBride RG, Kirby AC (1989) Calcium regulation of skeletal myogenesis. I Cell content critical to myotube formation. In Vitro Cell Dev Biol 25:830–838
Przybylski RJ, Szigeti V, Davidheiser S, Kirby AC (1994) Calcium regulation of skeletal myogenesis. II Extracellular and cell surface effects. Cell Calcium 15:132–142
Rao PK, Kumar RM, Farkhondeh M, Baskerville S, Lodish HF (2006) Myogenic factors that regulate expression of muscle-specific microRNAs. Proc Natl Acad Sci U S A 103:8721–8726
Raynaud F, Carnac G, Marcilhac A, Benyamin Y (2004) m-Calpain implication in cell cycle during muscle precursor cell activation. Exp Cell Res 298:48–57
Richard I, Broux O, Allamand V, Fougerousse F, Chiannilkulchai N, Bourg N, Brenguier L, Devaud C, Pasturaud P, Roudaut C et al (1995) Mutations in the proteolytic enzyme calpain 3 cause limb-girdle muscular dystrophy type 2A. Cell 81:27–40
Rosenberg MI, Georges SA, Asawachaicharn A, Analau E, Tapscott SJ (2006) MyoD inhibits Fstl1 and Utrn expression by inducing transcription of miR-206. J Cell Biol 175:77–85
Ryten M, Dunn PM, Neary JT, Burnstock G (2002) ATP regulates the differentiation of mammalian skeletal muscle by activation of a P2X5 receptor on satellite cells. J Cell Biol 158:345–355
Sakuma K, Nakao R, Aoi W, Inashima S, Fujikawa T, Hirata M, Sano M, Yasuhara M (2005) Cyclosporin A treatment upregulates Id1 and Smad3 expression and delays skeletal muscle regeneration. Acta Neuropathol (Berl) 110:269–280
Sakuma K, Nishikawa J, Nakao R, Watanabe K, Totsuka T, Nakano H, Sano M, Yasuhara M (2003) Calcineurin is a potent regulator for skeletal muscle regeneration by association with NFATc1 and GATA-2. Acta Neuropathol (Berl) 105:271–280
Salzberg S, Mandelboim M, Zalcberg M, Shainberg A, Mandelbaum M (1995) Interruption of myogenesis by transforming growth factor beta 1 or EGTA inhibits expression and activity of the myogenic-associated $(2' - 5')$ oligoadenylate synthetase and PKR Exp Cell Res 219:223–232
Schulze M, Belema-Bedada F, Technau A, Braun T (2005) Mesenchymal stem cells are recruited to striated muscle by NFAT/IL-4-mediated cell fusion. Genes Dev 19:1787–1798
Schutzle UB, Wakelam MJ, Pette D (1984) Prostaglandins and cyclic AMP stimulate creatine kinase synthesis but not fusion in cultured embryonic chick muscle cells. Biochim Biophys Acta 805:204–210
Serrano AL, Murgia M, Pallafacchina G, Calabria E, Coniglio P, Lomo T, Schiaffino S (2001) Calcineurin controls nerve activity-dependent specification of slow skeletal muscle fibers but not muscle growth. Proc Natl Acad Sci U S A 98:13108–13113
Shainberg A, Yagil G, Yaffe D (1969) Control of myogenesis in vitro by Ca 2 + concentration in nutritional medium. Exp Cell Res 58:163–167
Shen W, Li Y, Tang Y, Cummins J, Huard J (2005) NS-398, a cyclooxygenase-2-specific inhibitor, delays skeletal muscle healing by decreasing regeneration and promoting fibrosis. Am J Pathol 167:1105–1117
Shen W, Prisk V, Li Y, Foster W, Huard J (2006) Inhibited skeletal muscle healing in cyclooxygenase-2 gene-deficient mice: the role of PGE2 and PGF2alpha. J Appl Physiol 101:1215–1221
Shin KS, Park JY, Ha DB, Chung CH, Kang MS (1996) Involvement of K(Ca) channels and stretch-activated channels in calcium influx, triggering membrane fusion of chick embryonic myoblasts. Dev Biol 175:14–23
Sokol NS, Ambros V (2005) Mesodermally expressed Drosophila microRNA-1 is regulated by Twist and is required in muscles during larval growth. Genes Dev 19:2343–2354
Sotiropoulos A, Ohanna M, Kedzia C, Menon RK, Kopchick JJ, Kelly PA, Pende M (2006) Growth hormone promotes skeletal muscle cell fusion independent of insulin-like growth factor 1 up-regulation. Proc Natl Acad Sci U S A 103:7315–7320
Stiber JA, Tabatabaei N, Hawkins AF, Hawke T, Worley PF, Williams RS, Rosenberg P (2005) Homer modulates NFAT-dependent signaling during muscle differentiation. Dev Biol 287:213–224
Stockholm D, Barbaud C, Marchand S, Ammarguellat F, Barritault D, Richard I, Beckmann J, Martelly I (1999) Studies on calpain expression during differentiation of rat satellite cells in primary cultures in the presence of heparin or a mimic compound. Exp Cell Res 252:392–400

Szigeti GP, Szappanos H, Deli T, Cseri J, Kovacs L, Csernoch L (2007) Differentiation-dependent alterations in the extracellular ATP-evoked calcium fluxes of cultured skeletal muscle cells from mice. Pflugers Arch 453:509–518

Takaesu G, Kang JS, Bae GU, Yi MJ, Lee CM, Reddy EP, Krauss RS (2006) Activation of p38{alpha}/{beta} MAPK in myogenesis via binding of the scaffold protein JLP to the cell surface protein Cdo. J Cell Biol 175:383–388

Trama J, Go WY, Ho SN (2002) The osmoprotective function of the NFAT5 transcription factor in T cell development and activation. J Immunol 169:5477–5488

Trama J, Lu Q, Hawley RG, Ho SN (2000) The NFAT-related protein NFATL1 (TonEBP/NFAT5) is induced upon T cell activation in a calcineurin-dependent manner. J Immunol 165:4884–4894

Vandenburgh HH, Shansky J, Solerssi R, Chromiak J (1995) Mechanical stimulation of skeletal muscle increases prostaglandin F2 alpha production, cyclooxygenase activity, cell growth by a pertussis toxin sensitive mechanism. J Cell Physiol 163:285–294

Williams NG, Zhong H, Minneman KP (1998) Differential coupling of alpha1-, alpha2-, beta-adrenergic receptors to mitogen-activated protein kinase pathways and differentiation in transfected PC12 cells. J Biol Chem 273:24624–24632

Winslow MM, Pan M, Starbuck M, Gallo EM, Deng L, Karsenty G, Crabtree GR (2006) Calcineurin/NFAT signaling in osteoblasts regulates bone mass. Dev Cell 10:771–782

Woo SK, Dahl SC, Handler JS, Kwon HM (2000) Bidirectional regulation of tonicity-responsive enhancer binding protein in response to changes in tonicity. Am J Physiol Renal Physiol 278: F1006–F1012

Xu Q, Yu L, Liu L, Cheung CF, Li X, Yee SP, Yang XJ, Wu Z (2002) p38 Mitogen-activated protein kinase-, calcium-calmodulin-dependent protein kinase-, calcineurin-mediated signaling pathways transcriptionally regulate myogenin expression. Mol Biol Cell 13:1940–1952

Yamauchi J, Nagao M, Kaziro Y, Itoh H (1997) Activation of p38 mitogen-activated protein kinase by signaling through G protein-coupled receptors. Involvement of Gbetagamma and Galphaq/11 subunits. J Biol Chem 272:27771–27777

Yoshida N, Yoshida S, Koishi K, Masuda K, Nabeshima Y (1998) Cell heterogeneity upon myogenic differentiation: down-regulation of MyoD and Myf-5 generates 'reserve cells'. J Cell Sci 111(Pt 6):769–779

Zalin RJ (1977) Prostaglandins and myoblast fusion. Dev Biol 59:241–248

Zalin RJ (1987) The role of hormones and prostanoids in the in vitro proliferation and differentiation of human myoblasts. Exp Cell Res 172:265–281

Zetser A, Gredinger E, Bengal E (1999) p38 mitogen-activated protein kinase pathway promotes skeletal muscle differentiation. Participation of the Mef2c transcription factor. J Biol Chem 274:5193–5200

Zhang Z, Ferraris JD, Brooks HL, Brisc I, Burg MB (2003) Expression of osmotic stress-related genes in tissues of normal and hyposmotic rats. Am J Physiol Renal Physiol 285:F688–F693

Zhao Y, Samal E, Srivastava D (2005) Serum response factor regulates a muscle-specific microRNA that targets Hand2 during cardiogenesis. Nature 436:214–220

Zimowska M, Szczepankowska D, Streminska W, Papy D, Tournaire MC, Gautron J, Barritault D, Moraczewski J, Martelly I (2001) Heparan sulfate mimetics modulate calpain activity during rat Soleus muscle regeneration. J Cell Physiol 188:178–187

CHAPTER 8

MUSCLE REGENERATION IN ANIMAL MODELS

BRUCE M. CARLSON
Institute of Gerontology, 300 North Ingalls Building
University of Michigan, Ann Arbor, MI 48109, USA

1. INTRODUCTION

As emphasis on research in skeletal muscle regeneration has increased during the past four decades, there has been a corresponding increase in the number of animal models used to study the phenomenon. Not all animals are the same, nor are all animal models the same. The goal of this chapter is to outline the major animal models used in experimental studies of skeletal muscle regeneration and to point out the advantages and limitations of each model.

2. NON-MAMMALIAN MODELS

2.1 The Amphibian Limb and Tail

The limbs and tail of urodele amphibians are capable of regenerating after amputation by a process known as epimorphic regeneration Wallace, 1981; Tsonis, 1996). After amputation, the amputation surface is covered by a wound epidermis, which interacts with the underlying tissues of the extremity to stimulate a process known as dedifferentiation. The dedifferentiative process produces mesenchyme-like cells that accumulate distally to form a regeneration blastema from which a new limb or tail regenerates. The end product of epimorphic regeneration is virtually identical to that of the originally amputated part.

Skeletal muscle, along with other limb tissues, is involved in the dedifferentiative process. The big question is what dedifferentiation means. At the histological level, there is clearly a loss of the structure of the differentiated mesodermal tissues at the end of the stump. What is less clear are the dynamics of dedifferentiation at the cellular level. This topic is covered in detail in Chapter 9 of this book.

S. Schiaffino and T. Partridge (eds.), Skeletal Muscle Repair and Regeneration, 163–179.

Amphibian limb muscle is also capable of regenerating by another mode, called tissue regeneration, which follows the same pathway and appears to operate under the same types of controls as muscle regeneration in mammals (Carlson, 1970, 1979). The urodele amphibian limb is one of the few models in which the regeneration of an entire muscle can be brought about by two completely different mechanisms. This affords investigators the opportunity to study interactions between the tissue and epimorphic modes of muscle regeneration. This can be done by mincing the muscles in a limb and then amputating the limb at a level that transects the minced muscle. The issue is whether the minced muscle at the distalmost end of the limb stump will regenerate by the tissue or epimorphic mode. When this is done, a blastema forms at the end of the limb, and it is clear that at the end of the limb stump epimorphic regeneration in the form of blastema formation takes precedence over tissue regeneration, but the mechanisms underlying the interactions between these forms of muscle regeneration are totally unknown.

2.2 Jellyfish Muscle

A very interesting model for studying dedifferentiation and transdifferentiation of skeletal muscle is the muscle found in the umbrella of jellyfish medusae (Schmid, 1992; Schmid and Reber-Müller, 1995). When portions of the umbrella containing muscle cells are cultured in association with their normal extracellular matrix (the mesoglea), the muscle cells remain phenotypically stable for prolonged periods. However, if the matrix is digested away, the muscle becomes destabilized, and the cells lose most of their myofibrils. They then proliferate and transdifferentiate into new cell types, principally smooth muscle and sensory nerve cells. Of interest in this model is that the pattern of *Msx* expression is seemingly reciprocal to that which has been reported for dedifferentiating vertebrate skeletal muscle (Galle et al., 2005).

3. MAMMALIAN MODELS

The choice of an animal model for studying mammalian muscle regeneration is largely dependent upon the questions being asked. Some questions concern the overall success of regeneration of a muscle as a whole and interactions between a regenerating muscle and other tissues in its local or general environment. Other questions involve overall molecular or biochemical analysis at specific stages of muscle regeneration, and homogeneity and synchronicity at the tissue level are important. Yet other approaches, directed at the individual cell level, require the ability to focus on specific regions and specific times. Models already exist for most of these needs.

3.1 Models Involving the Complete or Near-complete Degeneration of an Entire Muscle

3.1.1 The minced muscle model

This earliest model of whole muscle regeneration had a bizarre origin. During the Lysenko era in the Soviet Union, a dominant theory in Lysenkoist biology was the "new cell theory" of Lepeshinskaya (1945), which posited that under certain developmental and post-traumatic conditions cells can arise *de novo* from a proteinaceous soup called "living substance." In an era when mammalian skeletal muscle was thought not to be able to regenerate, Studitsky decided to test the theory by chopping a muscle into one mm^3 pieces and reimplanting the mince back into the muscle bed. Amazingly, new muscle did regenerate, and Studitsky (1953) wrote that this experiment confirmed the validity of the new cell theory. Nevertheless, Studitsky's laboratory subsequently conducted much important early fundamental research on the biology of regenerating skeletal muscle (Studitsky, 1959).

Despite its unusual origin, this model has proven to be quite useful for the study of certain aspects of muscle regeneration (see Carlson, 1972 for general review). First, there is no question but that all muscle fibers in the minced muscle have been destroyed. This is basically an ischemia model. One of its strengths was that at the histological level it shows nicely the correlation between the course of revascularization and the progression of muscle regeneration from satellite cell activation to muscle fiber differentiation. Because of the developmental gradient, all stages can be seen in a single regenerating muscle. This model has proven to be very important in understanding the basis for morphogenesis of a muscle regenerating by the tissue mode, because at the time of implantation the mince is grossly shapeless and totally disorganized in its internal architecture (Carlson, 1972, Chapter 6). This model was used to considerable advantage in early studies of muscle-nerve interactions in regeneration (see below).

Major disadvantages of the minced muscle model are that in the rat, at least, large amounts of connective tissue adhesions ultimately form. This reduces the functional capacities of minced muscle regenerates. Although the first careful studies on contractile properties of regenerating muscle were done on regenerates derived from this model (Carlson and Gutmann, 1972), it was not an ideal one for such studies. In addition, the success of this model depends greatly upon the size of the muscle to be minced. In the rat, mincing the extensor digitorum longus (EDL) or soleus muscles alone did not produce acceptable regenerates (Carlson, unpublished). In muscles larger than the rat gastrocnemius regeneration is limited by the speed of revascularization, and a muscle larger than the rat gastrocnemius also does not regenerate well by this method. I know of one case where mincing was done on a badly injured human tibialis anterior muscle, and the results were disastrous. A final disadvantage of the mincing model is a counterpoint of one of its advantages. As mentioned above, there is a spatiotemporal gradient of regeneration in a regenerating mince. Although very advantageous for histological studies, it makes whole-muscle biochemical analysis very difficult because different stages of both degeneration and regeneration are present in the same specimen.

3.1.2 Free muscle grafting

In its simplest form free muscle grafting consists of removing a muscle completely from its bed and replacing it orthotopically. In what is called a standard graft, the cut tendons are reconnected with the stumps of the host, but revascularization and reinnervation are allowed to occur spontaneously. This model was initially devised in order to allow better physiological analysis of regenerating small fast and slow muscles like the EDL and soleus (Carlson and Gutmann, 1975). Independently, free muscle grafting in humans was initiated at the same time (Freilinger et al., 1981). This is basically an ischemia model. The center of the graft falls into a state of ischemic necrosis, and over a few days regeneration occurs along a centripetal gradient.

The level of overall success in free grafting models is a function of the completeness of reinnervation rather than of revascularization. In a standard graft of the rat EDL muscle, maximum tetanic force typically returns to slightly over 1/3 of control levels. If the motor nerve is implanted into the grafted muscle, the level of functional recovery is about 50%, but if the motor nerve to the muscle is allowed to remain intact, the graft, which still degenerates and regenerates, recovers 100% of its original mass and about 85–90% of its maximum contractile force (Carlson et al., 1981).

Free grafting is size limited, and there is some species variation in the response of the muscle to grafting. In rats, the rectus femoris muscle regenerates after free grafting, but in adult rats, it takes 6–7 weeks before the center of the graft has become revascularized. Free grafting has also been successful after grafting in cats (Faulkner et al., 1976), but in monkeys, the center of a free graft becomes filled with a dense core of collagenous connective tissue, with a concentric rim of regenerated muscle fibers surrounding that (Markley et al., 1978).

A major advantage of the free grafting model is that it produces little scar tissue in mice and rats and it is very useful for physiological analysis. It is excellent for the analysis of motor units in regenerating muscle, as well as for routine contractile properties. Like the minced muscle model, a major disadvantage is the presence of a gradient of regenerative maturity in the early transplants (first 2–3 weeks in small muscles). Another disadvantage in whole transplant models is the survival of a thin rim of original muscle fibers around the periphery of an early graft. This considerably complicates gross biochemical or molecular analysis.

Because it is basically an ischemia model, free muscle grafting is size-limited. In my hands, the upper limit in rats is a muscle of 0.5–1.0 gm. Free grafting of the entire rat gastrocnemius muscle has usually been unsuccessful for me. Human free muscle grafting was faced with the same limitations, and once microvascular suturing methods became available, surgeons switched to that method, which produced much better clinical results that were not limited by size. However, transplantation with vascular anastomoses results in survival of the original muscle fibers, so it is no longer a regeneration model.

3.1.3 Cross-transplantation models

A valuable variation on whole muscle grafting has been the cross-transplantation of muscles. Cross-transplantation models function best in distinguishing between intrinsic properties and environmental influences on regenerating muscles. This was first done between fast and slow muscles (rat EDL and soleus) to investigate the trophic effect of the nerve on the functional differentiation of regenerating muscle (Gutmann and Carlson, 1975). In the premolecular era of muscular dystrophy research, minced muscles were transplanted between normal and dystrophic mice as a test of myogenic vs. neurogenic theories of muscular dystrophy (Salafsky, 1971; Neerunjun and Dubowitz, 1975), and it has also been used to sort out intrinsic vs. environmental differences in muscles regenerating in normal and diabetic rats (Gulati and Swamy, 1991). Finally, a cross-age transplantation model showed that old muscles grafted into young rats regenerated as well as same-age young muscles, whereas young muscles grafted into old hosts regenerated as poorly as old muscles (Carlson and Faulkner, 1989).

3.1.4 Parabiosis models

Parabiosis models can be useful for studying humoral influences upon regenerative processes. Although historically such studies have been used most frequently in studies involving mechanisms of compensatory hypertrophy of the liver, there has been one report of the use of old-young parabiotic pairs in the study of muscle regeneration. Conboy et al. (2005) created parabiotic mice, with old and young members of the pairs and were able to establish that a parabiotic connection with a young mouse restored activation of Notch signaling and also increased the proliferation and regenerative capacity of aged satellite cells. Carlson (unpublished) has had considerable experience with old-young parabiotic rats. Such pairs are produced by removing an oval-shaped area of skin on the lateral side of the trunk on opposite sides of each member of the pair and then suturing the skin around the oval. In most cases, there are no untoward side effects of the operation. Conboy et al. (2005) maintained their mice as parabiotic pairs for 5 weeks before injuring a muscle.

Carlson (unpublished) transplanted muscles at the time of operation and then waited 2 months before physiological analysis of the grafts. Over the two months, the parabiotic rats lost weight, possibly due to increased stress, but were otherwise in good shape. Analysis of the grafts was unsuccessful because of limitations inherent in the surgical model, when contractile properties of the regenerated muscles were used as the endpoint. The principal difficulty was mechanical. In a side-by-side pair, there appeared to be more mechanical stress placed on the outer limb of each member of the pair. Therefore an exercise effect was introduced. Although there was an improvement in regeneration of grafted muscles of the old members of the pairs, the results were not statistically significant.

If properly controlled, the parabiosis model can be a very powerful one, but it is important to understand the conditions associated with the model that could skew the interpretations of the data. Other then the mechanical example noted above, one of the most important is the potential for long-term stress effects on the rats to influence the processes under consideration.

3.1.5 Injection of myotoxic agents

One of the most popular methods for producing massive amounts of muscle damage is to inject any of a variety of myotoxic agents into a muscle. This is a seemingly easy method for producing muscle damage, but unless one takes into account the patterns of pathology produced by these agents, it is possible to misinterpret the results produced by analysis of the tissues. One advantage of many of these toxins is that they are not neurotoxic, which produces a cleaner experimental system. These agents will be loosely grouped into three categories – local anesthetics, toxins and other chemical agents.

3.1.5.1 Local anesthetics Most of the traditional local anesthetics are myotoxic to varying degrees (Benoit and Belt, 1970; Burke et al., 1972). However, the one that is most commonly used for experimental studies is bupivacaine (Marcaine). Reasons for Marcaine's popularity are the large number of muscle fibers destroyed by a single injection and the fact that injected Marcaine produces a non-ischemic lesion, resulting in a spatially and temporally homogeneous field of degenerating and regenerating muscle fibers. In contrast, some of the other local anesthetics produce ischemic lesions (Foster and Carlson, 1980), which introduces considerable variability into whole muscle analysis.

Important considerations in the Marcaine-injection model are concentration and site of injection. In the rat, 0.75% is the most effective concentration, but sometimes batches of this concentration have to be specially made up because clinical application in recent years has tended to use 0.5% or lower concentrations. Shelf life is important. After 6 months to a year, Marcaine loses its potency. The most common injection model is an injection into the tibialis anterior muscle. In the rat, a satisfactory introduction of the anesthetic can be made with a # 28 needle about 25–30 mm long. The easiest technique for introduction is to insert the needle through the skin at the distal end of the tibialis anterior and push the needle to the proximal end. Then as the needle is withdrawn, anesthetic is injected along the needle path. For larger rats, two or three injections into different areas of the muscle provides better coverage. For better placement, or for injecting deeper muscles, such as the EDL, injection under direct surgical exposure is preferable. Normally it is best to inject as much anesthetic as the muscle can hold, but for larger muscles, that amount of anesthetic can cause the death of the rat.

One of the disadvantages of this technique is that with simple injections, it is extremely rare to cause degeneration of all of the muscle fibers in the rat. Therefore for gross analysis one must recognize that there is a mix of normal and degenerating or regenerating muscle fibers in the sample. Another disadvantage is variability of the extent of the lesion. However for studies involving tissue sections, this is not a serious limitation. In fact the presence of some obviously normal muscle fibers can serve as an intrinsic control. For destroying all of the muscle fibers in a rat muscle without destroying the internal architecture, one of the surest techniques is to remove the muscle, soak it in Marcaine for a few minutes and then graft the muscle back into the host (Carlson and Gutmann, 1976). The Marcaine diffusing into the muscle kills the muscle fibers at the periphery that would otherwise survive in a muscle graft. In the rat EDL muscle, usually only 2–5 muscle fibers out of $\sim$4,500 survive this procedure.

A highly effective technique for producing massive muscle degeneration and regeneration in rat extraocular muscles is injection of Marcaine into the retrobulbar space (Carlson and Rainin, 1985). In primate muscle such injections must be made directly into the extraocular muscles in order to be effective (Carlson et al., 1992).

Species differences in susceptibility to Marcaine are important. Whereas it is highly myotoxic in the rat, it is less so in primates and even less so in the mouse. For this reason, other myotoxic agents are often preferable in these species.

3.1.5.2 Toxins One of the most widespread models for producing muscle fiber degeneration and regeneration is the application of venom myotoxins (rev. Gutiérrez and Ownby, 2003; Harris, 2003). The many myotoxins can be placed into three main groups – cardiotoxins, small polypeptides from new world vipers; cobra venom motoxins, 60–65 amino acid polypeptides; and phospholipase A_2 toxins, derived from a wide variety of snakes. Some of these toxins act locally and are injected directly into a muscle. Other act systemically and can be injected in the vicinity of a muscle. In models involving direct injection, issues of sparing of some muscle fibers are very similar to those encountered by those who use local anesthetics (d'Albis et al., 1988).

An issue of importance with any myotoxic agent is whether the agent preferentially destroys specific muscle fiber types. For example, slow muscle fibers may have a greater susceptibility than fast fibers to the toxic effects of notexin, a commonly used myotoxic agent (Harris et al., 1975). Another variable that must be taken into account is the vascular damage induced by some toxins. This can change the character of the muscle lesion that is produced.

Myotoxins are not limited to snake venoms. Venom from bees and tarantulas can also cause myonecrosis (Ownby and Odell, 1983; Ownby et al., 1997). A newer model of myotoxicity is injection of the bacterial *Clostridium sordellii* lethal toxin (Barbier et al., 2004). This toxin also causes degenerative changes in the neuromuscular junction, so it introduces a more complex array of pathologies than some of the other toxins.

3.1.5.3 Other chemical agents A wide variety of chemical agents cause muscle damage. The spectrum includes compounds such as organophosphate pesticides (Dettbarn, 1984), antiphlogistic substances (Řeřábková, 1983), neuroleptic drugs (Svendsen, 1983), formocresol (Standish, 1964), and clenbuterol (Burniston et al., 2002). Most of these agents, however, produce sporadic patterns of muscle fiber damage, and none has been adopted as a standard model for studying muscle regeneration.

3.1.6 Ischemia models

Physiologists use a wide variety of ischemia models to study the effects of temporary hypooxygenation on skeletal muscle (e.g. Paoni et al., 2002). Typically, these involve ligation or compression of a major artery to a limb or to a specific muscle. Most of these, however, do not produce complete degeneration of the muscle fibers. Depending upon the severity of restriction of blood flow or the time of total ischemia, the effect on muscle fibers is greater or less (Karpati et al., 1974). Studies on the rat (Phillips et al., 1987) showed that both myonuclear and satellite cell survival is extinguished by 4 hours of ischemia. Both mincing and free transplantation are the most commonly used models that involve ischemia.

3.2 Models Involving Partial Destruction of a Muscle

A number of surgical models produce localized lesions, rather than causing destruction of the entire muscle. Although they may be more difficult to analyze in the laboratory, they approximate more closely than other models injuries that occur in real life to humans.

3.2.1 Transection

Transection of a muscle occurs most frequently during the course of surgery, where the incision penetrates one or more layers of muscle to gain access to some other organ. Transection produces a clean lesion, which is humans is usually followed by scarring at the site. On either side of the incision, the transected muscle fibers often undergo abortive attempts at regeneration, and the regenerated ends commonly become embedded in connective tissue (Äärimaa et al., 2004). Earlier studies on muscle fiber regeneration often applied the term continuous regeneration or budding to this response (Field, 1960).

3.2.2 Crushing or contusion

Many studies have involved inflicting local mechanical trauma to muscles (e.g. Järvinen and Sorvari, 1975; Minamoto et al., 1999; Stauber et al., 1990). Most of these models employ some sort of mechanical apparatus to produce the lesion. A problem inherent in all is to produce reproducible lesions from one animal to the next, especially if analysis of regeneration includes contractile properties of the damaged and regenerating muscle. It is also important to recognize that fast and slow muscles react differently in terms of both their degeneration and regeneration, to crush injury (Bassaglia and Gautron, 1995). In a comparison of regeneration after notexin-induced or crush injury, Fink et al. (2003) noted the recovery of oxidative capacity in the regenerating soleus muscle was diminished after crush injury as compared with after notexin-induced injury.

3.2.3 Amputation

Two early studies reported that muscles amputated close to the origin or near the distal end can regenerate the missing tissue (Dimitrova, 1959; Litver et al., 1961). Carlson (unpublished) was unsuccessful in repeating the experiment of Litver et al., (1961), but was told by a Russian colleague that it really did work. The experiment was repeated (Carlson, 1974). All of the gastrocnemius muscle was removed except for a roughly 5 mm long proximal stump. The nature of the response depended upon whether or not the regenerating Achilles tendon made contact with the stump of the muscle. If no contact was made, minimal regeneration attached the stump to the underlying tissues. On the other hand, if the Achilles tendon regenerated far enough to make contact with the proximal muscle stump, much more regeneration ensued, and a tongue of regenerating muscle grew out of the stump. This shows the importance of mechanical tension on the growth of regenerating muscle. The same concept has more recently been shown to apply to muscles being engineered to grow in vitro (Vandenberg, 1982).

3.2.4 Heat or cold injury

Thermal injury models – either heat or cold – have been used to replicate situations that can damage muscles in humans (Toader-Radu, 1978). Injury is commonly produced by applying hot or cold pieces of metal onto the skin or directly onto a muscle. Another use of freezing injury is to eliminate the possibility of any surviving cells remaining in a muscle, while yet not mechanically destroying the integrity of the muscle. This model has been used to investigate the possibility of an exogenous cellular contribution to a regenerating muscle or to serve as a neutral base for the regeneration of muscle from implanted cells (Morgan et al., 1987; Schultz et al., 1986). The principal concern in freezing models is to be certain that the freezing technique used does, in fact, kill all of the cells in the original muscle. Removing the muscle entirely from the body and subjecting it to repeated freezing and thawing before re-implantation is the method most certain to produce this result.

3.2.5 Innervation models

A variety of animal models have been devised to investigate various aspects of muscle-nerve relationships during muscle regeneration. Some have been focused at the whole muscle level, whereas others have allowed analysis at the level of the single muscle fiber.

3.2.5.1 Denervation models The most drastic perturbation of innervation consists of eliminating all innervation to the regenerating muscle. Classic whole muscle denervation models consist of transecting the sciatic nerve, tying off the stumps to prevent reinnervation and then grafting or mincing a hindlimb muscle. In all cases tried, early muscle fiber regeneration proceeds normally, but there is a significant denervation effect on later stages of regeneration. In both mice (Mufti, 1977) and frogs (Hsu, 1974) the regenerating muscle fibers begin to undergo massive degenerative changes during the third week, and by the end of the third week, almost all muscle fibers within the regenerate have disappeared. Regenerating rat muscle responds somewhat differently to denervation. Early muscle fiber regeneration occurs normally in the absence of nerves, but this is followed by progressive atrophy, rather than degeneration of muscle fibers, starting in the third week (Mong, 1977; Zhenevskaya, 1962).

The response of individual regenerating muscle fibers to denervation was studied in cutaneous pectoris muscle of frogs (Burden et al., 1979). In this model, the motor nerve supplying a thin strip of damaged muscle fibers was cut. Early muscle fiber regeneration occurred and a normal distribution of acetylcholine receptors was found below the persisting basal laminae of the original muscle fibers.

3.2.5.2 Elimination of specific nerve components In early research, Zhenevskaya (rev. 1974) conducted experiments in which specific functional components of the nerves leading to regenerating muscles were eliminated, mainly by tying off components of the spinal roots leading to limb muscles. She established clearly that regenerating mammalian muscle requires motor innervation, but that it would occur in the absence of either sensory or autonomic innervation.

3.2.5.3 Quantitative innervation models Free grafting of the EDL muscle in rats has been used to look at the relationship between different amounts of innervation and the overall course of muscle regeneration. The standard grafting model, which relies upon spontaneous reinnervation, produces regenerates with only about 35% of contractile force and a 45% decrease vs. control in the number of motor units (Côté and Faulkner, 1984). In contrast, the nerve-intact model of grafting (Carlson et al., 1981) results in near normal restoration of mass and contractile function. This is associated with a normal number of motor units in the regenerate (Cederna et al., 2001).

3.2.5.4 Motor endplate-less (MEPless) models Normally regenerating motor nerve fibers regenerate back to the neuromuscular junction region during both nerve and muscle fiber regeneration (Marshall et al., 1977). Even if muscle fibers are prevented from regenerating by x-irradiation, regenerating nerve fibers grow back to the original endplate region of the persisting basal laminae of the original muscle fibers (Sanes et al., 1978). Nevertheless, if the motor endplate zone is surgically removed, nerve fibers regenerating into a regenerating mammalian muscle will form neuromuscular junctions *de novo*. This was shown by Womble (1986), who removed the endplate zone of the rat soleus muscle and surgically connected the two ends of the muscle before grafting it. In this model, the soleus is preferable to the EDL muscle because in the soleus, the MEP zone extends in a straight line across the muscle, in contrast to the EDL in which the MEP zone is distributed in a long "V" shape.

3.2.6 Miscellaneous muscle regeneration models

A number of other animal models are useful for studying various aspects of muscle degeneration and regeneration. Exercise injury models and models involving non-muscle stem cells and genetic manipulation are treated separately in this volume and will not be covered here.

3.2.6.1 Hormonal models One of the classical models of a testosterone-sensitive muscle is the levator ani of the rat. One of the most convenient ways of producing damage in situ is to crush the muscle along a desired length with a hemostat (Gutmann and Carlson, 1978). This muscle is divided in the midline by a raphe, and each half is separately innervated. This allows a number of types of manipulations. In addition to regeneration in situ, the levator ani muscle can be grafted into the bed of a limb muscle. This model was used to determine whether the hormonal sensitivity was neurogenic or myogenic (Carlson et al., 1979).

3.2.6.2 Cellular implantation models The implantation of normal myoblasts into a dystrophic regenerating muscle has been viewed as a potential treatment for muscular dystrophy or other myopathies (Partridge, 2002; Skuk and Tremblay, 2003). The general principle behind these models is that introducing exogenous myoblasts into a diseased or regenerating muscle will provide the basis for and increased mass of functional new muscle and/or normal genes that will replace muscle cells deficient in a gene, such as dystrophin.

The basic technique consists of growing a population of myoblasts in culture and then injecting a defined number of these cells into a diseased muscle or into a lesion in an otherwise normal muscle. Systemic injection techniques have not proven to be effective. Because of the limited migration of the injected cells, it is necessary to make injections into many areas of large muscles, and even then, the injected cells tend to remain in the needle tracks. One of the major issues in myoblast transplantation is the rapid, massive death of the vast majority of implanted cells (Hodgetts et al., 2000). On a longer term basis, immunological rejection is an issue unless autologous cells are transplanted. It is clear that incorporation of transplanted myoblasts is much more effective in areas of regenerating than in normal muscle. A variant on this theme is the implantation of a single muscle fiber into a muscle. If cells of the implant can be distinguished from those of the host, this model allows one to estimate the number of progeny that can be derived from the small number of satellite cells that are associated with the implanted fiber.

3.2.6.3 Hamster cheek pouch model The hamster cheek pouch has been used as transplantation site for several types of tissue, including regenerating skeletal muscle (Faulkner et al., 1983). The strength of this model is the ability to view living tissue in a special chamber introduced into the cheek pouch by intravital microscopy. This model has proven to be very useful in studying the dynamics of revascularization.

3.2.6.4 Diffusion chambers The implantation of pieces of muscle into Millipore diffusion chambers has been used to study some aspects of regeneration under more controlled cellular environmental conditions. This technique was used in mice by Yarom et al. (1982), who reported enhanced regeneration of pieces of human muscle when they were associated with autogenous bone marrow cells. Carlson (unpublished) has found diffusion chambers to be of limited value in rat models because the chambers quickly become surrounded by a capsule of dense connective tissue, which soon causes the degeneration of the cellular contents of the chamber.

3.2.6.5 Micropuncture lesions Karpati and Carpenter (1982) devised a model for producing very fine lesions in individual muscle fibers. Muscles were exposed, and by the use of a hand-operated micromanipulator, tiny puncture lesions were made in single exposed muscle fibers. This model proved useful in studying calcium flow dynamics in an immediate post-injury period and for producing areas of segmental necrosis within single muscle fibers.

3.2.6.6 The fruit-bat web model One early animal model of muscle regeneration was the web musculature of the fruit bat (Church, 1970). These large bats have numerous thin strips of muscle arranged in parallel rows along much of the length of the wing. In addition, the neurovascular bundle is highly accessible. Church (1970) applied crush lesions to some of the muscle bands and studied their subsequent regeneration. On the negative side of this excellent anatomical model are the remote locations of the fruit-bat colonies and the vicious nature of the bats.

3.2.6.7 Loading and unloading models Limb muscles can be made to regenerate under conditions of increased or decreased load. A classical loading model for the soleus muscle is to ablate the synergistic gastrocnemius and plantaris muscles, thus forcing the soleus to assume the entire load of extending the ankle. When these muscles were removed a week after orthotopically transplanting the soleus muscle, the regenerating soleus underwent significant hypertrophy (Coan and Tomanek, 1981).

The converse model is unweighting (Adams et al., 2003). The most common laboratory technique for producing unweighting is the hindlimb suspension model, in which a rat is partially suspended by its tail so that the hindlimbs cannot touch the floor of the cage. The postural soleus muscle is particularly susceptible to unweighting. Not only does it regenerate more poorly than control muscle in the unweighted condition, but upon return to normal gravitational conditions after space flight or to weight-bearing after hindlimb suspension, even the normal soleus shows signs of muscle fiber damage.

4. EVALUATION OF ANIMAL SPECIES AS MODELS FOR MUSCLE REGENERATION STUDIES

Many animal species have been used in experimental studies on muscle regeneration, and each has advantages and disadvantages. This section will outline some of them.

Urodele amphibians are the species of choice for studies of epimorphic muscle regeneration, because they are among the few vertebrates that are capable of doing so. This remains the system of choice for studying muscle fiber dedifferentiation. Because urodele muscle also regenerates by the tissue mode, it remains a good object for comparing the tissue and epimorphic regeneration of muscle. A complication of urodele muscle is the disposition of the satellite cells, which in newts are separated from the muscle fibers by basal lamina material. Whether this constitutes a significant problem or just a minor variation on a theme remains to be seen.

Among the mammals, the rat has been the traditional species for creating surgical models of regeneration. A major advantage of the rat is the tremendous amount of collected background information already available on muscle regeneration in that species. Although convenient for experimental surgical models, the rat is also not ideal for models involving human surgical models because of its small size. For application of muscle regeneration techniques to humans, dimensionality is very important. Especially in free grafting models, much of what works in a rat is ineffective in humans or even larger mammals because both revascularization and reinnervation are not very successful in freely grafted large muscles. Another disadvantage of the rat is its tendency to produce much more scarring after trauma than smaller rodents.

The biggest disadvantage of the mouse is its small size, but the huge number of disease and genetic models in the mouse more than compensates for the size disadvantage. However, as the mouse is becoming more commonly used for regeneration studies, techniques are being adapted to fit the mouse. Mice are not ideal for studies of denervation and reinnervation because of their small size. It is much more difficult to maintain mouse muscle in a denervated condition than it is for rats, rabbits or larger animals. Because of their small size, ischemia models are not convenient in the mouse. Scarring is not as great a problem in the mouse as it is in the rat. Although now the use of cellular markers, such as green fluorescent protein, is now routine in the mouse, it is noteworthy that early myotubes of mice and rats are easily distinguishable by histology alone. This has been used in rat-to-mouse transplantation studies (Phillips et al., 1987).

The dog has seen limited use as a muscle regeneration model, despite the fact that there is a dystrophic strain. Nevertheless, earlier studies have shown that of all the laboratory animals, muscle regeneration in the dog is closer to that of the human than in any other non-primate mammal (Studitsky and Ignatieva, 1961).

Because of the expense, primates have been little used in studies of muscle regeneration. However, when available, they model the time course and distribution of damage from agents, such as local anesthetics, in a way much more comparable to the human situation than that from other laboratory animals (Komorowski et al., 1990). Similarly, primates are important models in experimental studies of myoblast transplantation (Skuk and Tremblay, 2003).

REFERENCES

Äärimaa V, Kääriäinen M, Vaittinen S, Tanner J, Järvinen T, Best, T, Kalim H (2004) Restoration of myofiber continuity after transection injury in the rat soleus. Neuromusc Disord 14:421–428

Adams GR, Caiozzo VJ, Baldwin KM (2003) Skeletal muscle unweighting: spaceflight and ground-based models. J Appl Physiol 95:2185–2201

Barbier J, Popoff MR, Molgó J (2004) Degeneration and regeneration of murine skeletal neuromuscular junctions after intramuscular injection with a sublethal dose of Clostridium sordellii lethal toxin. Infect Immun 72:3120–3128

Bassaglia Y, Gautron J (1995) Fast and slow rat muscles degenerate and regenerate differently after whole crush injury. J Muscle Res Cell Motil 16:420–429

Benoit PW, Belt WD (1970) Destruction and regeneration of skeletal muscle after treatment with a local anesthetic, bupivacaine (Marcaine). J Anat 107:547–556

Burden SJ, Sargent PB, McMahan UJ (1979) Acetylcholine receptors in regenerating muscle accumulate at original synaptic sites in the absence of the nerve. J Cell Biol 82:412–425

Burke GW, Fedison JR, Jones CR (1972) Muscle degeneration produced by local anesthetics. Virginia Dent J 49:33–37

Burniston JG, Ng Y, Clark WA, Colyer J, Tan L-B, Goldspink DF (2002) Myotoxic effects of clenbuterol in the rat heart and soleus muscle. J Appl Physiol 93:1824–1832

Carlson BM (1970) Relationship between the tissue and epimorphic regeneration of muscles. Am Zool 10:175–186

Carlson BM (1972) The Regeneration of Minced Muscles. Basel: Karger, 128 pp

Carlson BM (1974) Regeneration from short stumps of the gastrocnemius muscle. Experientia 30: 275–276

Carlson BM (1979) The relationship between the tissue and epimorphic regeneration of muscle. In Muscle Regeneration, ed., A. Mauro. New York: Raven Press, pp. 57–71

Carlson BM, Emerick S, Komorowski TE, Rainin EA, Shepard BM (1992) Extraocular muscle regeneration in primates: Local anesthetic-induced lesions. Ophthalmology 99:582–589

Carlson BM, Faulkner JA (1989) Muscle transplantation between young and old rats: Age of host determines recovery. Am J Physiol 256 (Cell Physiol 25):C1262–C1266

Carlson BM, Gutmann E (1972) Development of contractile properties of minced muscle regenerates in rats. Exp Neurol 36:239–249

Carlson BM, Gutmann E (1975) Regeneration in free grafts of normal and denervated rat muscles. Contractile properties. Pflügers Arch 353:215–225

Carlson BM, Gutmann E (1976) Free grafting of the extensor digitorum longus muscle in the rat after Marcaine pretreatment. Exp Neurol 53:82–93

Carlson BM, Herbrychová A, Gutmann E (1979) Retention of hormonal sensitivity in free grafts of the levator ani muscle. Exp Neurol 63:94–107

Carlson BM, Hník P, Tuèek S, Vejsada R, Bader DM, Faulkner JA (1981) Comparison between grafts with intact nerves and standard free grafts of the rat extensor digitorum longus muscle. Physiol Bohemoslovaca 30:505–513

Carlson, BM, Rainin EA (1985) Rat extraocular muscle regeneration: Repair of local anesthetic-induced damage. Arch Ophthalmol 103:1373–1377

Cederna PS, Asato H, Gu X, van der Meulen J, Kuzon WM, Carlson BM, Faulkner JA (2001) Motor unit properties of nerve-intact extensor digitorum longus muscle grafts in young and old rats. J Gerontol:Biol Sci 56A:B254–B258

Church JCT (1970) Cell quantitation in regenerating bat web muscle. In Regeneration of striated muscle, and myogenesis, eds. Mauro A, Shafiq SA, AT. Milhorat. Amsterdam: Excerpta Medica, pp. 101–117
Coan MR, Tomanek RJ (1981) The growth of regenerating soleus muscle transplants after ablation of the gastrocnemius muscle. Exp Neurol 71:278–294
Conboy IM, Conboy MJ, Wagers AJ, Girma ER, Weissman IL, Rando TA (2005) Rejuvenation of aged progenitor cells by exposure to a young systemic environment. Nature 433:760–764
Côté C, Faulkner JA (1984) Motor unit function in skeletal muscle autografts of rats. Exp Neurol 84:292–305
d'Albis A, Couteaux R, Janmot C, Roulet A, Mira J-C (1988) Regeneration after cardiotoxin injury of innervated and denervated slow and fast muscles of mammals. Euro J Biochem 174:103–110
Dettbarn W-D (1984) Pesticide induced muscle necrosis: Mechanisms and prevention. Fundamen Appl Toxicol 4:S18–S26
Dimitrova A (1959) Regeneration of amputated muscles in mammals (Bulgarian). Bulgar Akad Nauk 3:165–173
Faulkner JA, Maxwell LC, Mufti SA, Carlson BM (1976) Skeletal muscle fiber regeneration following heterotopic autotransplantation in cats. Life Sci. 19:289–95
Faulkner JA, Weiss SW, McGeachie JK (1983) Revascularization of skeletal muscle transplanted into the hamster cheek pouch: Intravital and light microscopy. Microvasc Res 26:49–64
Field EJ (1960) Muscle regeneration and repair. In Structure and function of muscle, Vol. 3, ed. GH. Bourne. New York: Academic Press, pp. 139–170
Fink E, Fortin D, Serruriere B, Ventura-Clapier R, Bigard AX (2003) Recovery of contractile and metabolic phenotypes in regenerating slow muscle after notexin-induced or crush injury. J Muscle Res Cell Motil 24:421–429
Freilinger G, Holle J, Carlson BM (1981) Muscle Transplantation. Wien: Springer Verlag, 311 pp
Foster AH, Carlson BM (1980) Myotoxicity of local anesthetics and regeneration of the damaged muscle fibers. Anesth Analg 59:727–736
Galle S, Yanze N, Seipel K (2005) The homeobox gene Msx in development and transdifferentiation of jellyfish striated muscle. Interenat J Devel Biol 49:961–967
Gulati AK, Swamy MS (1991) Regeneration of skeletal muscle in streptozotocin-induced diabetic rats. Anat Rec 229:298–304
Gutiérrez JM, Ownby CL (2003) Skeletal muscle degeneration induced by venom phospholipases A_2: insights into the mechanisms of local and systemic myotoxicity. Toxicon 42:915–931
Gutmann E, Carlson BM (1975) Contractile and histochemical properties of regenerating cross-transplanted fast and slow muscles in the rat. Pflügers Arch 353:227–239
Gutmann E, Carlson BM (1978) The regeneration of a hormone-sensitive muscle (levator ani) in the rat. Exp Neurol 58:535–548
Harris JB (2003) Myotoxic phospholipases A_2 and the regeneration of skeletal muscles. Toxicon 42: 933–945
Harris JB, Johnson MA, Karlsson E (1975) Pathological responses of rat skeletal muscle to a single subcutaneous injection of a toxin isolated from the venom of the Australian tiger snake, Notechis scutatus. Clin Exp Pharmacol Physiol 2:383–404
Hodgetts SI, Beilharz MW, Scalzo AA, Grounds MD (2000) Why do cultured transplanted myoblasts die in vivo? DNA quantification shows enhanced survival of donor male myoblasts in host mice depleted of CD4+ and CD8+ cells or NK1.1+ cells. Cell Transplant 9:489–502
Hsu L (1974) The role of nerves in the regeneration of minced skeletal muscles in adult Anurans. Anat Rec 179:119–136
Järvinen M, Sorvari T (1975) Healing of a crush injury in rat striated muscle. Acta Path Microbiol Scand, Sec. A 83:259–265
Karpati G, Carpenter S (1982) Micropuncture lesions of skeletal muscle cells: a new experimental model for the study of muscle cell damage, repair, and regeneration. In Disorders of the motor unit, ed. DL. Schotland. New York: John Wiley & Sons, pp. 517–533
Karpati G, Carpenter S, Melmed C, Eisen AA (1974) Experimental ischemic myopathy. J Neurol Sci 223:129–161

Komorowski TE, Shepard B, Okland S, Carlson BM (1990) An electron microscopic study of local anesthetic-induced skeletal muscle fiber degeneration and regeneration in the monkey. J Orthop Res 8:495–503
Lepeshinskaya OB (1945) The origin of cells from living substance and the role of living substance in the organism (Russian). Moscow: Izdatel Akad Nauk SSSR, 231 pp
Litver GM, Dampel NN, Simelson IB, Kostkin VB (1961) Organic regeneration of skeletal muscles in rats (Russian). Byull Eksp Biol Med 52:101–105
Markley JM, Faulkner JA, Carlson BM (1978) Regeneration of skeletal muscle after grafting in monkeys. Plastic Reconstr Surg 62:415–422
Marshall LM, Sanes JR, McMahan UJ (1977) Reinervation of original synaptic sites on muscle fiber basement membrane after disruption of the muscle cells. Proc Natl Acad Sci USA 74:3073–3077
Minamoto AB, Grazziano CR, Salvini TDF (1999) Effect of single and periodic contusion on the rat soleus muscle at different stages of regeneration. Anat Rec 254:281–287
Mong FSF (1977) Histological and histochemical studies on the nervous influence on minced muscle regeneration of triceps surae of the rat. J Morphol 151:451–462
Morgan JE, Coulton GR, Partridge TA (1987) Muscle precursor cells invade and repopulate freeze-killed muscles. J Muscle Res Cell Motil 8:386–396
Mufti SA (1977) Regeneration following denervation of minced gastrocnemius muscles in mice. J Neurol Sci 33:251–266
Neerunjun JS, Dubowitz V (1975) Identification of regenerated dystrophic minced muscle transplanted in normal mice. J Neurol Sci 24:33–38
Ownby CL, Odell GV (1983) Pathogenesis of skeletal muscle necrosis induced by tarantula venom. Exp Molec Path 38:283–296
Ownby CL, Powell JR, Jiang MS, Fletcher JE (1997) Melittin and phospholipase A_2 from bee (Apis mellifera) venom causes necrosis of murine skeletal muscle tissue in vivo. Toxicon 35:67–80
Paoni NF, Peale F, Wang F, Errett-Baroncini C, Steinmetz H, Toy K, Bai W, Williams PM, Bunting S, Gerritsen ME, Powell-Braxton L (2002) Time course of skeletal muscle repair and gene expression following acute hind limb ischemia in mice. Physiol Genom 11:263–272
Partridge T (2002) Myoblast transplantation. Neuromusc Disord 12:S3–S6
Phillips GD, Lu D, Mitashov VI, Carlson BM (1987) Survival of myogenic cells in freely grafted rectus femoris and extensor digitorum longus muscles. Am J Anat 180:365–372
Řeřábková L (1983) Poškození kosterního svalu intramuskulárním podáním protozánětlivých latek pyrazolidinového typu. Sbornik Lékař 85:161–166
Salafsky B (1971) Functional studies of regenerated muscles from normal and dystrophic mice. Nature 229:270–273
Sanes JR, Marshall LM, McMahan UJ (1978) Reinnervation of muscle fiber basal lamina after removal of myofibers. J Cell Biol 78:176–198
Schmid V (1992) Transdifferentiation in medusae. Internat Rev Cytol 142:213–261
Schmid V., Reber-Müller S (1995) Transdifferentiation of isolated striated muscle of jellyfish in vitro: the initiation process. Semin Cell Biol 6:109–116
Schultz E, Jaryszak DL, Gibson MC, Albright DJ (1986) Absence of exogenous satellite cell contribution to regeneration of frozen muscle. J Muscle Res Cell Motil 7:361–367
Skuk D, Tremblay JP (2003) Myoblast transplantation: the current status of a potential therapeutic tool for myopathies. J Muscle Res Cell Motil 24:285–300
Standish SM (1964) Striated muscle regeneration after chemical injury. Arch Pathol 77:330–339
Stauber WT, Fritz VK, Dahlmann B (1990) Extracellular matrix changes following blunt trauma to rat skeletal muscles. Exp Molec Path 52:69–86
Studitsky AN (1953) Types of new formations of cells from living substance in processes of histogenesis and regeneration (Russian). Zhur Obshch Biol 14:177–197
Studitsky AN (1959) The Experimental Surgery of Muscles (Russian). Moscow: Izdatel Akad Nauk SSSR, 338 pp
Studitsky AN, Ignatieva ZP (1961) Restoration of muscles in higher mammals (Russian). Moscow: Izdatel Akad Nauk SSSR, 192 pp

Svendsen O (1983) Local muscle damage and oily vehicles: A study on local reactions in rabbits after intramuscular injection of neuroleptic drugs in aqueous or oily vehicles. Acta Pharmacol Toxicol 52:298–304

Toader-Radu M (1978) Dynamics of regeneration in skeletal muscle following localized heat injury. Rev Roum Morphol Embryol Physiol, Morphol-Embryol 24:69–73

Tsonis PA (1996) Limb Regeneration. Cambridge: Cambridge Univ Press, 241 pp

Vandenberg HH (1982) Dynamic mechanical orientation of skeletal muscle myofibers in vitro. Devel Biol 93:438–443

Wallace H (1981) Vertebrate Limb Regeneration. Chichester: John Wiley & Sons, 276 pp

Womble MD (1986) The clustering of acetylcholine receptors and formation of neuromuscular junctions in regenerating mammalian muscle grafts. Am J Anat 176:191–205

Yarom R, Meyer S, Carmy O, Ghidoni B, More R (1982) Enhancement of human muscle growth in diffusion chambers by bone marrow cells. Virchows Arch (Cell Pathol) 41:171–180

Zhenevskaya RP (1962) Experimental histologic investigation of striated muscle tissue. Rev Canad Biol 21:457–470

Zhenevskaya RP (1974) Neurotrophic regulation of plastic activity of muscular tissue (Russian), Moscow: Izdatel Nauka, 239 pp

CHAPTER 9

SKELETAL MUSCLE RECONSTITUTION DURING LIMB AND TAIL REGENERATION IN AMPHIBIANS: TWO CONTRASTING MECHANISMS

ELLY M. TANAKA
Max-Planck Institute of Molecular Cell Biology and Genetics, Pfotenhauerstrasse 108, 01307 Dresden, Germany

"Eye of newt and toe of frog"........the potent properties of amphibian tissues was recognized by William Shakespeare in his dark drama Macbeth. Indeed the amphibian models have provided important contributions and insight to muscle regeneration. Most obviously, the satellite cell was first described by Mauro (1961) in frog muscle (Mauro, 1961). A further dramatic aspect of the amphibians is their ability to regenerate not only muscle, but entire limb and tail structures (Fig. 1). The anurans (frogs and toads) display limb and tail regeneration during tadpole stages but lose this ability upon metamorphosis whereas the caudata (salamanders such as newts and axolotls) retain the ability to regenerate their limbs and tails throughout life.

During limb and tail regeneration the wholesale growth and reformation of muscle tissue is coordinated with the growth and patterning of other tissues such as bone, nerve and skin. The regeneration of muscle under these conditions results in a much closer replica to the original structure compared to normal muscle regeneration, which results in muscle that is functional but has noticeable deficits or distinctions from the original, such as centrally rather than peripherally located myonuclei (see (Carlson, 2003)). An important question is whether the satellite cell activation used in normal muscle tissue repair is deployed and is sufficient to support wholesale regeneration of muscle during limb regeneration, which would suggest that extracellular factors solely determine the mode of muscle regeneration, or whether limb and tail regeneration involve activation of muscle progenitors by a distinct mechanism compared to normal muscle regeneration. While the answers

S. Schiaffino and T. Partridge (eds.), Skeletal Muscle Repair and Regeneration, 181–197.

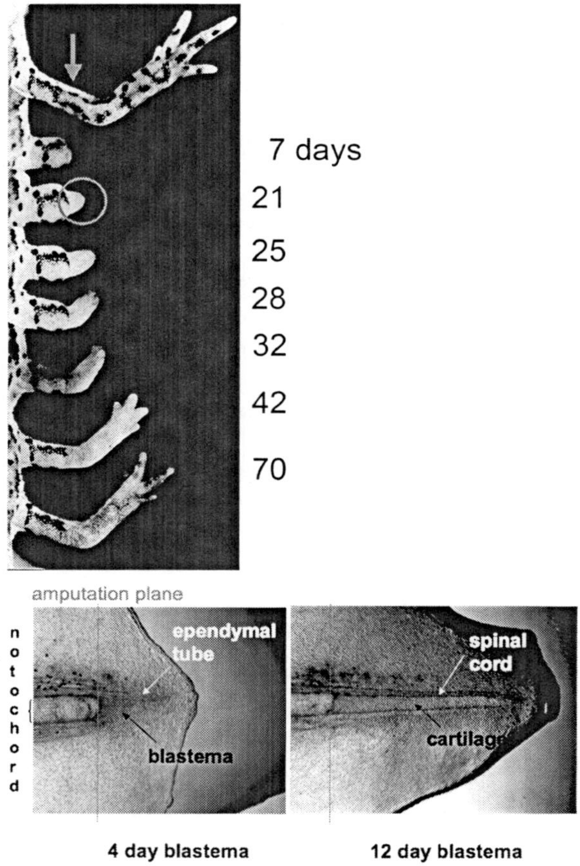

Figure 1. A. The progression of newt limb regeneration. Amputation of the newt limb in the upper arm is followed by wound closure, then by blastema formation by 21 days (red circle). In the subsequent weeks, the limb undergoes growth and patterning to reconstitute a patterned limb structure by 70 days. (From Goss, 1969). B. Tail regeneration in larval axolotl. Amputation of the tail results in outgrowth of the spinal cord as an ependymal tube, and blastema formation. By 12 days, formation of the ventral cartilage rod is visible. Later myotomes will form along the tail.

to such questions are not yet completely resolved, a number of recent experiments in different amphibian models are beginning to address these fascinating questions. Most intriguingly, it now appears that frogs and salamanders may employ rather different mechanisms of recruiting muscle progenitor cells into regenerating appendages. Frogs appear to exclusively rely on satellite cells, while salamanders additionally use the unique mechanism of muscle fiber dedifferentiation. A question is whether the cellular mechanism of dedifferentiation accounts for the life-long ability of salamanders to regenerate in contrast to the temporally limited ability of frogs to undertake regeneration.

Considering the distinctions between frogs and salamanders, in this chapter I will describe our understanding of how muscle cells are recruited to appendage regeneration separately for salamanders and frogs. Then the chapter will close with some speculation concerning the relationship between the amphibian regeneration to mammalian regeneration.

1. DEDIFFERENTIATION OF SKELETAL MUSCLE FIBERS VERSUS ACTIVATION OF SATELLITE CELLS DURING SALAMANDER LIMB AND TAIL REGENERATION

In contrast to muscle repair that relies largely on satellite cells of the myogenic lineage, regeneration of an entire limb or tail proceeds by forming a complex pool of progenitor cells called the blastema at the surface of the cut appendage (Fig. 2A). For example the limb blastema, which morphologically resembles the developing limb bud, harbors the progenitors to form the different cell types of the limb including muscle, tendon, cartilage, dermis, nerve sheath, and blood vessels. This blastema lies directly underneath, and depends on signals from a specialized epithelial structure called the wound epidermis that provides essential signals that support regeneration (see (Stocum, 1995; Tanaka, 2003). An important question is how does the mature tissue produce these progenitor cells? Are the progenitor cells committed to the individual lineages such as muscle, or do they have a broader multipotency?

The genesis of muscle progenitors from muscle tissue during salamander appendage regeneration has long been controversial due to the proposal by a number of researchers that dedifferentiation occurs, a mechanism completely separate from satellite cell activation (Fig. 2A) (Hay, 1959; Lentz, 1969). Interestingly, histological examination of newt muscle many years ago revealed no classical satellite cells beneath the muscle basal lamina but rather an interstitial cell outside the basal lamina was found (Popiela, 1976). This finding suggested that perhaps newt muscle was unique in lacking a reserve satellite cell and thus perhaps depended on an alternative mechanism such as dedifferentiation. While the detailed, histological descriptions of dedifferentiation were quite compelling, such studies were unable to directly track the fate of the muscle fibers in the regenerating tissue, and thus could not determine if such "dedifferentiating cells" were viable and truly contributed to the blastema (Hay, 1959; Lentz, 1969).

Recent studies employing new cellular and molecular tools have broken through some of these roadblocks and have come to a surprising situation. On one hand a number of cell tracking studies confirm that muscle dedifferentiation does occur during regeneration in salamanders. On the other hand, it has recently been shown that newt muscle tissue does contain satellite cells that are molecularly similar to satellite cells from other organisms. Given the apparent existence of both progenitor cell mechanisms in salamander tissue, the exciting question that remains is what is the relative contribution of muscle dedifferentiation versus satellite cell activation? Do the progenitor cells that arise from dedifferentiation versus

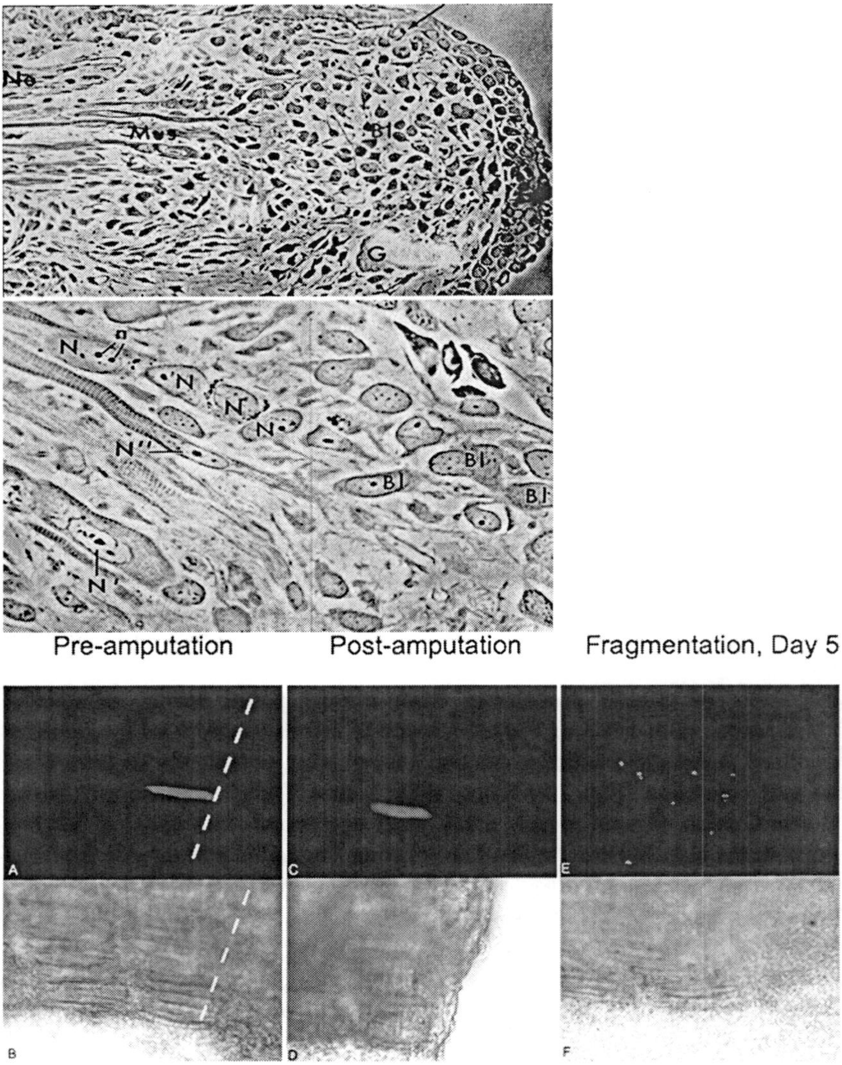

Figure 2. A. Longitudinal section of a regenerating newt limb blastema. Upper panel: The blastema cells (Bl) that are the regeneration progenitor cells, is covered with an epidermal cap (Ep). The blastema also contains giant cells (G). On the left is the mature portion of the limb including muscle fibers (Mus) and nerve (Ne). Lower panel: higher zoom of the upper panel highlighting the transition zone between mature tissue and blastema. N" represents nuclei still contained in a muscle fiber. N', nuclei proposed by Hay to be in fragmenting myofibers. N, nuclei proposed to have budded off of muscle fibers and are in the process of forming blastema cells (Bl). Courtesy of (Hay, 1959). B. Lineage tracing of muscle fibers in the axolotl tail. A muscle fiber was injected with rhodamine dextran (left panel) and then the tail cut adjacent to this fiber (middle panel). After 4 days, the fiber fragmented into mononucleate cells that then proliferated and spread at 5 days (right panel). A,C,E fluorescent images of muscle cell; B,D,F fluorescence overlayed with DIC images. Courtesy of (Echeverri et al., 2001)

satellite cell activation harbor distinctive properties or fates? Below are described the studies documenting muscle cell dedifferentiation and satellite cell activation in salamanders.

1.1 Salamander Muscle Dedifferentiation

Several different lineage marking methods have been employed in recent years to establish the phenomenon of muscle dedifferentiation in salamanders. Lo et al. initially probed muscle dedifferentiation by implanting in vitro-formed myotubes that had been lineage labelled via microinjection of rhodamine dextran (Lo et al., 1993). These studies utilized a non-clonal myogenic cell line derived from newt limb muscle that readily forms myotubes under low serum conditions (Ferretti and Brockes, 1988). The myotubes were then sieved away from residual myoblasts and replated prior to injection with fluorescent lineage tracer. One week after implantation of such myotubes into the regenerating limb, histological examination revealed fluorescent tracer in mononucleate cells of the blastema. Later time points showed a larger number of labeled mononucleate cells, indicating the cells had proliferated. At later stages of regeneration, the cell tracer was found in muscle fibers and in rare cases in newly forming cartilage, indicating that the implanted cells had differentiated into two different cell types. These studies also implemented tritiated thymidine as a second tracer in order to confirm that the label recovered in the regenerating tissue truly derived from the implanted cells. Both labels were observed in mononucleate cells after implantation. These groundbreaking results left open the question of whether the observed phenomenon was limited to cultured, implanted myotubes, or whether endogenous, fully differentiated muscle fibers could undergo this transition.

To track endogenous muscle dedifferentiation, Echeverri et al. injected muscle fibers in larval axolotl tails and imaged the fluorescent cells in live animals (Echeverri et al., 2001). Under such conditions, the fibers were observed to fragment into mononucleate cells at 4 or 5 days after tail amputation (Fig. 2B). These cells doubled in number and entered into the blastema but the long-term fate of the cells could not be observed due to dilution of lineage tracer. However since endogenous fibers were being followed information concerning the quantitative contribution of dedifferentiated muscle to the tail blastema could be inferred from these experiments. Based on the frequency of dedifferentiation, and the number of muscle fibers at the amputation plane, it was estimated that approximately 17% of the blastema could arise from muscle dedifferentiation. A highest frequency of 40% of muscle fibers was observed to undergo dedifferentiation suggesting that all fibers may not undergo dedifferentiation after tail amputation. This is consistent with histological studies of the regenerating limb that described muscle dedifferentiation in the surface layer of muscle but not in underlying muscle cells (Lentz, 1969).

Muscle dedifferentiation involves the breakdown of muscle-specific myofibrils, the "budding" of nuclei surrounded by a cytoplasm from the syncitium and the reacquisition of proliferative capacity. The events of budding and cell cycle re-entry represent two distinguishable, independent pathways. In vivo observations of dedifferentiating muscle fibers and implanted myotubes showed that nuclei can already re-enter S-phase while in the muscle fiber, as assayed by uptake of tritiated thymidine or BrdU, prior to cellularization (Hay and Fischman, 1961; Kumar et al., 2000). Velloso et al. further asked whether progression through S-phase was required for subsequent budding by inhibiting progression through S-phase in cultured myotubes via X-irradiation or by microinjection of expression plasmids encoding the cell cycle inhibitor CDKN2, $p16^{INK4}$ prior to implantation (Velloso et al., 2000). Under these conditions, implanted myotubes still formed mononucleate cells indicating that cell cycle progression was not required for budding. Interestingly, the number of labeled mononucleate cells that were recovered in these experiments was lower than from untreated myotubes, presumably due to their inability to proliferate and expand.

What are the signals that trigger muscle dedifferentiation, and what intracellular pathways are involved in this event? Surprisingly little is still known about this phenomenon on a molecular level. The in vivo tracking of dedifferentiating muscle fibers indicated that muscle cells had to be clipped by the injury in order to induce efficient muscle dedifferentiation (Echeverri et al., 2001). Such observations suggest that release of muscle fibers from neighboring contacts, and/or induction of muscle contraction may be one important trigger for dedifferentiation. Both the budding and the muscle cell cycle re-entry have been reconstituted in in vitro cell culture assays but in neither case has the key extracellular factor yet been identified.

1.2 In vitro Reconstitution of Myotube Cell Cycle Re-entry

Cell cycle re-entry of differentiated myotubes has been studied using the newt A1 myogenic cell culture (Tanaka et al., 1997). These newt myoblasts withdraw from the cell cycle and differentiate upon serum starvation, similar to myogenic cells from mouse and other organisms. In contrast to mouse cells, however, the newt myotubes re-enter the cell cycle upon re-addition of serum. Since serum represents the clotted fraction of blood, this attribute indicated a direct link between responses to wounding and the initiation of muscle dedifferentiation. This connection was strengthened by the observation that addition of active thrombin to serum generated a higher level of stimulatory activity within serum (Tanaka and Brockes, 1998; Tanaka et al., 1999). A series of experiments demonstrated that thrombin proteolysis was involved in generating an active molecule in serum that was termed SPRF (Fig. 3). The SPRF activity is a glycosylated protein that circulates in serum as a high-molecular entity (Straube et al., 2004). The molecular identity of the factor is still unknown. The ability of newt myotubes to respond to the SPRF activity is inhibited by contact with other cells (Tanaka et al., 1997). This is likely an in vitro correlate for the requirement to release muscle fibers from their neighbors to induce dedifferentiation in vivo, as described by Echeverri et al. (2001).

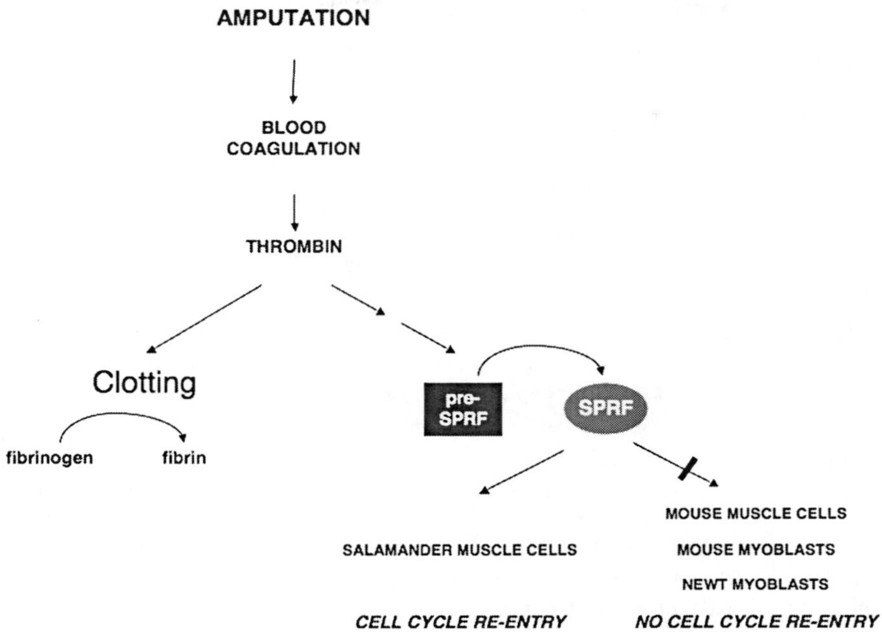

Figure 3. A model for induction of muscle cell cycle re-entry in response to limb amputation. Injury induces clotting of blood that involves activation of the thrombin protease. Thrombin activity also generates an active factor, termed SPRF, that can induce cell cycle re-entry in newt myotubes but not in mouse myotubes

While the in vivo role of the thrombin-stimulated serum factor on muscle dedifferentiation has not yet been tested, the role of thrombin in triggering dedifferentiation has been tested in the lens regeneration system. After removal of the newt lens, regeneration occurs via the dedifferentiation and transdifferentiation of the pigmented epithelial cells of dorsal iris. This again, occurs via the cell cycle re-entry of pigmented cells followed by loss of pigmentation and epithelial character. Imokawa et al. injected the irreversible thrombin inhibitor, Phe-Pro-Arg-Chloromethylketone into the anterior chamber of the eye just after lens removal and observed that S-phase was strongly depressed in the dorsal iris (Imokawa and Brockes, 2003). In vitro assays additionally showed that the pigmented epithelial cells undergo S-phase in response to the thrombin-activated serum (Simon and Brockes, 2002). Therefore the SPRF activity appears to be involved in multiple contexts of dedifferentiation and thus may be a central initiator of regeneration.

The myotube budding activity has also been reported in vitro. McGann et al. reported that a protein extract derived from early regenerating newt limb tissue could induce 16% of cultured newt cultured myotubes to undergo cellularization (McGann et al., 2001). 25% of the myotubes also underwent DNA synthesis. Strikingly, the

Table 1.

Cell cycle re-entry	
SPRF	Newt
Blastema Extract	Newt and Mouse
Cellularization	
Blastema extract	Newt and Mouse
Msxl	Newt* and Mouse

* Depletion of MSXl inhibits cellularization

extract was also active on mouse C2C12 myotubes that showed a similar cell cycle re-entry and cellularization. Addition of extract also apparently caused a decrease in muscle proteins such MyoD, myogenin and troponinT in 15–18% of the myotubes. No further work on the identity of the budding initiating factor has so far been reported.

One unresolved issue is the activity of the extract versus serum on mouse myotubes. Tanaka et al. observed that the serum factor initiated cell cycle re-entry in newt but not mouse myotubes, whereas McGann et al. reported that blastema extract could induce cell cycle re-entry in mouse myotubes (McGann et al., 2001; Tanaka et al., 1997). It is unclear whether the cell cycle re-entry factor in the two preparations is distinctive, or whether the blastema extract causes an preliminary dedifferentiation of mouse myotubes that then makes them receptive to serum cues. The species specific profile of cell cycle re-entry and cellularization in response to different cues is summarized in Table 1. While cultured myotubes require a stimulus from blastema extract, myofibers isolated directly from the newt were observed to spontaneously undergo budding in culture through the formation of "cauliflower" structures that seemed to form colonies of proliferative cells (Kumar et al., 2004). This suggests that the mechanical process of isolation may have been sufficient in these fibers to induce the dedifferentiation response in these cells.

2. INTRACELLULAR COMPONENTS OF THE DEDIFFERENTIATION PATHWAY

The signaling pathway that is induced in dedifferentiating muscle cells is an important question. The transcription factor *msx1* is the only intracellular factor putatively involved in muscle dedifferentiation that has been so far characterized. *msx1* expression was noticed to be associated with regenerative zones; *msx1* is found not only in the limb blastema, but also in the tip of the mouse digit that is the sole region of the mouse limb that is able to undergo regeneration (Koshiba et al., 1998; Reginelli et al., 1995). On a cellular level, constitutive expression of *msx1* was known to block myogenic differentiation of mouse C2C12 cells (Song et al., 1992). Based on these properties, Odelberg et al. asked whether induction of *msx1* in C2C12 myotubes would reverse the differentiated state (Odelberg et al., 2000). Up to 50% of *msx1*-expressing myotubes were observed to have reduced expression of muscle differentiation factors, such as MRF4, myogenin, MyoD and p21. The authors also found that 8.8% of myotubes underwent fragmentation and apparently made proliferative clones, as assayed by following individual myotubes under the light microscope day by day. These colonies could be pushed toward osteogenic, adipogenic and chondrocyte fates when cultured under the appropriate media conditions. These results indicated that *msx1* could be a central regulator of muscle cell dedifferentiation.

Thus far the only further insight into the role of *msx1* in muscle dedifferentiation was provided by Kumar et al. in their dedifferentiating axolotl muscle fiber assay (Kumar et al., 2004). In situ hybridization for *msx1* transcripts in isolated fibers showed positive signal in fibers that were actively forming buds, while "quiescent" fibers that did not have buds were negative for the signal. In the limb blastema, in situ hybridization against *msx1* transcripts had typically been observed in the blastema but not in dedifferentiating muscle fibers (Koshiba et al., 1998). It is not known if this discrepency between the two results is due to the sensitivity of the in situ hybridization technique. To test the function of *msx1* in axolotl myofibers, Kumar et al. scrape loaded the muscle fibers with anti-*msx1* morpholinos and observed a reduced percentage of fibers that underwent the budding process thus implicating *msx1* in the budding process. In contrast, when Schnapp and Tanaka electroporated anti-*msx1* morpholinos into tail muscle fibers in vivo, no negative effect on muscle dedifferentiation was observed (Schnapp and Tanaka, 2005). The two studies utilized morpholinos targeting different *msx1* sequences; Kumar to the 5′ end of the *msx1* sequence described by Koshiba et al. (1998), while Schnapp observed that the axolotl *msx1* sequence likely begins at a methionine further 5′ to the Koshiba sequence, which would make the axolotl sequence more similar in size to other vertebrate *msx1* sequences and thus targeted the starting codons in this region with their morpholino. Morpholinos normally must be targeted to the beginning of the coding region or to splice junctions for effective translational repression. The divergent results could result from the differences in the target sequences.

3. SATELLITE CELLS ARE ASSOCIATED WITH NEWT MUSCLE FIBERS

While dedifferentiation has been a predominant focus of salamander regeneration studies, a recent study of newt muscle fibers unveiled that satellite cells indeed exist in newt muscle. Morrison et al. isolated newt limb muscle fibers to study their in vitro dedifferentiation potential, and to determine whether satellite cells are associated with the fibers (Morrison et al., 2006). In contrast to the Kumar studies, microinjection of lineage tracer into cultured newt muscle fibers did not reveal a significant level of budding and mononucleate cell formation from fibers. A distinction that may explain the discrepency between the two studies was the attachment of the muscle fibers in Morrison's study to Matrigel, compared to the floating, non-attached state of the fibers in Kumar's conditions. The fibers were also isolated from different species of salamander that could respond differently to the culture conditions.

In Morrison's study, careful histological characterization of the newt fibers revealed the presence of Pax7^{+}, M-cadherin^{+} mononucleate cells that are tightly associated with the muscle fiber. Unlike mammalian fibers, a collagen IV matrix lies between the fiber and the satellite cells (Fig. 4). Plating of the individual muscle fibers resulted in the outgrowth of Pax7^{+} and MyoD^{+} mononucleate cells from the fiber. These cultured cells could be induced toward the adipogenic and osteogenic pathways. Additionally the in vivo potential of the satellite cells during regeneration was investigated by pre-labeling culture satellite cells via BrdU incorporation followed by implantation into the limb. Implanted cells contributed to muscle, cartilage and wound epidermis. The contribution to wound epidermis was unexpected, as previous in vivo labeling studies strongly indicate that the wound epidermis derives solely from epidermal rather than internal tissues of the limb (Hay and Fischman, 1961). The use of long-term BrdU incorporation as a labeling technique, which could disrupt normal DNA structure and thus gene expression could have allowed the implanted cells to acquire fates that they do not normally form in vivo. Therefore, while the demonstration that newt muscle harbors satellite cells is an important contribution to our knowledge of the newt limb regeneration system, we still do not know the true contribution of satellite cells versus dedifferentiating muscle fibers to the regeneration blastema. Clearly an essential issue is to

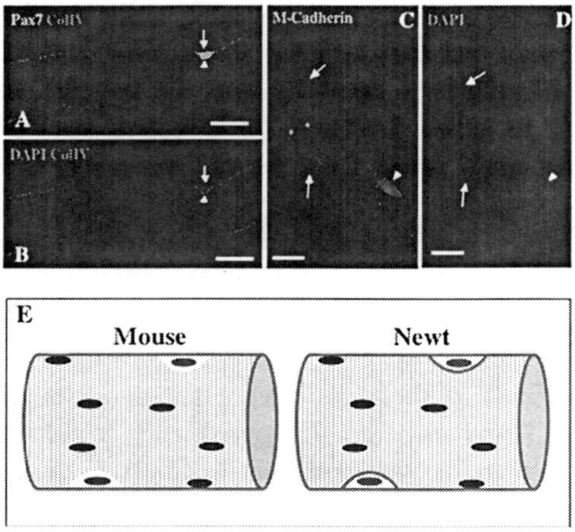

Figure 4. Molecular characterization of newt muscle satellite cells. A. Immunohistochemical staining shows that Pax7^{+} satellite cells lie next to newt muscle fibers but are completely surrounded by a Collagen IV containing basement membrane. C. The satellite cells are also positive for M-cadherin. B,D. DAPI staining of myofibers to highlight all nuclei. E. Schematic outlining the differences between mouse and newt muscle fibers. Courtesy of (Morrison et al., 2006)

track the in vivo contribution of satellite versus dedifferentiating muscle fibers to the regenerate.

4. MUSCLE REGENERATION OCCURS FROM SATELLITE CELLS DURING FROG TAIL REGENERATION

The implicit assumption was that regeneration mechanisms between animals relatively close in evolution such as two major branches of the amphibians–frogs and salamanders–would occur by the same mechanisms. However, a number of observations have accumulated documenting profound differences in limb and tail regeneration between salamanders and frogs. The most obvious manifestation of such a difference can be observed during Xenopus versus axolotl tail regeneration. During larval stages, both Xenopus and axolotls still harbor a notochord along the tail axis (Fig. 1B). Upon amputation of the tail the Xenopus regenerates back the notochord while the axolotl forms a rod of cartilage instead of notochord. In axolotl, the spatial arrangement of the cartilage rod and myotomes are dependent on the orientation of the spinal cord, and indeed the whole process of tail regeneration requires the presence of the spinal cord (Holtzer, 1956). In contrast, Xenopus tail regeneration can occur in the absence of the spinal cord. Finally, the Xenopus spinal cord does not appear to regenerate an exact replica of the original, and is missing some key neuronal cell types (Goss, 1969).

The biology of muscle regeneration is also emerging as one of the distinctions between frogs and salamanders. In an elegant series of experiments Slack and colleagues as well as Ryffel et al. have traced the origin and mechanism of regenerating muscle during Xenopus tail regeneration. Their studies suggest that in contrast to the salamander system where muscle dedifferentiation is implicated as a significant contributor to the blastema, muscle dedifferentiation apparently does *not* contribute to tail regeneration in Xenopus. Rather the satellite cell is likely the main contributor to muscle.

The fate of Xenopus muscle during tail regeneration has been mapped by several means that have come to the conclusion that muscle dedifferentiation does not occur during Xenopus tadpole tail regeneration. Ryffel et al. generated loxP reporter animals where muscle specific expression of cre-recombinase resulted in persistent expression of YFP selectively in muscle cells (Ryffel et al., 2003). When these larval tails were amputated, YFP+ cells were not found in the regenerating blastema until day 16 post amputation. After this day, YFP+ muscle cells formed, which presumably represented newly differentiated muscle in which the muscle specific recombination occurred. This work indicated that muscle fibers do not contribute cells to the regeneration blastema, and that there is no transdifferentiation of muscle cells during Xenopus regeneration. Gargioli and Slack arrived at similar conclusions based on analogous cre/loxP experiments (Gargioli and Slack, 2004).

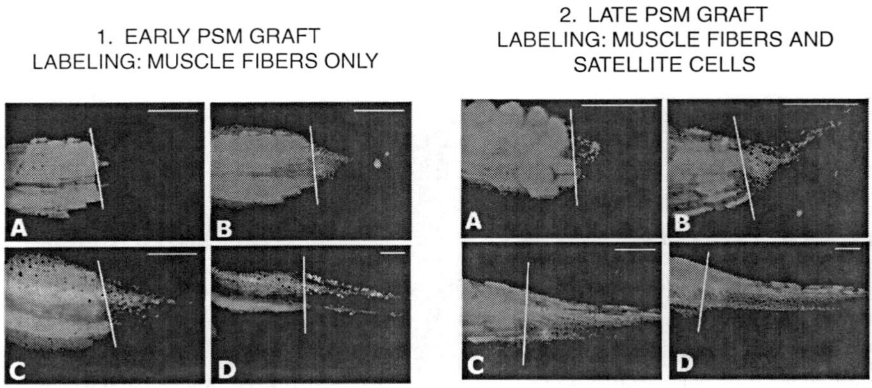

Figure 5. Tracking of muscle fibers and satellite cells during Xenopus tail regeneration. (1) Embryonic Stage 13 transplants of posterior presomitic mesoderm (PSM) from GFP+ donors into unlabeled hosts results in muscle fiber labeling. Amputation of such tails results in no GFP labeling in the tail regenerate. A. Tail regenerate at 3 days post-amputation. B. 5 dpa. C. 10 dpa. D. 20 dpa. White line represents amputation plane. (2) In contrast, embryonic stage 17 grafts of PSM result in muscle fiber and satellite cell labeling. In this case, labeled muscle is observed in the tail regenerate. A. Tail regenerate at 3 days post-amputation. B. 5 dpa. C. 10 dpa. D. 20 dpa. White line represents amputation plane. Courtesy of (Gargioli and Slack, 2004)

Gargioli and Slack were additionally able to test the role of satellite cells in tail regeneration by transplanting the embryonic region that generates satellite cells in the Xenopus tail from a GFP+ donor into a normal host (Gargioli and Slack, 2004). Transplantation of medial presomitic mesoderm at embryonic stage 13 produced labeling exclusively in muscle cells (Fig. 5). Tail amputation of such animals resulted in no GFP+ cells in the regenerating tissue, further confirming that muscle dedifferentiation does not contribute to the tail blastema. In contrast, transplantation of medial presomitic mesoderm at stage 17 resulted in GFP-labeling of both muscle fibers and satellite cells. In such animals, abundant GFP+ cells were observed in the regenerating tissue, and ultimately labeled muscle fibers in the regenerate (Fig. 5). These results have lead to the conclusion that satellite cell activation rather than muscle dedifferentiation occurs during Xenopus tail regeneration.

In a further set of experiments, Chen et al. suggest that satellite cells are likely the main cellular source for muscle regeneration in the Xenopus tail by interfering with Pax7 function during regeneration (Chen et al., 2006). Pax7 function was disrupted using transgenic animals expressing a Pax7EnR fusion construct under the heat shock promoter. Induction of Pax7EnR expression during regeneration resulted in the reduction in satellite cell number close to the amputation plane and in the regenerate. Histological analysis indicated that satellite cells may be undergoing increased apoptosis and not decreased proliferation under these conditions. Interestingly, a normal tail regenerated after the first amputation, but if this regenerated tail was re-amputated (with concomitant induction of Pax7EnR expression during the second regeneration), the secondary regenerates harbored reduced or no muscle. These results suggest that normal Pax7 function is required to maintain the satellite cell pool, and when such a pool is compromised, muscle regeneration in the tail regenerate is impaired. Because the Pax7EnR construct was driven in all cells, and throughout regeneration, it cannot be ruled out that the depletion of muscle arose from effects beyond derangement of muscle satellite cells. It is possible that another cell type failed, or that muscle progenitors in the regenerate itself also failed to contribute to this phenotype. Nonetheless, taking all the Xenopus experiments together, the picture emerges that muscle dedifferentiation is unlikely to occur, and activation of satellite cells is likely the primary source of muscle progenitors in Xenopus tail regeneration.

5. PLASTICITY OF MUSCLE-DERIVED CELLS DURING THE PROCESS OF REGENERATION

The primary focus of the above studies has been which cell within muscle tissue contributes to the regeneration blastema during epimorphic regeneration. A further question is what happens within the blastema to muscle-derived cells—do they acquire plasticity and contribute to other cell fates? Again, the results in the salamander and in frog appear to differ. The clear conclusion from embryonic transplantation work in Xenopus is that each tissue type renews itself but does not contribute to other tissue types (Gargioli and Slack, 2004). This included the work from muscle transplants which gave rise only to GFP+ muscle in the tail regenerate. The situation seems to differ in salamander regeneration. Implantation of lineage marked myotubes resulted in dedifferentiation at early stages and at later stages, redifferentiation into not only muscle fibers but also at a low frequency, incorporation into cartilage (Lo et al., 1993). Likewise, implantation of cultured satellite cells resulted in population of cartilage and epidermis, although in this case, it could not be excluded that the cultured satellite cell preparation did not have some contaminating cell types within it prior to transplantation, and the labeling method could have rendered the cells artificially more plastic (Morrison et al., 2006). Thus, the results indicate that muscle-derived blastema cells in the salamander may have a broader spectrum of fates open to them compared to comparable cells in Xenopus. It should be noted however, that the regenerating larval Xenopus tail harbors no cartilage structures, so the lack of observed plasticity may be due to the lack of cartilage.

6. EVOLUTIONARY CONSIDERATIONS FOR MUSCLE REGENERATION

Amphibians have provided an interesting evolutionary perspective on the potential of muscle regeneration. On one hand, Xenopus employs satellite cells in the process of epimorphic regeneration, and thus provides a perspective that this cell type that is found in all vertebrates can in some contexts participate in the regenerating functional, complex body structures, atleast at early stages in the life cycle.

On the other hand, the implementation by the salamander of dedifferentiation in addition to satellite cell activation suggests that robust, lifelong regeneration could correlate with the unique cellular behavior. In addition to dedifferentiation itself, the cell tracking studies that have been performed so far suggest that muscle and satellite cell-derived blastema cells have a broader potential beyond muscle, and could potentially achieve a plasticity similar to mesenchymal stem cells. More extensive cell lineage tracking experiments of muscle fibers and satellite cells in vivo are imperative to solidify the apparent plasticity of the salamander cells.

If such plasticity of differentiation and cell lineage is truly a prerequisite for bonafide regeneration of complex body structures, the critical issue is whether such plasticity can be induced in mammals. The intriguing results from Keating and colleagues (McGann et al., 2001; Odelberg et al., 2000) would imply that mammalian cells have the potential for plasticity, and it is the environmental cues that are distinctive, however much work must be done to confirm and extend these results. The idea that mouse muscle cells are not irreversibly committed to a differentiated state and that differentiation is a continually maintained state has a concrete basis from work with viral oncoproteins that can induce S-phase re-entry, and also through work with heterokaryons (Blau et al., 1985a; Blau et al., 1985b; Iujvidin et al., 1990; Tiainen et al., 1996). Interestingly, when newt/mouse myotube heterokaryons were produced and challenged with serum it was found that cell cycle re-entry in response to serum is a dominant trait (Velloso et al., 2001). Again, this suggests that the mouse muscle cell may ultimately have the potential to access dedifferentiation and plasticity beyond what they normally achieve in normal muscle repair.

7. WHAT DISTINGUISHES A PROGENITOR CELL INVOLVED IN TISSUE REPAIR VERSUS REGENERATION OF AN ENTIRE LIMB?

As Carlson has pointed out, even in the salamander, the champion of regeneration, injury of muscle results in imperfect repair of muscle while limb amputation results in apparently flawless reproduction and structuring of muscle (Carlson, 2003). This would suggest that the mode of injury can have a large influence on how the progenitor cells rebuild muscle. Limb amputation has a key geometric distinction from lateral limb wounds that induce muscle repair. Complete severing of the limb in the salamander is followed by the migration of epidermis and dermis from around the circumference of the limb toward the center of the cut surface. The congruence of cells from with anterior, posterior, dorsal and ventral identities is an essential trigger for limb blastema formation (Endo et al., 2004). This was shown by grafting cuffs of skin that artificially produced positional discontinuity along the side of the limb, which resulted in the formation of ectopic, supernumerary limbs at the site of positional discontinuity (Bryant and Iten, 1976; Carlson, 1967; Maden and

Turner, 1978). This meeting of cells from different limb areas triggers a series of inductive events that assigns the blastema progenitor cells positional identities along the limb axis. It is unknown whether all progenitor cells (deriving from muscle, cartilage, dermis, nerve) acquire such patterning information in order to execute the complex morphogenetic events of limb regeneration, or whether only selected cell types acquire positional identity, and other cell types are passive passengers that follow the patterning of a few cell types. Preliminary results from our lab examining HoxA gene expression as an indicator of positional identity during regeneration indicates that muscle derived progenitor cells do indeed acquire axial positional identity during limb regeneration (Kragl and Tanaka, unpublished). This leads us to the hypothesis that the distinction between muscle repair and epimorphic regeneration is the acquisition during bonafide regeneration of positional cues for correctly patterning the new structure. This provides the muscle progenitor cells with the correct information to form structures according to developmental programs that are reactivated in these cells. Such developmental programs may not be reinitiated after simple muscle injury. When extrapolated to mammals, then induction of perfect regeneration would necessitate the re-creation of positional identity cues that may no longer be present in adult mammalian tissue. In summary, the amphibian systems, through on the one hand their conservation with other vertebrates, but on the other hand, their unique regenerative abilities provide an important source of information on how to potentially induce better muscle regeneration in other vertebrates.

While satellite cells appear to be universally important for muscle repair, studies in caudates (salamanders) a different, rather unique mechanism of progenitor cell production has been described during bonafide limb and trail regeneration; the dedifferentiation of mature muscle cells.

ACKNOWLEDGEMENTS

I would like to thank Werner Straube for his contributions and ideas to newt muscle dedifferentiation. The work in the lab on this topic was supported by the Max-Planck Gesellschaft, DFG, VW Foundation and the BMBF.

REFERENCES

Blau HM, Chiu CP, Pavlath GK, Webster C (1985a) Muscle gene expression in heterokaryons. Adv Exp Med Biol 182:231–247

Blau HM, Pavlath GK, Hardeman EC, Chiu CP, Silberstein L, Webster SG, Miller SC, Webster C (1985b) Plasticity of the differentiated state. Science 230:758–766

Bryant SV, Iten LE (1976) Supernumerary limgs in amphibians: experimental production in Notophthalmus viridescens and a new interpretation of their formation. Dev Biol 50:212–234

Carlson BM (1967) Studies on the mechanism of implant-induced supernumerary limb formation in Urodeles. I. The histology of supernumerary limb formation in the adult newt, Triturus viridescens. J Exp Zool 164:227–242

Carlson BM (2003) Muscle regeneration in amphibians and mammals: passing the torch. Dev Dyn 226:167–181

Chen Y, Lin G, Slack JM (2006) Control of muscle regeneration in the Xenopus tadpole tail by Pax7. Development 133:2303–2313
Echeverri K, Clarke JD, Tanaka EM (2001) In vivo imaging indicates muscle fiber dedifferentiation is a major contributor to the regenerating tail blastema. Dev Biol 236:151–164
Endo T, Bryant SV, Gardiner DM (2004) A stepwise model system for limb regeneration. Dev Biol 270:135–145
Ferretti P, Brockes JP (1988) Culture of newt cells from different tissues and their expression of a regeneration-associated antigen. J Exp Zool 247:77–91
Gargioli C, Slack JM (2004) Cell lineage tracing during Xenopus tail regeneration. Development 131:2669–2679
Goss RJ (1969) Principles of regeneration. Academic Press, New York
Hay ED (1959) Electron Microscopic Observations of Muscle Dedifferentiation in Regenerating Amblystoma Limbs. Developmental Biology 1:555–585
Hay ED, Fischman DA (1961) Origin of the blastema in regenerating limbs of the newt Triturus viridescens. An autoradiographic study using tritiated thymidine to follow cell proliferation and migration. Dev Biol 3:26–59
Holtzer S (1956) The inductive activiy of the spinal cord in urodele tail regeneration. Journal of Morphology 99:1–39
Imokawa Y, Brockes J P (2003) Selective activation of thrombin is a critical determinant for vertebrate lens regeneration. Curr Biol 13:877–881
Iujvidin S, Fuchs O, Nudel U, Yaffe D (1990) SV40 immortalizes myogenic cells: DNA synthesis and mitosis in differentiating myotubes. Differentiation 43:192–203
Koshiba K, Kuroiwa A, Yamamoto H, Tamura K, Ide H (1998) Expression of Msx genes in regenerating and developing limbs of axolotl. J Exp Zool 282 :703–714
Kumar A, Velloso CP, Imokawa Y, Brockes JP (2000) Plasticity of retrovirus-labelled myotubes in the newt limb regeneration blastema. Dev Biol 218:125–136
Kumar A, Velloso, CP, Imokawa Y, Brockes, JP (2004) The regenerative plasticity of isolated urodele myofibers and its dependence on MSX1. PLoS Biol 2:1168–1176
Lentz, TL (1969) Cytological studies of muscle dedifferentiation and differentiation during limb regeneration of the newt Triturus. Am J Anat 124:447–479
Lo DC, Allen F, Brockes JP (1993) Reversal of muscle differentiation during urodele limb regeneration. Proc Natl Acad Sci U S A 90:7230–7234
Maden M, Turner RN (1978) Supernumerary limbs in the axolotl. Nature 273:232–235
Mauro A (1961) Satellite cell of skeletal muscle fibers. J Biophys Biochem Cytol 9:493–495
McGann CJ, Odelberg SJ, Keating MT (2001) Mammalian myotube dedifferentiation induced by newt regeneration extract. Proc Natl Acad Sci U S A 98:13699–13704
Morrison JI, Loof S, He P, Simon A (2006) Salamander limb regeneration involves the activation of a multipotent skeletal muscle satellite cell population. J Cell Biol 172:433–440
Odelberg SJ, Kollhoff A, Keating MT (2000) Dedifferentiation of mammalian myotubes induced by msx1. Cell 103:1099–1109
Popiela H (1976) Muscle satellite cells in urodele amphibians: faciliatated identification of satellite cells using ruthenium red staining. J Exp Zool 198:57–64
Reginelli AD, Wang YQ, Sassoon D, Muneoka K (1995) Digit tip regeneration correlates with regions of Msx1 (Hox 7) expression in fetal and newborn mice. Development 121:1065–1076
Ryffel GU, Werdien D, Turan G, Gerhards A, Goosses S, Senkel S (2003) Tagging muscle cell lineages in development and tail regeneration using Cre recombinase in transgenic Xenopus. Nucleic Acids Res 31:e44–e57
Schnapp E, Tanaka EM (2005) Quantitative evaluation of morpholino-mediated protein knockdown of GFP, MSX1, and PAX7 during tail regeneration in Ambystoma mexicanum. Dev Dyn 232: 162–170
Simon A, Brockes JP (2002) Thrombin activation of S-phase reentry by cultured pigmented epithelial cells of adult newt iris. Exp Cell Res 281:101–106

Song K, Wang Y, Sassoon D (1992) Expression of Hox-7.1 in myoblasts inhibits terminal differentiation and induces cell transformation. Nature 360:477–481

Stocum D (1995) "Wound repair, regeneration, and artificial tissues." Springer-Verlag, New York

Straube WL, Brockes JP, Drechsel DN, Tanaka EM (2004) Plasticity and reprogramming of differentiated cells in amphibian regeneration: partial purification of a serum factor that triggers cell cycle re-entry in differentiated muscle cells. Cloning Stem Cells 6:333–344

Tanaka EM (2003) Regeneration: if they can do it, why can't we? Cell 113:559–562

Tanaka EM, Brockes JP (1998) A target of thrombin activation promotes cell cycle re-entry by urodele muscle cells. Wound Repair Regen 6:371–381

Tanaka EM, Drechsel DN, Brockes JP (1999) Thrombin regulates S-phase re-entry by cultured newt myotubes. Curr Biol 9:792–799

Tanaka EM, Gann AA, Gates PB, Brockes JP (1997) Newt myotubes reenter the cell cycle by phosphorylation of the retinoblastoma protein. J Cell Biol 136:155–165

Tiainen M, Spitkovsky D, Jansen-Durr P, Sacchi A, Crescenzi M (1996) Expression of E1A in terminally differentiated muscle cells reactivates the cell cycle and suppresses tissue-specific genes by separable mechanisms. Mol Cell Biol 16:5302–5312

Velloso CP, Kumar A, Tanaka EM, Brockes JP (2000) Generation of mononucleate cells from post-mitotic myotubes proceeds in the absence of cell cycle progression. Differentiation 66:239–246

Velloso CP, Simon A, Brockes JP (2001) Mammalian postmitotic nuclei reenter the cell cycle after serum stimulation in newt/mouse hybrid myotubes. Curr Biol 11:855–858

CHAPTER 10

MUSCLE FIBRE REGENERATION IN HUMAN SKELETAL MUSCLE DISEASES

GEORGE KARPATI[1] AND MARIA J. MOLNAR[1,2]

[1] *Neuromuscular Research Group, Montreal Neurological Institute, McGill University, Montreal, Quebec, Canada*

[2] *Molecular Medicine Division, National Institute of Psychiatry and Neurology, Budapest*

1. INTRODUCTION

1.1 Terms of Reference

The subject of this Chapter is largely restricted to the description and illustration of the salient microscopic and histochemical features of skeletal muscle fibre regeneration in human diseases and brief references to the essential modern concepts of the underlying cellular and molecular mechanisms. For more details, the reader is referred to other chapters in this book, which describe basic cellular and molecular aspects of various forms of myogenic precursor cells as well as molecular signals in muscle fibre growth and development in animal models.

1.2 Terminology

Certain unique features of skeletal muscle fibres are particularly relevant to regeneration. These include large size, elongated shape and multinuclearity (Allbrook, 1981). Definition of the terminology is essential. *Regeneration* is a process of reconstitution of a skeletal muscle fibre following necrosis (or acute cell death) which in muscle fibres is usually segmental. Thus, regeneration must be distinguished from varied types of muscle fibre *repair* that follows different forms of muscle fibre damage but not necrosis. For that reason, designating necrosis as "degeneration" is discouraged. In some instances, actively *regenerating muscle fibres* can be distinguished from muscle fibres that have completed the regenerative

S. Schiaffino and T. Partridge (eds.), Skeletal Muscle Repair and Regeneration, 199–215.

process (*regenerated fibres*). There are precise microscopic distinguishing features of these cells that permit their identification. Since necrosis is a common feature of human myopathies, particularly the dystrophies and certain inflammatory myopathies, regeneration, which usually follows necrosis, is also a common myopathological finding (Carpenter and Karpati, 2001). Skeletal muscle regeneration in mature muscle has many common features with embryonic myogenesis (Zhao and Hoffman, 2004).

2. DESCRIPTION OF THE ESSENTIAL FEATURES OF MUSCLE FIBRE REGENERATION

2.1 Necrosis

Muscle fibre necrosis is one of the two commonest severe forms of damage to muscle fibres, the other being muscle fibre atrophy. Apoptosis is not a proven way of extinction of muscle fibres in human muscle diseases.

Skeletal muscle fibres are multinucleated cells and necrosis almost always involves only a segment of the fibre (Fig. 1). The length of a necrotic segment can greatly vary which implies that the successfully regenerating segment can also vary accordingly (Grounds, 1991). Necrosis entails rapid cell death during which stereotyped cellular changes develop including, loss of the plasmalemma and myonuclear dissolution being the earliest. Gradual dissolution of the contractile material and cellular organelles will follow which leads to the conversion of the necrotic fibre segment into amorphous debris. In about six-eight hours after the to-be-necrotic fibre passed the "point of no return", phagocytic macrophages invade the necrotic segment and the necrotic debris is rapidly removed. An efficacious phagocytic process is essential for optimal regeneration to take place. Two items do not get destroyed in the necrotic segment: the basal lamina and the satellite cells. Both are of utmost importance in the regenerative process (vide infra).

Muscle fibre necrosis is the cardinal destructive process in several forms of muscular dystrophies and inflammatory myopathies such as Duchenne/Becker muscular dystrophy (DMD), most forms of limb girdle dystrophies, congenital dystrophies as well as dermatomyositis or polymyositis (Karpati, 2002). However, in several major diseases in these categories, such as myotonic dystrophy or facioscapulohumeral dystrophy or sporadic inclusion body myositis, necrosis is rare and it is atrophy that leads to progressive reduction of muscle mass. Accordingly, regeneration is only present in diseases where muscle fibre necrosis occurs (Table 1).

The trigger factors for muscle fibre necrosis are varied and in many instances they are not fully elucidated. The best understood situation is in DMD where the central problem is a genetically determined deficiency of the cytoskelatal molecule dystrophin (Anderson, 2002; Nowak and Davies, 2004). This leads to a major secondary reduction of several of the dystrophin-associated proteins (including the dystroglycans and sarcoglycans) as well as an abnormal intracellular redistribution of nNOS. The lack of dystrophin has several deleterious consequences. It

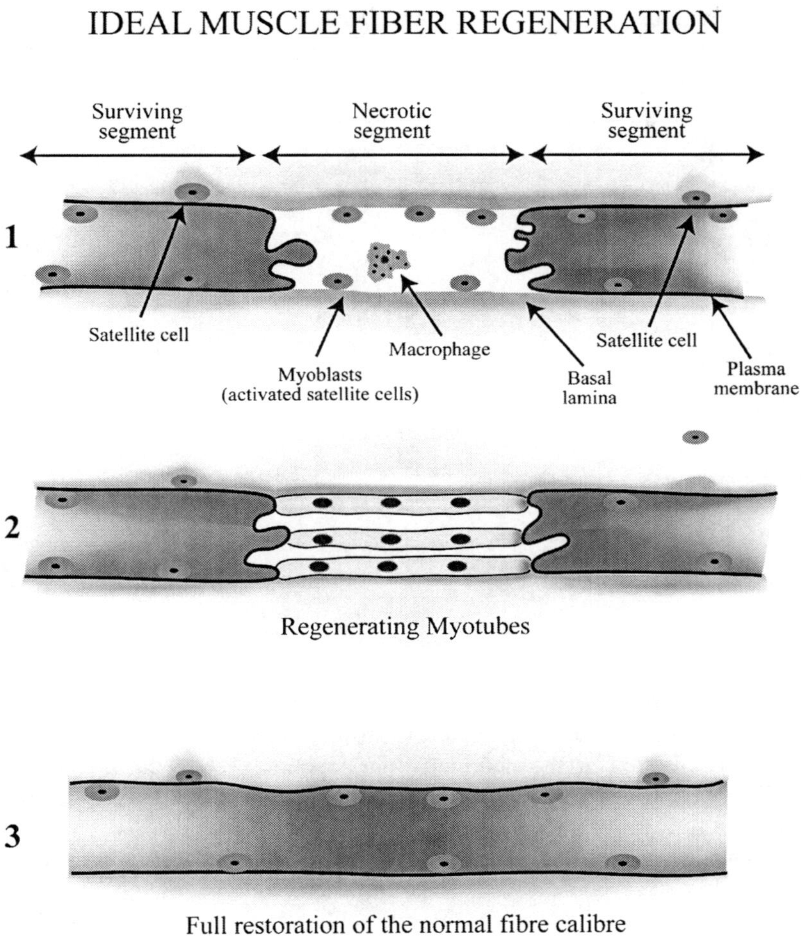

Figure 1. Schematic illustration of regeneration in a necrotic muscle fibre segment. The dimensions are not drawn to scale

leads to an impaired mechanical stability of the surface membrane making muscle fibres particularly vulnerable to lengthening contractions. Furthermore, dystrophin deficiency is suspected to cause perturbed molecular signalling reaching the muscle fibres. The displacement of nNOS from the surface of the muscle fibres is believed to cause reduced NO and a disturbed microcirculation of muscle. Finally, the dystrophin-associated protein deficiencies could cause additional deleterious effects corresponding to their normal physiological functions, most of which are presently unknown. Dystrophin deficiency also causes a relatively mild upregulation of the amount of extrasynaptic utrophin in DMD muscle fibres (Karpati, 1997). Utrophin,

Table 1. Major human skeletal muscle diseases with or without significant muscle regeneration

Necrotic muscle diseases with regeneration	Non-necrotic muscle diseases without regeneration
Dystrophies	
- Duchenne	- Myotonic Dystrophy Types I and II
- Becker	
- Congenital Dystrophies	- Facioscapulohumeral Dystrophy
- Most Limb Girdle Dystrophies	
Inflammatory Muscle Diseases	
- Dermatomyositis	- Inclusion Body Myositis
- Polymyositis	
Metabolic Muscle Diseases	
- McArdle's Disease	- Mitochondrial Diseases
- Malignant Hyperthermia Crisis	
Congenital Myopathies	
- Carnitine Palmytoil Transferase Deficiency (CPT)	- Nemaline Myopathy
	- Central Core Disease

being a close paralogue of dystrophin, could conceivably mitigate the deleterious effects of dystrophin deficiency, provided it is present in sufficient amount outside of its normal localisation at the neuromuscular junction. While the amount of extrasynaptic utrophin in DMD may cause some mitigation of the severity of the disease, it cannot fully prevent the ultimate dismal outcome of the disease. Utrophin is normally expressed in satellite cells, but there is no evidence that its upregulation in transgenic mdx mice specifically influences regenerative capacity.

One could presume that the predisposing molecular alterations for muscle fibre necrosis would have an impact on the myogenic precursor cells and therefore would adversely affect regeneration. However, clear evidence for this has not emerged. For example, the absence of dystrophin does not affect the activation and proliferation of the in situ myogenic precursor cells, as dystrophin is not expressed normally in satellite cells and myoblasts. The impaired regeneration that one sees in DMD is due to other circumstances (vide infra).

2.2 Satellite Cells (Dhawan and Rando, 2005; Seale and Rudnicki, 2000; Zammit and Beauchamp, 2001)

Satellite cells are small, spindle-shaped, dormant mononuclear cells situated between the basal lamina and plasma membrane (Figs. 2, 3). Their developmental origin is still not fully elucidated, but most likely they represent a subpopulation of embryonic myoblasts (Anderson, 2006). The satellite cell prevalence is usually

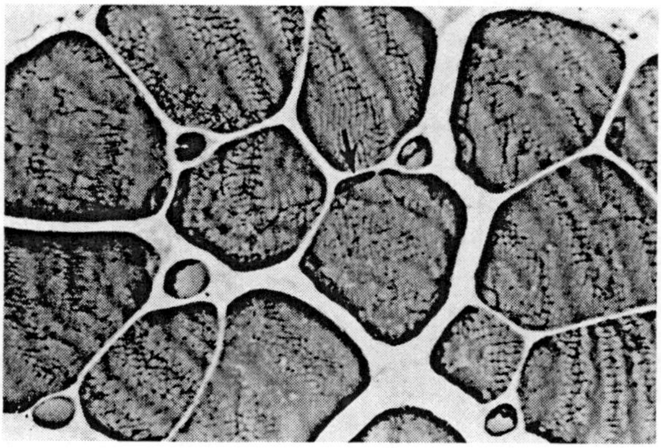

Figure 2. Arrow points to a resting satellite cell in a normal human muscle. Resin-embedded semi-thin section, X1370

expressed by the ratio of the number of satellite cell nuclei and myonuclei in a given fibre segment (rS/M) (Bischoff, 1994).

The most convenient method of displaying resting satellite cells on cryostat sections is by demonstrating their N-CAM or m-cadherin or CD34 reactivity (Fig. 4). Satellite cells have a major role in at least three important processes in skeletal muscle fibres: natural growth, work hypertrophy and regeneration. In this chapter, the focus is on the role of the satellite cells in human muscle fibre regeneration. In postnatal natural muscle fibre growth, they fuse into their parent muscle fibres and thus, their nuclei become myonuclei (Schultz and McCormick, 1994). As a result, the rS/M tends to decrease with age (Snow, 1977). This decrease is, however, mitigated by the fact that satellite cells divide before their fusion into their parent fibre (self-renewal) (Deasy et al., 2005; Zammit et al., 2004). A process similar to the one described in the natural growth of muscle fibres also takes place during work hypertrophy (Collins et al., 2005). There are many important characteristics of satellite cells, along with other potential myogenic cells, that are discussed in

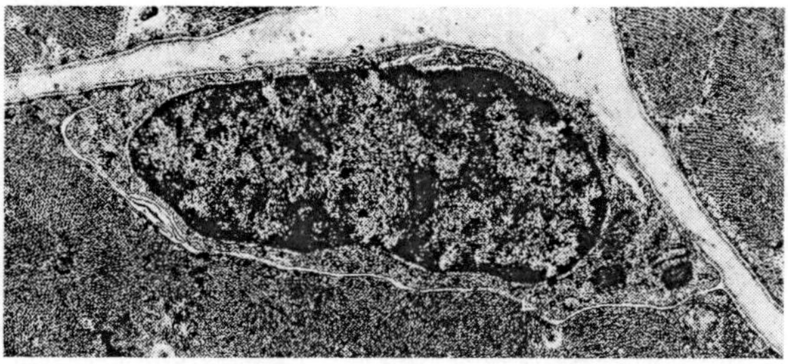

Figure 3. Electron micrograph of a normal human muscle showing a resting satellite cell in a muscle fibre. X18.800

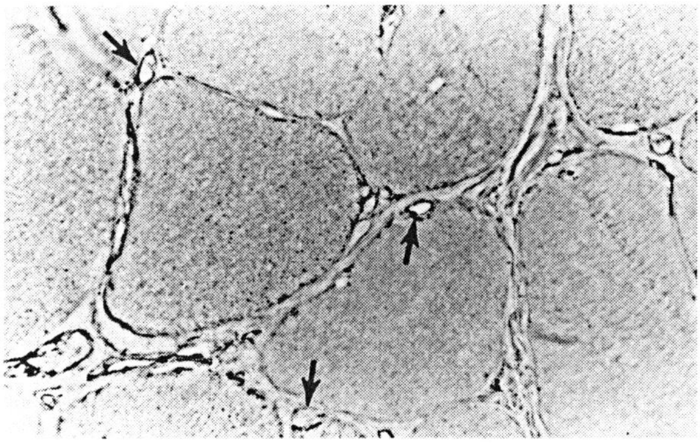

Figure 4. Resting satellite cells in normal human muscle fibres. The surface membrane of these cells shows N-CAM immunoreactivity, X350

various chapters of this volume, including their embryonic derivation and their various biochemical and molecular markers, and whether there are subpopulations of true stem cells (Qu-Petersen et al., 2002; Chen and Goldhamer, 2003) among them "side population" and CD34+ cells.

From the point of view of the satellite cells' role in regeneration (Sherwood and Wagers, 2006), two characteristics are of utmost importance:

a. The nature of the molecular/biochemical signal(s) that trigger the proliferative and differentiating processes in resting satellite cells leading to their transformation into myoblasts/myotubes in necrotic muscle fibre segments (Sherwood and Wagers, 2006);

b. A numerical threshold of their mitotic activity which could set a limit on their regenerating potential in case of recurrent necrosis of regenerated fibre segments (Renault et al., 2000).

It is unclear whether in skeletal muscles in situ, extramuscular myogenic cells can contribute to regeneration. In artificial (experimental) circumstances, blood-vessel-derived mesangioblasts and blood-born hematopoetic stem cells showed myogenic potential (Anversa et al., 2004; Tavian et al., 2005).

Section 2.2 as indicated above and further elaborated upon later, satellite cell activation entails vigorous mitogenic stimulation. This is necessary to generate enough myonuclei, through an adequate number of satellite cells, in the regenerated fibre segments. It appears that, at least, the majority of satellite cells have a limited number of possible mitotic cycles after which they become senescent and not capable of further regenerative action (Mouly et al., 2005). The total limit of possible mitotic cycles of satellite cells has been estimated to be about 60. The stem cell contingent of the myogenic cell population obviously does not have such constraints (Collins et al., 2007).

2.3 Microscopic Features of Human Muscle Regeneration (Fig. 1) (Carpenter and Karpati, 2001)

The following distinct sequential steps characterise the ideal muscle fibre regenerative process, all occurring within the basal laminar tube of the necrotic muscle fibre:

a. Appearance of activated satellite cells /myoblasts adhering to the basal lamina (Fig. 5). At that stage, phagocytosis and regeneration take place pari passu (Fig. 6) but the phagocytic macrophages can be distinguished from the activated myoblasts by their acid phosphatase positivity. Activated satellite cells, now called regenerating myoblasts, undergo multiple cell divisions. It is of interest that satellite cell activation may occur in normal segments of muscle fibres adjacent to the necrotic/regenerating segment (Fig. 5).

b. Fusion of the myoblasts to form strips of multinucleated myotubes still adhering to the basal lamina.

c. Rapid growth of the girth of the regenerative myotubes due to generation of myofibrils.

d. As the girth of myotubes increases, they will touch each other side by side and undergo lateral fusion as well as fusion to the surviving stump of the fibre to form the beginning of a multinucleated regenerating muscle fibre. These regenerating fibres (Figs. 7, 8, 9) are characterised by large, clear myonuclei and prominent nucleoli indicating vigorous transcriptional activity and blue cytoplasm by H&E due to abundant ribosomal RNA indicating vigorous translational activity. Myoblasts are richly endowed by mitochondria and glycogen reflecting vigorous

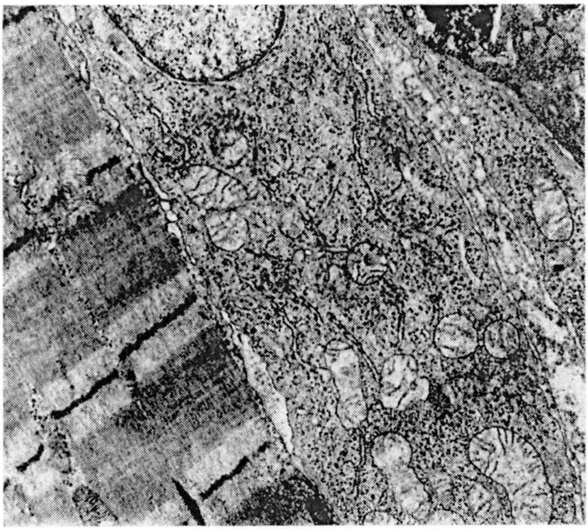

Figure 5. Electron micrograph of an activated satellite cell in a normal segment of a muscle fibre adjacent to a necrotic/regenerating segment, X10.000

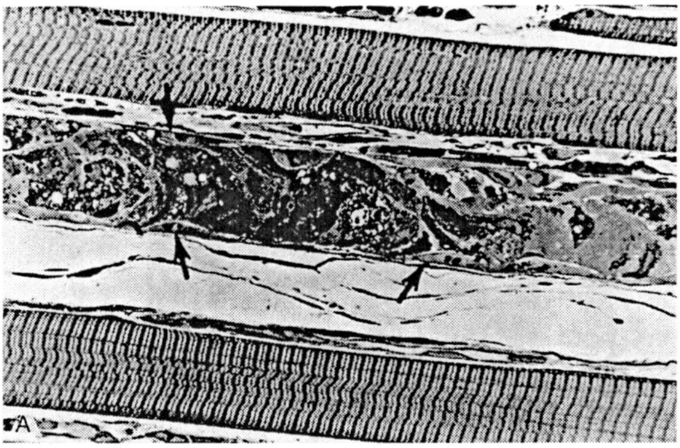

Figure 6. A necrotic fibre segment in which satellite cells (arrows) became activated. This, so called "regen-degen" fibre, has phagocytic macrophages and myoblasts pari passu. Resin-embedded semi-thin section, inflammatory myopathy, X550

energy metabolism. A number of new molecules as markers of regeneration can be displayed by microscopic cytochemistry including cytoplasmic desmin (Fig. 10) and nCAM, developmental myosin heavy chain, diffuse class I MHC, sarcolemmal dystrophin and related molecules as well as utrophin. In human regenerating muscle fibres, the myonuclei are eventually located at the periphery. By contrast, in rodent muscles, a substantial number of myonuclei in regenerating and regenerated fibres remain central indefinitely. This has been a useful feature for measuring a cumulative dystrophic activity in animal models (mdx mice). The regenerated muscle fibre segment will have the same histochemical type as is that of the surviving stumps of the fibre. If the necrosis involves a segment of the fibre harbouring the motor end-plate, the remainder of the fibre

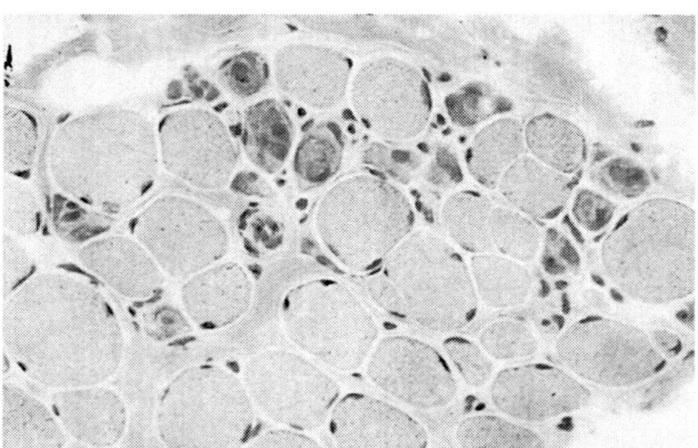

Figure 7. A cluster of small calibre regenerating fibres show basophilic cytoplasm with large, clear myonuclei and prominent nucleoli. Duchenne muscular dystrophy (DMD), Hematoxylin & eosin, X350

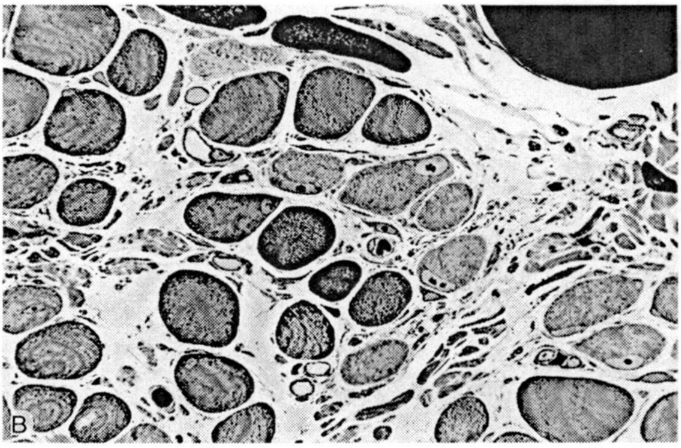

Figure 8. Epon-embedded semi-thin section showing several regenerating fibres with features similar to those described in Fig. 7, DMD, X400

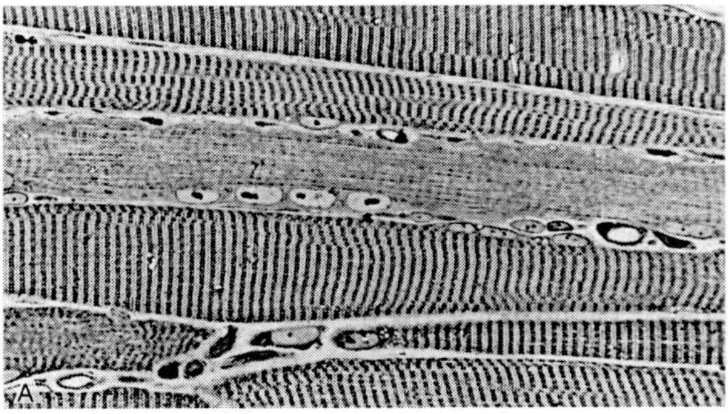

Figure 9. A regenerating muscle fibre similar to the one shown in Fig. 8 is displayed in longitudinal orientation. DMD, X550

behaves as if it were denervated until the regenerated segment with is newly formed endplate re-establishes the normal nerve muscle contact.

e. In ideal regeneration, the calibre of the regenerating fibres gradually increases and attains the girth of the fibre that was there before necrosis. When that happens, the regenerative markers disappear.

2.4 The Impact of the Basic Pathologic Process of a Given Muscle Disease upon Features of Regeneration

As mentioned before, regeneration of muscle fibres occurs only in those muscle diseases in which necrosis is present. Major examples of these diseases are listed in Table 1 along with those in which necrosis is scantly or does not occur.

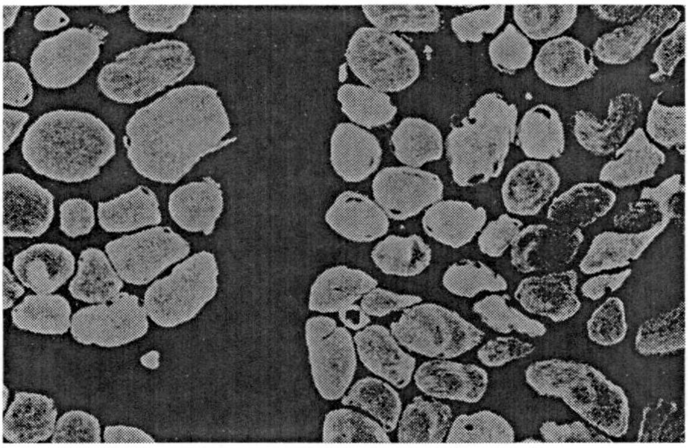

Figure 10. A cluster of regenerating fibres showing strong diffuse desmin immuno-reactivity DMD, X350

Among these diseases, we shall focus on Duchenne muscular dystrophy (DMD) (Anderson, 2002) and dermatomyositis (Hohlfeld, 2002).

In DMD the relentless repeated cycles of necrosis induces repeated regenerative activity, but as noted in Section 2.6, the proliferative capacity of an increasing number of satellite cells becomes gradually exhausted and, as a result, eventually regeneration fails. It is surmised, that the adult myogenic stem cell population, which is supposedly immortal, is not sufficient to maintain an adequate number of myogenic progenitor cells to compensate for the relentless cycles of muscle fibre necrosis and prevent muscle fibre loss.

In DMD progressive accumulation of endomysial connective tissue is a prominent feature and its impact on regeneration has been debated (Bernasconi et al., 1995).

One theory maintains that the major cause of the connective tissue accumulation is that at the site of regeneration failure, the skeins of empty basal lamina (Fig. 12) will serve as a nidus of progressive collagen deposition. According to another theory, the endomysial connective tissue accumulation is a result of the heightened activity of endomesial fibroblasts secondary to a high concentration of fibrogenic cytokines in the endomysial space (Bernasconi et al., 1995). Presumably, the putative fibrogenic cytokines originate from macrophages and other inflammatory cells in the dystrophic muscle. There is no reason to exclude the possibility that both processes operate. Since the excessive endomysial fibrosis is believed to be deleterious for muscle function, inhibition of fibrogenic cytokines could have therapeutic value (Bernasconi et al., 1995).

In dermatomyositis, the basic pathological process is muscle ischemia due to humoral immunity against capillaries and larger vessels of the muscle tissue (Karpati and Carpenter, 1993). As a result of the ischemia, either scattered necrotic fibres or various sized groups of necrotic muscle fibres (infarcts) are observed. The ischemia does not seem to interfere with the activation of satellite cells but the efficiency of phagocytosis is conspicuously impaired. This is reflected by the paucity of phagocytic macrophages in the necrotic fibres. This gives rise to the undue persistence of the necrotic debris and an impaired ability of the regenerative myoblasts to vigorously fuse with each other and then the fused myotubes to undergo lateral fusion (Karpati and Carpenter, 1993). As a result, many of the regenerated fibres appear as forked fibres (see Fig. 13). It is unclear as to what extent the forking of muscle fibres compromises the force generating capacity of the muscle. The only way to minimise this aberration of regeneration in dermatomysositis is to implement appropriate immunotherapy to combat blood vessel destruction and consequent ischemia.

2.5 Peculiar Features of Muscle Fibre Regeneration in Rodents

Since for many human muscle diseases there are suitable rodent animal models, peculiar features of regeneration in these models is of importance. The best example is the mdx mouse in which a spontaneous mutation in the dystrophin gene causes dystrophin deficiency and as such it is a useful model of DMD. Necrosis and regeneration in most muscles of the mature animal is widespread with one striking feature: in the regenerated fibres that a large proportion of myonuclei remain central in position indefinitely (Karpati et al., 1988). This is useful in identifying and measuring of all prior necrosis that had taken place in an mdx muscle. The permanent centronucleation of regenerated muscle fibres is also observed in rats, hamsters and guinea pigs but it contrasts with regenerated human muscle fibres in which myonuclei are peripheral. i.e.: sarcolemmal. The precise cellular mechanism(s) for the persistent central position of most myonuclei in regenerated rodent muscle fibres is undetermined (Turk et al., 2005). It is postulated that myonuclei remain in their normal peripheral position due an anchoring cytoskeletal system which, for some reason, in regenerated rodent muscle failed to be re-established. In contradistinction, all other regenerating features of the regenerated rodent fibres disappear (Figarella-Branger et al., 1999). There is no evidence that the central position of myonuclei in regenerated rodent fibres compromises contractile activity.

2.6 Distinctive Aberrations of Regeneration (Fig. 11)

In Section 2.3 characteristic features of ideal regeneration were described. However, as intimated in Section 2.4, various myopathological influences of the underlying disease may alter the ideal regenerative process causing distinctively aberrant regenerative fibres. These include the following:

a. ***The girth of the regenerated fibres may fall below*** what it should normally be for that muscle at that age. Otherwise, these fibres no longer express the regenerative marker molecules. This aberration is usually due to an insufficient number of myoblasts giving rise to a smaller than necessary number of myonuclei in the regenerated segment (Webster and Blau, 1990). This, in turn, will lead to a reduced cell volume to maintain a standard ratio of myonuclear number and cytoplasmic volume. This is probably the reason of widespread occurrence of small calibre fibres in Duchenne muscular dystrophy (DMD) where the

ABERRATIONS OF MUSCLE FIBER REGENERATION

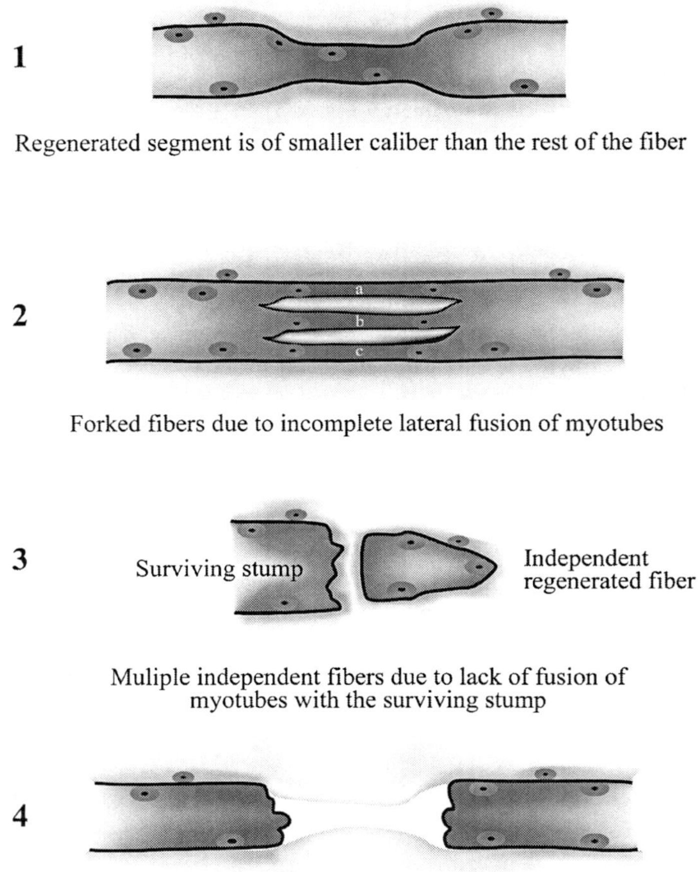

Figure 11. Schematic illustration of four possible types of aberrant regeneration in human muscle diseases. The dimensions are not drawn to scale

recurrent cycles of necrosis progressively reduce the number of satellite cells and their "regenerative vigour" available for regeneration. Therefore, it is probably incorrect to label small calibre fibres in DMD as being "atrophic".

b. ***Regeneration may completely fail*** and the necrotic fibre is not replaced by a regenerated one. These occurrences are ascertainable by visualising an empty ruffled basal laminar tube on epon histology or electron microscopy (Fig. 12). This aberration is due to the same circumstances in an extreme degree as explained under **a**. This occurs in muscles in DMD and is responsible for the

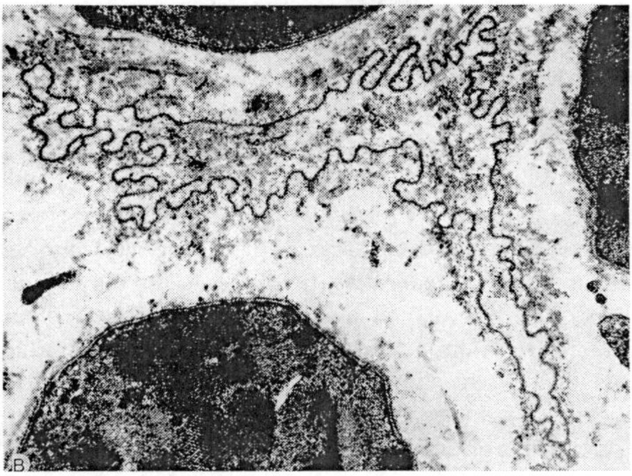

Figure 12. Electron micrograph showing a skein of empty basal lamina indicating a total lack of regeneration of a necrotic fibre (segment). DMD, X20000

progressive loss of muscle fibres and the severe clinical phenotype (Mouly et al., 2005).

c. ***Forking of muscle fibres*** occurs when the lateral fusion of the regenerating myotubes does not take place (Chou and Nonaka, 1977) usually because the poor removal of the necrotic debris by sluggish phagocytosis (Fig. 13). This is most common in ischemic myopathies such as dermatomyositis. In such a situation where there was a single muscle fibre of a certain calibre before necrosis, at the end of regeneration several smaller diameter muscle fibres will be present. The fusion of the myotubes to the surviving stump is not disturbed, hence

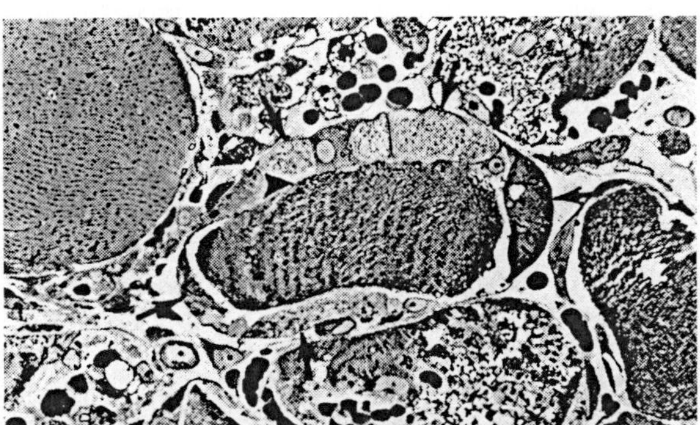

Figure 13. A regenerating muscle fibre contains prominent peripheral myotubes but there is a large central portion of persisting necrotic debris due to poor phagocytosis. This will probably result in forked fibres. Ischemic myopathy, resin-embedded semi-thin section, X550

the appearance of "forking" of the regenerated segment. Each small calibre forked fibre will be surrounded by its own individual basal lamina. While the functional consequence of this aberration of regeneration is unknown, it is possible that the contractile function of the muscle may be compromised. The forked muscle fibres in a cluster are of the same histochemical type, hence the designation of "myopathic type grouping" (Fig. 14) as distinct from the more common neuropathic type grouping in which the number of grouped fibres is usually higher. The most common cause of forking of regenerated muscle fibres is impaired lateral fusion of the regenerating myotubes. This may occur if the phagocytosis is sluggish and removal of the necrotic debris is slow or incomplete. Such an occurrence is common in dermatomyositis where muscle ischemia reduces the number of available blood-born macrophages that would be destined to become phagocytes.

d. Another rare but deleterious form of aberration is when the *regenerating myotubes fail to fuse with one or both surviving stumps*. In such instances, the continuity of the muscle fibre is interrupted that would negatively affect force generation by such fibre even if the regenerated segment attained a normal calibre.

There is no evidence that in a denervated muscle fibre the expected regenerative process is significantly disturbed. However, there is probable evidence that in old age, muscle fibre regeneration, at least in certain muscles, is suboptimal probably due to the fact that more and more satellite cells become senescent as their regenerative potential is used up for the restitution of muscle fibres that become necrotic from normal usage involving lengthening contractions (Grobler et al., 2004).Other factors that seem to adversely affect muscle fibre regeneration include ischemia and protein malnutrition of the host.

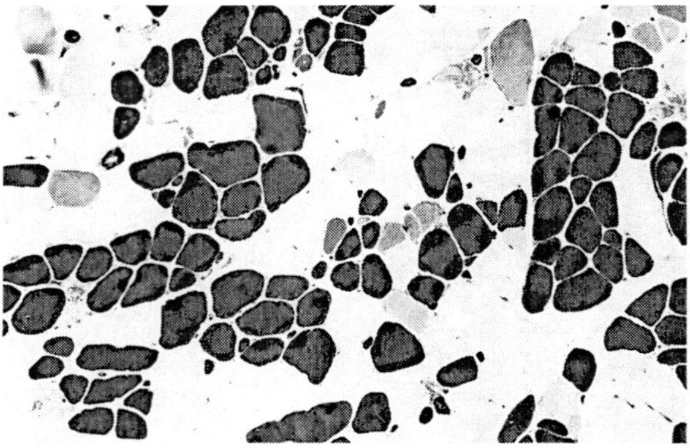

Figure 14. Relatively small clusters of muscle fibres of the same histochemical type displayed with the myosin ATPase reaction at pH4.6. This "myopathic type grouping" is a reflection of forking of regenerated fibres. DMD, X350

3. THERAPEUTIC OPPORTUNITIES

As noted above, aberrant regeneration is fairly common in human myopathies, the most serious ones being a complete failure of a necrotic segment to regenerate or regenerate to significantly lower-than-normal calibre fibres. Aberrant regeneration can substantially aggravate the severity of muscle dysfunction and clinical phenotype in a given disease such as DMD. Understanding the molecular and cellular basis of aberrant regeneration makes it possible to institute measures that eliminate or reduce the degree and prevalence of aberrant regeneration. The administration of molecules (or transferring genes encoding them) that could stimulate satellite cell proliferation and/or differentiation, such as IGF 1 or LIF could produce therapeutic effects (Lefaucheur and Sebille, 1995). A molecule that could override the mitotic ceiling of satellite cells could be of significant benefit, but such molecule, thus far, has not been identified. Suppression of myostatin was also suggested to be an enhancer of regeneration (McCroskery et al., 2005; Wagner et al., 2005).

Another possible therapeutic approach exploiting knowledge pertaining to regeneration is myoblast or stem cell transfer (Cossu and Biressi, 2005; Qu et al., 1998). If large amounts of purified normal myogenic cells are introduced by direct injection or by vascular dissemination into immuno-suppressed hosts (patients) and the introduced cells survive in the host muscles, they may fuse with the host fibres or with each other. In the former case the may act as cell-mediated therapeutic gene transfer agents, while in the latter situation they can act as tissue replacement agents. The use of cultured myoblasts derived from satellite cells of a partially HLA-matched middle aged donor (such as the fathers of DMD patients) has several drawbacks. One is that many donor satellite cells had already undergone several divisions in the host muscles and many of them may be on the verge senescence. This problem is avoided if myogenic stem cells are used. Embryonic stem cells could theoretically be ideal (notwithstanding ethical issues), but no studies have been published about the feasibility of such approach even in animal models. In summary, thus far, the therapeutic efficiency and safety of cell therapies have been suboptimal in published human trials (Partridge, 2000).

4. CONCLUSIONS

a. Skeletal muscle fibres normally have excellent regenerative capacity due the on-site strategic presence of large numbers of myogenic precursor cells (satellite cells or muscle stem cells).

b. Most satellite cells, however, cannot be considered stem cells as their mitotic capacity is not unlimited.

c. The molecular background of satellite cell differentiation and the triggering of their proliferative and differentiating behaviour is complex and by no means fully elucidated.

d. In various pathological states, the ideal regenerative process may be subverted and various aberrant regenerative patterns may arise aggravating the clinical phenotype of the disease.

e. Understanding the molecular/biochemical factors that cause or contribute to aberrant regeneration can be therapeutically helpful in many myopathies, such as DMD.

ACKNOWLEDGEMENTS

Some of the Figures shown have been prepared by Stirling Carpenter, MD.

REFERENCES

Allbrook D (1981) Skeletal muscle regeneration. Muscle Nerve 4:234–245

Anderson LVB (2002) Dystrophinopathies. In: Karpati G (ed) Structural and molecular basis of skeletal muscle diseases. International Society of Neuropathology Press, Los Angeles, pp 6–23.

Anderson JE (2006) The satellite cell as a companion in skeletal muscle plasticity: currency, conveyance, clue, connector and colander. J Exp Biol 209:2276–2292

Anversa P, Kajstura J, Leri A (2004) Circulating progenitor cells: search for an identity. Circulation 110:3158–3160

Bernasconi P, Torchiana E, Confalonieri P, Brugnoni R, Barresi R, Mora M, Cornelio F., Morandi L, Mantegazza R (1995) Expression of transforming growth factor-beta 1 in dystrophic patient muscles correlates with fibrosis. Pathogenetic role of a fibrogenic cytokine. J Clin Invest 96:1137–1144

Bischoff R (1994) The satellite cell and muscle regeneration. In: Engel AG, Frazini-Armstrong C (eds) MyologyMcGraw-Hill, New York, pp 97–118.

Carpenter S, Karpati G (2001) Skeletal Muscle Pathology. New York: Oxford University Press

Chen JC, Goldhamer DJ (2003) Skeletal muscle stem cells. Reprod Biol Endocrinol 1:101

Chou SM, Nonaka I (1977) Satellite cells and muscle regeneration in diseased human skeletal muscles. J Neurol Sci 34:131–145

Collins CA, Olsen I, Zammit PS, Heslop L, Petrie A, Partridge TA, Morgan JE (2005) Stem cell function, self-renewal, and behavioral heterogeneity of cells from the adult muscle satellite cell niche. Cell 122:289–301

Collins CA, Zammit PS, Ruiz AP, Morgan JE, Partridge TA (2007) A population of myogenic stem cells that survives skeletal muscle aging. Stem Cells 25:885–894

Cossu G, Biressi S (2005) Satellite cells, myoblasts and other occasional myogenic progenitors: possible origin, phenotypic features and role in muscle regeneration. Semin Cell Dev Biol 16:623–631

Deasy BM, Gharaibeh BM, Pollett JB, Jones MM, Lucas MA, Kanda Y, Huard J (2005) Long-term self-renewal of postnatal muscle-derived stem cells. Mol Biol Cell 16:3323–3333

Dhawan J, Rando TA (2005) Stem cells in postnatal myogenesis: molecular mechanisms of satellite cell quiescence, activation and replenishment. Trends Cell Biol 15:666–673

Figarella-Branger D, Pellissier JF, Bianco N, Karpati G (1999) Sequence of expression of MyoD1 and various cell surface and cytoskeletal proteins in regenerating mouse muscle fibers following treatment with sodium dihydrogen phosphate. J Neurol Sci 170:151–160

Grobler LA, Collins M, Lambert MI, Sinclair-Smith C, Derman W, St Clair GA, Noakes TD (2004) Skeletal muscle pathology in endurance athletes with acquired training intolerance. Br J Sports Med 38:697–703

Grounds MD (1991) Towards understanding skeletal muscle regeneration. Pathol Res Pract 187:1–22

Hohlfeld R (2002) Polymyositis and Dermatomyositis. In: Karpati G (ed) Structural and molecular basis of skeletal muscle diseases. ISN Neuropath Press, Basel, pp 221–227

Karpati G (2002) General Pathological, Immunopathological, and Genetic Background of Skeletal Muscle Disorders. In: Karpati G (ed) Structural and molecular basis of skeletal muscle diseases. ISN Neuropath Press, Basel, pp 1–3

Karpati G (1997) Utrophin muscles in on the action. Nat Med 3:22–23

Karpati G, Carpenter S (1993) Pathology of the inflammatory myopathies. Baillieres Clin Neurol 2: 527–556

Karpati G, Carpenter S, Prescott S (1988) Small-caliber skeletal muscle fibers do not suffer necrosis in mdx mouse dystrophy. Muscle Nerve 11:795–803

Lefaucheur JP, Sebille A (1995) Muscle regeneration following injury can be modified in vivo by immune neutralization of basic fibroblast growth factor, transforming growth factor beta 1 or insulin-like growth factor I. J Neuroimmunol 57:85–91

McCroskery S, Thomas M, Platt L, Hennebry A, Nishimura T, McLeay L, Sharma M, Kambadur R (2005) Improved muscle healing through enhanced regeneration and reduced fibrosis in myostatin-null mice. J Cell Sci 118:3531–3541

Mouly V, Aamiri A, Bigot A, Cooper RN, Di Donna S, Furling D, Gidaro T, Jacquemin V, Mamchaoui K, Negroni E, Perie S, Renault V, Silva-Barbosa SD, Butler-Browne GS (2005) The mitotic clock in skeletal muscle regeneration, disease and cell mediated gene therapy. Acta Physiol Scand 184:3–15

Nowak KJ, Davies KE (2004) Duchenne muscular dystrophy and dystrophin: pathogenesis and opportunities for treatment. EMBO Rep 5:872–876

Partridge T (2000) The current status of myoblast transfer. Neurol Sci 21:S939–S942

Qu Z, Balkir L, van Deutekom JC, Robbins PD, Pruchnic R, Huard J (1998) Development of approaches to improve cell survival in myoblast transfer therapy. J Cell Biol 142:1257–1267

Qu-Petersen Z, Deasy B, Jankowski R, Ikezawa M, Cummins J, Pruchnic R, Mytinger J, Cao B, Gates C, Wernig A, Huard J (2002) Identification of a novel population of muscle stem cells in mice: potential for muscle regeneration. J Cell Biol 157:851–864

Renault V, Piron-Hamelin G, Forestier C, DiDonna S, Decary S, Hentati F, Saillant G, Butler-Browne GS, Mouly V (2000) Skeletal muscle regeneration and the mitotic clock. Exp Gerontol 35:711–719

Schultz E, McCormick KM (1994) Skeletal muscle satellite cells. Rev Physiol Biochem Pharmacol 123:213–257

Seale P, Rudnicki MA (2000) A new look at the origin, function, and "stem-cell" status of muscle satellite cells. Dev Biol 218:115–124

Sherwood RI, Wagers AJ (2006) Harnessing the potential of myogenic satellite cells. Trends Mol Med 12:189–192

Snow MH (1977) The effects of aging on satellite cells in skeletal muscles of mice and rats. Cell Tissue Res 185:399–408

Tavian M, Zheng B, Oberlin E, Crisan M, Sun B, Huard J, Peault B (2005) The vascular wall as a source of stem cells. Ann N Y Acad Sci 1044:41–50

Turk R, Sterrenburg E, de Meijer E, van Ommen G, den Dunnen J, 't Hoen P (2005) Muscle regeneration in dystrophin-deficient mdx mice studied by gene expression profiling. BMC Genomics 13:98

Wagner KR, Liu X, Chang X, Allen RE (2005) Muscle regeneration in the prolonged absence of myostatin. Proc Natl Acad Sci U S A 102:2519–2524

Webster C, Blau HM (1990) Accelerated age-related decline in replicative life-span of Duchenne muscular dystrophy myoblasts: implications for cell and gene therapy. Somat Cell Mol Genet 16: 557–565

Zammit PS, Beauchamp JR (2001) The skeletal muscle satellite cell: stem cell or son of stem cell? Differentiation 193–204

Zammit PS, Golding JP, Nagata Y, Hudon V, Partridge TA, Beauchamp JR (2004) Muscle satellite cells adopt divergent fates: a mechanism for self-renewal? J Cell Biol 166:347–357

Zhao P, Hoffman EP (2004) Embryonic myogenesis pathways in muscle regeneration. Dev Dyn 229: 380–392

CHAPTER 11

SKELETAL MUSCLE REPAIR AFTER EXERCISE-INDUCED INJURY

TERO A.H. JÄRVINEN[1], MINNA KÄÄRIÄINEN[1], VILLE ÄÄRIMAA[2], MARKKU JÄRVINEN[1] AND HANNU KALIMO[3,4]

[1] *Medical School, University of Tampere, Tampere, Finland; and Departments of Orthopaedic and Plastic Surgery, Tampere University Hospital, Tampere, Finland*

[2] *Department of Orthopedic Surgery, University and University Central Hospital of Turku, Turku, Finland*

[3] *Department of Pathology, University and University Central Hospital of Helsinki and Turku, Helsinki and Turku, Finland*

[4] *Department of Pathology, University and University Hospital of Uppsala, Uppsala, Sweden*

The muscle injury induced by either excessive or injurious exercise commonly afflicts normal i.e. basically healthy muscle. The exercise that injures diseased muscle (e.g. in muscular dystrophies, myositides or McArdle's disease) needs not be very excessive, because the muscle is itself diseased and vulnerable. This underlines the importance of appropriate physical therapy to patients with a chronic neuromuscular disease. Regeneration in human myopathies is dealt with in Chapter 10, and only some specific aspects on the effects of exercise on "myopathic muscles" are discussed here. Even though the experimental studies have provided most of the information presented in this chapter, the focus is on human aspects in the regeneration of muscle injuries.

1. MECHANISMS OF SKELETAL MUSCLE INJURY

Muscle injuries are one of the most common injuries occurring in sports, their frequency varying from 10 to 55% of all the sustained injuries. Muscle injuries can be of *shearing type* (caused by *contusion, strain or laceration*), in which the muscle fibers and their basal lamina and mysial sheaths rupture. In the other type of injury, *in situ necrosis (or rhabdomyolysis)*, the myofibers are necrotized while the basal

S. Schiaffino and T. Partridge (eds.), Skeletal Muscle Repair and Regeneration, 217–242.

lamina and mysial sheaths remain intact (Järvinen et al., 2005; Huard et al., 2002; Beiner & Jokl, 2001; Best & Hunter, 2000; Garrett, 1996). Over 90% of all sports-related injuries are either contusions or strains, whereas muscle lacerations are uncommon injuries in sports (Järvinen et al., 2005). Muscle contusion occurs when a muscle is subjected to sudden, heavy extrinsic compressive force, such as a direct blow, i.e. the injury is not a consequence of the intrinsic force of the exercise itself. If the external force also causes an open wound, the injury is called a laceration, in which the affected muscle can be torn or sharply cut depending on the cause of the laceration. In strains, the myofibers are exposed to an excessive intrinsic tensile force. Their severity vary from very mild strain injury like delayed onset muscle soreness (DOMS) to "real" strains, shearing type of muscle injuries, in which myofibers and the associated connective tissue structures including blood vessels are ruptured. "Real" muscle strains induced by exercise do not pathobiologically differ in a significant way from those above mentioned shearing type of injuries, contusions or lacerations, caused by external force.

2. CLINICAL CLASSIFICATION OF MUSCLE STRAINS

The clinical picture of a muscle strain depends on the extent and nature of the muscle destruction and the hematoma that develops at the site of the injury. In exercise induced strains the hematoma is most often *intramuscular*. The extravasated blood within the intact muscle fascia increases intramuscular pressure, which subsequently compresses the bleeding blood vessels and thereby eventually limits the size of the hematoma. In a severe strain the epimysium of the injured muscle may also rupture and then an intermuscular hematoma develops.

DOMS may be considered the mildest form of strain injury, but since muscle fibers are not torn in DOMS, many do not regard it as a strain injury. Even the name is actually just a symptom, but since it is well known and widely used, we also employ it as a surrogate for the lacking pathogenetic term. Based on the clinical impairment muscle injuries may be classified as mild, moderate or severe (Kalimo et al., 1997; Jackson & Feagin, 1973). *Mild* (first degree) strain represents a tear of only few muscle fibers with minor swelling and discomfort accompanied with no or only minimal loss of strength and restriction of the movements (ability to mobilize). *Moderate* (second degree) strain, in turn, is a greater damage of the muscle with a clear loss of function (ability to contract). *Severe* (third degree) strain occurs when a tear extends across the entire cross-section of the muscle (a very rare consequence of excessive intrinsic force alone) and thus results in a virtually complete loss of muscle function.

3. DELAYED ONSET MUSCLE SORENESS

General. The mildest type of muscle injury due to intrinsic sporting exercise is DOMS, an injury that all active sports people must have experienced. DOMS is commonly a consequence of an overenthusiastic exercise of untrained muscle, which is tolerated while engaged in that activity, but followed by muscle soreness 1–3 days after the exercise. This phenomenon strikes especially if the exercise includes eccentric work, i.e. lengthening of contracted muscles like in running downhill or squatting with weights. DOMS has been extensively studied both experimentally and also in humans using volunteers for physically quite demanding performances and muscle biopsies. Since the experimental animals do not complain of their sore muscles, experimental studies have been designed to correspond to human eccentric work, namely forceful extension of contracted muscle. This has been implemented by electrical stimulation of the motor nerve to induce tetanic contraction of the muscle, which is then forcefully lengthened (e.g. Lieber et al., 1994). The findings are essentially concordant in humans and experimental animals (e.g. Friden & Lieber, 2002), though some differences have been encountered and new views on the interpretation of the results have been presented (Yu et al., 2002, 2004).

Clinical aspects. The symptoms of stiffness, soreness and tenderness with palpation develop during the first 1–2 days with a peak on days 2 or 3, and they disappear usually with no treatment by days 5–7. The pain is aggravated by passive stretch of the sore muscle and the strength of the muscle is decreased. This is usually associated with a rise in serum creatine kinase (CK), which is usually modest but sometimes up to 20-fold. CK peaks around days 3 to 6 and returns to normal during the first or occasionally second week after the eccentric exercise. Inflammatory reaction has been reported in both experimental animals and in humans, even though it has been rarely analysed in human DOMS (MacIntyre et al., 2001). The pain in DOMS is mediated by type III and IV nociceptors, which in DOMS are most likely stimulated by factors (such as bradykinin, prostaglandins and serotonin) released from the inflammatory cells. Nonsteroidal anti-inflammatory drugs (NSAIDs) have been used to reduce the pain, but the relatively mild inflammation does not actually need any alleviation by treatment with NSAIDs.

Pathogenesis of DOMS. In human DOMS develops after eccentric work excessive for the fitness level of the muscle. In animal experiments the eccentric muscle contraction is repeated many times. But it has been shown that even a single eccentric stretch in rabbits may be sufficient to result in reduced biomechanical capacity (i.e. the load that causes failure of the muscle upon stretching is reduced) and to stimulate the dormant satellite cells to divide (Fig. 1a and b; Äärimaa et al., 2004a). Such an injury is, however, very mild since the offspring of the activated satellite cells did not seem to mature further into myoblasts expressing muscle specific proteins nor fuse with the parent myofiber (Äärimaa et al., 2004a).

Even though DOMS is associated with CK rise, which must indicate some degree of sarcolemmal damage inducing leak of sarcoplasmic proteins, it has been demonstrated that in DOMS usually no frank necrosis of myofibers ensues (Yu & Thornell, 2002; Yu et al., 2002, 2003, 2004; Äärimaa et al., 2004a). The main structural finding has been focal loss of the myofibrillar (sarcomeric) structures. The early loss of desmin has been considered a key feature of the lesions in

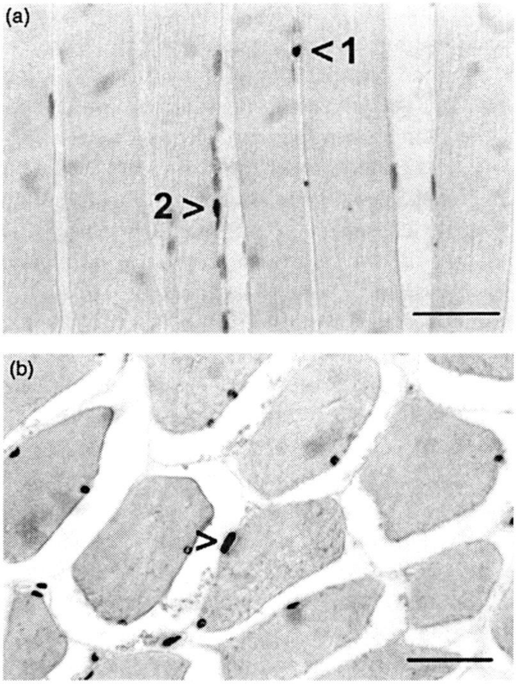

Figure 1. The nuclei of proliferating cells have incorporated bromodeoxyuridine (BrDU; arrowheads) 1 day after a single eccentric stretch of rabbit anterior tibialis muscle. (a) The rounded nucleus 1 is located in the endomysium being most likely a fibroblast nucleus. The elongated nucleus 2 closely apposed to a myofiber is most probably a satellite cell nucleus. Longitudinal section. (b) The BrDU immunopositive nucleus closely apposed to a myofiber is most probably a satellite cell nucleus. Cross section. Anti-BrdU and hematoxylin counterstain. Bar 50 μm

experimental animals (Lieber et al., 1994; Friden & Lieber, 2002). Interestingly, this may be a feature of injury due to eccentric work in experimental animals, but human studies using both high resolution frozen section immunofluorescence and electron microscopy have led to divergent interpretation of the results (Yu & Thornell, 2002; Yu et al., 2002, 2003, 2004). Remarkably the DOMS induced by eccentric work while running downstairs disclosed structural changes, which Thornell's team interpret as remodeling of myofibrils in the affected fibers rather than actual injury (Yu & Thornell, 2002; Yu et al., 2002, 2003, 2004). They noticed loss of immunostaining for the main component of Z-discs, α-actinin and the two giant molecules (important in keeping myofibrils in register), titin and nebulin whereas desmin and actin actually increased in the regions of sarcomeric "blurring". They propose that α-actinin, titin and nebulin are rather released from their normal location than being degraded as previously suggested. Subsequently actin and desmin molecules are recruited to the site of remodeling to provide the basic molecules needed for the formation of new sarcomeres. Thereafter the organizing molecules α-actinin, titin and nebulin are reintegrated to initiate the organization of actin filaments into orderly new sarcomeres, the number of which is at the same time increased (supernumerary sarcomeres). This sequence of events is similar to that observed in myofibril formation during muscle development. The outcome of remodeling, i.e. the increase in the number of sarcomeres could be one of the factors that make the muscle more resistant to future eccentric work (Yu & Thornell, 2002; Yu et al., 2002, 2003, 2004). The sequence of these changes is described in the schematic Fig. 2.

Even though the evidence for remodeling with sarcomerogenesis in DOMS is well supported by the analyses of Thornell's team, some kind of injury must also occur, since serum CK elevation is a constant finding and inflammatory cells also invade the muscles suffering from DOMS.

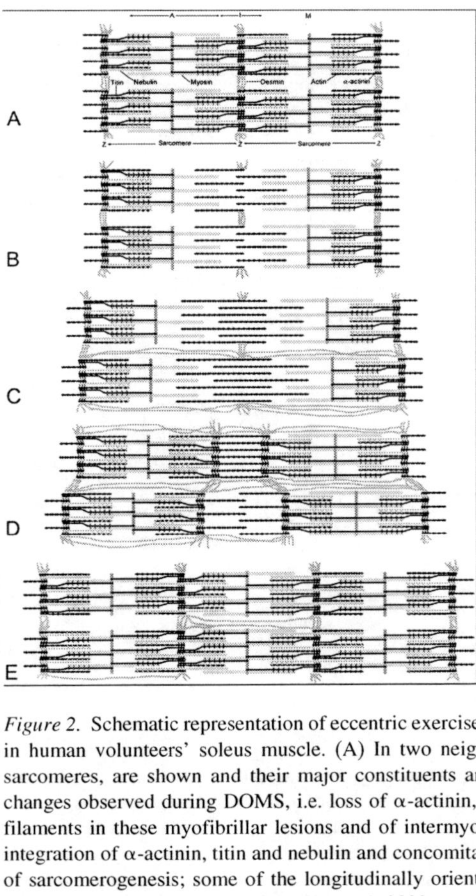

Figure 2. Schematic representation of eccentric exercise-induced myofibril remodeling based on findings in human volunteers' soleus muscle. (A) In two neighboring myofibrils, each one composed of two sarcomeres, are shown and their major constituents are labelled. (B) The first stage of the structural changes observed during DOMS, i.e. loss of α-actinin, titin and nebulin. (C) Localised growth of actin filaments in these myofibrillar lesions and of intermyofibrillar desmin intermediate filaments. (D) Re-integration of α-actinin, titin and nebulin and concomitant formation of new sarcomeres. (E) Completion of sarcomerogenesis; some of the longitudinally oriented desmin filaments still persist for some time. Reproduced with permission from Yu et al. 2003

4. THE PATHOBIOLOGY OF MUSCLE STRAINS

In strains, the myofibers are exposed to such an excessive intrinsic tensile force that fullblown shearing injury occurs, i.e. not only the myofibers rupture but also their basal lamina as well as mysial sheaths and blood vessels running in the endo- or perimysium are torn. The rupture is most commonly located close to the myotendinous junction (MTJ), the level of the last T-tubules being described as the weakest site. This site is often not situated next to the proper tendons, since in muscles tapering towards their tendons, numerous MTJs are formed within the muscle belly, where the myofibers attach to the intramuscular fascia within and round the muscle.

The healing of a strain injury follows a fairly constant pattern, which is similar as in contusions or lacerations (Järvinen et al., 2005). As the regeneration process of injured skeletal muscle is described elsewhere in this book, we will present this process only briefly as a necessary background for describing the role of exercise in the repair. (Fig. 3).

Three phases have been identified in this process (Hurme et al., 1991a; Kalimo et al., 1997) (Fig. 3): (1) Destruction phase, (2) Repair phase and (3) Remodeling phase.

1) *Destruction phase*. The ruptured myofibers become necrotized only over a short distance (Figs. 4a and b). The propagation of the necrosis is halted by a "fire door", a contraction band formed within a couple of hours, in the shelter of which

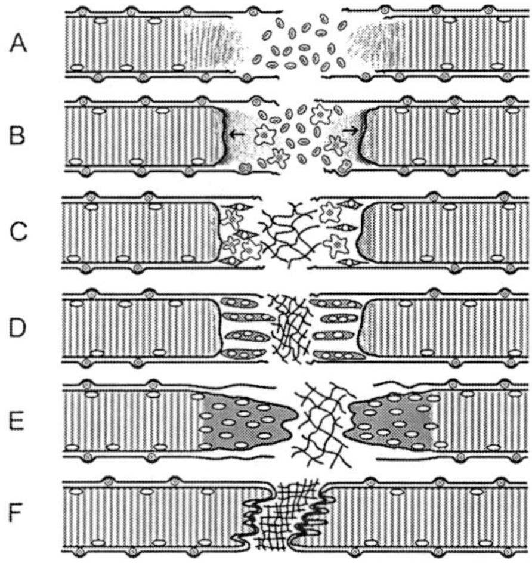

Figure 3. The regeneration of a shearing injury. (A) Torn myofiber and BL. (B) Contraction band and demarcation membrane seal the torn fiber ends. Satellite cells (SC) begin to proliferate and inflammation reaction begins. (C) SCs differentiate into myoblasts and fibroblasts begin to produce collagens and form scar tissue. (D) Myoblasts fuse into myotubes. (E) Myotubes fuse with the surviving parts of the torn fibers and start to form new MTJs. (F) Fully regenerated fiber with organized scar tissue and MTJs attached to it. Courtesy of Dr. Samuli Vaittinen

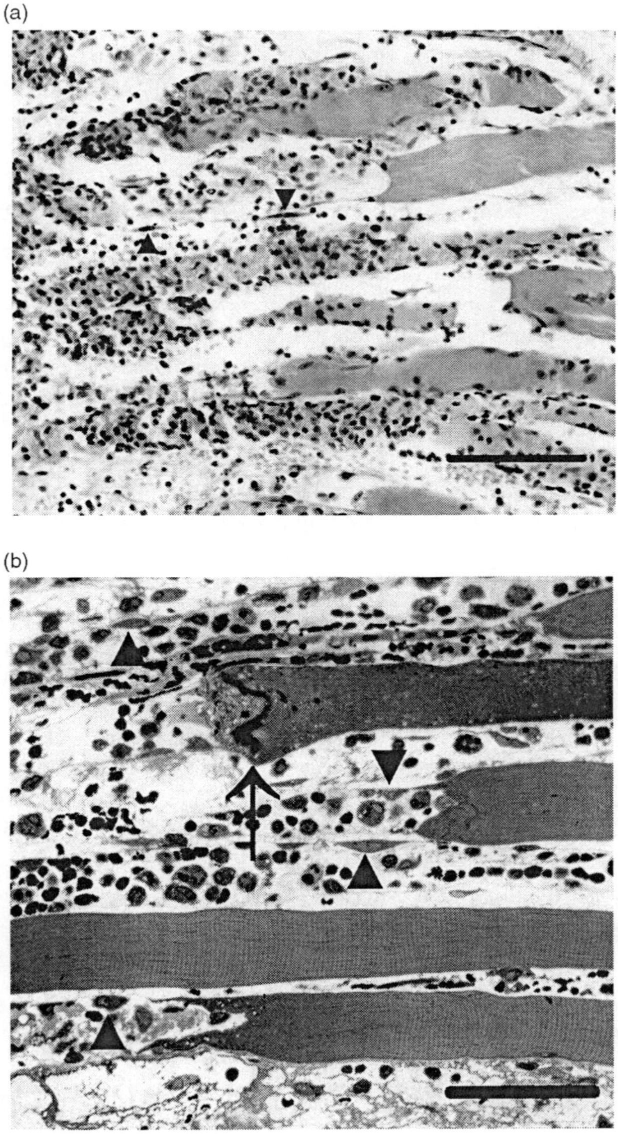

Figure 4. Transected myofibers 2 days after shearing (sharp transection) injury of rat gastrocnemius muscle. The necrotic part of the fibers is phagocytosed by macrophages visible within the preserved basal lamina cylinders. There are inflammatory cells also between the fibers. In one fiber there is still discernible a contraction band (arrow in b), which seals off the transected myofiber preventing the extension of the necrosis beyond a couple of millimeters. Satellite cells within the basal laminas have differentiated into myoblasts (arrowheads). (a) Herovici staining. (b) Epon semithin section and toluidine blue staining. Bar 100 μm

the rupture is sealed by a new sarcolemma (Fig. 4b). The ruptured myofibers contract and the gap between the stumps is filled by a hematoma. The injury induces a brisk inflammatory cell reaction (Figs. 4a and b; see Chapter 12).

2) *Repair phase*. This begins with phagocytosis of the necrotized tissue by blood derived monocytes (Figs. 4a and b). The myogenic reserve cells, satellite cells, are activated and begin the reparation of the breached myofiber. Firstly, committed satellite cells begin to differentiate into myoblasts: desmin positivity may be detected in their sarcoplasm already by 6 hours after injury (cf. Fig. 5b; Vaittinen et al., 2001). Secondly, undifferentiated stem satellite cells begin to proliferate by 24 hours (Fig. 5a) and thereafter contribute to the formation of myoblasts (Fig. 5b), at the same time providing new satellite cells by asymmetric cell division for future needs of regeneration (Rantanen et al., 1995a, Kuang et al., 2007). The myoblasts arising from the committed and stem satellite cells then fuse to form myotubes within a couple of days (Rantanen et al., 1995a). More recently stem cells from non muscle origin have been suggested to contribute to regeneration of myofibers (Chargé & Rudnicki, 2004, see Chapter 4). Within 5–6 days the necrotized part of the ruptured myofiber inside the remaining old basal lamina is replaced by the regenerating myofiber, which then begins to penetrate into the connective tissue scar between the stumps of the ruptured myofibers (Fig. 6). The injury site is also revascularized by ingrowing capillaries.

3) *Remodeling phase*. This is the period of maturation of the regenerating myofibers, which includes formation of a mature contractile apparatus and attachment of the ends of the regenerated myofibers to the intervening scar by newly formed MTJs (Figs. 7a–c). The retraction of the scar pulls the ends closer to each other, but they appear to stay separated by a thin layer of connective tissue to which the ends remain attached by newly formed MTJs (Figs. 7b and c; Vaittinen et al., 2002). Restoration of the motor innervation of the denervated parts of the ruptured myofibers (i.e. abjunctional muscle stumps separated from the original neuromuscular junction by scar tissue, Fig. 8) is also a prerequisite for the recovery of the functional capacity of the muscle (Rantanen et al., 1995b).

5. IMMOBILIZATION AND REMOBILIZATION IN MUSCLE HEALING

A short period of immobilization following a shearing type of muscle injury is mandatory and beneficial and certainly desired by the patient. The immobilization allows the scar tissue connecting the injured muscle stumps to gain the required strength to withstand the contraction-induced forces applied on the regenerating tissue without a rerupture. However, immobilization should be restricted to last for less than a week, so that the adverse effects of immobility *per se* are limited to minimum (Järvinen et al., 2005; Jarvinen 1975, 1976a, 1976b). Reruptures at the site of the original muscle trauma are common if active mobilization is begun immediately after the injury. By placing the injured muscle at rest for the first

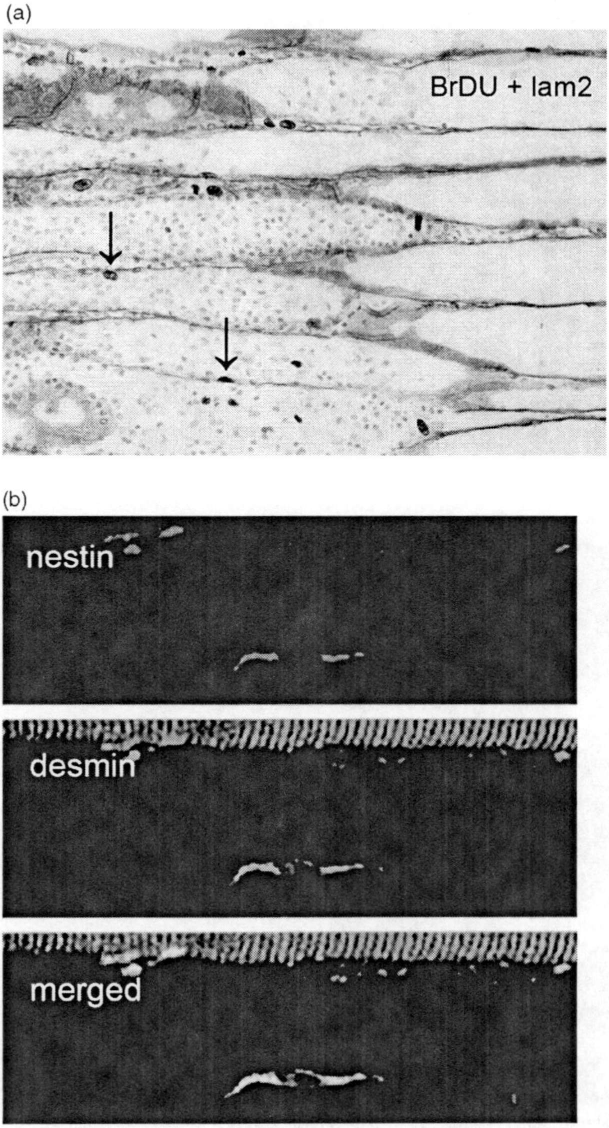

Figure 5. Committed satellite cells can differentiate directly into desmin positive myoblasts already by 6 hours, whereas stem satellite cells first divide beginning around 24 hours post injury (arrows in Figure 5a) and presumably one from each mitosis differentiates into a myoblast and begins to express intermediate filaments desmin and nestin (Figure 5b), while the other is left in reserve. Rat gastrocnemius muscle 24 hours after toxin-induced injury. (a) Anti bromodeoxyuridine and laminin2 + hematoxylin counterstain. (b) Anti nestin and desmin confocal immunofluorescence. Reproduced with permission from Vaittinen et al. 2001

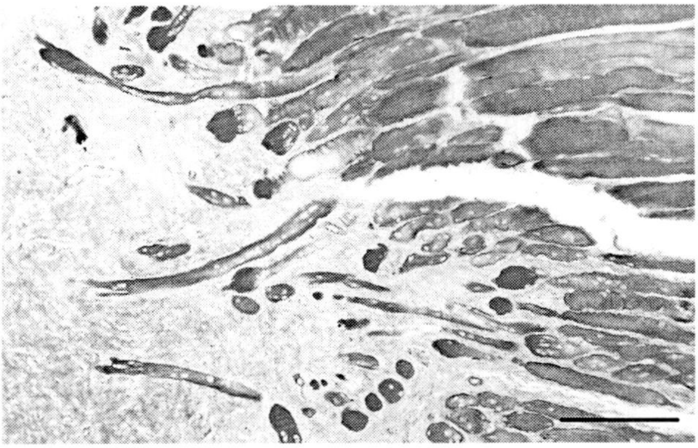

Figure 6. The regenerating myotubes have filled the old basal lamina cylinders (by about day 5) and thereafter they begin to penetrate into the still relatively wide intervening scar on the left. Rat gastrocnemius muscle 7 days after shearing injury. Anti-desmin. Bar 150 μm

(a)

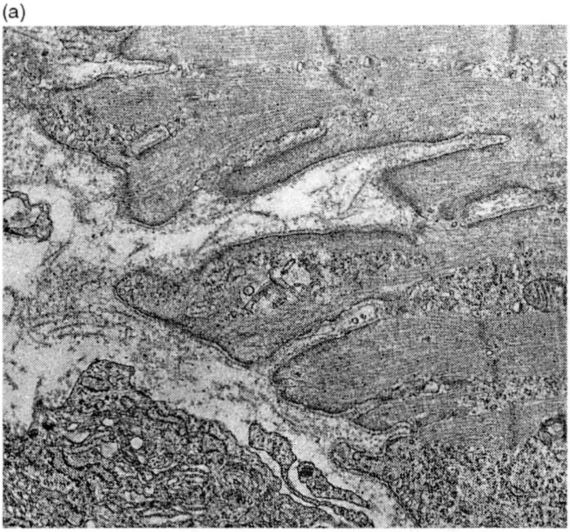

Figure 7. (a) An early (21 days after shearing injury of rat gastrocnemius) neoformed myotendinous junction (MTJ) through which the stumps of the transected myofibers attach to the intervening scar. The myofilaments extend to the sarcolemma, which forms characteristic finger-like projections. (b) By 9 months the stumps of the transected myofibres have become juxtaposed due to contraction of the scar, to which the stumps are attached by integrin **α**7 positive neoformed MTJs but not fused with the stumps from the opposite direction (stumps from the left: arrows 1 and from the right: arrows 2). (c) Two juxtaposed, neoformed MTJs at 9 months after the transection. MTJs are immunopositive for integrin α7 subunit and perijunctional sarcoplasm immunostains for nestin. Despite their close proximity, the regenerated myofibres approaching each other from opposite directions have not fused, but retained their neoformed MTJs. (b) and (c). Shearing injury of rat soleus. (a) An electron micrograph. (b) Anti-α7 integrin and hematoxylin counterstain. (c) Anti-α7 integrin and anti-nestin confocal immunofluorescence. Figure 7b and c reproduced with permission from Vaittinen et al. 2001

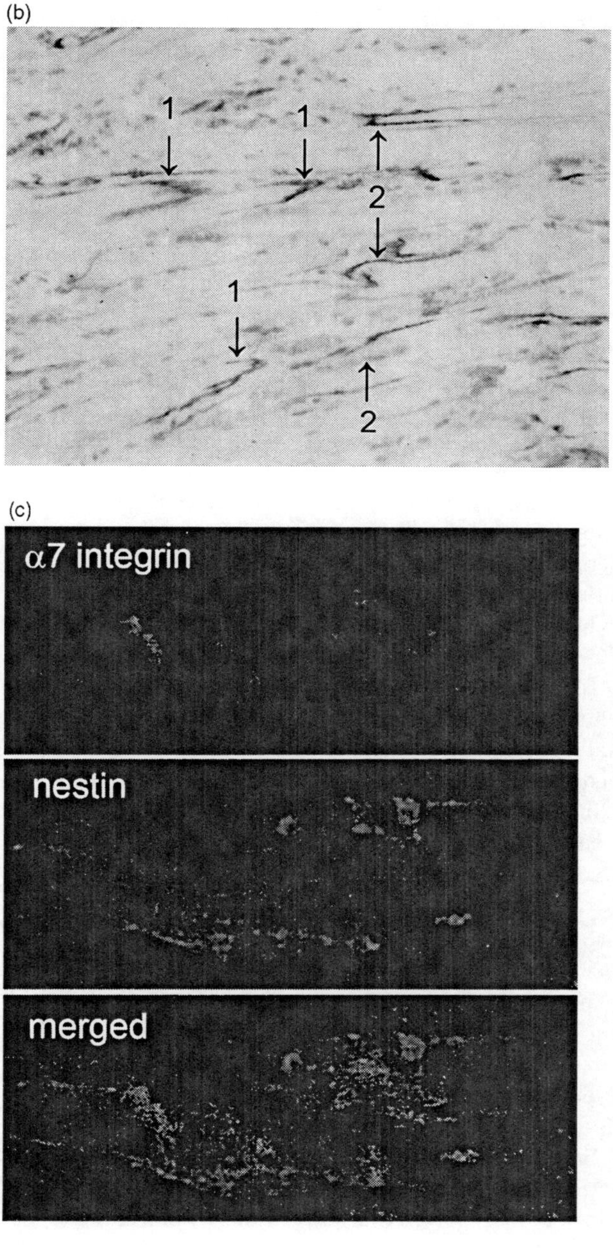

Figure 7. (Continued)

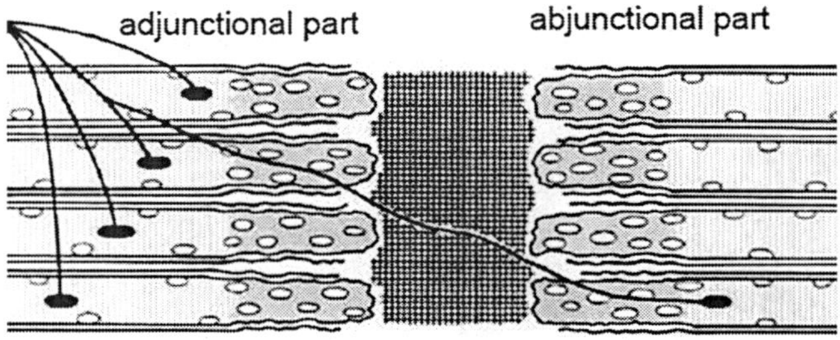

Figure 8. Since each myofiber has only a single neuromuscular junction (NMJ) the abjunctional stumps of asymmetrically transected myofibers become myogenically denervated. Nerve sprouts are able to pierce through the scar and form a new NMJ on the abjunctional stump (one sprout depicted)

4–6 days after the injury, the excessive scar formation and reruptures at the injury site can be best prevented (Järvinen & Lehto, 1993; Jarvinen 1975, 1976a, 1976b). Avoiding reruptures is important, as it has been shown that the reruptures are actually the most severe skeletal muscle injuries which cause the longest lost time from sport activity (Brooks et al., 2006).

The initial immobilization must be followed by active rehabilitation. Early mobilization as the acute treatment of muscle trauma was first recommended – largely based on his vast personal experience in sports medicine – by Dr. Woodard in 1953. Today, this empirical notion is supported by a considerable amount of experimental evidence (Järvinen et al., 2005; Kannus et al., 2003; Buckwalter, 1995; Jarvinen, 1975, 1976a, 1976b). Thus, the mobilization of the injured skeletal muscle should be started gradually (i.e. within the limits of pain) as soon as possible as the early mobilization has been shown to best expedite and intensify the regeneration phase of the injured skeletal muscle. The exercise is also crucial for the induction of appropriate molecules and correct orientation of the regenerating myofibers (see below; Jarvinen 1975, 1976a, 1976b; Buckwalter, 1995; Kääriäinen et al., 2001; Kannus et al., 2003; Järvinen et al., 2005).

6. CLINICAL PRACTICE IN REHABILITATION OF INJURED MUSCLE

As highlighted above, the most crucial decision in the treatment of patients with injured skeletal muscle is to decide when the injured muscle can be remobilized without causing a rerupture. In clinical practice the decision must be based on all the information about the injurious event as well as the site, quality and severity of the injury. Mobilization after 3 to 7 days is commonly recommended. Experimentally, the connective tissue scar between the muscle stumps has been found to be the weakest point in the traumatized muscles until day 10 after injury (Järvinen, 1976b; Crisco et al., 1994; Kääriäinen et al., 1998). After day 10 the failure in the regenerating muscles occurs within the myofibers either near the injury site or the intact MTJs, which is also the rupture site in healthy uninjured muscles (Crisco et al., 1994). Thus, the remobilization is begun already at the time, when there is still a considerable risk of rerupture and hence it must be done with caution. On the other hand, if immobilization is prolonged, for example from 14 days onwards, the traumatized muscles become significantly atrophied and the orientation of the regenerated myofibers is markedly more disordered (Jarvinen 1975). Only a single prospective randomized study exists on different treatment regimens for hamstring strains (Sherry & Best, 2004). Exercise protocol consisting of progressive agility and trunk stabilization yielded a significantly better outcome than a regimen focusing only on stretching and strengthening of the injured hamstring muscles, indicating that all emphasis in the rehabilitation should not be placed only on the injured muscle (Sherry & Best, 2004).

Largely based on our experimental findings, we have adopted the following practice for the treatment of our athletes with an acute muscle injury: The required *relative immobility* can be achieved simply by applying a firm adhesive taping or alike over the injured muscle. Cast is naturally not needed. We highly recommend the use of crutches to athletes with the most severe lower extremity muscle injuries, as well as when the injury is located at a site where adequate immobilization is otherwise difficult to attain, such as the groin area (Woodard, 1954). We also instruct the athlete to be very cautious for the first three to seven days after the injury to prevent the injured muscle from any kind of stretching. After this period of relative immobility, more active use of the injured muscle can be started gradually within the limits of pain.

It is of particular importance that all physical rehabilitation activities should always start with an adequate warming-up of the injured muscle (Petersen & Hölmich, 2005; Noonan et al., 1993; Safran et al., 1989; Safran et al., 1988), as it has been shown to reduce muscle viscosity and relax muscles neurally. When warming up is combined with stretching, the elasticity of muscle is increased. Stimulated, warm muscles absorb more energy than unstimulated muscles and, can thus better withstand loading without a risk of failure/rerupture. (Noonan et al., 1993; Safran et al., 1989; Safran et al., 1988). The other purpose of stretching is to distend the maturing scar in a phase when it is still plastic, but already has the required strength to prevent a functionally disabling retraction of the muscle stumps. Painless elongation of the maturing scar can be achieved by gradual stretching, beginning with bouts of 10 to 15 seconds at a time and then proceeding up to a period of 1 minute. Stretching should also involve repeated stretches of the same muscle because repeated elongation has been shown to decrease the (counter) resistance of the muscle to stretching (Petersen, 2005).

7. ADHESION OF REGENERATING MYOFIBERS

The complex attachment of myofibers to extracellular matrix (ECM) is established mainly by dystrophin- and integrin-associated complexes (Fig. 9; Ervasti 2007; Grounds et al., 2005; Kääriäinen et al., 2000). These complexes play a key role in transmitting the mechanical force of myofibers to ECM structures. Most of the contractile forces are transmitted through the MTJs, which are fingerlike projections

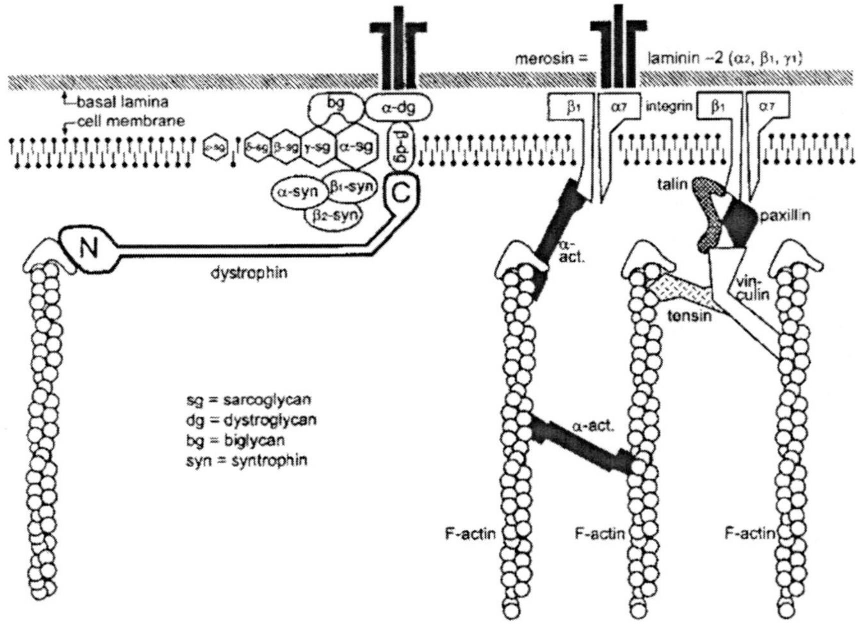

Figure 9. A schematic picture of the dystrophin- and integrin-associated complexes by which the myofibers attach to the extracellular matrix

into the connective tissue at the ends of myofibers (Fig. 7a). These sites are enriched with molecules of dystrophin and integrin complexes, which are also located at sarcolemma along the whole myofiber. Integrins are transmembrane cell adhesion molecules found in many tissues. Integrins consist of two subunits α and β. The α-unit determines the type of extracellular ligand and the β-unit the intracellular site of adhesion. In skeletal muscle the most important type of integrin is α7β1-type which binds via the α7 unit to extracellular laminin-2 (merosin) and via the β1 unit to intracellular proteins, including α-actinin, talin, vinculin, paxillin and tensin, which further connect to the cytoskeleton and the contractile apparatus of the myofiber (von der Mark et al., 2002; Burkin & Kaufman, 1999). Dystrophin is a large subsarcolemmal molecule which forms a dystrophin glycoprotein complex (DGC) with several transmembrane glycoproteins including α- and β-dystroglycan, sarcoglycans, and syntrophins. Intracellularly dystrophin binds to actin, and extracellularly the DGC binds via α-dystroglycan to laminin-2 (merosin). Mutations in various components of the DGC and integrin complex cause muscular dystrophies (see below; Davies & Nowak, 2006).

When myofibers are breached, the continuity of the tendon-muscle-tendon unit is disrupted at the rupture site, and the contractile force cannot be transmitted across the gap formed between the ruptured stumps (Fig. 3). Instead, the stumps are simply pulled further apart during contraction. Thus, it is, of course, essential

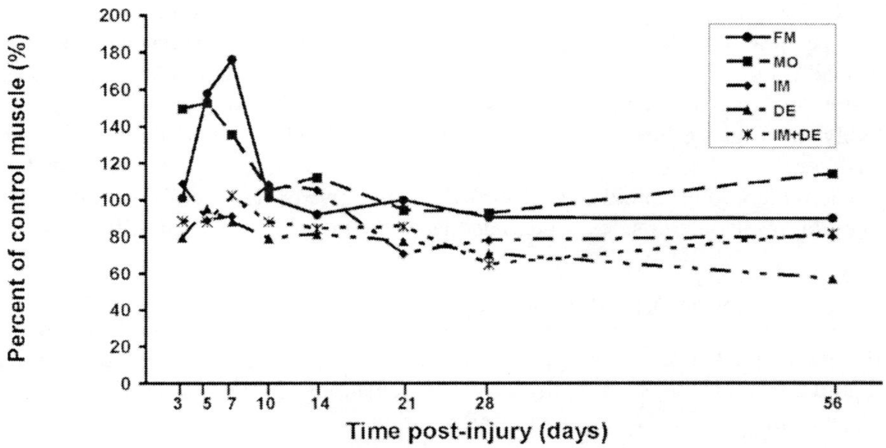

Figure 10. The intensity of sarcolemmal α7 integrin immunoreactivity in the preserved (intact) parts of myofibers after a shearing (sharp transection) injury of rat soleus. At the early stage of regeneration the intensity was clearly increased in both actively mobilized (MO) and freely moving (FM) rats (earlier in the MO rats). This indicates reinforced attachment to the extracellular matrix along the lateral aspects of ruptured myofibers as an attempt to compensate for the loss of attachment at the ruptured end. In immobilized (IM), denervated (DE) and immobilized + denervated (IM + DE) rats a decrease or only a modest increase was recorded indicating that exercise is needed to induce upregulation of α7 integrin. The results are expressed as percentages of the control values in the intact contralateral muscles. Reproduced with permission from Kääriäinen et al. 2001

for functional recovery to restore this continuity. The ends of the regenerating myofibers, attempting to penetrate through the scar tissue, maintain a growth cone-appearance for 10–14 days in experimental animals (Hurme et al., 1991a; Hurme & Kalimo, 1992; Kääriäinen et al., 2000). During this time period the ends cannot yet firmly attach to the scar. Instead, the regenerating myofibers reinforce their adhesion to the ECM on their lateral aspects both in the intact and regenerating parts of the myofibers by "ectopic" accumulation of extra α7β1-integrin along the lateral sarcolemma of the regenerating muscle fibers (Fig. 10). This reinforced lateral adhesion both reduces the movement of the stumps and the pull on the still fragile scar, reducing the risk of rerupture while allowing some use of the injured muscle before the healing is completed (Kääriäinen et al., 2000, 2001, 2002). Interestingly, mechanical stress appears to be a prerequisite for the reinforced lateral adhesion as the increase of α7β1-integrin protein along the lateral sarcolemma does not occur in the absence of mechanical stress (Fig. 10; Kääriäinen et al., 2001).

Later during the regeneration process, when the regenerating muscle fibers have extended out of the old BL cylinders and penetrated into the scar tissue, the formation of new MTJs at the tips of the regenerating myofibers takes place (Hurme et al., 1991a). Exactly the same adhesion molecules as in the normal MTJs accumulate in these neoformed MTJs (Fig. 7a–c) with simultaneous normalization of the α7β1-integrin expression in the lateral sarcolemma MTJs (Fig. 10; Kääriäinen et al., 2000, 2001, 2002). It is conceivable that when regenerating myofibers are still penetrating into the scar their tips cannot yet be firmly attached and they instead reinforce their lateral adhesion. Later, when the penetration of the regenerating fibers ceases and firm adhesion is needed, $\alpha 7\beta 1$-integrin molecules most likely move along the sarcolemma to neoformed MTJs. In fact, it is known that $\beta 1$-integrins are able to move along cell surface, a mechanism possibly guided by ligand binding (Felsenfeld et al., 1996). An alternative explanation is that the excess integrin on the lateral sides is degraded and new integrin is formed at the tips.

8. THE INTERVENING SCAR

Immediately after an injury to the skeletal muscle, a gap formed between the ruptured muscle fibers is filled with a hematoma. Blood derived fibrin and fibronectin cross-link to form early granulation tissue, an initial ECM that acts as a scaffold and anchorage site for the invading inflammatory cells and subsequently for fibroblasts (Hurme et al., 1991; see Chapters 12 and 13). Fibroblasts then start synthesizing the proteins and proteoglycans of the ECM to restore the integrity of the connective tissue framework (Goetsch et al., 2003; Hurme et al. 1991b; Lehto et al., 1985a–c, 1986).

Among the first synthesized ECM proteins are fibronectin and tenascin-C (TN-C) (Hurme et al., 1991b, 1992b; Lehto et al., 1985a).Fibronectin first turns into multimeric fibrils and then forms the superfibronectin with greatly enhanced adhesive properties (Wierbicka-Patynowski et al., 2003; Morla et al., 1994). Both fibronectin and TN-C contain numerous fibronectin type III repeats that can be unfolded (stretched) by the mechanical loading. Thus, both of these ECM molecules possess elastic properties, being capable of stretching to several times their resting length in response to mechanical loading exposed on them (Erickson, 2002;Oberhauser et al., 1998) and they are thought to provide strength and elasticity for the early granulation tissue in injured skeletal muscle (Järvinen et al., 2000, 2003a;b; Hurme & Kalimo, 1992b). The expression of fibronectin and TN-C are soon followed by type III collagen (Lehto et al., 1985a; Hurme et al., 1991b), while the production of type I collagen is initiated a couple of days later remaining thereafter elevated for several weeks (Yan et al., 2003; Hurme et al., 1991b; Lehto et al., 1985a,b). The initially abundant granulation tissue between the surviving muscle stumps subsequently condenses into a remarkably small connective tissue mass composed mainly of type I collagen (Hurme et al., 1991a, 1991b; Lehto et al., 1985a, 1986).

Despite the best efforts of elastic superfibronectin and TN-C in providing the mechanical strength to the early granulation tissue, the scar at the injury site is the weakest point of the injured skeletal muscle early after trauma, (Kääriäinen et al., 1998; Järvinen et al., 1976a) but its tensile strength increases considerably with the production of type I collagen (Kääriäinen et al., 1998; Lehto et al., 1985a,c, 1986). The mechanical stability of the collagen, in turn, is attributable to the formation of intermolecular cross-links between collagen molecules during the maturation of the scar tissue (Lehto et al., 1985c). Approximately ten days after the trauma, the maturation of the scar has reached the point at which it no longer is the weakest link of the injured muscle, but rather, if loaded to failure, the rupture usually occurs within the muscle tissue adjacent to the neoformed MTJs. However, a relatively long time is still needed until the strength of muscle is completely restored to the pre-injury level (Kääriäinen et al., 1998, Crisco et al., 1994; Lehto et al., 1985a; Jarvinen, 1975, 1976a).

On the scar side of the neoformed MTJs, TN-C is abundant throughout the early granulation tissue, but disappears when granulation tissue condenses and turns into a permanent collagen type I positive scar. However it re-emerges in the neoformed MTJs forming bands, identical to those seen in the normal MTJ (Hurme, 1992b). As an elastic and adhesive molecule, TN-C absorbs the forces created by muscle contractions (Järvinen et al., 2003a,b, Hurme, 1992b). Interestingly, α7β1-integrin forms clusters adjacent to TN-C in the neoformed MTJs. α7β1-integrin can bind to the most elastic long form of TN-C (Mercado et al., 2004), and it is highly likely that it does so also in the neoformed MTJs in regenerating skeletal muscle. Together these molecules provide not only strong adhesion but also elasticity to the skeletal muscle to bear the forces created by the muscle contractions.

The specificity of the sequence of the molecular events described above is remarkable during the regeneration of the injured skeletal muscle. Different splicing variants of the α7β1-integrin subunits with different functional capabilities are expressed according to the functional requirements of the healing tissue. During the early phases when a dynamic adhesion of the growing myofibers to the extracellular matrix (ECM) is needed, the expression of α7A and β1A splice variants predominate, whereas in completely regenerated, mature, fully functional skeletal muscle, where the tensile forces require firm adhesion to the ECM, α7A/β1A splice variants are replaced by α7B/β1D splice variants (Kääriäinen et al., 2002). This expression pattern is consistent with the fact that β1D-integrin subunit interacts more strongly with the actin cytoskeleton than the β1A (Belkin et al., 1997) and can bind to the laminin of basal lamina (Vachon et al., 1997). This isoform is also capable of activating its extracellular domain for ligand binding through inside-out signaling (Vachon et al., 1997). α7B-subunit, in turn, has a cytoplasmic domain that has a potential actin-binding sequence as well as other postulated binding motifs (Song et al., 1993). Thus, the α7B/β1D splice variant is most suitable for implementation of firm adhesion all the way from the cytoskeleton to the ECM at the neoformed MTJs, when the muscle has recovered enough to allow normal muscle contractions (Kääriäinen et al., 2002).

Our experimental studies in rats have suggested that the opposing ends of the ruptured myofibers remain permanently separated by a thin layer of connective tissue to which the ends are attached by the neoformed MTJs (Fig. 7b and c; Vaittinen et al., 2002). Thus the original (preinjury) tendon–myofiber-tendon unit becomes replaced by two successive tendon–myofiber–neoformed MTJ units separated by the scar. This replacement is not likely to cause excessive functional disturbances during exercise, because these two successive units appear to contract synchronously, as both units become innervated by the same motor nerve (Rantanen et al., 1995b). Axon sprouts from the nerve innervating those tendon-myofiber-neoformed MTJ units that retained their motor end plates (the adjunctional units) when the myofiber ruptured, are capable of penetrating through the intervening scar and form new motor end plates on the tendon-myofiber-neoformed MTJ units beyond the scar (abjunctional units) (Fig. 8; Rantanen et al., 1995b).

In the very severe cases of muscle injuries, operative treatment may be of benefit, if the gap (diastasis) between the ruptured stumps is long. In such cases the axon sprouts may not be able to extend all the way to the myogenically denervated abjunctional tendon-myofiber-MTJ units and these remain permanently denervated and are atrophied. By surgically bringing the stumps closer to each other the chances for reinnervation are improved and the development of extensive scar tissue within the muscle can be avoided (Äärimaa et al., 2004b; Menetrey et al., 1999; Almekinders, 1991).

During natural repair process the interposed scar gradually diminishes in size, thereby bringing the stumps closer together until the myofibers finally become interlaced. However, the stumps most likely do not reunite, because of the neoformed MTJs by which the regenerated ends of breached myofibers attached to the intervening scar (Fig. 7a–c; Vaittinen et al., 2002). The fusion of the ruptured muscle stumps and restoration of the myofiber continuity is possible only if close surgical apposition of the stumps is performed immediately after the injury, i.e. at the cellular stage of myotubes and primitive myofibers, which are naturally capable of fusing with each other (Fig. 11a and b; Äärimaa et al., 2004b).

9. EXERCISE OF DISEASED MUSCLE

9.1 General

In healthy individuals the best intervention to improve physical (and mental) health is regular exercise. Aerobic exercise improves cardiorespiratory function, increases mitochondrial oxidative enzyme activity, improves muscle endurance and reduces obesity. Strength (progressive resistance exercise) training increases lean body mass, muscle protein mass, contractile force and power and improves physical function (e.g. Pate et al., 1995, Zinna & Yarasheski, 2003). If the individual suffers from a neuromuscular disease the benefit of and response to exercise may be remarkably different. Strength training and aerobic exercise training in these patients may slow down the deterioration of the physical performance and prevent permanent deformations (as well as improve the cardiorespiratory function, not to be discussed in this chapter). However, the disease process, for example in muscular dystrophies, may be such that over-exercise might accelerate the damage caused by the disease. Our evidence based knowledge of the pros and cons of exercise in different neuromuscular diseases is still scanty, because the pathogenesis of many *hereditary* muscle diseases has been clarified only recently. In

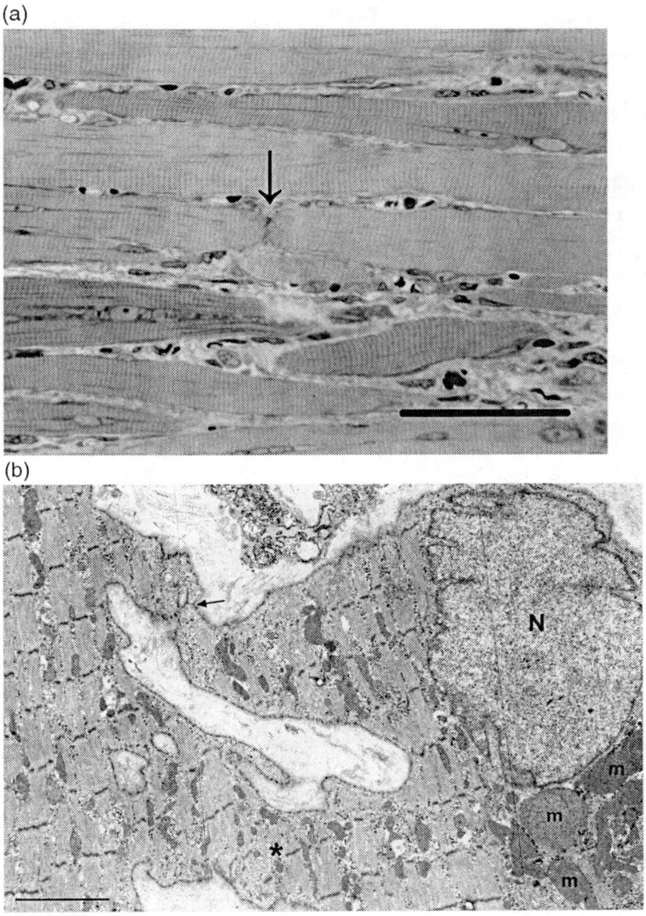

Figure 11. (a) Fusion of the stumps of transected myofibers which were surgically closely apposed immediately after injury. Note the continuity of several fibers and the apparent ongoing fusion right at the site of injury (arrowhead). (b) In EM two thin bridges of sarcoplasm are joining two myofibers. In one of them (arrow) the fusion appears to be occurring, and in the other (asterisk) the fusion is complete. The uniting myofibers have a somewhat irregular structure including myofibrils out of register together with an excessive number of large, sometimes gigantic mitochondria (m) and increased number of ribosomes between the myofibrils. 25 days after a shearing injury of rat soleus. N = nucleus, m = mitochondria. (a) Semithin epon section and toluidine blue staining. Bar 50 μm. (b) Electron micrograph. Bar 3 μm. Reproduced with permission from Äärimaa et al. 2002

many earlier studies the genetic defect was not known and the patients tested most likely suffered from different diseases and, thus, the results are not applicable to patients with a specific muscle disease. In addition to the ethical problems of treating patients with potentially harmful regime of excessive exercise, many hereditary muscle diseases are so rare that recruiting a large enough group of patients for statistical significances has been difficult, often impossible. Neither are *acquired* inflammatory myopathies common diseases and, besides, they have highly variable courses, which is further modified, for example in dermatomyositis and polymyositis, by the often effective immunosuppressive treatment (e.g. Briani et al., 2006). Thus, performing randomized controlled studies is difficult, and accordingly, in a Cochrane analysis from 2005 only two articles were considered to fulfill all inclusion criteria of the study (van der Kooi et al., 2005). Nevertheless some guidelines for different neuromuscular diseases have been presented. In this context we discuss the effects of exercise in some of the more common muscle diseases (see also Chapter 10).

9.2 Muscular Dystrophies

Muscular dystrophies are a pathogenetically heterogeneous group of hereditary muscle diseases, the cardinal feature of which is muscle fiber necrosis. A great majority of dystrophies are associated with defects in sarcolemmal and extracellular matrix molecules, which bind the contractile elements across the sarcolemma to the extracellular matrix [especially to laminin-2 (merosin) of basal lamina], discussed above and depicted in Fig. 8. However, there are dystrophies with gene defects in several other molecules, e.g. sarcomeric molecules and sarcoplasmic enzymes). The necrotizing character of muscular dystrophies has understandably called for caution in administering exercise therapy. Especially in those dystrophies in which the pathogenic mutation affects molecules associated with sarcolemma (Fig. 8) it is very logical that heavy exercise may aggravate the sarcolemmal damage and, thus, exacerbate the disease.

9.3 Animal Experiments

The most common muscular dystrophy, dystrophinopathy (Duchenne and Becker phenotypes), has animal models (e.g. *mdx* mice and golden retriever dogs with muscular dystrophy, GRMD) in which different effects of exercise have been analysed (Carter et al., 2002). The increased fragility of dystrophinopathic sarcolemma was verified in GRMD dogs, in which eccentric contractions induced greater force deficits (43.3% vs. 25.0%) than in healthy controls and even concentric contractions induced force deficits, though the difference did not reach significance (Childers et al., 2002). In experimental studies in *mdx* mice the potential therapy to restore the missing dystrophin molecules by transplantation of healthy myogenic precursor cells (mpc) into dystrophinopathy patients has been tested. A prerequisite for transplanted mpcs to fuse with myofibers is that myofibers are damaged, "preconditioned" to receive transplanted mpcs. By using exercise (swimming) as a mild additional damage in *mdx* mice ("for induction of fiber breaks"), the number of dystrophin positive fibers increased and thus grafting was more successful (Bouchentouf et al., 2006).

There are a few more natural animal species with muscular dystrophies, e.g. dy/dy mice with defect in laminin-2 and a hamster with a defect in δ-sarcoglycan (see below). In addition, many transgenic mouse models of various muscle diseases have been developed, for example transgenic mouse models for all four sarcoglycanopathies have been developed (Carter et al., 2002). It must, however, be emphasized that results with animal models cannot be directly applied to human diseases, since even in genetically homologous animal models (like *mdx* mice with defect in dystrophin gene) there are marked phenotypic differences. For example, mdx mice have only a bout of extensive myofiber necroses at the age of 2–4 weeks. Thereafter only lesser though continuous necroses and regeneration prevail, the mice show only minimal muscle weakness and their life span is normal. On the other hand, hamsters with δ-sarcoglycanopathy has mainly a cardiomyopathy with no obvious physical disability and only segmental necroses followed by active regeneration in skeletal muscle. A comprehensive review on the effects of exercise training in different animal models was published by Carter et al., 2002.

9.4 Dystrophinopathy

In human studies on dystrophinopathy patients it has been demonstrated in a 3-year-old boy with coincident dystrophinopathy and spina bifida that the dystrophic changes were markedly less severe in the lower extremities immobilized by the spina bifida. On these bases the authors recommended avoiding of excessive exercise (Kimura et al., 2006). It appears that there is a fairly good consensus that in dystrophinopathy, as well as in other necrotizing muscular dystrophies (see below) strenuous strength exercise should be avoided (Ansved 2003). On the other hand, aerobic low intensity exercise has been reported to be beneficial. In 1990's this was implemented by electrical stimulation simulating such low intensity exercise. Several studies reported positive effects of low frequency electrical stimulation (e.g. Zupan et al., 1993), the change in fiber type from more susceptible type IIB fibers to type I fibers being one of the explanations. Still in 2007 further systematic analyses were considered mandatory before recommendations to prescribe exercise for Duchenne type of dystrophinopathy can be given (Grange & Call, 2007).

9.5 Other Muscular Dystrophies

The group of limb girdle muscular dystrophies (LGMD) comprises many different entities with defects in sarcolemmal, sarcomeric and sarcoplasmic molecules and both autosomal dominant and recessive pattern of inheritance (Table). This has made it even more difficult to recruit patient groups to examine the effect of exercise. Furthermore, even when the sarcolemmal defect is located in the same dystrophin-glycoprotein complex, the pathogenic mechanism differ between defects in dystrophin vs δ-sarcoglycan, the defective molecule in LGMD2F (Milner & Kaufman, 2007). Yet, at present the same principle as for dystrophinopathy seems plausible for those LGMDs with sarcolemmal defects, i.e. high resistance training should be avoided (Ansved, 2003). Beneficial effects of endurance exercise has been reported in LGMD2I, which is caused by a defect in fukutin-related protein, a sarcoplasmic enzyme that glycosylates α-dystroglycan needed for appropriate binding of DGC to laminin-2 (Fig. 8; Sveen et al., 2007). Another reason for caution is the clinical phenotypes occasionally seen in some LGMDs, for example β-sarcoglycanopathy may manifest as exercise-induced myoglobinuria. In other LGMDs with non-sarcolemmal defects the recommendation could be at most an educated guess.

9.6 Facioscapulohumeral Muscular Dystrophy

The pathogenetic mechanism of this dystrophy (FSHD) is still open, the gene defect being a reduction of 3.3. kb DNA repeat sequences (D4Z4) at the telomeric end of chromosome 4q. This makes all predictions for the outcome of exercise treatment futile. However, FSHD is so common that sufficient numbers of patients have been recruited for *ex iuvantibus* treatment. This approach has been made easier by the fact that FSHD is usually much milder than dystrophinopathy or many LGMDs. Both low intensity aerobic exercise (Olsen et al., 2005) and strength training with or without albuterol (van der Kooi et al., 2004) has been shown to be beneficial, at least temporarily, without adverse effects on the disease.

9.7 Inflammatory Myopathies

Until early 1990's patients with inflammatory myopathies were encouraged to avoid physical activity and exercise, since those were suspected to aggravate the inflammatory process, and even until recently few studies have examined the effects of exercise in patients with myositides. Reasons for this were considered to be not only the relative rarity of these diseases, but also the lack of valid reliable outcome measures (Alexanderson & Lundberg, 2005). More recent studies have, however, shown reduced disability in patients with chronic dermatomyositis (caused by complement-mediated microangiopathy) or polymyositis (CD8 positive T-cell-mediated immunoreaction against an unknown myogenic antigen; Briani et al., 2006) following aerobic endurance training or mild/moderate up to intensive resistive muscular training. No signs of increased muscle inflammation were noted (Alexanderson et al., 2007). In the third idiopathic inflammatory myopathy, inclusion body myositis (another CD8 positive T-cell-mediated immunoreaction against an unknown myogenic antigen, but without response to immunosuppressive therapy) the results have been similarly beneficial (Arnardottir et al., 2003).

REFERENCES

Alexanderson H & Lundberg IE. (2005) The role of exercise in the rehabilitation of idiopathic inflammatory myopathies. Curr Opin Rheumatol 17:164–167.

Alexanderson H, Dastmalchi M, Esbjornsson-Liljedahl M, Opava CH, Lundberg IE. (2007) Benefits of intensive resistance training in patients with chronic polymyositis or dermatomyositis. Arthritis Rheum 57:768–777.

Almekinders LC. (1991) Results of surgical repair versus splinting of experimentally transected muscle. J Orthop Trauma 5:173–176.

Ansved T. (2003) Muscular dystrophies: influence of physical conditioning on the disease evolution. Curr Opin Clin Nutr Metab Care 6:435–439.

Arnardottir S, Alexanderson H, Lundberg IE & Borg K. (2003) Sporadic inclusion body myositis: pilot study on the effects of a home exercise program on muscle function, histopathology and inflammatory reaction. J Rehabil Med 35:31–35.

Beiner JM & Jokl P. (2001) Muscle contusion injuries: current treatment options. J Am Acad Orthop Surg 9:227–237.

Belkin, AM. et al. (1997) Muscle beta1D integrin reinforces the cytoskeleton-matrix link: modulation of integrin adhesive function by alternative splicing. J Cell Biol 139:1583–1595.

Best TM & Hunter KD. (2000) Muscle injury and repair. Physical Medical Rehabilitation Clinics in North America 11:251–266.

Bouchentouf M, Benabdallah BF, Mills P & Tremblay JP (2006) Exercise improves the success of myoblast transplantation in mdx mice. Neuromusc Disord 16:518–529.

Briani C, Doria A, Sarzi-Puttini P & Dalakas MC. (2006) Update on idiopathic inflammatory myopathies. Autoimmunity 39:161–170.

Brooks JHM, Fuller CW, Kemp SPT & Reddin DB. (2006) Incidence, risk and prevention of hamstring muscle injuries in professional rugby union. Am J Sports Med 34:1297–1306.

Buckwalter JA. (1995) Should bone, soft tissue, and joint injuries be treated with rest or activity? J Orthop Res 13:155–156.

Burkin DJ & Kaufman SJ. (1999) The α7β1 integrin in muscle development and disease. Cell Tissue Res. 296:183–190

Carter GT, Abresch RT & Fowler WM Jr. (2002) Adaptations to exercise training and contraction-induced muscle injury in animal models of muscular dystrophy. Am J Phys Med Rehabil 81(11 Suppl):S151–S161.

Chargé SBP & Rudnicki MA. (2004) Cellular and molecular regulation of muscle regeneration. Physiol Rev. 84:209–238.

Childers MK, Okamura CS, Bogan DJ, Bogan JR, Petroski GF, McDonald K & Kornegay JN (2002) Eccentric contraction injury in dystrophic canine muscle. Arch Phys Med Rehabil 83:1572–1578.

Crisco JJ, Jokl P, Heinen GT, et al. (1994) A muscle contusion injury model. Biomechanics, physiology, and histology. Am J Sports Med 22:702–710.

Davies KE & Nowak KJ. (2006) Molecular mechanisms of muscular dystrophies: old and new players. Nat Rev Mol Cell Biol. 7:762–773.

Ekstrand J & Gillquist J. (1983) Soccer injuries and their mechanism: a prospective study. Med and Sci Sports & Exerc 15:267–270.

Erickson HP. (2002) Stretching fibronectin. J Muscle Res Cell Motil 23:575–580.

Ervasti JM. (2007) Dystrophin, its interactions with other proteins, and implications for muscular dystrophy. Biochim Biophys Acta 1772:108–117.

Felsenfeld DP, Choquet, D & Sheetz, MP. (1996) Ligand binding regulates the directed movement of beta1 integrins on fibroblasts. Nature 383:438–440.

Friden J & Lieber RL. (2002) Tendon transfer surgery: clinical implications of experimental studies. Clinical Orthop Relat Res S163–S170

Garrett WE. (1996) Muscle strain injuries. Am J Sports Med 24: S2–S8.

Goetsch SC, Hawke TJ, Gallardo TD et al. (2003) Transcriptional profiling and regulation of the extracellular matrix during muscle regeneration. Physiol. Genomics 14:261–271.

Grange RW & Call JA. (2007) Recommendations to define exercise prescription for Duchenne muscular dystrophy. Exerc Sport Sci Rev 35:12–17.

Grounds MD, Sorokin L & White J. (2005) Strength at the extracellular matrix-muscle interface. Scand J Med Sci Sports 15:381–391.

Huard J, Li Y & Fu FH. (2002) Muscle injuries and repair: current trends in research. Journal of Bone & Joint Surgery 84-A:822–832.

Hurme T, Kalimo H, Lehto M & Järvinen M. (1991a) Healing of skeletal muscle injury. An ultrastructural and immunohistochemical study. Med Sci Sports Exerc 23:801–810.

Hurme T, Kalimo H, Sandberg M, et al. (1991b) Localization of type I and III collagen and fibronectin production in injured gastrocnemius muscle. Lab Invest 64:76–84.

Hurme T & Kalimo H. (1992a) Activation of myogenic precursor cells after muscle injury. Med Sci Sports Exerc 24:197–205.

Hurme T & Kalimo H. (1992b) Adhesion in skeletal muscle regeneration. Muscle Nerve 15:482–489.

Jackson DW & Feagin JA. (1973) Quadriceps contusions in young athletes: relation of severity of injury to treatment and prognosis. J Bone & Joint Surg 55-A:95–105.

Järvinen M. (1975) Healing of a crush injury in rat striated muscle. 2. A histological study of the effect of early mobilization and immobilization on the repair processes. Acta Pathol Microbiol Scand 83A:269–282.

Järvinen M. (1976a) Healing of a crush injury in rat striated muscle. 4. Effect of early mobilization and immobilization on the tensile properties of gastrocnemius muscle. Acta Chir Scand 142:47–56.

Järvinen M. (1976b) Healing of a crush injury in rat striated muscle. 3. A microangiographical study of the effect of early mobilization and immobilization on capillary ingrowth. Acta Pathol Microbiol Scand 84A:85–94.

Järvinen TA, Kannus P, Jarvinen TL, Jozsa L, Kalimo H, Jarvinen M (2002) Tenascin-C in the pathobiology and healing Process of musculoskeletal tissue injury. Scand J Med Sci Sports 10:376–382.

Järvinen M & Lehto MUK. (1993) The effect of early mobilization and immobilization on the healing process following muscle injuries. Sports Med 15:78–89.

Järvinen TAH, Józsa L, Kannus P, Järvinen TLN, Hurme T, Kvist M, Pelto-Huikko M, Kalimo H & Järvinen M. (2003a) Mechanical loading regulates the expression of tenascin-C in the myotendinous junction and tendon but does not induce *de novo*-synthesis in the skeletal muscle. J Cell Sci 116:857–866.

Järvinen TAH, Järvinen TLN, Kannus P & Kalimo H. (2003b) Ectopic expression of tenascin-C. J Cell Sci 116:3851–3853

Järvinen TAH, Järvinen TLN, Kääriäinen M, Kalimo H & Järvinen M. (2005) Biology of muscle trauma. Am J Sports Med 33:745–766.

Kalimo H, Rantanen J & Järvinen M. (1997) Muscle injuries in sports. Baillière's Clinical Orthop 2:1–24.

Kannus P, Parkkari J, Järvinen TLN, Järvinen TAH & Järvinen M. (2003) Basic science and clinical studies coincide: active approach is needed in the treatment of sports injuries. Scand J Med Sci Sports 13:150–154.

Kimura S, Ikezawa M, Nomura K, Ito K, Ozasa S, Ueno H, Yoshioka K, Yano S, Yamashita T, Matuskura M, Miike T (2006) Immobility reduces muscle fiber necrosis in dystrophin deficient muscular dystrophy. Brain Dev 28:473–476.

Kuang S, Kuroda K, Le Grand F & Rudnicki MA. (2007) Asymmetric self-renewal and commitment of satellite stem cells in muscle. Cell 129:999–1010.

Kääriäinen M, Kääriäinen J, Järvinen TLN, et al. (1998) Correlation between biomechanical and structural changes during the regeneration of skeletal muscle after laceration injury. J Orthop Res 16:197–206.

Kääriäinen M, Kääriäinen J, Järvinen TLN, et al. (2000) Integrin and dystrophin associated adhesion protein complexes during regeneration of shearing-type muscle injury. Neuromusc Disord 10:121–134.

Kääriäinen M, Liljamo T, Pelto-Huikko M, et al. (2001) Regulation of $\alpha 7$ integrin and merosin by mechanical stress during skeletal muscle regeneration. Neuromusc Disord 11:360–369.

Kääriäinen M, Nissinen L, Järvinen M, et al. (2002) Expression of $\alpha 7\beta 1$ integrin splicing variants during skeletal muscle regeneration. Am J Pathol 161:1023–1031.

Lehto M, Järvinen M & Nelimarkka O. (1986) Scar formation after skeletal muscle injury. A histological and autoradiographical study in rats. Arch Orthop Trauma Surg 104:366–370.

Lehto M, Duance VC & Restall D. (1985a) Collagen and fibronectin in a healing skeletal muscle injury. An immunohistochemical study of the effects of physical activity on the repair of injured gastrocnemius muscle in the rat. J Bone Joint Surg 67B:820–828.

Lehto M & Järvinen M. (1985b) Collagen and glycosaminoglycan synthesis of injured gastrocnemius muscle in rat. Eur Surg Res 17:179–185.

Lehto M, Sims TJ & Bailey AJ. (1985c) Skeletal muscle injury – molecular changes in the collagen during healing. Res Exp Med 185:95–106.

Lieber RL, Schmitz MC, Mishra DK & Friden J. (1994) Contractile and cellular remodeling in rabbit skeletal muscle after cyclic eccentric contractions. J Appl Physiol 77:1926–1934.

MacIntyre NJ, Bhandari M, Blimkie CJ, Adachi JD & Webber CE. (2001) Effect of altered physical loading on bone and muscle in the forearm. Can J physiol pharmacol 79:1015–1022.

Menetrey J, Kasemkijwattana C, Fu FH, et al. (1999) Suturing versus immobilization of a muscle laceration. A morphological and functional study in a mouse model. Am J Sports Med 27:222–229.

Mercado ML, et al. (2004) Neurite outgrowth by the alternatively spliced region of human tenascin-C is mediated by neuronal alpha7beta1 integrin. J Neurosci 24:238–247.

Milner DJ & Kaufman SJ. (2007) Alpha7beta1 integrin does not alleviate disease in a mouse model of limb girdle muscular dystrophy type 2F. Am J Pathol 170:609–619.

Morla A, Zhang Z & Ruoslahti E. (1994) Superfibronectin is a functionally distinct form of fibronectin. Nature 367:193–196.

Noonan TJ, Best TM, Seaber AV & Garrett WE Jr. (1993) Thermal effects on skeletal muscle behavior. Am J Sports Med 21:517–522.

Oberhauser AF, Marszalek PE, Erickson HP, Fernandez JM (1998) The molecular elasticity of the extracellular matrix Protien tenascin. Nature 393:181–185.

Olsen DB, Orngreen MC & Vissing J. (2005) Aerobic training improves exercise performance in facioscapulohumeral muscular dystrophy. Neurology 64:1064–1066.

Pate RR, Pratt M, Blair SN, Haskell WL, Macera CA, Bouchard C, Buchner D, Ettinger W, Heath GW, King AC, et al. (1995) Physical activity and public health. A recommendation from the Centers for Disease Control and Prevention and the American College of Sports Medicine. JAMA 273:402–407.

Petersen J & Hölmich P. (2005) Evidence based prevention of hamstring injuries in sports. Br J Sports Med 39:319–323.

Rantanen J, Ranne J, Hurme T, et al. (1995a) Satellite cell proliferation and expression of myogenin and desmin in regenerating skeletal muscle: evidence for two different populations of satellite cells. Lab Invest 72:341–347.

Rantanen J, Ranne J, Hurme T, et al. (1995b) Denervated segments of injured skeletal muscle fibres are reinnervated by newly formed neuromuscular junctions. J Neuropath Exp Neurol 54:188–191.

Safran MR, Garrett WE Jr, Seaber AV, Glisson RR & Ribbeck BM. (1988) The role of warm-up in muscular injury prevention. Am J Sports Med 16:123–129.

Safran MR, Seaber AV & Garrett WE Jr. (1989) Warm-up and muscular injury prevention. Sports 8:239–249.

Sherry MA & Best TM. (2004) A comparison of 2 rehabilitation programs in the treatment of acute hamstring strains. J Orthop Sports Phys Therapy 34:116–125.

Song WK, Wang W, Foster RF, et al. (1993) β36-α7 is a novel integrin alpha chain that is developmentally regulated during skeletal myogenesis. J Cell Biol 117:643–657.

Sveen ML, Jeppesen TD, Hauerslev S, Krag TO & Vissing J. (2007) Endurance training: an effective and safe treatment for patients with LGMD2I. Neurology 68:59–61.

Vachon PH, Xu H, Liu L, Loechel F, Hayashi Y, Arahata K, Reed JC, Wewer UM & Engvall E. (1997) Integrins (α7β1) in muscle function and survival. Disrupted expression in merosin-deficient congenital muscular dystrophy. J Clin Invest 100:1870–1881.

Vaittinen S, Lukka R, Sahlgren C, et al. (2001) The expression of intermediate filament protein nestin as related to vimentin and desmin in regenerating skeletal muscle. J Neuropathol Exp Neurol 60:588–59.

Vaittinen S, Hurme T, Rantanen J, et al. (2002) Transected myofibres may remain permanently divided in two parts. Neuromuscul Disord 12:584–587.

van der Kooi EL, Vogels OJ, van Asseldonk RJ, Lindeman E, Hendriks JC, Wohlgemuth M, van der Maarel SM & Padberg GW. (2004) Strength training and albuterol in facioscapulohumeral muscular dystrophy. Neurology 63:702–708.

van der Kooi EL, Lindeman E & Riphagen I. (2005) Strength training and aerobic exercise training for muscle disease. Cochrane Database Syst Rev (1):CD003907

von der Mark H, Williams I, Wendler O, Sorokin L, von der Mark K. & Poschl E. (2002) Alternative splice variants of α7 β1 integrin selectively recognize different laminin isoforms. J Biol Chem 277:6012–6016

Wierzbicka-Patynowski I, Schwarzbauer JE (2003) The ins and outs of fibronectin matrix assembly. J cell Sci 116:3269–3276.

Woodard C. (1954) What is active treatment? In Woodard C (eds) Sports Medicine. London: Max Parrish & Co, pp 1–14.

Yan Z, Choi S, Liu X, Zhang M, Schageman JJ, Lee SY, Hart R, Lin L, Thurmond FA & Williams RS. (2003) Highly coordinated gene regulation in mouse skeletal muscle regeneration. J Biol Chem 278:8826–8836.

Yu JG & Thornell LE. (2002) Desmin and actin alterations in human muscles affected by delayed onset muscle soreness: a high resolution immunocytochemical study. Histochem Cell Biol 118:171–9.

Yu JG, Malm C & Thornell LE. (2002) Eccentric contractions leading to DOMS do not cause loss of desmin nor fibre necrosis in human muscle. Histochem Cell Biol 118:29–34.

Yu JG, Carlsson L & Thornell LE. (2004) Evidence for myofibril remodeling as opposed to myofibril damage in human muscles with DOMS: an ultrastructural and immunoelectron microscopic study. Histochemistry and Cell Biology 121:219–227.

Yu JG, Furst DO & Thornell LE. (2003) The mode of myofibril remodelling in human skeletal muscle affected by DOMS induced by eccentric contractions. Histochemistry and Cell Biology 119:383–393.

Zinna EM & Yarasheski KE. (2003) Exercise treatment to counteract protein wasting of chronic diseases. Curr Opin Clin Nutr Metab Care 6:87–93.

Zupan A, Gregoric M, Valencic V & Vandot S. (1993) Effects of electrical stimulation on muscles of children with Duchenne and Becker muscular dystrophy. Neuropediatrics 24:189–192.

Äärimaa V, Rantanen J, Best T, Schultz E, et al. (2004a) Mild eccentric stretch injury in skeletal muscle causes transient effects on tensile load and cell proliferation. Scand. J Med Sci Sports 14:367–372.

Äärimaa V, Kääriäinen M, Vaittinen S, et al. (2004b) Restoration of myofiber continuity after transection injury by surgical suturing. Neuromuscul Disord 3:421–428.

CHAPTER 12

INFLAMMATION IN SKELETAL MUSCLE REGENERATION

JAMES G. TIDBALL
Departments of Physiological Science and Pathology and Laboratory Medicine, University of California, Los Angeles, Los Angeles CA

1. INTRODUCTION

Functional interactions between the immune system and skeletal muscle have been little explored. Until recently, most investigations of muscle inflammation consisted of histological assessments of acute muscle injury or diagnoses of inflammatory myopathies. However, few experimental studies addressed the mechanisms through which immune cells and skeletal muscle affect one another's differentiation, growth, function or viability. Nevertheless, hints have been provided in the scientific literature for decades to suggest that immune cells may play regulatory roles in muscle development, repair or regeneration. For example, 50 years ago, the sequential invasion of skeletal muscle by morphologically-distinct leukocyte populations was identified as a consequence of muscle injury (Godman, 1957). Furthermore, the elevated numbers of distinct leukocyte populations during muscle repair and regeneration suggested that leukocytes could play a role in those processes (Godman, 1957). Several investigators also noted the presence of leukocytes in elevated numbers and close proximity to skeletal muscle fibers during embryonic development (Abood and Jones, 1991; Nishikawa et al., 1998). Perhaps the unexplored regulatory roles of leukocytes in myogenesis are also reflected in their roles in muscle repair and regeneration.

This chapter focuses on evaluating recent discoveries concerning the functions of leukocytes that invade muscle following injury or modified muscle use, or during muscle disease. Emphases are put on leukocyte populations that can affect muscle repair and regeneration, and on the specific molecules that mediate interactions between muscle cells and leukocytes that affect muscle regeneration. These recent

S. Schiaffino and T. Partridge (eds.), Skeletal Muscle Repair and Regeneration, 243–268.

findings provide the basis for a model that is presented to represent potential functions and interactions between inflammatory cells that can affect repair and regeneration of skeletal muscle.

2. FUNCTIONS OF LEUKOCYTES IN MUSCLE REPAIR AND REGENERATION THAT FOLLOW ACUTE TRAUMA OR MODIFIED MUSCLE USE

Acute muscle trauma or modified muscle use results in a stereotypic inflammatory response that is characterized by the sequential invasion of muscle by specific myeloid cell populations, comprised almost entirely of neutrophils and specific subpopulations of macrophages (Fig. 1). This sequential response suggests that leukocytes may play multiple roles in regulating repair and regeneration of muscle, but they may also promote muscle damage through their capacity as cytolytic cells. The sequential invasion by myeloid cell populations also suggests that each successive stage of the invasion process may depend on interactions between the injured muscle and myeloid cells that invaded previously.

2.1 Do Neutrophils Influence Skeletal Muscle Regeneration?

Neutrophils are rapid and early invaders of skeletal muscle following muscle injury, with their concentrations in muscle becoming tremendously elevated within hours of muscle damage. Neutrophil activation in response to injury is highly-structured, and dominated by the respiratory burst, in which neutrophils rapidly release high concentrations of free radicals that can target cellular debris for removal by phagocytosis. Neutrophils also release proteases that can degrade tissue debris or extracellular matrix (ECM), and secrete pro-inflammatory cytokines that may further promote tissue inflammation (Wiedow and Meyer-Hoffert, 2005). Thus, neutrophils have the capacity to ready injured muscle for regeneration by removing damaged tissue,

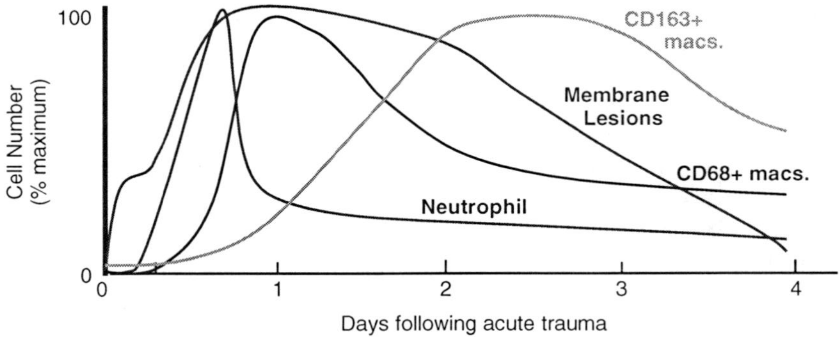

Figure 1. Relative time-course of myeloid cell invasion and muscle membrane injury following modified muscle use. Macs. = macrophages

to prepare the ECM for tissue remodeling and to signal the invasion of subsequent myeloid cell populations that may play their own roles in regeneration. However, many neutrophil-derived products, such as free radicals, are potentially cytotoxic and may cause muscle damage.

Several observations have provided the basis for speculations that the neutrophil-mediated phagocytosis facilitates muscle regeneration. For example, mouse strains with low rates of phagocytosis by leukocytes showed slower muscle regeneration following injury (Grounds, 1987) and old animals in which phagocytosis is slower experienced slower muscle regeneration (Zacks and Sheff, 1982; Grounds, 1987). In addition, depletion of phagocytic neutrophils and monocytes from mice prior to muscle injury by toxin injection resulted in greater quantities of cellular debris at the injury sites, and slower muscle regeneration (Teixeira et al., 2003). Similarly, depletion of phagocytic cells by intravenous administration of liposomes containing clondronate resulted in attenuated muscle inflammation, prolonged the removal of cellular debris and slowed muscle regeneration following an experimentally-imposed freeze injury (Summan et al., 2006). Although the clodronate-treatment caused reductions in macrophage numbers (Summan et al., 2006), it is feasible that the treatment also reduced the concentration of phagocytic neutrophils at the peak of their invasion of the injured muscle. Other indirect evidence is consistent with the possibility that neutrophils promote muscle regeneration. Treatment of rabbits with the non-steroidal, anti-inflammatory drug flurbiprofen, which can affect the innate immune response including neutrophil function, slowed recovery of muscle function following injury (Mishra et al., 1995). Collectively, these investigations suggest a possible role for neutrophils and neutrophil-mediated phagocytosis in facilitating muscle regeneration, although conclusive data have not yet been produced.

Neutrophils also serve a positive role in tissue regeneration by signaling the invasion of other myeloid cells, such as macrophages, that could then mediate direct, positive effects on repair or regeneration. Although this neutrophil function has not yet been proven for injured muscle, it has been demonstrated in non-muscle models. For example, neutrophil depletion from rats prior to experimentally-induced lung inflammation caused a reduction in the number of macrophages that later invaded the injured tissue (Janardhan et al., 2006).

2.2 Macrophages Promote Skeletal Muscle Regeneration Following Acute Injury or Modified Muscle Use

2.2.1 Changes in macrophage phenotype during muscle's response to modified loading indicate that macrophages play more than one regulatory role in injured muscle

Recent investigations using a model of modified muscle use that produces mild muscle damage followed by inflammation, regeneration and growth have demonstrated a role for macrophages in promoting muscle repair, regeneration and growth in vivo. In this manipulation, rodent hindlimbs are elevated for a period of time above the cage floor so that they are no longer weight-bearing which causes a rapid loss of mass in some muscles as they adapt to the unloaded condition. The soleus muscle is most rapidly affected and can lose 30 to 40% of its mass in 10 days through mechanisms that resemble those that occur in the muscles of patients subjected to prolonged bedrest (Thomason and Booth, 1990). Subsequent reloading of the hindlimbs by normal weight-bearing and ambulation produces a predictable sequence of mild muscle injury, inflammation, regeneration and growth. Muscle reloading produces an increase in central-nucleated fibers, renewed expression of developmental genes and a rapid increase in fiber cross-sectional area, so that the fibers return to their pre-suspension size within 4 to 7 days of reloading (Thomason and Booth, 1990; Krippendorf and Riley, 1993; St. Pierre and Tidball, 1994). This predictable, stereotypic response of the rodent soleus muscle to hindlimb suspension followed by reloading (HS/Rel) has permitted the in vivo analysis of the contributions of specific leukocyte populations to injury, regeneration and growth that result from increased muscle loading.

Muscle inflammation in the HS/Rel model is characterized by a rapid invasion by neutrophils followed by at least two subpopulations of phenotypically-distinct macrophages (Krippendorf and Riley, 1993; St. Pierre and Tidball, 1994). The first sub-population is phagocytic and expresses CD68 (also known as ED1 antigen or macrosialin) (Rabinowitz and Gordon, 1991; Ramprasad et al., 1995) (Fig. 2). These characteristics of the early invading macrophage population are emblematic of macrophages that dominate Th1 immune responses. The Th1 response in other models of tissue injury and disease is characterized by the presence of interleukin-1 (IL-1), IL-2, interferon-γ (IFN-γ), and tumor necrosis factor (TNF), which can promote cellular immunity and are generally pro-inflammatory, leading to classical activation of macrophages (Gordon, 2003) (Fig. 3). Among these pro-inflammatory cytokines, stimulation by IFN-γ is essential for classical activation of the Th1 phenotype. The Th1 response is typically the initial response following trauma or injury of any tissue and leads to the rapid removal of foreign materials or tissue debris by phagocytosis (Mills et al., 2000; Anderson and Mosser, 2002; Gordon, 2003).

Skeletal muscle experiencing injury in the HS/Rel model is subsequently invaded by a sub-population of macrophages that express CD163 (also called ED2 antigen) and are not phagocytic and persist in injured muscle throughout the period of muscle regeneration and repair (Krippendorf and Riley, 1993; St. Pierre and Tidball, 1994). These characteristics are typical of macrophages that are present in the Th2 immune response. The Th2 response in other models of tissue injury and disease involves elevated levels of IL-4, IL-5, IL-6, IL-10 and IL-13, which can have anti-inflammatory effects and can cause de-activation of classically-activated, Th1 macrophages and induce alternative activation of macrophages (Gordon, 2003). Deactivation is associated with the down-regulation of pro-inflammatory cytokines and reduction of macrophages expressing CD68. IL-10 is a potent deactivating cytokine that contributes to the termination of the inflammatory response and can enable macrophages to shift from a classically-activated phenotype to an alternatively-activated phenotype (Lang et al., 2002; Mosser, 2003). This shift is an

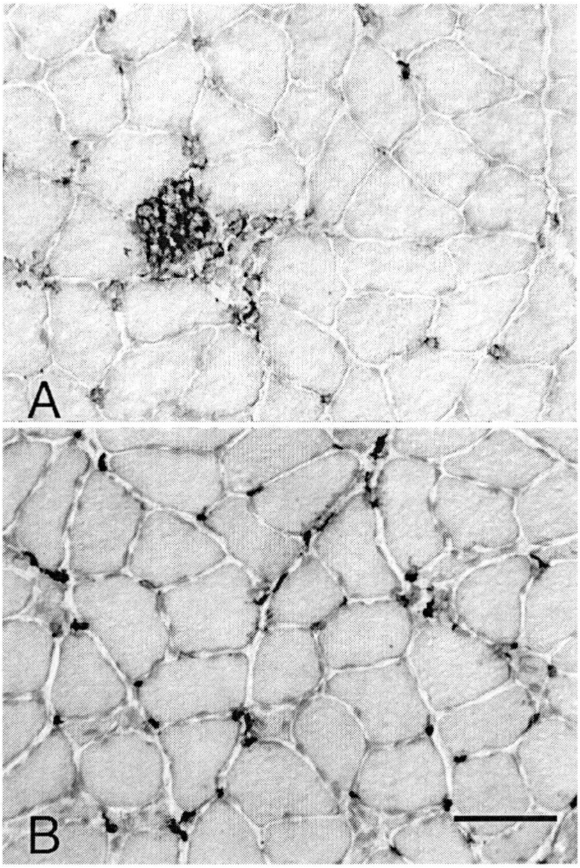

Figure 2. Section of soleus muscle from rat subjected to 10 days of hindlimb unloading followed by 1 day of muscle reloading. Red cells are labeled with anti-CD68. Note muscle fiber near the center of the micrograph that is heavily invaded with CD68+ macrophages. B. Soleus muscle section from rat subjected to 10 days of hindlimb suspension followed by 2 days of muscle reloading. Red cells are labeled with anti-CD163. Bars = 80 μm

important response to tissue injury because alternative activation promotes tissue repair and involves increased expression of genes that are involved in connective tissue remodeling and fibrosis, such as procollagens I and III, metalloproteinases and arginase (Gordon, 2003; Anderson and Mosser, 2002).

2.2.2 CD68+ *macrophages may target and remove cellular debris caused by oxidative damage*

The early-invading, phagocytic, CD68+/CD163− macrophages reach their highest concentration in muscle at about 24 hours following the onset of injury or reloading

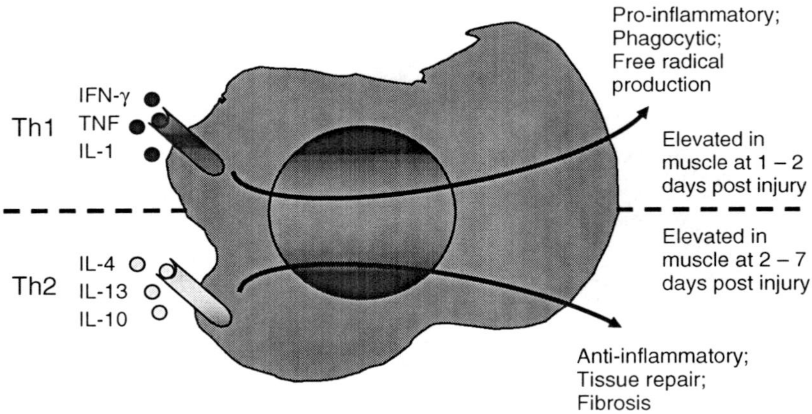

Figure 3. Representation of differing results of macrophage activation by Th1 cytokines (upper) or Th2 cytokines (lower) that can affect muscle repair and regeneration

and then rapidly decline (St. Pierre and Tidball, 1994; Frenette et al., 2000) (Fig. 1). Although there are few functional studies of CD68, it is known to function as a receptor for oxidized low density lipoproteins (oxLDLs) (Ottnad et al., 1995; Ramprasad et al., 1995; Van Velzen et al., 1997). CD68+/CD163− cells are at their highest concentrations during acute, Th1 cytokine-mediated inflammation in many tissues. Although most CD68 is located in late endosomes, nearly 15% is located on the surface of macrophages, and rapid shuttling between the surface and intracellular pools occurs (Ramprasad et al., 1996; Kurushima et al., 2000). The affinity of CD68 for oxLDLs, the rapid shuttling of CD68 between the macrophage cell surface and the endosomal compartment supports speculations that CD68 functions in the recognition of oxidatively-injured cells and cellular debris that are then targeted for phagocytosis.

Phagocytosis of injured cells by CD68+/CD163− macrophages would be particularly relevant to understanding the inflammatory response to injured muscle, in which oxidative damage is a major cause of muscle injury and removal of damaged muscle fibers is an early and rapid event. Recent findings have shown that the majority of damage to the plasma membrane of muscle in HS/Rel results from a superoxide and myeloperoxidase (MPO)-dependent mechanism, and is caused by neutrophils that invade reloaded muscle in advance of extravasation of CD68+/CD163− macrophages (Nguyen and Tidball, 2003; Nguyen et al., 2005). Null mutation of $gp91^{phox}$, the catalytic subunit of NADPH oxidase that is necessary for superoxide production by neutrophils, prevents most membrane damage in reloaded muscle, which suggests that superoxide or a downstream derivative may cause oxidative damage to the muscle membrane in this model (Nguyen and Tidball, 2003). Similarly, null mutation of MPO prevents most muscle membrane lysis during reloading (Nguyen et al., 2005). MPO is a neutrophil specific enzyme that catalyses the conversion of hydrogen peroxide to the highly-oxidative and cytolytic free radical, hypochlorous acid. Lipoproteins at the cell surface can be oxidized by MPO products to target cells for CD68 binding (Zouaoui Boudjeltia et al., 2004). Thus, oxidative modification of membrane lipoproteins by superoxide, hypochlorous acid or other free radicals may target muscle cells or fragments of muscle cells for phagocytic removal by CD68+/CD163− macrophages. Several findings support this possibility. Not only do oxidized lipoproteins bind CD68+/CD163− macrophages, but red blood cells experiencing oxidative stress also bind macrophages via CD68 (Nishikawa et al., 1990). Furthermore, exposure of skeletal muscle to oxidative stress in I/R injury leads to an increase in oxLDL at the muscle plasma membrane (Grisotto et al., 2000) and oxidative modification of LDLs by MPO increases their binding by macrophage receptors (Hazell and Stocker, 1993; Podrez et al., 2000; Carr et al., 2000).

2.2.3 *Non-phagocytic,* CD68−/CD163+ *macrophages accumulate in muscle during repair, regeneration and growth*

The CD68−/CD163+ macrophages that dominate the second population of leukocytes to invade muscle in the HS/Rel model and following acute injury reach their peak at 2 to 4 days of reloading (St. Pierre and Tidball, 1994; Frenette et al., 2000). Many observations suggest that CD68−/CD163+ macrophages may be associated with muscle repair or regeneration. For example, CD68−/CD163+ cells accumulate in reloaded muscle just prior to the onset of expression of proteins that are indicators of muscle repair and regeneration and they are present in highest numbers near regenerative fibers (St. Pierre and Tidball, 1994; McLennan, 1993). In addition, CD163+ peritoneal macrophages release factors in vitro that can promote myoblast proliferation (Massimino et al., 1997). This likely role of CD68−/CD163+ cells in promoting tissue repair is not muscle specific; they accumulate in numerous tissues during the healing phase following injury (Schaer et al., 2001; Schaer et al., 2002).

CD163 is a transmembrane glycoprotein that has 9 SRCR domains (i.e., scavenger receptor cysteine rich domains) that is constitutively expressed by monocytes and macrophages (Schaer et al., 2001). However, its expression levels are greatly affected by cytokines in patterns that are consistent with its role in tissue healing. For example, pro-inflammatory Th1 cytokines such as TNF-α down-regulate CD163 expression (Buechler et al., 2000), while the anti-inflammatory cytokine IL-10 is a powerful inducer of CD163 expression (Buechler et al., 2000; Sulahian et al., 2000; Schaer et al., 2001; Schaer et al., 2002). Similarly, anti-inflammatory agents such as glucocorticoids can cause an over 10-fold induction of CD163 expression (Schaer et al., 2002). IL-6, the expression of which is greatly elevated by increased muscle loading, does not affect CD163 expression, but does cause a translocation of CD163 from intracellular compartments to the cell surface (Buechler et al., 2000). IL-6 also causes large increases in IL-10 expression in muscle, and may thereby indirectly elevate CD163 levels (Buechler et al., 2000).

CD163 may play an active role in tissue repair and regeneration, and is not merely a marker of the healing phase of inflammation. CD163 is a receptor for complexed hemoglobin and haptoglobin, enabling these complexes to be internalized and degraded by macrophages (Kristiansen et al., 2001; Schaer et al., 2001). This process may be important in tissue repair following injury because hemolysis that is associated with injury causes rapid, local elevations of hemoglobin that can be toxic to cells (Moestrup and Moller, 2004). In addition, the iron in hemoglobin can catalyze production of iron-derived hydroxyl radicals by the Fenton reaction (Sadrzadeh et al., 1984; Lim et al., 1998), which can promote free radical mediated damage to the muscle cells. This, also, would be attenuated by CD163 binding and internalization of hemoglobin (Philippidis et al., 2004). There are further, downstream benefits of internalization of hemoglobin / haptoglobin complexes. Following internalization, the breakdown of heme is catalyzed by hemeoxygenase to yield bilirubin, free iron and carbon monoxide (CO); CO in turn can then decrease the expression of pro-inflammatory Th1 cytokines and increase expression of anti-inflammatory Th2 cytokines, to promote healing (Otterbein, 2000; Otterbein and Choi, 2000). In addition, recent findings indicate a feed-forward system through which CD163 binding can amplify its anti-inflammatory role; hemoglobin / haptoglobin binding by CD163 increases expression of hemeoxygenase and IL-10, which would further promote the anti-inflammatory effects (Philippidis et al., 2004).

2.2.4 Macrophages promote muscle repair, regeneration and growth following modified muscle use and acute injury

The ability of macrophages to attenuate inflammation and promote tissue repair in several models of injury and disease, and the coincidence of their time of arrival in muscle with the onset of muscle repair, regeneration, growth suggests that they may promote muscle regeneration. Several investigations support this likely role for these macrophages. For example, activated monocytes and macrophages released a factor that increased the proliferation rates of satellite cells in vitro (Cantini and Carraro, 1995). In addition, conditioned medium from macrophage cell line cultures (J774) increased the proportion of cultured myoblasts that express MyoD, and increased the proportion of muscle cells that express myogenin when applied at later stages of differentiation (Cantini et al., 2002), indicating that macrophage-derived factors could have a positive effect on both proliferation and differentiation of muscle cells. However, other investigators observed that the co-culturing of muscle cells with macrophages increased muscle cell numbers and increased the proportion of muscle cells that expressed MyoD, but decreased the proportion of cells expressing myogenin (Merly et al., 1999). Thus, macrophages apparently promoted activation but inhibited differentiation of muscle in these other studies. Although the differences in the effects of macrophages on muscle differentiation in vitro could reflect differences in experimental design, perhaps the distinctions are attributable to differences in the differentiation or activation state of the macrophages (i.e., classical vs. alternative). However, in a previous study (Massimino et al., 1997), CD163+ macrophages were selectively enriched prior to co-culturing with muscle cells, after which they produced both an increase in muscle cell proliferation and an increase in the number of myonuclei per myotube in the cultures. Thus, CD163+ macrophages may be able to stimulate proliferation and differentiation of muscle cells.

In vivo studies also support a role for macrophages in promoting growth, regeneration and repair following muscle injury or modified use. For example, when mice were irradiated to deplete populations of proliferative cells such as monocytes and macrophages, muscle regeneration following whole muscle engraftment was impaired (Lescaudron et al., 1999). Furthermore, null mutation of cyclooxygenase-2 (COX-2) or COX-2 specific inhibition produced reductions in muscle inflammation that were associated with reductions in MyoD expression and slower muscle growth (Bondesen et al., 2004). In vitro studies had previously implicated COX-2 in myoblast proliferation (Zalin, 1987), differentiation (Schutzle et al., 1984) and fusion (David and Higginbotham, 1981; Entwistle et al., 1986; Horsley et al., 2003), and the later in vivo findings in which COX-2 signaling was ablated showed that COX-2 could mediate these effects after muscle injury in vivo (Bondesen et al., 2004). However, those findings do not necessarily mean that the loss of inflammatory cells, especially macrophages, in the COX-2 deficient mice was responsible for the defects in muscle regeneration. In a more recent study (Bondesen et al., 2006), myofiber growth and satellite cell fusion were significantly reduced during periods muscle reloading in the HS/Rel model in mice in which COX-2 signaling was ablated. However, the magnitude of the treatment effect was similar in muscles in which there was extensive inflammation and those with little inflammation, leading the investigators to conclude that the positive effects of COX-2 signaling in muscle regeneration were not solely associated with inflammatory cells.

Recent, in vivo studies of the role of inflammatory cells in muscle regeneration following injury or modified muscle use support the distinct roles of early-invading, Th1 macrophages and later invading Th2 macrophages in the regenerative process. Depletion of phagocytic macrophages by the systemic application of liposomes that contained clodronate prior to freeze injury resulted in slower clearing of cellular debris, but did not significantly affect the growth of regenerating muscle fibers (Summan et al., 2006). In contrast, antibody depletion of the late-invading population of macrophages that are primarily CD68 − /CD163+, non-phagocytic cells did not affect the influx of early-invading macrophages that are primarily CD68 + /CD163−, but produced significantly slower myofiber growth in the HS/Rel model (Tidball and Wehling-Henricks, 2007). This finding indicates that CD68 − /CD163+ macrophages promote muscle growth in vivo. In addition, depletion of the late-invading population of macrophages slowed muscle fiber regeneration that was indicated by reductions in fiber centronucleation. Surprisingly, depletion of the late-invading macrophage population also slowed the repair of muscle membrane lesions that were caused by muscle reloading, and this was also reflected by persistent elevation of dysferlin (Tidball and Wehling-Henricks, 2007), a protein that has been implicated in the repair of damaged cell membranes (Bansal et al., 2003). Finally, macrophage-depleted muscles also showed a prolonged elevation of MyoD+ cells, suggesting that CD68 – /CD163+ macrophages may affect muscle differentiation in vivo (Tidball and Wehling-Henricks, 2007), as shown previously in vitro (Massimino et al., 1997).

2.2.5 *What signals the shift in macrophages from a classically-activated,* CD68+ *phenotype to an alternatively-activated,* CD163+ *phenotype during muscle repair and regeneration?*

IL-4, IL-10 and IL-13 are Th2 cytokines that are capable of inducing alternative activation of macrophages, suggesting that the expression or secretion of at least one of these molecules induces the shift in macrophage activation state that occurs during muscle regeneration. Although systemic IL-10 levels are elevated following intense exercise over a time-course that would correspond to the shift from classical to alternative activation of macrophages in injured muscle (Smith et al., 2000), no direct data establish this function for IL-10 in muscle. However, IL-4 expression in muscle is significantly elevated in skeletal muscle during regeneration that results from acute freeze damage (Horsley et al., 2003). Although Th2 lymphocytes are the most prominent and well-characterized source of IL-4, they do not invade muscle following acute injury or modified use. Instead, myotubes themselves appear to be the primary and perhaps exclusive source of IL-4 in regenerating muscle (Horsley et al., 2003). Following acute injury, the growth of regenerating myotubes and myofibers is significantly slowed in mice that are systemic null mutants for either IL-4 or the IL-4-α receptor (Horsley et al., 2003). Interestingly, the reduction of fiber growth that occurs in the regenerating muscles of mice that are deficient in IL-4 signaling (Horsley et al., 2003) is similar to the reduction observed when late-invading macrophages are depleted in the HS/Rel (Tidball and Wehling-Henricks, 2007). Although it is feasible that the reduced fiber growth could partially reflect loss of the pro-regenerative, pro-growth effects that are mediated by alternatively activated macrophages, direct effects of IL-4 on muscle cell growth and proliferation have been clearly demonstrated. Muscle cells from IL-4 or IL-4α receptor null mutant mice produce smaller myotubes with fewer myonuclei in vitro than formed by wild-type muscle, indicating a defect in myoblast fusion (Horsley et al., 2003). Furthermore, defects in IL-4 –mediated signaling are associated with slower directed migration of myoblasts in vitro (Jansen and Pavlath, 2006), and exogenous IL-4 increases migratory activity of transplanted myoblasts within host skeletal muscles (Lafreniere et al., 2006).

2.2.6 *How do macrophages promote muscle growth, repair or regeneration following muscle injury?*

2.2.6.1 Leukemia inhibitory factor (LIF) Although the presence of CD68 – /CD163+ macrophages has been associated with muscle growth and differentiation in vivo and in vitro, the factor that is produced by CD163+ macrophages to mediate these effects is unknown. However, leukemia inhibitory factor (LIF) is an attractive candidate molecule because it is expressed by macrophages (Abe et al., 1991), its expression in injured muscle is elevated, it can increase muscle growth in vivo and stimulate satellite cell proliferation in vitro (Barnard et al., 1994; Vakakis et al., 1995; Finkelstein et al., 1996; Kurek et al., 1996, 1997).

LIF is a glycosylated cytokine that is secreted by immune cells and a variety of other cell types. The broad biological functions of LIF include inducing osteoblast proliferation, regulating hematopoiesis and promoting cardiac hypertrophy (Verfaillie and McGlave, 1991; Metcalf et al., 1990; Pan et al., 1999). LIF has also been implicated in tissue repair. Axotomy of sympathetic neurons induced LIF at the lesion site (Banner and Patterson, 1994; Carlson et al., 1996) and LIF promoted neural repair (Finkelstein et al., 1996) and Schwann cell survival (Dowsing et al., 1999). Exogenous LIF also promotes muscle growth and regeneration (Kurek et al., 1996). LIF administration to crushed muscle by an osmotic pump increased fiber size and regeneration (Barnard et al., 1994). LIF administration to muscle during reinnervation also produced a significant increase in muscle fiber cross-sectional area (Finkelstein et al., 1996). Furthermore, LIF expression is elevated during muscle regeneration (Kurek et al., 1996). LIF null mutants also repaired crushed muscle more slowly than wild type mice (Kurek et al., 1997), further supporting the role of LIF in muscle repair and regeneration.

Many of the LIF-mediated effects on muscle growth and regeneration may reflect direct actions of the cytokine on muscle. Skeletal muscle cells express the LIF receptor (LIFR) (Bower et al., 1995) and LIF promotes myoblast proliferation in vitro (Vakakis et al., 1995; Bower et al., 1995) by signaling through the JAK2-STAT3 pathway (Spangenburg and Booth, 2002) and may inhibit proliferation through activation of the ERK signaling pathway (Jo et al., 2005). However, there may be multiple sources of LIF in injured muscle in vivo that may activate these processes. Careful studies that employed LIF cDNA probes for in situ hybridization for LIF mRNA localization showed that there may be low levels of LIF expression in injured muscle fibers (Kurek et al., 1996; Kami and Senba, 1998), but that most LIF mRNA is located in mononucleated cells in injured or diseased muscle. Thus, the elevation of muscle LIF that occurs during increased muscle loading or following muscle injury (Kurek et al., 1996; Sakuma et al., 2003) may be attributable to inflammatory cells that invade the muscle.

Numerous investigations indicate that LIF expression by myeloid cells or mesenchymal cells is increased by Th1 cytokines. Both TNF-α and IL-1β can greatly increase LIF expression (Alexander et al., 1992; Gough and Williams, 1989; Hamilton et al., 1993; Chabaud et al., 1998; Campbell et al., 1993), which suggests that LIF expression would be highest in Th1 inflammatory responses. However, other findings indicate that elevated LIF expression may either promote the Th1 response, by increasing IL-1β expression and TNF-α (Villiger et al., 1993; Zhu et al., 2001) or attenuate classical activation and promote the Th2 inflammatory response, by decreasing IL-1β and TNF-α expression (Banner et al., 1998; Weber et al., 2005) and increasing IL-10 expression (Weber et al., 2005). The most parsimonious interpretation of current reports is that Th1 cytokines can increase LIF

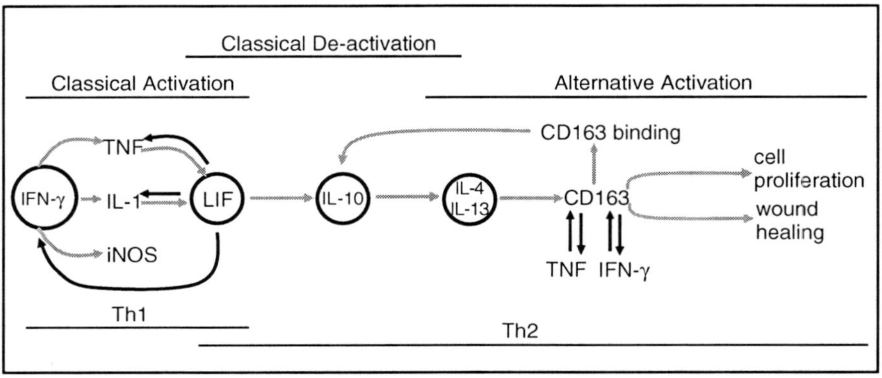

Figure 4. Schematic of hypothetical model of key molecules involved in macrophage response to muscle acute injury or modified loading (circled in blue). Green arrows indicate a stimulatory effect on molecule's expression or activity. Red arrows indicate an inhibitory effect on expression or activity. Black arrows indicate that both stimulatory and inhibitory effects on expression or activity have been demonstrated experimentally

expression, and LIF may in turn provide positive or negative feedback on the Th1 response that may be determined by tissue type or other factors in the inflamed tissues (Fig. 4).

2.2.6.2 Tumor necrosis factor-alpha (TNF-α) TNF-α has proven to serve broad and apparently antagonist roles in muscle viability, growth and function. Following acute muscle injury, inflammatory cells are the primary source of TNF-α. For example, most TNF-α in muscle experiencing acute trauma is located in CD11b+ cells (Warren et al., 2002) that include neutrophils and macrophages (Collins and Grounds, 2001). At later stages of muscle response to acute injury, the muscle fibers themselves begin to display a slight immunoreactivity to anti-TNF-α (Warren et al., 2002). However, TNF-α levels in muscle can also be influenced significantly by TNF-α synthesis in other tissues. For example, TNF-α expression by a transgene in the lung not only increased TNF-α concentration in muscle, but also caused an increase in TNF-α expression by the muscle itself (Langen et al., 2006).

TNF-α was originally named "cachexin" because it appears at high levels in the sera of cancer patients who experienced muscle wasting and because it could induce muscle wasting when administered to experimental animals (Beutler et al., 1985; Tracey et al., 1988). Subsequent, in vitro studies showed that exogenous TNF-α would reduce the quantity of protein in myotube cultures without decreasing the DNA content, indicating that it induced a catabolic state and did not merely cause cell death (Li et al., 1998). The apparent increase in proteolysis was attributable, at least in part, to the activation of NFκB binding to DNA that corresponded to increased degradation of the inhibitor of NFκB, IκB-α (Li et al., 1998). These findings supported a specific and direct effect of TNF-α in promoting muscle wasting. However, application of TNF-α to myoblast cultures increased their DNA content, indicating a mitogenic effect (Li, 2003), and inhibited myoblast differentiation and fusion (Guttridge et al., 2000; Langen et al., 2001; 2004). Interestingly, the TNF-α that promoted cell proliferation in myoblast cultures (6 ng / ml) was the same as the concentration that increased catabolism in myotube cultures (Li et al., 1998) and the same as the concentration in the serum of some cachectic cancer patients (Nakashima et al., 1995). Collectively, in vitro studies show that TNF-α can promote muscle cell death, proliferation, growth or wasting; the effects may largely depend on the state of differentiation of the cells.

In vivo studies further illustrate the difficulties in predicting the effect of TNF-α on muscle cell growth or viability. Although systemic elevations of TNF-α caused muscle wasting (Tracey et al., 1988) and inhibited muscle regeneration (Coletti et al., 2005), intraperitoneal injection of TNF-α into healthy mice increased the number of satellite cells that were mitotic (BrDU+) during the subsequent 16 hours (Li, 2003). Because no BrDU+ satellite cells were found in control mouse muscles, the latter results indicated that elevating systemic levels of TNF-α activates satellite cells, as well as increasing their proliferation. The conflicting outcomes on muscle viability caused by systemic administration of TNF-α to healthy muscle are also mirrored in the effects of ablating TNF-α signaling in muscle that is subjected to acute trauma. Although null mutation of the TNF receptors I and II (TNF-R) showed no histologically-obvious differences in the inflammatory and regenerative responses to acute freeze injury compared to wild-type mice, the concentration of MyoD mRNA was lower in TNF-R null mice and in anti-TNF-α treated mice 3 days post-injury (Warren et al., 2002). Furthermore, the rate of strength recovery following injury was slower (Warren et al., 2002), showing that TNF-α plays a positive role in muscle regeneration following acute muscle injury. However, when mice expressing a TNF-α transgene in their lungs that caused a secondary increase in muscle TNF-α were subjected to HS/Rel, there was a slowing of muscle growth and a reduction in the expression of perinatal MyHC in the reloaded muscles of the transgenic mice (Langen et al., 2006). These findings support a negative regulatory role of TNF-α in muscle regeneration following acute injury. In contrast to those findings, regeneration and repair following acute muscle injury caused by cardiotoxin (CTX) injection into the muscle was diminished by null mutation of TNF-R (Chen et al., 2005). Null mutants showed less regeneration that was assessed by fiber centronucleation, as well as a slower recovery of force production, a more persistent inflammation, reduced activation of MEF-2 and lower levels of MyoD (Chen et al., 2005).

The unpredictability of the effects on muscle of perturbing TNF-α signaling in vitro or in vivo do not necessarily reflect conflicts in the findings. Initially, the differences in the experimental outcomes were attributed to differences in the concentration of TNF-α used in the experimental studies. However, even at the same TNF-α concentrations, the effect on muscle growth or regeneration can vary with the state of differentiation of the muscle cells. Indeed, muscle inflammation following acute injury in vivo may provide a system in which the levels of leukocyte-derived TNF-α production are coordinated with the stage of muscle differentiation so that the net effect is to promote regeneration. The high levels of expression of TNF-α in neutrophils and early invading macrophages (Collins and Grounds, 2001) correspond with the initial, rapid Th1 inflammatory response that follows muscle trauma. This relatively high TNF-α concentration can activate satellite cells, stimulate their proliferation and inhibit their differentiation. The subsequent increase in Th2 macrophages would attenuate the Th1 response and reduce TNF-α concentrations to levels that would permit differentiation and fusion of myogenic cells, but below levels that would cause a negative protein balance. This model is consistent with most experimental observations.

2.2.6.3 TNF-like weak inducer of apoptosis (TWEAK) TWEAK, a member of the TNF-α superfamily shows promise as a macrophage-derived factor that can play a significant role in affecting the response of skeletal muscle to injury. Although TWEAK exists primarily as a transmembrane protein, it can be proteolytically cleaved and released as a soluble cytokine (Chicheportiche et al., 1997) and subsequently act on target cells via binding its receptor, fibroblast growth factor-inducible14 (Fn14) (Chicheportiche et al., 1997; Marsters et al., 1998). Because skeletal muscle is one of the tissues that expresses high levels of TWEAK and Fn14 (Chicheportiche et al., 1997; Meighan-Mantha et al., 1999; Feng et al., 2000) and signaling via the TWEAK/Fn14 has been implicated in the differentiation and repair of non-muscle cells and tissues (Jakubowski et al., 2005; Polek et al., 2003), the possible role of TWEAK in muscle regeneration has been recently explored, illuminating another potential mechanism through which macrophages may influence muscle regeneration (Dogra et al., 2006; Girgenrath et al., 2006).

Although muscle expresses TWEAK, inflammatory cells appear to be the major source of the cytokine following muscle injury caused by CTX injection and the inflammatory cells are probably macrophages. Cell fractionation of injured muscle 3 days after CTX injection showed that the highest levels of TWEAK expression were in the Mac-1+ cell fraction (Girgenrath et al., 2006). Although both neutrophils and macrophages express Mac-1, the majority of leukocytes in muscle at 3 days post-injury would likely be macrophages. TWEAK-mediated signaling was implicated in muscle regeneration by the finding that null mutation of Fn14 reduced the number of centrally-nucleated muscle fibers in CTX-injected muscles. However, whether the defect in regeneration in Fn14 null mice is a consequence of disruption of direct effects of TWEAK on muscle cells or indirect effects caused by perturbing other functions of leukocytes in muscle regeneration in vivo is uncertain. Because ablation of Fn14 caused a great reduction of the numbers of F4/80+ macrophages and Gr-1+ neutrophils in the injured muscle and a slower removal of cellular debris, some of the impaired regeneration could be attributable to impaired phagocytosis of debris and reduction of other macrophage-derived factors that can promote regeneration.

Although perturbed signaling through TWEAK/Fn14 may affect muscle regeneration in vivo through direct and indirect effects on muscle, in vitro findings show clearly that TWEAK can have direct effects on skeletal muscle cells. Application of TWEAK to cultured muscle cells increased cell proliferation that was associated with elevated expression of the cell cycle regulator, cyclin D1 (Dogra et al., 2006). In addition, differentiation was further inhibited by TWEAK-mediated activation of NFkB, which led to degradation of MyoD (Dogra et al., 2006). Although the net effect of TWEAK treatment of myoblasts in vivo was to greatly reduce the formation of myotubes (Dogra et al., 2006; Girgenrath et al., 2006), signaling via the TWEAK/Fn14 pathway is not essential for muscle regeneration in vivo because Fn14 slows, but does not prevent muscle recovery from acute injury (Girgenrath et al., 2006).

2.2.7 *How do muscle cells and inflammatory cells communicate following muscle injury?*

The chemical signals that are exchanged between muscle and immune cells to initiate the inflammatory process provide a potentially important control point in the regulation of muscle healing and growth following injury. Perhaps the first direct evidence for muscle-derived signals influencing the inflammatory response was provided in a study in which minced muscle tissue that had been previously crush-injured in vivo was shown to be chemoattractive to neutrophils and macrophages in vitro (Robertson et al., 1993). Interestingly, macrophages but not neutrophils released a substance that was chemoattractive for muscle cells in vitro (Robertson et al., 1993), showing reciprocal signaling functions between muscle and macrophages following injury.

Numerous molecules have subsequently been identified that are capable of chemoattracting inflammatory cells to muscle, where they could then participate in modulating muscle repair and regeneration. In at least one model of injury, chemoattraction of leukocytes into the injured muscle occurred through activation of the complement system, and did not require the de novo synthesis or secretion of signaling molecules or their receptors. During the reloading phase of the HS/Rel model, neutrophil and CD68+/CD163− macrophages invasion into muscle was significantly reduced by treating animals with the soluble complement receptor, sCR1 (Frenette et al., 2000). The systemic treatment with sCR1 inhibited complement activation by both the classical pathway and alternative pathways. Although the mechanism by which complement activation would occur following muscle injury or modified use was not identified, studies in other experimental models of acute injury suggest interesting possibilities. For example, antibody-independent activation of the classical pathway can be caused by the release of intermediate filament proteins from injured cells (Linder et al., 1979) and several investigators have noted that loss of the intermediate filament protein desmin from muscle is an early consequence of muscle injury in vivo (Lieber et al., 1994).

Muscle cells also actively release substances that can activate leukocytes and attract them to muscles. In vitro, satellite cells release chemokine ligand-22 (CCL22, also called MP-derived chemokine [MPC]), monocyte chemoattractant protein-1 (MCP-1, also called CCL2), fractalkine, and urokinase type plasminogen-activator receptor (μPAR) (Chazaud et al., 2003). Each of these cytokines contributed significantly to the chemotaxis of monocytes and macrophages across a barrier of endothelial cells in vitro. Interestingly, expression of these cytokines in vitro was highest soon after the satellite cells were activated to re-enter the cell cycle, and then declined at about the time when the cells withdrew from the cell cycle to undergo differentiation and fusion to form myotubes (Chazaud et al., 2003), suggesting that the time-course of release of chemoattractants from muscle could potentially regulate the time-course of inflammation.

The increase in MCP-1 expression by activated satellite cells could be significant in the context of muscle injury and repair, in which satellite cells could signal the rapid invasion of leukocytes to promote muscle regeneration. Indeed, recent findings show that null mutation of MCP-1 slows muscle regeneration following I/R injury, with delayed removal of cellular debris and slower growth of regenerative fibers following injury (Shireman et al., 2007). As predicted in the in vitro model (Chazaud et al., 2003), null mutation of MCP-1 or its receptor CCR2 also produced a reduction in macrophage invasion into the injured muscle (Shireman et al., 2007; Contreras-Shannon et al., 2007). Defects in morphological indices of regeneration were also reflected in impaired recovery of function following muscle injury. Recovery of muscle strength following an acute freeze injury was slower in CCR2 null mutants than in wild-type mice (Warren et al., 2004). However, satellite cells may not be the primary source of MCP-1 in injured muscle, at least in the I/R injury model. Although numerous cell types can express MCP-1, including muscle cells, endothelial cells, neutrophils and macrophages (Kumar et al., 1997; Stark et al., 1997; Hilgers et al., 2000; Reyes-Reyna and Krolick, 2000; Summan et al., 2003), MCP-1 was primarily located in macrophages and endothelial cells in muscle subjected to ischemic injury (Shireman et al., 2006).

Several recent findings indicate that the plasminogen system also provides mechanisms through which muscle and inflammatory cells interact to promote muscle regeneration. The plasminogen system consists of urokinase plasminogen activator (μPA) and its substrate, (plasminogen), the activated serine protease (plasmin) and the μPA receptor (μPAR) and inhibitor (PAI-1). The best-characterized function of the plasminogen system is the proteolysis of extracellular matrix proteins, which can facilitate the migration of cells involved in tissue repair and regeneration, including skeletal muscle (Lluis et al., 2001). Experimental perturbations of the plasminogen system support the view that a significant portion of its role in regeneration is mediated through inflammatory cells. Null mutants for plasminogen experienced significantly less muscle inflammation and slower muscle regeneration following an acute injury (Suelves et al., 2002) that corresponded to an increase in connective tissue in the more slowly regenerating muscle. Although these defects in regeneration could be attributed to a reduction in the removal of cellular debris by leukocytes, thereby leading to slowed regeneration, other findings suggest that more direct roles of leukocytes in the regenerative process were deficient. For example, null mutation of PAI-1 yields significantly elevated invasion of macrophages into muscle following an acute injury by toxin injection, and the increase in inflammation is accompanied by elevated expression of markers of muscle regeneration, such as MyoD, and improved recovery of muscle function (Koh et al., 2005).

3. FUNCTIONS OF LEUKOCYTES IN MUSCLE REPAIR AND REGENERATION DURING MUSCLE DISEASE

The experimental dissection of the role of inflammatory cells in muscle regeneration following acute trauma or modified use has proven much simpler than evaluating the role of leukocytes in regeneration of diseased muscle. The greater complexity of analyzing leukocyte function in diseased muscle is largely attributable to the simultaneous involvement of both innate and acquired immune responses to the diseased muscle. While muscle trauma leads exclusively to a highly-structured innate immune response that is dominated by myeloid cells in which there is a shift from a Th1 to a Th2 inflammatory response over a predictable time course, muscle diseases typically involve cellular or humoral responses in which there is an ancillary innate response to the muscle damage that occurs as a consequence of the disease. Thus, in the case of polymyositis, the most common inflammatory myopathy, a T-cell mediated cellular immune response is the major cause of muscle damage (Christopher-Stine and Plotz, 2004), but it is unknown whether other leukocyte populations play a significant role in promoting regeneration of muscle fibers that are damaged by auto-aggressive T-cells. Similarly, muscle cell death in dermatomyositis is driven by a humoral immune response (Christopher-Stine and Plotz, 2004), but the possibility that other leukocyte populations could simultaneously drive regeneration has not been unexplored.

A role for the immune system in muscle regeneration is also unclear in dystrophinopathies. Dystrophinopathies are progressive, muscle-wasting diseases that are caused by mutations in the dystrophin gene, which encodes a membrane-associated cytoskeletal protein. Loss of dystrophin leads to Duchenne muscular dystrophy (DMD) in humans and mdx dystrophy in mice. Muscle fibers that lack dystrophin are mechanically weaker and experience lesions during high intensity muscle contractions, leading to a leakage of cytosolic proteins into the extracellular space. However, this mechanical damage to dystrophic muscle fibers stimulates an immune response to the damaged tissue that consists of a strong, innate immune response that is dominated by macrophages, and a less obvious but functionally-significant cellular immune response, in which cytotoxic T-lymphocytes induce apoptosis of muscle fibers (Spencer et al., 1997).

Unlike macrophage function in muscle experiencing acute trauma, macrophages in dystrophic muscle cause cytolysis rather than regeneration, at least at the peak of muscle pathology in 4-week-old mdx mice (Wehling et al., 2001). Macrophages isolated from 4-week-old mdx muscle lyse myotubes in vitro, and depletion of macrophages from mdx mice reduces muscle fiber damage in vivo by approximately 80% (Wehling et al., 2001). However, this early-invading population of macrophages is of the classically-activated Th1 phenotype; whether macrophages of the Th2 phenotype are also present and promote muscle fiber regeneration has not been tested. Although the depletion of macrophages by injection of the pan-macrophage antibody, anti-F4/80, caused a reduction in the number of regenerative fibers in the mdx muscle (Wehling et al., 2001), this effect is presumably attributable to a reduction in fiber damage by classically-activated macrophages, rather than a decrease in regenerative capacity.

A role for inflammatory cell-derived TNF-α in regeneration of dystrophic muscle is similarly unclear, and reflects both the pleiotrophic functions of TNF-α and the more complex immune response to dystrophin-deficiency. Null mutation of TNF-α in dystrophin-deficient mdx mice showed effects on the dystrophic pathology that varied with the stage of the disease and the muscle that was analyzed (Spencer et al., 2000). For example, no histologically-discernible effect of the TNF-α mutation was observed in the quadriceps of 4-week-old mdx mice, but the TNF-α mutation caused a significant *increase* in pathology in the mdx diaphragm. However, during the regenerative stage of the mdx pathology at 8 weeks of age, TNF-α ablation caused a reduction of histopathology in the quadriceps, while increasing the numbers of regenerative fibers displaying central nucleation (Spencer et al., 2000). These findings suggest that TNF-α promotes regeneration in at least the mdx quadriceps, although at the cost of promoting diaphragm pathology during the acute onset of the disease. Despite the negative effect of the TNF-α mutation on diaphragm histology early in the disease, the mutation yielded functional improvements in the diaphragm, including increased force production and improved ventilation in response to hypercapnia in 7-month-old mice (Gosselin et al., 2003). Whether these long-term improvements relate to reducing damage, increasing regeneration or modifying the functional capacity of surviving fibers is unknown. However, blocking TNF-α signaling by the systemic administration of a soluble TNFR caused a reduction in histopathology of mdx tibialis anterior muscle in 3-week-old mice that was followed by a reduction in central nucleated fibers in 4-week-old mdx mice, which suggests that the beneficial effect of ablation of TNF-α mediated signaling occurs by reducing damage rather than promoting regeneration, at least in the tibialis anterior at this stage of the pathology (Hodgetts et al., 2006).

4. CONCLUSIONS

Our understanding of the roles played by leukocytes in regulating skeletal muscle's response to acute injury or modified use now supports a model in which macrophages play a central regulatory role. There is good agreement between in vitro observations and in vivo studies using several model of acute muscle damage that show that macrophages release soluble substances that influence the proliferation, differentiation, growth, repair and regeneration of muscle. Further, the model is supported by in vivo findings that there is a shift from a Th1-responsive macrophage population that stimulates satellite cell activation and proliferation, while inhibiting differentiation, to a Th2-responsive population that promotes differentiation, growth and repair. Although it is likely that the regulatory influences of macrophages on muscle regeneration will be affected by redundant and overlapping signaling pathways, several pathways of importance have been identified. In particular, TNF-mediated effects have now been shown to be important in affecting muscle regeneration, although the underlying mechanisms remain enigmatic. Recent discovery of a role for the TWEAK/Fn14 pathway in muscle regeneration indicates that further investigations of this system will be essential for a complete understanding of how macrophages can influence muscle regeneration.

Unfortunately, much less is known concerning how inflammatory cells can affect muscle regeneration in muscle diseases, such as idiopathic inflammatory myopathies and the muscular dystrophies. Indeed, there are no definitive findings to show that immune cells can directly affect regeneration of diseased muscle. Based on our recently acquired knowledge of interactions between macrophages and acutely injured muscle, similar roles are likely to be played by these myeloid cells present in diseased muscle, although the more complex inflammatory infiltrate of diseased muscle will complicate in vivo analysis.

ACKNOWLEDGEMENTS

The author received support from grants from the National Institutes of Health (AR40343, AR47721) and the Muscular Dystrophy Association during the preparation of this chapter.

REFERENCES

Abe T, Ohno M, Sato T, Murakami M, Kajiki M, Kodaira R (1991) "Differentiation induction" culture of human leukemic myeloid cells stimulates high production of macrophage differentiation inducing factor. Cytotechnology 5 Suppl 2:S75–S93.

Abood EA, Jones MM (1991) Macrophages in developing mammalian skeletal muscle: evidence for muscle fibre death as a normal developmental event. Acta Anat (Basel) 140(3):201–212.

Alexander HR, Wong GG, Doherty GM, Venzon DZ, Fraker DL, Norton JA (1992) Differentiation factor/leukemia inhibitory factor protection against lethal endotoxemia in mice: synergistic effect with interleukin 1 and tumor necrosis factor. J Exp Med 175:1139–1142.

Anderson CF, Mosser DM (2002) A novel phenotype for an activated macrophage: the type 2 activated macrophage. J Leukoc Biol 72:101–106.

Banner LR, Patterson PH (1994) Major changes in the expression of the mRNAs for cholinergic differentiation factor/leukemia inhibitory factor and its receptor after injury to adult peripheral nerves and ganglia. Proc Nat Acad Sci 91:7109–7113.

Banner LR, Patterson PH, Allchorne A, Poole S, Woolf CJ (1998) Leukemia inhibitory factor is an anti-inflammatory and analgesic cytokine. J Neurosci 18:5456–5462.

Bansal D, Miyake K, Vogel SS, Groh S, Chen CC, Williamson R, McNeil PL, Campbell KP (2003) Defective membrane repair in dysferlin-deficient muscular dystrophy. Nature 423:168–172.

Barnard W, Bower J, Brown MA, Murphy M, Austin L (1994) Leukemia inhibitory factor (LIF) infusion stimulates skeletal muscle regeneration after injury: injured muscle expresses LIF mRNA. J Neurol Sci 123:108–113.

Beutler B, Greenwald D, Hulmes JD, Chang M, Pan YC, Mathison J, Ulevitch R, Cerami A (1985) Identity of tumour necrosis factor and the macrophage-secreted factor cachectin. Nature 316(6028):552–554.

Bondesen BA, Mills ST, Kegley KM, Pavlath GK (2004) The COX-2 pathway is essential during early stages of skeletal muscle regeneration. Am J Physiol Cell Physiol 287:C475-C483.

Bondesen BA, Mills ST, Pavlath GK (2006) COX-2 pathway regulates growth of atrophied muscle via multiple mechanisms. Am J Physiol Cell Physiol 290(6):C1651–1659.

Bower J, Vakakis N, Nicola NA, Austin L (1995) Specific binding of leukemia inhibitory factor to murine myoblasts in culture. J Cell Physiol 164:93–98.

Buechler C, Ritter M, Orso E, Langmann T, Klucken J, Schmitz G (2000) Regulation of scavenger receptor CD163 expression in human monocytes and macrophages by pro- and antiinflammatory stimuli. J Leukoc Biol 67:97–103.

Campbell IK, Waring P, Novak U, Hamilton JA (1993) Production of leukemia inhibitory factor by human articular chondrocytes and cartilage in response to interleukin-1 and tumor necrosis factor alpha. Arthritis Rheum 36:790–794.

Cantini M, Carraro U (1995) Macrophage-released factor stimulates selectively myogenic cells in primary muscle culture. J Neuropathol Exp Neurol 54:121–128.

Cantini M, Giurisato E, Radu C, Tiozzo S, Pampinella F, Senigaglia D, Zaniolo G, Mazzoleni F, Vittiello L (2002) Macrophage-secreted myogenic factors: a promising tool for greatly enhancing the prolieferative capacity of myoblasts in vitro and in vivo. Neurol Sci 23:189–194.

Carlson CD, Yuchen B, Jonakait GM, Hart RP (1996) Interleukin-1β increases leukemia inhibitory factor mRNA levels through transient stimulation of transcription rate. Glia 18:141–151.

Carr AC, Myzak MC, Stocker R, McCall MR, Frei B (2000) Myeloperoxidase binds to low-density lipoprotein: potential implications for atherosclerosis. FEBS Lett 487:176–180.

Chabaud M, Fossiez F, Taupin JL, Miossec P (1998) Enhancing effect of IL-17 on IL-1-induced IL-6 and leukemia inhibitory factor production by rheumatoid arthritis synoviocytes and its regulation by Th2 cytokines. J Immunol 161:409–414.

Chazaud B, Sonnet C, Lafuste P, Bassez G, Rimaniol A-C, Poron F, Authier F-J, Dreyfus PA, Gherardi RK (2003) Satellite cells attract monocytes and use macrophages as a support to escape apoptosis and enhance muscle growth. J Cell Biol 163:1133–1143.

Chen SE, Gerken E, Zhang Y, Zhan M, Mohan RK, Li AS, Reid MB, Li YP (2005) Role of TNF-alpha signaling in regeneration of cardiotoxin-injured muscle.Am J Physiol Cell Physiol 289(5): C1179–C1187

Chicheportiche Y, Bourdon PR, Xu H, Hsu YM, Scott H, Hession C, Garcia I, Browning JL (1997) TWEAK, a new secreted ligand in the tumor necrosis factor family that weakly induces apoptosis. J Biol Chem 272(51):32401–32410.

Christopher-Stine L, Plotz PH (2004) Adult inflammatory myopathies. Best Practice & Res Clin Rheum 18:331–344.

Coletti D, Moresi V, Adamo S, Molinaro M, Sassoon D (2005) Tumor necrosis factor-α gene transfer induces cachexia and inhibits muscle regeneration. Genesis 43:120–128.

Collins RA, Grounds MD (2001) The role of tumor necrosis factor alpha (TNF-alpha) in skeletal muscle regeneration. Studies in TNF-alpha (-/-) and TNF-alpha (-/-)/LT-alpha (-/-) mice. J Histochem Cytochem 49:989–1001.

Contreras-Shannon V, Ochoa O, Reyes-Reyna SM, Sun D, Michalek JE, Kuziel WA, McManus LM, Shireman PK (2007) Fat accumulation with altered inflammation and regeneration in skeletal muscle of CCR2 -/- mice following ischemic injury. Am J Physiol Cell Physiol. 292(2): C953–967.

David JD, Higginbotham CA (1981) Fusion of chick embryo skeletal myoblasts: interactions of prostaglandin E1, adenosine 3′:5′ monophosphate, and calcium influx. Dev Biol 82(2):308–316.

Dogra C, Changotra H, Mohan S, Kumar A (2006) Tumor necrosis factor-like weak inducer of apoptosis inhibits skeletal myogenesis through sustained activation of nuclear factor-kappaB and degradation of MyoD protein. J Biol Chem 281(15):10327–10336.

Dowsing BJ, Morrison WA, Nicola NA, Starkey GP, Bucci T, Kilpatrick TJ (1999) Leukemia inhibitory factor is an autocrine survival factor for Schwann cells. J Neurochem 73:96–104.

Entwistle A, Curtis DH, Zalin RJ (1986) Myoblast fusion is regulated by a prostanoid of the one series independently of a rise in cyclic AMP. J Cell Biol 103(3):857–866.

Feng SL, Guo Y, Factor VM, Thorgeirsson SS, Bell DW, Testa JR, Peifley KA, Winkles JA (2000) The Fn14 immediate-early response gene is induced during liver regeneration and highly expressed in both human and murine hepatocellular carcinomas. Am J Pathol 156(4):1253–1261.

Finkelstein DI, Bartlett PF, Horne MK, Cheema SS (1996) Leukemia inhibitory factor is a myotrophic and neurotrophic agent that enhances the reinnervation of muscle in the rat. J Neuroscience Res 46:122–128.

Frenette J, Cai B, Tidball JG (2000) Complement activation promotes muscle inflammation during modified muscle use. Am J Pathol 156:2103–2110.

Girgenrath M, Weng S, Kostek CA, Browning B, Wang M, Brown SA, Winkles JA, Michaelson JS, Allaire N, Schneider P, Scott ML, Hsu YM, Yagita H, Flavell RA, Miller JB, Burkly LC, Zheng TS (2006) TWEAK, via its receptor Fn14, is a novel regulator of mesenchymal progenitor cells and skeletal muscle regeneration. EMBO J 25(24):5826–5839.

Godman GC (1957) On the regeneration and redifferentiation of mammalian striated muscle. J Morphol 100:27–82.

Gordon S (2003) Alternative activation of macrophages. Nat Rev Immunol 3:23–35.

Gosselin LE, Barkley JE, Spencer MJ, McCormick KM, Farkas GA (2003) Ventilatory dysfunction in mdx mice: impact of tumor necrosis factor-alpha deletion. Muscle Nerve 28:336–343.

Gough NM, Williams RL (1989) The pleiotropic actions of leukemia inhibitory factor. Cancer Cells 1:77–80.

Grisotto PC, dos Santos AC, Coutinho-Netto J, Cherri J, Piccinato CE (2000) Indicators of oxidative injury and alterations of the cell membrane in the skeletal muscle of rats submitted to ischemia and reperfusion. J Surg Res 92:1–6.

Grounds MD (1987) Phagocytosis of necrotic muscle in muscle isografts is influenced by the strain, age, and sex of host mice. J Pathol 153:71–82.

Guttridge DC, Mayo MW, Madrid LV, Wang CY, Baldwin AS Jr. (2000) NF-kappaB-induced loss of MyoD messenger RNA: possible role in muscle decay and cachexia. Science 289(5488):2363–2366.

Hamilton JA, Waring PM, Filonzi EL (1993) Induction of leukemia inhibitory factor in human synovial fibroblasts by IL-1 and tumor necrosis factor-alpha. J Immunol 150:1496–1502.

Hazell LJ, Stocker R (1993) Oxidation of low-density lipoprotein with hypochlorite causes transformation of the lipoprotein into a high-uptake form for macrophages. Biochem J 290:165–172.

Hilgers KF, Hartner A, Porst M, Mai M, Wittmann M, Hugo C, Ganten D, Geiger H, Veelken R, Mann JF (2000) Monocyte chemoattractant protein-1 and macrophage infiltration in hypertensive kidney injury. Kidney Int 58(6):2408–2419.

Hodgetts S, Radley H, Davies M, Grounds MD (2006) Reduced necrosis of dystrophic muscle by depletion of host neutrophils, or blocking TNF-alpha function with Etanercept in mdx mice. Neuromuscul Disord 16(9–10):591–602.

Horsley V, Pavlath GK (2003) Prostaglandin F2(alpha) stimulates growth of skeletal muscle cells via an NFATC2-dependent pathway. J Cell Biol 161(1):111–118.

Horsley V, Jansen KM, Mills ST, Pavlath GK (2003) IL-4 acts as a myoblast recruitment factor during mammalian muscle growth. Cell 113:422–423.

Krippendorf BB, Riley DA (1993) Distinguishing unloading- versus reloading-induced changes in rat soleus muscle. Muscle Nerve 16:99–108.

Kumar AG, Ballantyne CM, Michael LH, Kukielka GL, Youker KA, Lindsey ML, Hawkins HK, Birdsall HH, MacKay CR, LaRosa GJ, Rossen RD, Smith CW, Entman ML (1997) Induction of monocyte chemoattractant protein-1 in the small veins of the ischemic and reperfused canine myocardium. Circulation 95(3):693–700.

Janardhan KS, Sandhu SK, Singh B (2006) Neutrophil depletion inhibits early and late monocyte/macrophage increase in lung inflammation. Front Biosci 11:1569–1576.

Jakubowski A, Ambrose C, Parr M, Lincecum JM, Wang MZ, Zheng TS, Browning B, Michaelson JS, Baetscher M, Wang B, Bissell DM, Burkly LC (2005) TWEAK induces liver progenitor cell proliferation. J Clin Invest 115(9):2330–2340.

Jansen KM, Pavlath GK (2006) Mannose receptor regulates myoblast motility and muscle growth. J Cell Biol 174(3):403–413.

Jo C, Kim H, Jo I, Choi I, Jung SC, Kim J, Kim SS, Jo SA (2005) Leukemia inhibitory factor blocks early differentiation of skeletal muscle cells by activating ERK. Biochim Biophys Acta 1743(3):187–197.

Kami K, Senba E (1998) Localization of leukemia inhibitory factor and interleukin-6 messenger ribonucleic acids in regenerating rat skeletal muscle. Muscle Nerve 21(6):819–822.

Koh TJ, Bryer SC, Pucci AM, Sisson TH (2005) Mice deficient in plasminogen activator inhibitor-1 have improved skeletal muscle regeneration. Am J Physiol Cell Physiol 289(1):C217–223.

Kristiansen M, Graversen JH, Jacobsen C, Sonne O, Hoffman HJ, Law SK, Moestrup SK (2001) Identification of the haemoglobin scavenger receptor. Nature 409:198–201.

Kumar AG, Ballantyne CM, Michael LH, Kukielka GL, Youker KA, Lindsey ML, Hawkins HK, Birdsall HH, MacKay CR, LaRosa GJ, Rossen RD, Smith CW, Entman ML (1997) Induction of monocyte chemoattractant protein-1 in the small veins of the ischemic and reperfused canine myocardium. Circulation 95(3):693–700.

Kurek JB, Bower JJ, Romanella M, Koentgen F, Murphy M, Austin L (1997) The role of leukemia inhibitory factor in skeletal muscle regeneration. Muscle Nerve 20:815–822.

Kurek JB, S Nouri S, Kannourakis G, Murphy M, Austin L (1996) Leukemia inhibitory factor and interleukin-6 are produced by diseased and regenerating skeletal muscle. Muscle Nerve 19:1291–1301.

Kurushima H, Ramprasad M, Kondratenko N, Foster DM, Quehenberger O, Steinberg D (2000) Surface expression and rapid internalization of macrosialin (mouse CD68) on elicited mouse peritoneal macrophages. J Leukoc Biol 67:104–108.

Lafreniere JF, Mills P, Bouchentouf M, Tremblay JP (2006) Interleukin-4 improves the migration of human myogenic precursor cells in vitro and in vivo. Exp Cell Res 312(7):1127–1141.

Lang R, D Patel D, Morris JJ, Rutschmanand RL, Murray PJ (2002) Shaping gene expression in activated and resting primary macrophages by IL-10. J Immunol 169:2253–2263.

Langen RC, Schols AM, Kelders MC, Wouters EF, Janssen-Heininger YM (2001) Inflammatory cytokines inhibit myogenic differentiation through activation of nuclear factor-kappaB. FASEB J 15(7):1169–1180.

Langen RC, Van der Velden JL, Schols AM, Kelders MC, Wouters EF, Janssen-Heininger YM (2004) Tumor necrosis factor-alpha inhibits myogenic differentiation through MyoD protein destabilization. FASEB J 18:227–237.

Langen RC, Schols AM, Kelders MC, van der Velden JL, Wouters EF, Janssen-Heininger YM (2006) Muscle wasting and impaired muscle regeneration in a murine model of chronic pulmonary inflammation. Am J Respir Cell Mol Biol 35(6):689–696.

Lescaudron L, Peltekian E, Fontaine-Perus J, Paulin D, Zampieri M, Garcia L, Parrish E (1999) Blood borne macrophages are essential for the triggering of muscle regeneration following muscle transplant. Neuromusc Disorders 9:72–80.

Li YP (2003) TNF-alpha is a mitogen in skeletal muscle. Am J Physiol Cell Physiol 285(2):C370–C376.

Li Y-P, Schwartz RJ, Waddell ID, Holloway BR, Reid MB (1998) Skeletal muscle myocytes undergo protein loss and reactive oxygen-mediated NF-κB activation in response to tumor necrosis factor α. FASEB J 12:871–880.

Lieber RL, Schmitz MC, Mishra DK, Friden J (1994) Contractile and cellular remodeling in rabbit skeletal muscle after cyclic eccentric contractions. J Appl Physiol 77(4):1926–1934.

Lim SK, H. Kim H, Lim SK, bin Ali A, Lim YK, Wang Y, Chong SM, Costantini F, Baumman H (1998) Increased susceptibility in Hp knockout mice during acute hemolysis. Blood 92:1870–1877.

Linder E, Lehto VP, Stenman S (1979) Activation of complement by cytoskeletal intermediate filaments. Nature 278:176–178.

Lluis F, Roma J, Suelves M, Parra M, Aniorte G, Gallardo E, Illa I, Rodriguez L, Hughes SM, Carmeliet P, Roig M, Muñoz-Canoves P (2001) Urokinase-dependent plasminogen acivaton is required for efficient skeletal muscle regeneration in vivo. Blood 97:1703–1711.

Marsters SA, Sheridan JP, Pitti RM, Brush J, Goddard A, Ashkenazi A (1998) Identification of a ligand for the death-domain-containing receptor Apo3. Curr Biol 8(9):525–528

Massimino M, Rapizzi E, Cantini M, Libera L, Mazzoeni F, Arsian P, Carraro U (1997) ED2+ macrophages increase selectively myoblast proliferation in muscle cultures. Biochem Biophys Res Commun 235:754–759.

Merly F, Lescaudron L, Rouaud T, Crossin F, Gardahaut MF (1999) Macrophages enhance muscle satellite cell proliferation and delay their differentiation. Muscle Nerve 22:724–732.

McLennan IS (1993) Resident macrophages (ED2- and ED3-positive) do not phagocytose degenerating rat skeletal muscle fibers. Cell Tissue Res 272:193–196.

Meighan-Mantha RL, Hsu DK, Guo Y, Brown SA, Feng SL, Peifley KA, Alberts GF, Copeland NG, Gilbert DJ, Jenkins NA, Richards CM, Winkles JA (1999) The mitogen-inducible Fn14 gene encodes a type I transmembrane protein that modulates fibroblast adhesion and migration. J Biol Chem 274(46):33166–33176.

Metcalf D, Nicola NA, Gearing DP (1990) Effects of injected leukemia inhibitory factor on hematopoietic and other tissues in mice. Blood 76:50–56.

Mills CD, Kincaid K, Alt JM, Heilman MJ, Hill AM (2000) M-1/M-2 macrophages and the Th1/Th2 paradigm. J Immunol 164 :6166–6173.

Mishra DK, Friden J, Schmitz MC, Lieber RL (1995) Anti-inflammatory medication after muscle injury. A treatment resulting in short-term improvement but subsequent loss of muscle function. J Bone Joint Surg 77(10):1510–1519.

Moestrup SK, Moller HJ (2004) CD163: a regulated hemoglobin scavenger receptor with a role in the anti-inflammatory response. Ann Med 36:347–354.

Mosser DM (2003) The many faces of macrophage activation. J Leukoc Biol 73:209–212.

Nakashima J, Tachibana M, Ueno M, Baba S, Tazaki H (1995) Tumor necrosis factor and coagulopathy in patients with prostate cancer. Cancer Res 55(21):4881–4885.

Nguyen HX, Tidball JG (2003) Null mutation of gp91^{phox} reduces muscle membrane lysis during muscle inflammation in mice. J Physiol (Lond) 553:833–841.

Nguyen HX, Lusis AJ, JG Tidball (2005) Null mutation of myeloperoxidase in mice prevents mechanical activation of neutrophil lysis of muscle cell membranes in vitro and in vivo. J Physiol (Lond) 565:403–413.

Nishikawa A, Murata E, Akita M, Kaneko K, Moriya O, Tomita M, Hayashi H (1998) Roles of macrophages in programmed cell death and remodeling of tail and body muscle of *Xenopus laevis* during metamorphosis. Histochem Cell Biol 109(1):11–17.

Nishikawa K, Arai H, Inoue K (1990) Scavenger receptor-mediated uptake and metabolism of lipid vesicles containing acidic phospholipids by mouse peritoneal macrophages. J Biol Chem 265: 5226–5231.

Ottnad E, Parthasarathy S, Sambrano GR, Ramprasad MP, Quehenberger O, Kondratenko N, Green S, Steinberg D (1995) A macrophage receptor for oxidized low density lipoprotein distinct from the receptor for acetyl low density lipoprotein: partial purification and role in recognition of oxidatively damaged cells. Proc Natl Acad Sci U S A 92:1391–1395.

Otterbein LE, Bach FH, Alam J, Soares M, Tao Lu H, Wysk M, Davis RJ, Flavell RA, Choi AM (2000) Carbon monoxide has anti-inflammatory effects involving the mitogen-activated protein kinase pathway. Nat Med 6:422–428.

Otterbein LE, Choi AM (2000) Heme oxygenase: colors of defense against cellular stress. Am J Physiol Lung Cell Mol Physiol 279:L1029–L1037.

Pan J, Fukuda K, Saito M, Matsuzaki J, Kodama H, Sano M, Takahashi T, Kato T, Ogawa S (1999) Mechanical stretch activates the JAK/STAT pathway in rat cardiomyocytes. Circ Res 84: 1127–1136.

Philippidis P, Mason JC, Evans BJ, Nadra I, Taylor KM, Haskard DO, Landis RC (2004) Hemoglobin scavenger receptor CD163 mediates interleukin-10 release and heme oxygenase-1 synthesis: antiinflammatory monocyte-macrophage responses in vitro, in resolving skin blisters in vivo, and after cardiopulmonary bypass surgery. Circ Res 94:119–126.

Podrez EA, Febbraio M, Sheibani N, Schmitt D, Silverstein RL, DP Hajjar DP, Cohen PA, Frazier WA, Hoff HF, Hazen SL (2000) Macrophage scavenger receptor CD36 is the major receptor for LDL modified by monocyte-generated reactive nitrogen species. J Clin Invest 105:1095–1108.

Polek TC, Talpaz M, Darnay BG, Spivak-Kroizman T (2003) TWEAK mediates signal transduction and differentiation of RAW264.7 cells in the absence of Fn14/TweakR. Evidence for a second TWEAK receptor. J Biol Chem 278(34):32317–32323.

Rabinowitz SS, Gordon S (1991) Macrosialin, a macrophage-restricted membrane sialoprotein differentially glycosylated in response to inflammatory stimuli.J Exp Med 174(4):827–836.

Ramprasad MP, Fischer W, Witztum JL, Sambrano GR, Quehenberger O, Steinberg D (1995) The 94- to 97-kDa mouse macrophage membrane protein that recognizes oxidized low density lipoprotein and phosphatidylserine-rich liposomes is identical to macrosialin, the mouse homologue of human CD68. Proc Natl Acad Sci U S A 92:9580–9584.

Ramprasad MP, Terpstra V, Kondratenko N, Quehenberger O, Steinberg D (1996) Cell surface expression of mouse macrosialin and human CD68 and their role as macrophage receptors for oxidized low density lipoprotein. Proc Natl Acad Sci U S A 93:14833–14838.

Reyes-Reyna SM, Krolick KA (2000) Chemokine production by rat myocytes exposed to interferon-gamma. Clin Immunol 94(2):105–113.

Robertson TA, Maley MA, Grounds MD, Papadimitriou JM (1993) The role of macrophages in skeletal muscle regeneration with particular reference to chemotaxis. Exp Cell Res 207:321–331.

Sadrzadeh SM, Graf E, Panter SS, Hallaway PE, Eaton JW (1984) Hemoglobin. A biologic Fenton reagent. J Biol Chem 259:14354–14356.

Sakuma K, Nishikawa J, Nakao R, Watanabe K, Totsuka T, Nakano H, Sano M, Yasuhara M (2003) Calcineurin is a potent regulator for skeletal muscle regeneration by association with NFATc1 and GATA-2. Acta Neuropathol (Berl) 105(3):271–280.

Schaer DJ, Boretti FS, Hongegger A, Poehler D, Linnscheid P, Staege H, Muller C, Schoedon G, Schaffner A (2001) Molecular cloning and characterization of the mouse CD163 homologue, a highly glucocorticoid-inducible member of the scavenger receptor cysteine-rich family. Immunogenetics 53:170–177.

Schaer DJ (2002) The macrophage hemoglobin scavenger receptor (CD163) as a genetically determined disease modifying pathway in atherosclerosis. Atherosclerosis 163:199–201.

Schaer DJ, Boretti FS, Schoedon G, Schaffner A (2002) Induction of the CD163-dependent haemoglobin uptake by macrophages as a novel anti-inflammatory action of glucocorticoids. Br J Haematol 119:239–243.

Schutzle UB, Wakelam MJ, Pette D (1984) Prostaglandins and cyclic AMP stimulate creatine kinase synthesis but not fusion in cultured embryonic chick muscle cells. Biochim Biophys Acta 805(2): 204–210.

Shireman PK, Contreras-Shannon V, Reyes-Reyna SM, Robinson SC, McManus LM (2006) MCP-1 parallels inflammatory and regenerative responses in ischemic muscle. J Surg Res 134(1):145–157.

Shireman PK, Contreras-Shannon V, Ochoa O, Karia BP, Michalek JE, McManus LM (2007) MCP-1 deficiency causes altered inflammation with impaired skeletal muscle regeneration. J Leukoc Biol 81(3):775–785.

Smith LL, Anwar A, Fragen M, Rananto C, Johnson R, Holbert D (2000) Cytokines and cell adhesion molecules associated with high-intensity eccentric exercise. Eur J Appl Physiol 82(1–2):61–67.

Spangenburg EE, Booth FW (2002) Multiple signaling pathways mediate LIF-induced skeletal muscle satellite cell proliferation. Am J Physiol Cell Physiol 283(1):C204–C211.

Spencer MJ, Walsh CM, Dorshkind KA, Rodriguez EM, Tidball JG (1997) Myonuclear apoptosis in dystrophic mdx muscle occurs by perforin-mediated cytotoxicity. J Clin Invest 99(11):2745–2751.

Spencer MJ, Marino MW, Winkler WM (2000) Altered pathological progression of diaphragm and quadriceps in TNF-deficient and dystrophin-deficient mice. Neuromusc Disorders 10: 612–619.

Stark VK, Hoch JR, Warner TF, Hullett DA (1997) Monocyte chemotactic protein-1 expression is associated with the development of vein graft intimal hyperplasia. Arterioscler Thromb Vasc Biol 17(8):1614–1621.

St. Pierre BA, Tidball JG (1994) Differential response of macrophage subpopulations to soleus muscle reloading following rat hindlimb suspension. J Appl Physiol 77:290–297.

Suelves M, Lopez-Alemany R, Lluis F, Aniorte G, Serrano E, Parra M, Carmeliet P, Munoz-Canoves P (2002) Plasmin activity is required for myogenesis in vitro and skeletal muscle regeneration in vivo. Blood 99(8):2835–2844.

Sulahian TH, Hogger P. Wahner AE, Wardwell K, Goulding NJ, Sorg C, Droste A, Stehling M, Wallace PK, Morganelli PM, Guyre PM (2000) Human monocytes express CD163, which is upregulated by IL-10 and identical to p155. Cytokine 12:1312–1321.

Summan M, McKinstry M, Warren GL, Hulderman T, Mishra D, Brumbaugh K, Luster MI, Simeonova PP (2003) Inflammatory mediators and skeletal muscle injury: a DNA microarray analysis. J Interferon Cytokine Res 23(5):237–245.

Summan M, Warren GL, Mercer RR, Chapman R, Hulderman T, Van Rooijen N, Simeonova PP (2006) Macrophages and skeletal muscle regeneration: a clodronate-containing liposome depletion study. Am J Physiol Regul Integr Comp Physiol 290(6):R1488–R1495.

Teixeira CFP, Zamuner SR, Zuliani JP, Fernandes CM, Cruz-Hofling MA, Fernandes I, Chaves F, Gutierrez JM (2003) Neutrophils do not contribute to local tissue damage, but play a key role in skeletal muscle regeneration, in mice injected with *Bothrops asper* snake venom. Muscle Nerve 28:449–459.

Thomason DB, Booth FW (1990) Atrophy of the soleus muscle by hindlimb unweighting. J Appl Physiol 68:1–12.

Tidball J, Wehling-Henricks M (2007) Macrophages promote muscle membrane repair and muscle fiber growth and regeneration during modified muscle loading in mice in vivo. J Physiol (Lond) 578:327–336.

Tracey KJ, Wei H, Manogue KR, Fong Y, Hesse DG, Nguyen HT, Kuo GC, Beutler B, Cotran RS, Cerami A, Lowry SF (1988) Cachectin/tumor necrosis factor induces cachexia, anemia, and inflammation. J Exp Med 167(3):1211–1227.

Vakakis N, Bower J, Austin L (1995) In vitro myoblast to myotube transformations in the presence of leukemia inhibitory factor. Neurochem Int 27:329–335.

Verfaillie C, McGlave P (1991) Leukemia Inhibitory Factor/human interleukin for DA cells: a growth factor that stimulates the in vitro development of multipotential human hematopoietic progenitors. Blood 77:263–270.

Van Velzen AG, Da Silva RP, Gordon S, Van Berkel TJ (1997) Characterization of a receptor for oxidized low-density lipoproteins on rat Kupffer cells: similarity to macrosialin. Biochem J 322: 411–415.

Villiger PM, Geng Y, Lotz M (1993) Induction of cytokine expression by leukemia inhibitory factor. J Clin Invest 91:1575–1581.

Warren GL, Hulderman T, Jensen N, McKinstry M, Mishra M, Luster MI, Simeonova PP (2002) Physiological role of tumor necrosis factor α in traumatic muscle injury. FASEB J 16:1630–1632.

Warren GL, O'Farrell L, Summan M, Hulderman T, Mishra D, Luster MI, Kuziel WA, Simeonova PP (2004) Role of CC chemokines in skeletal muscle functional restoration after injury. Am J Physiol Cell Physiol 286:C1031–C1036.

Weber MA, Schnyder-Candrian S, Schnyder B, Quesniaux V, Poli V, Stewart CL, Ryffel B (2005) Endogenous leukemia inhibitory factor attenuates endotoxin response. Lab Invest 85:276–284.

Wehling M, Spencer MJ, Tidball JG (2001) A nitric oxide synthase transgene ameliorates muscular dystrophy in mdx mice. J Cell Biol 155:123–131.

Wiedow O, Meyer-Hoffert U (2005) Neutrophil serine proteases: potential key regulators of cell signalling during inflammation. J Intern Med 257(4):319–328.

Zacks SI, Sheff MF (1982) Age-related impeded regeneration of mouse minced anterior tibial muscle. Muscle Nerve 5:152–161.

Zalin RJ (1987) The role of hormones and prostanoids in the in vitro proliferation and differentiation of human myoblasts. Exp Cell Res 172(2):265–281.

Zhu M, Oishi K, Lee SC, Patterson PH (2001) Studies using leukemia inhibitory factor (LIF) knockout mice and a LIF adenoviral vector demonstrate a key anti-inflammatory role for this cytokine in cutaneous inflammation. J Immunol 166:2049–2054.

Zouaoui Boudjeltia K, Moguilevsky N, Legssyer I, Babar S, Guillaume M, Delree P, Vanhaeverbeek M, Brohee D, Ducobu J, Remacle C (2004) Oxidation of low density lipoproteins by myeloperoxidase at the surface of endothelial cells: an additional mechanism to subendothelium oxidation. Biochem Biophys Res Commun 3252:434–438.

CHAPTER 13

COMPLEXITY OF EXTRACELLULAR MATRIX AND SKELETAL MUSCLE REGENERATION

MIRANDA D. GROUNDS

School of Anatomy & Human Biology, The University of Western Australia

1. INTRODUCTION: THE ENVIRONMENT OF CELLS IS OF MAJOR IMPORTANCE

Skeletal muscle regeneration occurs by the activation, proliferation and fusion of muscle precursor cells (myoblasts) that usually lie in a satellite cell position on the surface of the sarcolemma beneath the external lamina (basement membrane) of myofibres. Thus these myoblasts (satellite cells), as is the whole surface of the myfibre and the neuromuscular junctions (NMJ), are in intimate contact with the specialized components of the extracellular matrix (ECM) that compose the basement membrane (often referred to as basal lamina). Interactions between the surface of myofibres and the basement membrane ensure the effective transfer of the contractile force (generated in the sarcomeres) laterally across the cell surface to the ECM for full force generation. Beyond the intimate interactions of skeletal muscle cells with the basement membrane lies interstitial connective tissue that contains blood vessels, nerves and lymphatics. The ECM also contains a wealth of circulating and other proteins (Bortoluzzi et al., 2006). The importance of the ECM is often overlooked yet it is the environment of the cells (from myoblasts and myofibres to endothelium and nerves) that dictates what will occur.

During development, myoblasts migrate through the ECM and proliferate to form new muscle fibres initially in the absence of a basement membrane. This contrasts with the situation in mature muscle where the myoblasts (derived mainly from satellite cells) lie beneath the basement membrane of myofibres. Most myofibre damage resulting from exercise/contraction-induced injury, muscle disease or muscle transplantation, does not severely disrupt the basement membrane, thus the myoblasts become activated, proliferate and fuse within this niche environment.

S. Schiaffino and T. Partridge (eds.), Skeletal Muscle Repair and Regeneration, 269–301.

While there is strong evidence that the basement membrane persists as a scaffold after myofibre necrosis, it is not clear exactly which molecular components survive or are modified in the first week after damage and this is clearly critical to define the local micro-environment of the responding myoblasts. In more severe injuries in vivo such as major trauma or crush injury where the myofibre and basement membrane is completely disrupted, the myoblasts will be in intimate contact with interstitial ECM components and this will dictate their response. These in vivo issues are critical to consider when making comparisons with observations derived from tissue culture where myoblast behaviour is studied in an artificial environment often without any defined ECM components.

This review first outlines in general the main components of the ECM environment and their dynamic regulation with some reference to skeletal muscle: this is an enormous subject and thus only a hint of the complexity is possible. The roles of specific components in vivo during muscle regeneration are then discussed with respect to inflammation, revascularization, myoblast activation and proliferation, fusion of myoblasts and myotube formation, re-innervation and maturation to restore full function to the regenerated myofibre and, finally, fibrosis.

2. COMPONENTS OF THE ECM SURROUNDING SKELETAL MUSCLE FIBRES

2.1 Interstitial Connective Tissue

The interstitial connective tissue around muscle fibres accounts for about 1–10% of the muscle tissue, it provides a scaffold and basic mechanical support for blood vessels and nerves, it also provides elasticity and especially transfers the mechanical force generated within myofibres to move the skeleton (Kjaer, 2004). In skeletal muscle the connective tissue which surrounds a whole muscle fascicle is the *epimysium* and the connective tissue that encloses a whole bundle of myofibres is the *perimysium* (Fig. 1). The perimysium contains networks of blood vessels and nerves that extend into the *endomysium* that intimately surrounds each myofibre (i.e is in contact with the basement membrane). The tendons, that attach the ends of muscles to bones, are dense specialized connective tissue structures with much type I collagen designed to withstand tension. The myotendinous junctions and costameres on myofibres are specialized areas for force transmission and are enriched in sarcolemmal and basement membrane associated components (Tidball, 1991). Central to muscle function is the transfer of mechanical force from the sarcomeres within myofibres, laterally through the basement membrane to the ECM [reviewed in (Grounds et al., 2005)], initially to the endomysium and then integration and transmission of this force through the perimysium, epimysium and tendons to the bones. The ability of mechanical forces to modify ECM components (especially the expression of collagens, tenascin-C and metalloproteinases) and resultant mechanotransduction with associated intracellular signaling during normal skeletal muscle function has been reviewed extensively (Kjaer, 2004; Sarasa-Renedo and Chiquet, 2005) and is of considerable interest to sports medicine.

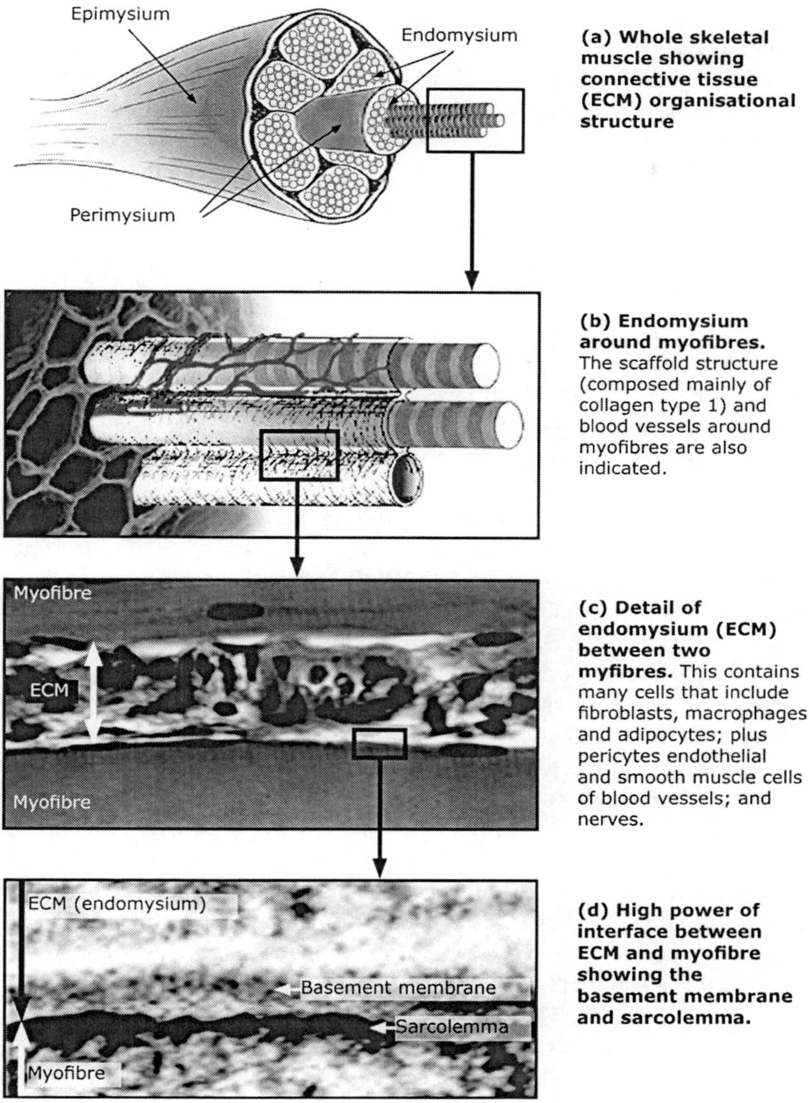

Figure 1. **The ECM structure of skeletal muscle**. The organization of the ECM is indicated (a) for the connective tissue around an entire anatomical muscle (epimysium) and surrounding muscle bundles (perimysium), through to that in close contact with individual muscle fibres (endomysium). The review focuses on the endomysium (b, c) that includes the specialized basement membrane (d) intimately associated with the surface of the myofibre. The endomysium forms a scaffold around each myofibre as shown by scanning electron microscopy (b) of the ECM structure (mainly collagen 1) remaining when myofibres have been removed [based on image courtesy of P Allingham, Queensland].The main molecular components of the endomysium are outlined in Table 1. The basement membrane (d) is shown by transmission electron microscopy (of a longitudinal section of mature muscle) as a fuzzy layer close to the sarcolemma: the basement membrane forms a complex ECM network linked by many molecular connections to transmembrane proteins at the myofibre surface

The composition of the interstitial connective tissue in muscle tissue is probably generally similar to that for any tissue, being composed of water and a complex mesh of interacting molecules some of which are structural and some that are sticky and more adhesive. The collagen fibrils (that provide strength) interact with a multitude of proteoglycans and many other ECM molecules e.g fibronectin and tenascin, along with metalloproteinases (that modify the ECM) and cytokines. The key interstitial ECM components are briefly outlined below and the complex roles of proteoglycans are discussed in more detail later (see Section 2.3). An overview of the main ECM components in the endomysium around a mature myofibre is presented in Table 1.

2.1.1 Collagens

Collagens are highly important structural proteins with many functions, they form a large superfamily all composed of 3 polypeptide α chains that are coiled into a triple helix rope like rod structure made possible by the presence of glycine, they are rich in (Gly-X-Y) repeats where X is often proline and Y is hydroxyproline [reviewed in (Gelse et al., 2003; Myllyharju and Kivirikko, 2004; Ricard-Blum and Ruggiero, 2005)]. The number and complexity of collagens is illustrated in vertebrates where at least 27 collagens (types I to XXVII) with 42 distinct α chains have already been identified, plus more than 20 proteins with collagen-like triple helix domains, around 20 isoenzymes of collagen modifying enzymes, and more than 1300 mutations have been reported for human collagen genes [reviewed in (Myllyharju and Kivirikko, 2004; Ricard-Blum and Ruggiero, 2005)]. Collagen synthesis involves extensive post-translational modifications that require numerous enzymes and most collagens are able to form supramolecular aggregates (Kjaer, 2004). Additional complexity arises from the fact that the non-collagenous domains of many collagens have important functions, for example the proteolytically derived fragments of the C-domain of collagen XVIII (known as endostatin), XV (restin) and IV all inhibit angiogenesis (Myllyharju and Kivirikko, 2004). Beyond their critical structural biomechanical role, collagens affect diverse cellular functions through their multitude of interactions with other proteins and the entrapment and storage of growth factors such as Transforming Growth Factor-β (TGF-β), and insulin-like Growth Factor (IGF) -I and -II (Gelse et al., 2003).

The classic fibrillar collagens are collagens I, II and III with minor types V and XI, with two additional novel members XXIV and XXVI being recently identified (Ricard-Blum and Ruggiero, 2005). Collagen I is the most abundant and widely studied. It is the major collagen in the interstitial connective tissue and tendons (representing from ~30% to 97% of total collagen) and of huge structural importance for the transfer of muscle force: muscles vary in the content and types of collagen (Kjaer, 2004). Collagens age over time with more intermolecular cross-links initially resulting in increased biomechanical strength: however, the slow turnover of collagen combined with glycation that modifies the cross-links and is dramatically accelerated in diabetes, leads to stiffness and reduced optimal function in aged tissues (Avery and Bailey, 2005; Kjaer, 2004). Based on their structure

Table 1. **Indication of the ECM (endomysium) structure and molecular composition around an individual adult skeletal myofibre** [shows a change in regenerating muscle]. This is a rapidly evolving field and these data are only indicative of the complexity, with many omissions (see text). The main distribution of these molecules in interstitial ECM and basement membrane of undamaged myofibres is indicated, but this is sometimes unclear *in vivo* and after damage more molecules may extend into the basement membrane to the cell surface

Molecule	Collagens	Fibronectin	Tenascin	Laminins	Glycosaminoglycans (GAGs)				Galetins	Proteeases	
Number of forms	27	3	4	16	Many				14	Many Controlled by complexity of activators and inhibitors	
Forms in skeletal muscle	**Coll** I, IV, VI, XV, XVIII, IX , XI		**Tenascin** (tn) C, X(Y), R, J1	**Laminins** (LM) 211, 221 (α2 chain) 411,421 511,521	**Chondroitin Sulphates** Biglycan Versiscan Decorin	**Dermatan sulphates**	**Heparan Sulphates** Perlecan Syndecans Glycipan-1 Agrin	**Hyaluronan**		**MMPs** (~28) −2,−7,−9, −14, −24,−25 also **RECK**, and **TIMPs**	**Plasmin Trypsin Thrombin calpain**
Interstitial ECM	**mainly coll I**	**fn**	[tnC normally low but increases after damage, loading]		**Rich e.g decorin**	**Rich**		**(hyaluronan)**		**MMP-2** [MMP-9 increases after injury: made by myoblasts]	
Basement membrane	**coll VI coll IV**	? [normally NOT in matureB/M but may be after damage]	? [perhaps after damage]	**Mainly LM-211** [LM-411 and 511 in myotubes]	**versican e.g biclycan**		**Rich e.g. perlecan, syndecans-1, 3, 4, glypican-1 agrin**	? [no information in vivo]	**gal-1**		

Notes:

(a) Many other molecules that are present (often in smaller amounts) but are not shown here, may be of critical importance.

(b) The composition of the interstitial ECM (e.g for GAGs) varies between endomysium (shown here), perimysium and epimysium.

(c) Different forms of molecules (e.g laminin- 221, 411, 421, 511 and 521) are present in specialized regions of the sarcolemma such as myotendinous and neuromuscular junctions (generally not shown).

(d) Sarcolemma associated proteins such as integrins (that bind ECM molecules) and ADAM12L are not included in this Table (see text).

and supramolecular organizations, collagens are grouped as such fibril-forming collagens, or as fibril-associated, network-forming, anchoring, basement-membrane, trans-membrane and other collagens (Gelse et al., 2003).

Collagen IV is the major structural component of basement membranes where it integrates laminins, nidogens and other molecules into a stable supramolecular aggregate. The interstitial connective tissue is anchored to the basement membrane through a microfilament network rich in collagen VI and defects in collagen VI result in several myopathies (Bonnemann and Laing, 2004). The multiplicity of linkages for the ECM is illustrated for collagen VI that binds the fibrillar collagens type I and II, basement membrane components like collagen IV and perlecan, cell membrane associated molecules like integrins, and proteoglycans like decorin and biglycan (Jenniskens et al., 2006; Lechner et al., 2006; Wiberg et al., 2002). Skeletal muscle cells make collagen I (and probably III), collagen IV, VI, and XV. Information is not yet available for other collagens such as XVIII, IX and XII that are found in skeletal muscle. Adhesion of cells to collagens is mediated by glycoproteins that modulate many cellular responses throughout the ECM (see Section 2.3).

2.1.2 *Fibronectin*

Fibronectin is a large fibril forming adhesive glycoprotein that exists in 3 forms: a soluble dimeric (plasma) form that circulates in the blood and other bodily fluids; oligomers of (cell-surface) fibronectin that are transiently attached to the cell surface, and highly insoluble (matrix) fibronectin fibrils in the ECM. Fibronectin binds to collagen, tenascin and other ECM molecules and is involved in adhesion and cell migration. Down-regulation appears essential for myotube formation, at least during embryogenesis and in culture (see Section 3.3.1). Fibronectin also undergoes alternative splicing to produce the extra domain-B (ED-B) containing isoform, which is exclusively expressed during embryogenesis, tissue repair and tumor angiogenesis (Khan et al., 2006).

2.1.3 *Tenascins*

Tenascins are also extracellular glycoproteins involved in cell adhesive events. There are 4 members of the tenascin family. Structurally these molecules consist of a large complex of six polypeptide chains (linked by disulphide bonds) projecting from a central core. In skeletal muscle, tenascin-C has a similar location to fibronectin but is normally present at low levels in uninjured muscle. Tenascin-C increases rapidly after mechanical loading (Sarasa-Renedo and Chiquet, 2005) and damage and is closely associated with inflammation (Chiquet and Fluck, 2003). Tenascin-C influences integrin and syndecan signaling; it is proposed that tenascin-C is anti-adhesive and serves to counterbalance the adhesive effects of fibronectin [reviewed in (Chiquet-Ehrismann, 2004)]. The tenascin-C null mouse has a surprisingly mild phenotype (Mackie and Tucker, 1999) although the impact of this defect on muscle regeneration has not been investigated.

Mammalian tenascin-X (and the similar avian tenascin-Y) is highly expressed in the connective tissue of developing skeletal muscles and influences myogenesis (Hagios et al., 1999). Various tenascins are associated with muscle innervation: tenascin-C seems to be required for reinnervation of muscle, tenascin-W is also in the connective tissue of skeletal muscle and both it and tenascins-R are involved in neural function (Chiquet-Ehrismann, 2004). In addition, the synthesis of tenascin (J1) by fibroblasts increases in response to muscle denervation (Gatchalian et al., 1989).

2.2 Basement Membrane and Interactions with Cell Surface Molecules of the Myofibre

Since the immediate environment has the most impact on the behaviour of muscle cells, attention is focussed on the basement membrane and associated molecules. Basement membranes are highly specialised ECM sheets that in skeletal muscle are intimately associated with the surface of myofibres and satellite cells (Fig. 1). The major components are laminins, collagen IV, nidogens and the heparan sulphate proteoglycan perlecan. Laminins and collagen IV self-assemble to form networks that are linked both directly and via binding to nidogen and perlecan; in addition, binding to cell surface molecules like the dystroglycan complex and integrins is crucial for the transfer of contractile forces from the myofibre to the ECM (Grounds et al., 2005). The basement membrane has structural and functional specialisations as illustrated by the distinct molecular composition at the neuromuscular junction (Jenniskens et al., 2006), that are not discussed in this review.

Laminins and integrins. Laminins are found only in basement membranes and are composed of 3 chains, α, β and γ, that each occur in multiple forms and combine into about 16 different heterotrimers: the α chains are the functional important portion of the laminin molecule. In mature skeletal muscle, laminins containing the $\alpha2$ chain are the main component of the basement membrane, with laminin-2 ($\alpha2\beta1\gamma1$) being the major form around the sarcolemma of mature myofibres, whereas laminin-4 ($\alpha2\beta2\gamma1$) occurs at myotendinous and neuromuscular junctions (NMJ) (Grounds et al., 2005). The new nomenclature for laminins refers to these as LM-211 and LM-221 respectively (Aumailley et al., 2005). The only other laminins in skeletal muscle cells are characterised by the presence of laminin $\alpha4$ and $\alpha5$ chains, combined with $\beta1\gamma1$ as laminin-8 (LM-411) and laminin-10 (LM-511) or with $\beta2\gamma1$ as LM-421 and LM-521 respectively, and have a more restricted localization at NMJ in mature muscle (Grounds et al., 2005). The vascular epithelium of blood vessels also expresses LM-411 and LM-511 (Ringelmann et al., 1999). The situation is different in developing and regenerating skeletal muscle where, although laminin $\alpha2$ is the dominant chain, laminin $\alpha4$ and $\alpha5$ chains are more broadly distributed [reviewed in (Grounds et al., 2005; Gullberg et al., 1999; Sorokin et al., 2000)]. Where laminin $\alpha2$ is absent, the other laminins $\alpha4$ and $\alpha5$ that are expressed on young myotubes cannot substitute for laminin

$\alpha 2$ on mature myofibres: it is emphasised that laminin-1 ($\alpha 1\beta 1\gamma 1$: LM-111) is not found in skeletal muscle tissue (Ringelmann et al., 1999; Sorokin et al., 2000).

Integrins are key receptors that link the ECM to the inside of the cell and the expression of specific integrins on myoblasts, myotubes and myofibres is associated with binding to different laminins (Grounds et al., 2005). Integrins are cell surface adhesion molecules that can signal bidirectionally across the plasma membrane. They are heterodimers composed of an α and β subunit and, while at least 18α and 8β chains have been characterised, only a restricted number of $\alpha\beta$ heterodimers have been identified (Belkin and Stepp, 2000). Integrins bind a huge array of ECM ligands including collagens, fibronectin, tenascin and laminin (and other molecules) and play a central role in translating mechanical and structural cues into intracellular molecular signals that affect many aspects of cell behaviour (Ingber, 2006; Kjaer, 2004; Larsen et al., 2006).

Beyond binding to integrins, laminins bind to the dystrophin glycoprotein complex (DGC) that is the main transmembrane link responsible for force transmission to the ECM (Grounds et al., 2005). The crucial role of the DGC in normal muscle function and strength is demonstrated by the wide range of clinical muscle disorders that result from defects of one of the many molecules associated with this complex (Kanagawa et al., 2005). For example, defective or absent dystrophin disrupts the DGC in Duchenne Muscular Dystrophy (DMD), with resultant cell membrane fragility leading to repeated myofibre necrosis that ultimately results in scar tissue when muscle regeneration fails. Binding of the ECM to the DGC occurs at α-dystroglycan that (in muscle) binds to laminin-2 (LM211) and perlecan that both contain laminin globular domains (Kanagawa et al., 2005) and also binds to biglycan (Lechner et al., 2006). The importance of post-translational processing has recently been recognized with glycosylation (linkage of sugars) being the crucial modification that modulates the function of dystroglycan as a receptor for these extracellular binding partners (Barresi and Campbell, 2006).

Carbohydrate accounts for roughly half the molecular weight of α-dystroglycan with many sites for O-linked glycosylation. It is now recognized that a range of mutations (that account for half of the genes involved in congenital muscular dystrophies) affect enzymes called glycosyl transferases, resulting in reduced glycosylation of α-dystroglycan that manifests as a group of diseases called dystroglycanopathies with a range of severity that affect skeletal muscle, the eye and brain [reviewed in (Barresi and Campbell, 2006; Martin, 2006)]. It seems that underglycosylation of α-dystroglycan inhibits the binding with extracellular matrix proteins such as laminin, effectively divorcing this important cell adhesion molecule from its extracellular environment.

2.3 Glycosaminoglycans

All extracellular proteins have evolved in the presence of linear polysaccharides (sugars) called glycosaminoglycans (GAGs). The most common sulphated GAGs in skeletal muscle are heparan sulphates (HS), chondroitin sulphates and dermatan sulphates. The expression of these GAGs and especially the role of HS in skeletal muscle development and physiology are the subject of an excellent recent review (Jenniskens et al., 2006). The perimysium is rich in chondroitin and dermatan sulphates, whereas the endomysium and basement membrane are rich in heparan sulphates. A non-sulphated member of this family, hyaluronan, is also found in skeletal muscle.

Heparan sulphate (HS) proteoglycans are found on the surface of all animal cells and are also a major component of the ECM: they are the major GAGs in the basement membrane of skeletal muscle. The HS chains are assembled intracellularly on the core proteins by enzymes in the Golgi, there is extraordinary structural heterogeneity and further processing occurs by sulphatases on the cell surface. Because of their high negative charge the HS chains bind to almost all ECM molecules and to a wide range of other proteins of diverse function, with the crucial importance of HS-protein interactions for regulating many aspects of biology including development, normal tissue function, metabolism, inflammation, regeneration and stem cells being increasingly recognized (Bishop et al., 2007; Coombe and Kett, 2005). The observation that HS binds to fibroblast growth factor-2 (FGF-2) catalysed a wealth of research that identified different GAG binding partners for various growth factors and also the requirement for specific HSs to bind and activate the FGF receptor. Interestingly, it was shown that FGF-2 binds to the HS chains of syndecans (present in proliferating myoblasts) but not to chondroitin sulphate chains or to other HS chains on different proteoglycans (Olwin and Rapraeger, 1992). An unexpected ligand for HS is the extracellular histone H1 that binds specifically to perlecan in the basement membrane of myotubes/myofibres (Henriquez et al., 2002). Heparin affin regulatory peptide (HARP) is another heparin binding growth factor that is involved in myogenesis and skeletal muscle regeneration (Caruelle et al., 2004). These GAG–protein interactions modulate many extracellular events ranging from sequestering and stabilizing growth factors, to mediating the internalization of ligands. The use of artificial mimetic GAGs (Barbosa et al., 2005) that improve myogenesis in vivo (Zimowska et al., 2001) confirms the wide importance of GAGs during muscle regeneration.

About 26 enzymes participate in the formation of HS chains in mammals (Bishop et al., 2007) and modification of the sulphation status of HS sidechains of GAGs influences their extracellular interactions with signalling molecules, for example to influence FGF-2 and Wnt signaling (Ai et al., 2006; Ai et al., 2003). These important modifications are effected by sulphatases that in muscle act on substrates such as perlecan, gylcipan and syndecans (Ai et al., 2006; Jenniskens et al., 2006).

Many studies into the important roles of HS proteoglycans in normal, myopathic and regenerating skeletal muscle in vivo were pioneered by Enrique Brandan and his colleagues in Chile [reviewed recently (Jenniskens et al., 2006)]. In adult skeletal muscle, the chondroitin sulfate proteoglycan, decorin, is very abundant: it is located mainly in the perimysium and is upregulated during regeneration (Brandan et al., 1991), whereas biglycan, another chondroitin sulfate proteoglycan, is mainly in the endomysium and NMJ and binds to α-dystroglcan on the sarcolemma (Casar et al., 2004b; Droguett et al., 2006). Biglycan and decorin are both small leucine-rich proteoglycans that interact with collagen and cytokines; like the pro-fibrotic cytokine TGF-β, they are both implicated in fibrosis in various disorders (see Section 3.5). The HS proteoglycans syndecan-3 and syndecan-4 are rich on the surface of skeletal muscle and satellite cells, where they play roles in cell signalling and cell adhesion and have many effects on myogenesis and regeneration (Casar et al., 2004a; Cornelison et al., 2004; Cornelison et al., 2001). Other HS proteoglycans of importance in skeletal muscle and myogenesis include the cell surface-associated syndecan-1, glypican (Velleman et al., 2004) and perlecan (Henriquez et al., 2002; Kanagawa et al., 2005) and the basement membrane-associated collagen VIII and agrin that is located at the NMJ (Jenniskens et al., 2006).

Hyaluronan (also known as hyaluronic acid) is the only GAG that lacks a protein core, it is found in the interstitial connective tissue (is very important in cartilage) and is of especial interest to inflammation (Hascall et al., 2004). Less widely recognized is the fact that hyaluronan is found within cells associated with many organelles especially the nucleus, often in cells undergoing mitosis (Hascall et al., 2004). In addition, hyaluronan rapidly accumulates in smooth muscle and other cells in response to endoplasmic reticulum (ER) stress where it forms cable-like structures that appear to originate in the perinuclear or ER region and extend outside the cell. These hyaluronan cable structures bind leukocytes and play an important role in the early inflammatory response (Hascall et al., 2004). Hyaluronan also influences myogenesis in tissue cultured cells where a decrease appears necessary for myotube formation (see 3.3.1).

Versican is a chondroitin sulfate proteoglycan of the ECM that participates in cell adhesion, proliferation, migration and angiogenesis, and the intracellular signaling pathways resulting from the engagement of versican by cell surface molecules is the subject of a recent review (Rahmani et al., 2006). Versican binds to ECM components such as hyaluronan, collagen I, fibulin, fibronectin, tenascin-R, selectins and chemokines as well as to cell surface molecules such as CD44 and integrin β1. Versican is a downstream target of the transcription factor Pax3 (that is an important regulator of the myogenic genes MyoD and Myf5) and transcription of versican is also regulated by p53.

The functions of these complex carbohydrates often involve recognition of specific sugar components by families of molecules called lectins. Galectins are a family of proteins that bind β-galactosidase (lactose) and have at least 14 family members in mice: they have extracellular and also intracellular functions that include roles in angiogenesis and inflammation (Houzelstein et al., 2004; Liu and Rabinovich, 2005). Galectin-1 is found in many species, generally has immunosuppressive effects and is implicated in myogenesis in several ways. Galectin-1 is present in developing and adult skeletal muscle (Watt et al., 2004). It is found within the cytoplasm of C2C12 myoblasts and secreted during terminal differentiation; the extracellular galectin-1 co-localises with laminin in vitro where it may inhibit binding of myoblasts to laminin (Cooper et al., 1991). A striking effect of galectin-1 is the conversion of non-myogenic cells, such as dermal fibroblasts and especially human mesenchymal stem cells, into myoblasts that can contribute to muscle regeneration (Chan et al., 2006). However, the precise role of galectin-1 in myogenesis during development and in regenerating muscles remains to be elucidated (Kami and Senba, 2005).

2.4 Proteases in the Extracellular Environment

The extraordinary complexity of molecular interactions that contribute to the formation and dynamic modulation of the ECM is daunting and can be barely touched upon within the scope of this review. Many ECM molecules and other proteins including enzymes are secreted and act outside the cell and the secretome of skeletal muscle has been described recently by computational analysis (Bortoluzzi et al., 2006). Proteolysis is a major contributor to dynamic modulation of the cell surface and ECM (Overall and Blobel, 2007). Matrix metalloproteinases (MMPs) are secreted enzymes that are especially important for specifically processing and degrading all components of the ECM, cellular receptors and cytokines. The MMPs are key modulators of many biological processes including skeletal muscle formation, angiogenesis, cellular migration, inflammation and wound healing. Twenty-eight members of the MMP family have been identified to date in humans (Bernal et al., 2005). MMPs are produced as inactive zymogens, some are activated during secretion whereas most are activated extracellularly often by proteases (e.g initiated by uPA/plasmin). In skeletal muscles, expression of MMP-2 is constitutive and the active form of MMP-2 is transiently increased after injury: MMP-2 can degrade denatured collagens I, II and III, native collagens type IV and V, elastin and also proteoglycans and fibronectin (Kherif et al., 1999). Cultured C2C12 mouse and human myoblasts secrete MMP-2 (Kherif et al., 1999). By 24 hours after damage MMP-9 is induced and levels are sustained for a few days: MMP-9 degrades similar substrates and is produced mainly by inflammatory cells (Kherif et al., 1999). The role of MMPs in skeletal muscle has been reviewed recently (Carmeli et al., 2004). Tissue culture studies of MMP-2, MMP-7, MMP-9 and MT1-MMP (membrane-type MMP also known as MMP-14) in myoblasts and myotubes show distinct changes during differentiation; MMP-2 is expressed constitutively by myoblasts and myotubes whereas MMP-9 is expressed only in pre-fusion myoblasts [see (Carmeli et al., 2004; Echizenya et al., 2005; Lluri and Jaworski, 2005)]. Other MMPs associated with skeletal muscle include MMP-25 and MMP-24 and difference in expression patterns between species was noted (Bernal et al., 2005).

Of additional interest to skeletal muscle is the gene *RECK* that codes for a glycoprotein that regulates at least 3 members of the MMP family, MMP-2, MMP-9 and MT1-MMP. RECK suppresses myoblast fusion/myotube formation possibly by inhibiting MMP-2 and MT1-MMP, the RECK promoter is down-regulated by MyoD (to allow myotube formation) and is activated by MRF4 (possibly to promote myofibril maturation) (Echizenya et al., 2005). A recent report links MT1-MMP with degradation of fibronectin (Ohtake et al., 2006). Further studies in MT1-MMP null mice indicate that MT1-MMP cleaves laminin $\alpha 2$ and is important in maintaining myofibre integrity (Ohtake et al., 2006).

Other important proteases that regulate ECM molecules are the ADAMs (A Disintegrin and Metalloprotease) family of transmembrane glycoproteins, with > 30 members. These molecules possess a MMP domain that processes and sheds/releases the ectodomain of membrane-anchored growth factors and receptors (Blobel, 2005; Lafuste et al., 2005). One of these, ADAM12, is highly conserved between species, it is a multidomain protein with protease, cell adhesion, fusion, and signaling activities. One of the 2 isoforms of ADAM12, the long membrane anchored form ADAM12-L is associated with skeletal muscle and fusion. This molecule was originally named meltrin-α (Wewer et al., 2005). ADAM-12L is highly expressed in developing muscles, is downregulated in adults but is re-expressed in regenerating muscles by fusing myoblasts and newly formed myofibres (Lafuste et al., 2005). The extracellular domains of ADAM12-L interact with syndecans (Iba et al., 2000) and integrins, specifically with the α7 chain (the major integrin α-chain on adult skeletal muscle) and also to integrin α9β1 that is fundamental for myoblast fusion (Lafuste et al., 2005). ADAM12-L also interacts with the IGF growth factor system, in particular to cleave IGF binding proteins-3 and -5. The shorter secreted soluble form, ADAM12-S, is inhibited by TIMP-3. Thus collectively these proteases regulate the bioavailability of IGFs. This is especially important during pregnancy where ADAM12-S is elevated in the placenta and serum (Loechel et al., 2000). The availability of the unbound form of IGF-1 has major implications since IGF-1 plays many roles in skeletal muscle related to myogenesis, apoptosis, protein metabolism and the balance between atrophy and hypertrophy (Shavlakadze and Grounds, 2006). The proteolytic processing and shedding of the ectodomains of membrane anchored growth factors, cytokines and receptors has many biological consequences for skeletal muscle formation and regeneration and ADAM12-related ectodomain shedding is a topic of increasing interest (Wewer et al., 2005).

Tissue inhibitors of metalloproteinases (TIMPs) are endogenous inhibitors of MMPs. The balance between MMPs/TIMPs exquisitely regulates the turnover and remodeling of ECM during normal development, angiogenesis and pathogenesis (Chirco et al., 2006) and it seems likely that dysregulation of MMP-9/TIMP-1 or MMP-2/TIMP-2 expression contributes to muscle pathogenesis. TIMP-1 expression is undetectable on myoblasts but increases during differentiation and is high in cultured myotubes (Carmeli et al., 2004). Tissue culture studies of C2C12 cells also support a role for TIMP-2 (a specific activator of proMMP-2) during myogenesis since TIMP-2 is very low on proliferating myoblasts but increases during differentiation: it is also suggested that TIMP-2 may exert MMP-independent functions (Lluri and Jaworski, 2005).

Additional players in the extracellular environment are a myriad potent proteases produced by inflammatory and other cell types. One very important player is the serine protease plasmin and the associated plasminogen activation (PA) system (Nagamine et al., 2005). In addition to degrading fibrin clots, plasmin degrades fibronectin and laminin and activates MMPs and several latent growth factors: it is implicated in ECM degradation and tissue remodeling in situations such as skeletal muscle regeneration (Suelves et al., 2005). The protease plasmin arises from activation of plasminogen (Plg) and this balance involves complex regulation by inhibitors and activators. The generation of plasmin is inhibited mainly by plasminogen activator inhibitor type 1 (PAI-1). The number of signals that activate PAI-1 is vast (and includes hypoxia and TGF-β). PAI-1 is rapidly inactivated by oxygen radicals produced by neutrophils and other cells (Dellas and Loskuttof, 2005). The role of PAI-1 in the development of skeletal muscle is of tremendous interest, as PAI-1 is stabilized by binding to the plasma protein vitronectin (Dellas and Loskuttof, 2005). PAI-1 binding to vitronectin inhibits IGF-1/insulin signaling by competing for vitronectin binding through the integrin α5β3 binding site (Maile et al., 2006). Activation of plasmin occurs by 2 different pathways: the tissue type plasminogen activator (t-PA) is mainly involved in dissolution of fibrin, whereas the urokinase type plasminogen activator (uPA) enhances activation of cell-bound plasminogen. A key molecule in this latter pathway is the urokinase receptor (uPAR) (Nagamine et al., 2005) that also binds to vitronectin and integrins to affect integrin-mediated adhesion to the ECM. Many tissue culture studies have demonstrated roles for components of the PA system during myogenesis [discussed in 3.3.1 and see (Suelves et al., 2005)]. Skeletal muscle regeneration induced by injury has been analyzed in mice deficient for Plg, PAI-1, tPA or uPA and has demonstrated the profound effects of these molecules on the inflammatory response which contributes to muscle repair, and also on fibrosis. In particular, uPA (but not tPA) dependent plasmin activation was required for efficient skeletal muscle regeneration (Suelves et al., 2005).

Extracellular proteases such as thrombin and trypsin, proteolytically activate PARs (Protease Activated Receptors) to directly initiate cell signaling (Mackie et al., 2002). Skeletal muscle cells express both PAR-1 and PAR-2 (Chinni et al., 2000). Many of these proteases are involved in excessive deposition of ECM and fibrosis (Hewitson et al., 2005; Mackie et al., 2002). Thrombin and plasmin are inhibited by extracellular serpins (SERine Protease INnhibitors, e.g PAI-1) to help control clotting, fibrosis and inflammation: the interaction between serpins and thrombin is accelerated by GAGs (Pike et al., 2005). Another protease is m-calpain that is secreted by cultured muscle cells into the ECM where it degrades fibronectin prior to myoblast fusion (Dourdin et al., 1999): it is noted that calpain levels are also regulated by GAGs (Zimowska et al., 2001). Proteases in the extracellular milieu can act on a wide range of potential substrates and intensive effort is now being directed at elucidating the substrate degradomes of proteases in vivo (these can differ to in vitro observations), i.e identifying which proteases cleave a particular substrate and which ECM substrates are cleaved by specific proteases (Overall and Blobel, 2007).

2.5 Regulation of ECM Gene Expression

The complex network of interactions within the ECM are of central importance for normal skeletal muscle function, with defects in many components of the dystrophin glycoprotein complex or of basement membrane associated proteins, e.g laminin-2 and collagen VI, leading to muscle breakdown or weakness. Dystrophic muscle of boys with DMD or the mdx mouse model have repeated cycles of endogenous muscle damage and necrosis that normally provoke muscle regeneration. Since each episode of regeneration results in deposition of ECM, repeated cycles of damage and repair result in an accumulation of interstitial connective tissue that over time distorts the environment of the cell, leading to increasing fibrosis and impaired new muscle formation that is a major problem in DMD.

Age results in changes to the molecular composition and amount of ECM [reviewed in (Grounds et al., 1998)]. Collagens are modified over time with a series of enzymatic and non-enzymatic alterations of intermolecular cross-links, initially increasing mechanical strength but later resulting in stiffness and sub-optimal muscle function (Avery and Bailey, 2005).

Mechanical signals influence many aspects of cell behaviour (Ingber, 2006) and the synthesis and breakdown of diverse ECM components is modified in response to muscle-generated forces and stresses resulting from gravity and compression [reviewed in (Kjaer, 2004; Sarasa-Renedo and Chiquet, 2005)]. These tensile forces are essential for maintaining full muscle strength and function and are of central interest to sports medicine and prevention of exercise-induced injury.

Gender has also been shown to influence the extent of fibrosis in dystrophic mdx mice, with fibrosis and production of collagen 1 being reduced in female mdx mice; this may relate to the extent of necrosis or to inflammation and appears to be oestrogen mediated (Salimena et al., 2004).

Fibroblasts produce many of the ECM molecules but specialized cells such as skeletal muscle also produce many components. Cooperation between the various cell types results in formation of the ECM. This is illustrated for the basement membrane where fibroblasts produce a proportion of some laminins e.g the laminin $\alpha 4$ chain in skeletal muscle (Frieser et al., 1997). Such cellular co-operation that is important in vivo needs to be considered when studying isolated pure myogenic cells in tissue culture.

The various factors outlined above, such as disease, age, mechanical load and gender that can each influence the composition of the ECM, need to be taken into account when evaluating the influence of ECM of skeletal muscle regeneration in vivo.

3. MUSCLE REGENERATION AND THE ECM

This discussion relates to new muscle formation that is preceded by myofibre necrosis. Other situations of muscle restoration such as hypertrophy that increases myofibre size after atrophy (Shavlakadze and Grounds, 2006) or re-innervation (that can occur after muscle disuse, atrophy, unloading or nerve damage) do not involve myofibre necrosis. Necrosis involves the breakdown of segments of the myofibre. Small breaks in the sarcolemma do not always lead to myofibre necrosis, since such damage may be repaired quickly by membrane resealing and this critical balance is important to consider. Interventions to increase mechanical strength at the sarcolemma or to increase sarcolemma resealing (both designed to reduce necrosis of dystrophic myofibres and thereby avoid the process of regeneration), have attracted much interest recently with the recognition that these can be influenced by various molecules both inside and outside the myofibre. The precise sequence of events that initiates sarcolemmal damage and the influx of Ca ions that can result in myofibre necrosis especially in dystrophic muscles, are still unclear (Whitehead et al., 2006). The involvement of $\alpha7\beta1$ integrin that binds laminin (Boppart et al., 2006), ADAM12 (Moghadaszadeh et al., 2003), inflammatory cells and inflammatory cytokines (especially TNFα) (Hodgetts et al., 2006; Radley and Grounds, 2006) emphasize the importance of extracellular factors in myofibre susceptibility to necrosis.

Once muscle damage has progressed to myofibre necrosis, the usual sequence of events for repair occurs (see Fig. 2) that includes inflammation (and phagocytosis) within the first few days, revascularization, myogenesis with formation of new muscle (usually completed within about 7–10 days after injury), and subsequent maturation, innervation and restoration of full function that can take months (Jarvinen et al., 2005). Transcription profiling of ECM gene expression during regeneration of mouse muscle (injured by cardiotoxin) reflects these main cellular events (Goetsch et al., 2003). Similarly, the close association during muscle breakdown and regeneration of changes in expression patterns of genes associated with inflammation, ECM composition and new muscle formation, is emphasised by microarray analysis of dystrophic muscles from young boys with DMD (Pescatori et al., 2007). The crucial role of ECM in all aspects of formation of new tissues has also attracted tremendous interest from the perspective of bioscaffolds for tissue engineering (Shin et al., 2003) with potential applications for construction of muscle tissue (Hill et al., 2006). The roles of various ECM components during the different aspects of muscle regeneration are outlined below.

3.1 Inflammation

It is essential to appreciate that if perturbation of a factor impairs inflammation and phagocytosis of necrotic tissue, this alone will inevitably result in impaired new muscle formation, without necessarily any direct effect of the factor on myogenesis *per se*. This dependent sequence of events is sometimes not considered when interpreting in vivo data. Degranulation of mast cells and rapid accumulation of neutrophils are early events after any muscle damage; these produce chemotactic signals that attract other inflammatory cells such as macrophages that have many functions including phagocytic removal of necrotic tissue and stimulation of angiogenesis [for a more detailed discussion of inflammation see Chapter 12 (Tidball, 2007). Many of the events of inflammation are influenced by ECM components.

One of the earliest events of inflammation involves neutrophils leaving the vasculature and entering damaged tissues, this relies on adhesion molecules and integrin signaling (Molteni et al., 2006). The expression of integrins that bind the basement membrane components laminin and collagen IV seem especially important, since reduced amounts of these ECM proteins are associated with the gaps through which neutrophils penetrate the basement membrane and transmigrate from the venules into skeletal muscle: it seems likely that proteases produced by the neutrophils

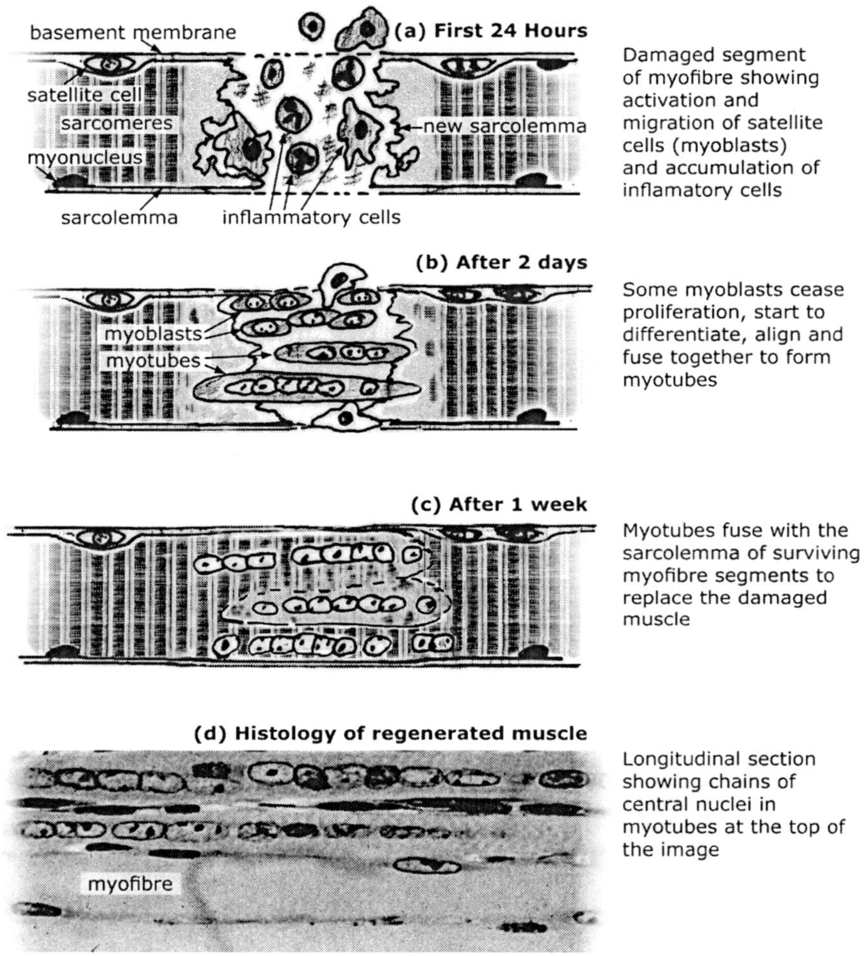

Figure 2. **Sequence of main cellular events during skeletal muscle regeneration** Myofibre necrosis that stimulates regeneration and new muscle formation is shown here in a longitudinal view as focal damage (where only a segment of the myofibre is injured): this can result from mechanical injury, excessive exercise or muscular dystrophy. (a) An early response to damage is inflammation, with the accumulation of neutrophils (within minutes) and macrophages that are conspicuous by 24 hours. During the first day after injury many ECM components are degraded by proteases; throughout the regenerative process there are many changes in ECM composition (discussed in the text). The satellite cells (myoblasts) on the surface of the myofibre become activated (rapidly) and start to synthesise DNA (by 18–24 hrs) for cell proliferation; myoblasts probably also start to migrate towards the damaged zone (in response to chemotactic signals). The sarcolemma is repaired (by 12 hrs) to re-seal the surviving myofibre segments. (b) Fusion of myoblasts into myotubes first occurs from 2.5 to 3 days after damage: this is preceeded by a switch from proliferation to differentiation and alignment of myoblasts. (c) Fusion between myotubes and the sarcolemma of the sealed surviving myofibre segments, to complete myofibre regeneration, does not occur until after one week (Based on Papdimitriou et al., 1990 and Robertson et al., 1993). The myofibre then matures and requires innervation for further growth and

(such as elastase) degrade the basement membrane to facilitate this transmigration (Wang et al., 2006).

Chemotactic signals produced by some ECM components (fibronectin, collagen type IV, laminin, entactin) or proteolytic fragments (of fibronectin and laminin) can act on a wide range of cells in vitro (Grounds and Davies, 1996; Hallmann et al., 2005), although in vitro chemotactic assays are notoriously difficult to extrapolate to the in vivo situation. In damaged muscles, proteases produced by macrophages, fibroblasts, endothelial cells and other cell types can rapidly break down ECM components into fragments which probably chemoattract leukocytes (and possibly myoblasts) to the injury site; these act in addition to chemoattractant cytokines.

The complexity of factors controlling many aspects of regeneration is illustrated by studies in mice where ECM protease activation systems are absent. For example, studies with knockout mice showed that uPA (but not tPA) dependent plasmin activation was required for efficient skeletal muscle regeneration and, accordingly, regeneration was accelerated in the absence of PAI-1. Since inflammation is impaired in the absence of uPA/plasmin this aspect alone would result in delayed/impaired myogenesis. Furthermore, plasmin is required for activation of MMPs and various growth factors and these aspects would also directly affect inflammation and angiogenesis and thus (indirectly) affect the kinetics of new muscle formation

Another factor to consider is hyaluronan, since this is deposited early in tissue repair and causes the attachment of leukocytes (Majors et al., 2003). However, the extent to which hyaluronan deposition might occur in skeletal muscle has not been investigated. The interacting role of other molecules that contribute to the inflammatory response including fibronectin, tenascin and the multitude of proteases and their inhibitors, is beyond the scope of this review.

3.2 Revascularization

3.2.1 Importance of revascularization and angiogenesis for new muscle formation

The ECM of skeletal muscle is filled with many blood vessels and this good blood supply is essential to provide the oxygen and energy required for muscle contraction. In minor injury there may be minimal disruption to this vasculature, whereas severe injury with extensive damage to the blood supply, or muscle transplantation, requires the formation of new blood vessels (angiogenesis) before regeneration can occur.

Figure 2.(Continued) functional contractilion. (d) Longitudinal section of recently regenerated skeletal muscle stained with haematoxylin and eosin, shows newly formed long myotubes with many central nuclei: these will mature into myofibres. Notes: (i) All of these cellular events will be delayed where the blood supply is damaged and re-vascularisation is required and (ii) myotube formation will be delayed/prevented when inflammation is impaired

Elegant studies in whole muscle grafts in mice, demonstrate that the inflammatory cells infiltrate damaged muscle and precede (slightly) the formation of new blood vessels (Roberts and McGeachie, 1990). Neovascularization is a limiting factor and, in the absence of revascularization, new muscle formation cannot progress. Extensive scar tissue formation results where revascularization is impaired, or large areas of muscle are damaged and ischaemia is prolonged: thus conditions that enhance new blood vessel growth are of considerable interest for effective muscle repair (Smythe et al., 2002). While the signals controlling angiogenesis e.g VEGF (Vascular Endothelial Growth Factor) have been extensively studied in many situations such as tumour formation and disease, there have been relatively few studies of the molecular control of angiogenesis in skeletal muscle. In old animals it seems that delayed revascularization of damaged muscles may result in part from sub-optimal stimulation by intrinsic signals produced by the aged muscle itself (Smythe & Grounds, unpublished data).

3.2.2 The endothelial cell

Basement membranes show distinct structural and functional differences due to isoforms of their molecular components that vary between different types of vascular endothelium e.g between capillaries and venules. In addition, other ECM molecules, such as SPARC, thrombospondins 1 and 2, fibronectin, nidogens 1 and 2, and collagen types VIII, XV, and XVIII are differentially expressed with the endothelium type and in various conditions (Hallmann et al., 2005). These regional variations need to be considered when evaluating data from different situations and assessing changes in the ECM related to new blood vessel formation associated with skeletal muscle regeneration.

3.2.3 ECM control of angiogenesis

A prerequisite for new blood vessel formation is localized breakdown of the basement membrane and the interstitial ECM, this precedes the proliferation and migration of capillary endothelial cells into the surrounding tissue and formation of new vessels. The cell-ECM interactions during blood vessel formation have been elegantly reviewed recently (Hallmann et al., 2005). Proteolytic degradation of the ECM is a critical component of endothelial invasion in angiogenesis, and therefore metalloproteinases and molecules that regulate MMPs and TIMPs are important players in neovascularisation (Carmeli et al., 2004). For example, heparanase is implicated in angiogenesis but since the enzyme is stored in a stable form intracellularly, whereas its heparan sulfate substrate is localized extracellularly, this suggests the existence of mechanisms that trigger heparanase secretion to control the ECM degradation that facilitates endothelial cell migration (Shafat et al., 2006). The balance between MMPs and TIMPs regulates the ECM turnover and remodeling, but TIMPs appear to play an additional much more complex role during angiogenesis and tumor progression (Chirco et al., 2006)

Heparan sulphate proteoglycans play critical roles in the growth of new blood vessels (demonstrated especially in tumours and cardiovascular diseases) since a plethora of growth factors, receptors, ECM molecules and enzymes bind to specific sites on the HS chains. For example, HS proteoglycans have profound effects on the bioactivity of the key angiogenic factor VEGF, affecting its diffusion, half-life and interaction with its tyrosine kinase receptors (Stringer, 2006). In contrast with all other known angiogenic factors, the proliferative action of VEGF seems to be restricted almost entirely to endothelial cells. Presumably similar events control angiogenesis in skeletal muscle repair. Some other ECM components that are implicated in angiogenesis include versican, hyaluronan, thrombospondin-1 (a potent inhibitor of angiogenesis), tenascin and molecules that interfere with integrin β1 (Hallmann et al., 2005). Semaphorins influence the affinity between integrins and their ECM ligands and secretion of class 3 semaphorin by non-endothelial tissues cooperates with VEGF to regulate endothelial cell vasculogenesis and also blood vessel sprouting through tissue boundaries (Bussolino et al., 2006).

3.3 ECM Impact on Myogenesis and Response to Muscle Damage

3.3.1 Tissue culture studies

Probably the earliest report to describe the influence of ECM on myogenesis was a tissue culture study by Hauschka and Konigsberg in 1966 who showed that a collagen substrate improved proliferation of embryonic chick myoblasts (Hauschka and Konigsberg, 1966). Since this time, cultured myoblasts have been routinely grown on type I collagen or denatured collagen (gelatin). Extrapolation of such observations to the in vivo situation should consider that, while during development myoblasts may be exposed to type I collagen in the interstitial connective tissue, this may not be the case in adult muscle after minor damage where the myoblasts lie in the niche environment beneath the basement membrane. Subsequently a basement membrane-like substrate called Matrigel was shown to enhance myoblast attachment (but not proliferation) and the later events of myogenesis especially the maintenance of myotubes (Hartley and Yablonka-Reuveni, 1990). It was suggested that mechanical tension contributed to the longevity of the myotubes, since this had also been reported in collagen gels [see (Hartley and Yablonka-Reuveni, 1990)]. Other groups had previously reported that laminins selectively enhance myoblast proliferation and differentiation but did not address the issue of myotube longevity [see (Hartley and Yablonka-Reuveni, 1990)]. Tissue culture studies are very useful to help define the role of specific molecules under defined conditions, yet myoblasts are often grown in the absence of fibroblasts although clearly both cell types (and others) contribute ECM components in vivo with important consequences. In vitro studies have identified various matrix components produced by muscle cells alone with an early tissue culture study showing that mouse G-8 skeletal muscle cells produced a protein that was probably fibronectin, in addition to basement membrane components (Beach et al., 1982). The role of mechanosignaling for maintenance of myotubes also needs to be considered (see Section 2.2).

Interactions between the ECM and growth factors were not usually appreciated in early tissue culture studies [for a recent review of GF during skeletal muscle regeneration, see Chapter 6 (Shefer and Yablonka-Reuveni 2007)]. There is now wide recognition of the critical importance of components of ECM binding for modulating GF action. For example HS proteoglycans interact with growth factors such as FGFs, Hepatocyte Growth Factor (HGF) and TGF-β. IGF-I and IGF-II are bound by a range of extracellular proteins that include vitronectin that is an important modulator of IGF-1 bioavailability (Kricker et al., 2003); the large family of IGF-1 binding proteins (IGF-BPs) in the ECM that vary with age and gender (Fiorotto et al., 2006); MMPs that modulate the interaction between these binding proteins and IGFs; and hormones that affect IGF-1 signalling. During pregnancy several MMPs are elevated that are proteases for IGF-BPs: one of these is PAPP-A (Pregnancy-Associated Plasma Protein–A) that cleaves IGFBP-2 but not IGFBP-3 to regulate IGF-1 bioavailability and growth (Kumar et al., 2005); another is ADAM12-S that cleaves IGFBP-5 and IGFBP-3 (Loechel et al., 2000). These observations raise intriguing questions about the impact of such additional proteases (PAPP-A and ADAM12-S that modify IGF-1 availability and myogenesis) on regeneration of injured muscles in pregnant mothers! Further complications come with ageing and changing levels of hormones, such as the anti-ageing hormone klotho that inhibits IGF-1 and insulin signaling (Kurosu et al., 2005). Clearly, ECM molecules combined with circulating proteins in the extracellular milieu play a major role in the availability of growth factors for receptor binding and activation of functional signaling.

A tissue culture study of the combined effects of various ECM substrates (gelatin, Matrigel, collagen IV, entactin-free laminin, fibronectin) with different GFs (FGF-2, PDGFs, TGF-β, LIF) on primary mouse muscle cultures found little impact of the ECM on myoblast proliferation (although laminin appears to be mitogenic). However, a marked influence of the ECM on myoblast fusion and myotube formation was observed, with Matrigel and fibronectin having the most benefit (Maley et al., 1995). This tissue culture study also emphasized strain-specific differences between the response of mouse myoblasts to the ECM substrates. Earlier studies using the MM14Dy adult mouse myogenic cell line or primary cultures of foetal mouse muscle grown on laminin (that lacked nidogen) showed enhanced myoblast elongation, motility and proliferation compared with cells grown on fibronectin or collagen (Ocalan et al., 1988). Different results between studies might be due to variations in the preparation and composition of the laminin substrates, the types of myogenic cells used and whether fibroblasts were present. Fibroblasts as well as muscle (and other) cells secrete a range of ECM components, yet more information is required regarding the profile of ECM molecules produced specifically by myoblasts and myotubes. Subsequent tissue culture studies in our laboratory (using C2C12 mouse myogenic cells and primary muscles cultures from C57Bl/Bl and Balb/c mice) with a wider range of ECM substrates (including laminin-1 and vitronectin) demonstrated enhanced attachment of myoblasts to collagen I, fibronectin and vitronectin, resulting in increased cell numbers despite no effect on proliferation (White, White and Grounds, unpublished). Fibronectin had previously been shown to promote the attachment and elongation of myoblasts and it is important to note that interactions between fibronectin and other ECM substrates result in conformational changes in fibronectin that influence the effects on C2C12 muscle cells (Garcia et al., 1999). Many studies support the idea that loss of fibronectin from the surface of myoblasts is required for fusion to occur: fibronectin degradation may be due to the protease m-calpain that is secreted into the ECM by muscle cells (Dourdin et al., 1997) or by MT1-MMP (Ohtake et al., 2006). The involvement of other MMP related molecules in fibronectin degradation and myoblast fusion is discussed in Section 2.4. It is concluded that the conditions required for myoblast differentiation and myotube formation appear to be controlled by 2 checkpoints, one that involves the muscle regulatory transcription factors (MRFs) and expression of myogenin, and a second that involves dynamic ECM remodelling (Ohtake et al., 2006). For example, decreased RECK results in elevated MT1-MMP to degrade fibronectin (also MMP-2 increases and TIMP-2 decreases) prior to myoblast fusion. A general trend is that many ECM components have little effect on myoblast proliferation, but do enhance myoblast attachment, myoblast differentiation, myotube formation and maturation.

Beyond the MMPs, many other ECM proteases influence myogenesis. For eample, ADAM12 mediates myoblast fusion through cell attachment involving integrin $\alpha9\beta1$ (Lafuste et al., 2005) and effects of components of the PA system on tissue cultured myoblasts have been intensively investigated [see Section 2.3 and refs in (Suelves et al., 2005)]: in brief, uPA and plasmin enhance differentiation and myogenesis and the integrated function of the complex PA system is required for myogenesis and also for growth factor (FGF-2, TGF-β and HGF) dependent proliferation and migration of satellite cells.

Early studies of GAGs on tissue cultured myoblasts (from chick embryos) identified a key role for hyaluronan as an inhibitor of myoblast differentiation, with no effect on myoblast attachment or proliferation (Elson and Ingwall, 1980; Kujawa et al., 1986). The importance of hyaluronan for differentiation and fusion was reflected by (transient) decreased synthesis prior to fusion in primary cultures of embryonic chick muscle (Yoshimura, 1985). Surprisingly, since this time hyaluronan seems to have received little attention, yet hyaluronon is rapidly synthesized by many cells in response to ER stress and accumulates during early events of inflammation (Hascall et al., 2004): this timing might correlate with inhibition of myoblast differentiation during regeneration of post-natal muscle. In order to affect myogenesis in adult muscle, the hyaluronan would need to be either made by the myogenic cells (the most likely scenario) or else available in the immediate environment of satellite cells/myoblasts located beneath the basement membrane in adult muscle, yet this crucial in vivo localization does not appear to have been examined.

Heparan sulphate proteoglycans have received much attention especially since recognition of their important regulation of GF activity, with myoblasts also demonstrating an absolute requirement for HS binding for high affinity interactions between FGF-2 and its receptor (Fuentealba et al., 1999; Rapraeger et al., 1991). Many in vitro studies using C2C12 myoblasts, primary cultures or isolated myofibres (Cornelison et al., 2001) have demonstrated the critical role of proteoglycan sulphation for normal myogenesis, with expression of perlecan, syndecan-1 and syndecan-3 being downregulated during myoblast differentiation and fusion, whereas glycipan-1 is upregulated in cultured C2C12 myoblasts [reviewed in (Casar et al., 2004a)]. The overall conclusion is that modulation of HS is essential for normal myogenesis to proceed and the many factors that can influence GAG composition and levels need to be considered.

Decorin appears to have no direct affect on myogenesis (as might be expected since it is not basement membrane associated) but it binds to TGF-β and myostatin to prevent function of these GF and thus can modulate myogenesis in vitro: whereas myostatin binds only to decorin, both TGF-β and TNFα bind decorin and bigylcan (Miura et al., 2006). The important modulation of TGF-β signaling by decorin, biglycan and betaglycan during myogenesis was demonstrated by studies on cultured C2C12 and decorin-null myoblasts (Droguett et al., 2006). Increased decorin and changes in the GAG sidechains of decorin are implicated in the switch of myogenic cells into the osteogenic lineage induced by bone morphogenic protein-2 (Gutierrez, 2006), illustrating the crucial role of such HS proteoglycans in cell lineage determination and probable stem cell niches.

Tissue culture is unable to fully reflect the much more complex in vivo situation that is the ultimate environment that we wish to understand and potentially manipulate. This can only be addressed by studying myogenesis in animal models where adult muscle is regenerating in response to experimental damage, muscle transplantation, or diseases like muscular dystrophy.

3.3.2 Muscle regeneration in animal models

Classical studies of muscle regeneration have used adult mice, rats and amphibians, but novel information is now being generated from smaller animals that are more readily manipulated genetically including zebra fish, worms (c.elegans) and fruit fly.

In adult mammalian skeletal muscle, most myoblasts are derived from satellite cells that lie on the sarcolemma beneath the basement membrane, these precursor cells are normally quiescent, but are activated in response to injury (and also growth). It is worth mentioning that since 1999, much attention was focused on the possible contribution of stem cells originating beyond satellite cells as an alternative source of myoblasts for muscle regeneration (Grounds et al., 2002). However, the enthusiasm for non-myogenic stem cells as a major player has abated and the satellite cells has returned to favour as the likely source of most myoblasts for repair of skeletal muscle (Zammit et al., 2006). The issues to address regarding myoblasts and muscle regeneration are; what are the factors that maintain the quiescent state of satellite cells, activate satellite cells, enhance satellite cell (myoblast) proliferation, or cause these myoblasts to cease proliferation and undergo differentiation that results in fusion to form myotubes? The in vivo experiments to address this fall into 3 main categories: descriptive histological and immunhistochemical studies of ECM changes during (i) normal muscle regeneration, or (ii) in genetically engineered mice that lack or over-express specific ECM components, and (iii) administration or inhibition/modification of various ECM molecules to influence the process of muscle regeneration.

Since the 1960s, immunohistochemistry has been used to describe the changing pattern of ECM components during skeletal muscle development and regeneration [see references in (Gulati et al., 1983a; Gulati et al., 1983b; Grounds et al., 1998; Huijbregts et al., 2001; Sorokin et al., 2000)]. The basic pattern that emerges (based on the above papers) is summarised briefly below and is focused on myogenic cells rather than revascularization or re-innervation. (see Table 1). It is noted that the precise timing of this pattern will differ with the nature/severity of the injury and also the species or strain of animal used.

Fibronectin is pronounced at all times in the interstitial connective tissue and is increased throughout the necrotic and regenerating zones of damaged muscles. Fibronectin is not normally a component of basement membranes, but it is not clear whether fibronectin penetrates beneath the basement membrane of damaged myofibres or whether myoblasts in vivo express fibronectin. Consequently, whether fibronectin must be decreased locally to facilitate myotube formation in regenerating adult muscles, as has been described for tissue cultured and developing muscle, does not seem to have been investigated. Similarly, the inhibitory role proposed for hyaluronan in myoblast fusion (see Section 3.3.1) has not been explored in regenerating adult skeletal muscle.

Little tenascin-C is present in uninjured muscles but the levels increase by about 3 days after damage, often associated with desmin positive cuffing cells presumed to be activated satellite cells (lying within the contour of the basement membrane of necrotic myofibres) and with myotube formation (Huijbregts et al., 2001). Tenascin-C is weak and patchy around very young myotubes becoming stronger and more continuous around older myotubes, but it is not present around mature myotubes and myofibres. Muscle regeneration has not been studied in tenascin-C null mice (that show apparently normal myogenesis during development (Mackie and Tucker, 1999)) and thus the precise importance of tenascin-C during the various in vivo events associated with new muscle formation in response to post-natal muscle damage remains to be defined.

While the basement membrane persists as a scaffold around damaged myofibres, molecules like collagen IV start to disappear by day 1 although some weak patchy distribution persists (Gulati et al., 1983b). (The degradation of these ECM components may be chemotactic to a wide range of cells including myoblasts). In newly formed myotubes, collagen IV is rapidly re-expressed and is pronounced but uneven by day 5 with more uniform distribution by 7 days after injury. Laminin $\alpha2$ in the basement membrane of mature myofibres also decreases after injury but is seen around young myotubes and increases with maturation, similar to collagen IV, as would be expected for formation of new basement membrane.

Increased production of collagen IV and laminins following myotube formation is confirmed by quantitation of matrix expression in primary cultures of mouse muscle (Huijbregts et al., 2001). Laminin α4 occurs weakly and transiently on some mature myotubes and laminin 5α appears to be cytoplasmic in young myotubes but it can be stronger (and transient) around older myotubes. Neither of these isoforms are present on mature myofibres even in the absence of laminin α2 (see Section 2.2). These basement membrane-associated molecules are clearly of critical important for normal mature muscle function (Grounds et al., 2005) but seem to be of lesser importance for myogenesis and myotube formation.

Proteoglycans in skeletal muscle have also received much attention (see Section 2.3 and 3.3.1) although some may play no direct role in myogenesis. For example, decorin is upregulated in the interstitial connective tissue of regenerating muscle and is of critical importance for GF interactions and for sequestering TGF-β but it is not normally in intimate contact with satellite cells/myoblasts (Brandan et al., 1991; Droguett et al., 2006). However, basement membrane and cell surface associated proteoglycans such as biglycan are very important for normal muscle function and it seems likely that they may play a direct role during myogenesis in vivo. Studies of muscle regeneration in normal and biglycan null mice showed a dramatic transient increase in biglycan during myoblast differenation and myotube formation supporting a direct role for it during myogenesis (Casar et al., 2004b). Yet biglycan is not necessary for skeletal muscle regeneration since myotube formation was not affected by the absence of biglycan, although small difference in myotube growth were noted and it seems likely that other proteoglycans may compensate for this defect.

In contrast, cell surface associated HS proteoglycans such as syndecan-4 and syndecan-3 appear to play a direct and important role in myogenesis in vivo: they are rich on the surface of myofibres and satellite cells (Casar et al., 2004a; Cornelison et al., 2001). Null mice that lack syndecan-4 or syndecan-3 have apparently normal muscle development but there are abnormalities in adult mice (being more severe for syndecan-3 nulls), striking differences between cultured myoblasts and isolated myofibres (with syndecan-4 null muscles showing major impairment of satellite cell activation) and a major disturbance of muscle regeneration in syndecan-4 null mice (Cornelison et al., 2004). It seems likely that other cell surface HS proteoglycans such as syndecan-1, glypican and perlecan are also directly involved in myogenesis in vivo during muscle regeneration (Casar et al., 2004a).

The important role of HS proteoglycans in modulating many aspects of muscle regeneration is further emphasised by systemic in vivo administration of synthetic proteoglycans called RGTAs (ReGeneraTing Agents) that raise the possibility of therapeutic enhancement of new muscle formation. The RGTAs appear to have remarkable benefits on the healing of many tissues following injury and ischaemia; including skeletal muscle (Barbosa et al., 2005; Zimowska et al., 2001). These synthetic HS proteoglycans are substituted dextrans that are extremely stable and probably enhance the bioavailibility of heparin binding growth factors in vivo (Meddahi et al., 2002). The proteoglycan decorin has also been administered locally into skeletal muscle with the aim of reducing fibrosis (Sato et al., 2003) as discussed below (see Section 3.5). These seem the main in vivo examples of ECM administration into skeletal muscles.

Proteases and their inhibitors undoubtedly play a very important role in myogenesis (see Section 2.4) and this is also true in regenerating muscles. The expression of MMP-2 and MMP-9 has been well described in regenerating and dystrophic mouse muscle (Kherif et al., 1999) where it was concluded that MMP-9 is related mainly to inflammation and probably the activation of satellite cells, whereas MMP-2 activation is high between 3 and 10 after injury, with maximal activity at 7 days, supporting a role for MMP-2 in myoblast proliferation and fusion (Kherif et al., 1999). The role of these and other proteases in tissue remodeling has been intensively investigated in mice that are deficient or over-express a specific protease. The application of this approach to skeletal muscle regeneration is illustrated for the plasminogen activation system using a range of null mice (Suelves et al., 2005): these in vivo studies demonstrate profound (probably indirect) effects on new muscle formation, with opposite roles for uPA and its inhibitor PAI-1. As emphasized, interpretation of the specific impact of protease activity on myogenesis in vivo is complicated by the crucial role of such proteases and of associated matrix degradation for other early events of tissue repair, such as inflammation and angiogenesis that precede new muscle formation.

3.4 Innervation, Maturation and Maintenance of Muscle Mass and Function

The function and maintenance of skeletal muscle mass requires innervation, and therefore any factors that impair re-innervation will impair the overall maturation and restoration of full function of regenerated muscle. It needs to be re-iterated that long-term observations (weeks and months) that demonstrate failure of muscle regeneration, may purely reflect inadequate innervation. Instead, it is sometimes wrongly inferred that impaired long-term regeneration reflects some problem with the efficiency of early events (during the first week) of inflammation, revacularisation and myogenesis *per se*.

In older animals, failure to maintain normal muscle innervation contributes to age-related muscle wasting (sarcopenia) and also to inefficiencies in muscle re-innervation (Shavlakadze and Grounds, 2003). Re-innervation of damaged muscle is further compromised in aged animals due to greater amounts of interstitial ECM that may physically interfere with nerve sprouts making effective contacts with regenerated myofibres, plus age-related changes in ECM composition that may retard the establishment and maintenance of new synapses (Grounds et al., 1998). Numerous studies have described ECM molecules associated with synapse formation during development and in various disorders but there has been relatively little work on ECM changes related to nerve sprouting and re-establishment of neuromuscular junctions in regenerating adult muscle. This is beyond the scope of this review but is discussed in Chapter 14 (Slater and Schiaffino, 2007).

3.5 Pathological Deposition of ECM and Fibrosis

With every cycle of regeneration it seems that the deposition of ECM, especially interstitial collagen, is increased slightly. Thus repeated injury and regeneration results in an accumulation of ECM and, over time, this altered environment probably adversely affects the events of regeneration and myogenesis and instead skews conditions to favour the deposition of fibrous (and fatty) tissue: this is pronounced in Duchenne muscular dystrophy.

Age also increases the deposition of ECM and it is well documented that increasing fibrosis occurs in regenerating muscles of older animals [reviewed in (Grounds et al., 1998)]. Accumulating molecular modifications to the cross-links of collagens (that turn over very slowly), with the dramatic acceleration of these changes in diabetes, leads to stiffness and impaired function in aged animals (Avery and Bailey, 2005). In skeletal muscles, there is also evidence that basement membrane can accumulate to form additional layers around myofibres and satellite cells in very old animals [reviewed in (Grounds et al., 1998)]: clearly this can create difficulties for signaling and also mechanical barriers for re-innervation.

Where inflammation or revascularization is delayed, or the process of myogenesis fails for other reasons, excessive fibrous connective tissue is deposited instead. Fibrosis (associated with fibroblast proliferation) is widely recognized as a severe complication of diseases in many tissues and intensive research is focused on interventions to reduce and ideally reverse this process: in brief, fibrosis is exacerbated by inflammation and proteases and one of the key cytokines that favours fibrosis is TGF-β (Chua et al., 2005). Levels of TGF-β are increased in many muscle diseases where fibrosis occurs and decorin is considered an anti-fibrotic agent (Zanotti et al., 2005). For example, in mouse skeletal muscles regenerating after experimental injury, decorin reduces the fibrosis that normally occurs by 2 weeks after injury (Sato et al., 2003). However, the situation is complicated and decorin does not seem to be effective as an anti-fibrotic agent in DMD, where increased decorin (and biglycan) are present yet fibrosis becomes pronounced (Fadic et al., 2006). Studies with the anti-fibrotic drug Pirfenidone resulted in a small but significant reduction in the level of hydroxyproline (a measure of collagen synthesis) in dystrophic muscles in mdx mice, and also revealed dissimilarities in collagen metabolism between functionally different skeletal muscles (Gosselin et al., 2007). Gamma interferon is also considered a potent anti-fibrotic agent (Foster et al., 2003). Disturbances of the plasminogen-activator inhibitor system increase fibrosis (Dellas and Loskuttof, 2005) and may contribute to the fibrosis in dystrophic muscle (Lopez-Alemany et al., 2003) and indirectly impact on myogenesis. In damaged mouse muscle where uPA/plasmin is decreased, the accumulation of fibrin impairs muscle regeneration, and the adverse effects on regeneration are reversed by the drug ancrod that reduces fibrinogen and hence fibrin accumulation (Suelves et al., 2005).

Other drugs can enhance fibrosis and impair muscle regeneration as has been demonstrated for the non-steroidal anti-inflammatory drug NS-398 (a cyclooxygenase-2-specific inhibitor) that resulted in delayed new muscle formation, associated with increased TGF-β1 and fibrosis and also reduced inflammation (Shen et al., 2005). Such studies emphasise the crucial and complex balance between ECM deposition and composition that influences the capacity to form new functional muscle in response to tissue damage.

4. CONCLUDING COMMENTS

To date the vast majority of investigations have focused on the impact of growth factors, MRFs and signaling cascades within muscle cells when evaluating factors that control myogenesis and skeletal muscle regeneration. Scant attention has been paid to the role of the ECM, with some notable exceptions. The extraordinary complexity and wealth of interacting systems in the extracellular in vivo environment that modulate cell behaviour is daunting to contemplate and many tools to investigate these systems have only recently become available. The numerous components that modulate the world of the ECM provide the opportunity for potential development of many new drugs and there seems little doubt that the importance of the exquisite control of cellular behaviour by ECM molecules will become increasing recognized.

ACKNOWLEDGEMENTS

I sincerely thank my various research colleagues for their support in unfunded explorations of the ECM and skeletal muscle over many years. In particular I thank Deirdre Coombe (Curtin University, Western Australia) and Lydia Sorokin (University of Muenster, Germany) for their critical reading of this manuscript, and Haslett Grounds for his assistance with the Figures.

REFERENCES

Ai X, Do AT, Kusche-Gullberg M, Lindahl U, Lu K, Emerson CP, Jr (2006) Substrate specificity and domain functions of extracellular heparan sulfate 6-O-endosulfatases, QSulf1 and QSulf2. J Biol Chem 281:4969–4976

Ai X, Do AT, Lozynska O, Kusche-Gullberg M, Lindahl U, Emerson CP, Jr (2003) QSulf1 remodels the 6-O sulfation states of cell surface heparan sulfate proteoglycans to promote Wnt signaling. J Cell Biol 162:341–351

Aumailley M, Bruckner-Tuder man L, Carter WG, Deutzmann R, Edgar D, Ekblom P, Engel J, Engvall E, Hohenester E, Jones JC, Kleinman HK, Marinkovich MP, Martin GR, Mayer U, Meneguzzi G, Miner JH, Miyazaki K, Patarroyo M, Paulsson M, Quaranta V, Sanes JR, Sasaki T, Sekiguchi K, Sorokin LM, Talts JF, Tryggvason K, Uitto J, Virtanen I, von der Mark K, Wewer UM, Yamada Y, Yurchenco PD (2005) A simplified laminin nomenclature. Matrix Biol 24:326–332

Avery NC, Bailey AJ (2005) Enzymic and non-enzymic cross-linking mechanisms in relation to turnover of collagen: relevance to aging and exercise. Scand J Med Sci Sports 15:231–240

Barbosa I, Morin C, Garcia S, Duchesnay A, Oudghir M, Jenniskens G, Miao HQ, Guimond S, Carpentier G, Cebrian J, Caruelle JP, van Kuppevelt T, Turnbull J, Martelly I, Papy-Garcia D (2005) A synthetic glycosaminoglycan mimetic (RGTA) modifies natural glycosaminoglycan species during myogenesis. J Cell Sci 118:253–264

Barresi R, Campbell KP (2006) Dystroglycan: from biosynthesis to pathogenesis of human disease. J Cell Sci 119:199–207
Beach RL, Burton WV, Hendricks WJ, Festoff BW (1982) Extracellular matrix synthesis by skeletal muscle in culture. Proteins and effect of enzyme degradation. J Biol Chem 257:11437–11442
Belkin AM, Stepp MA (2000) Integrins as receptors for laminins. Microscopy Res and Technique 51:280–301
Bernal F, Hartung HP, Kieseier BC (2005) Tissue mRNA expression in rat of newly described matrix metalloproteinases. Biol Res 38:267–271
Bishop JR, Schuksz M, Esko JD (2007) Heparan sulphate proteoglycans fine-tune mammalian physiology. Nature 446:1030–1037
Blobel CP (2005) ADAMs: key components in EGFR signalling and development. Nature 6:32–43
Bonnemann CG, Laing NG (2004) Myopathies resulting from mutations in sarcomeric proteins. Curr Opin Neurology 17:529–537
Boppart MD, Burkin DJ, Kaufman SJ (2006) Alpha 7 beta 1-integrin regulates mechanotransduction and prevents skeletal muscle injury. Am J Physiol Cell Physiol 290:C1660–1665
Bortoluzzi S, Scannapieco P, Cestaro A, Danieli GA, Schiaffino S (2006) Computational reconstruction of the human skeletal muscle secretome. Proteins 62:776–792
Brandan E, Fuentes ME, Andrade W (1991) The proteoglycan decorin is synthesized and secreted by differentiated myotubes. Europ J Cell Biology 55:209–216
Bussolino F, Valdembri D, Caccavari F, Serini G (2006) Semaphoring vascular morphogenesis. Endothelium 13:81–91
Carmeli E, Moas M, Reznick AZ, Coleman R (2004) Matrix metalloproteinases and skeletal muscle: a brief review. Muscle Nerve 29:191–197
Caruelle D, Mazouzi Z, Husmann I, Delbe J, Duchesnay A, Gautron J, Martelly I, Courty J (2004) Upregulation of HARP during in vitro myogenesis and rat soleus muscle regeneration. J Muscle Res Cell Motil 25:45–53
Casar JC, Cabello-Verrugio C, Olguin H, Aldunate R, Inestrosa NC, Brandan E (2004a) Heparan sulfate proteoglycans are increased during skeletal muscle regeneration: requirement of syndecan-3 for successful fiber formation. J Cell Sci 117:73–84
Casar JC, McKechnie BA, Fallon JR, Young MF, Brandan E (2004b) Transient up-regulation of biglycan during skeletal muscle regeneration: delayed fiber growth along with decorin increase in biglycan-deficient mice. Dev Biol 268:358–371
Chan J, O'Donoghue K, Gavina M, Torrente Y, Kennea N, Mehmet H, Stewart H, Watt DJ, Morgan JE, Fisk NM (2006) Galectin-1 induces skeletal muscle differentiation in human fetal mesenchymal stem cells and increases muscle regeneration. Stem Cells 24:1879–1891
Chinni C, de Niese MR, Jenkins AL, Pike RN, Bottomley SP, Mackie EJ (2000) Protease-activated receptor-2 mediates proliferative responses in skeletal myoblasts. J Cell Sci 113:4427–4433
Chiquet-Ehrismann R (2004) Tenascins. Int J Biochem Cell Biol 36:986–990
Chiquet M, Fluck M (2003) Ectopic expression of tenascin-C. J Cell Sci 116:3851–3853
Chirco R, Liu XW, Jung KK, Kim HR (2006) Novel functions of TIMPs in cell signaling. Cancer Metastasis Rev 25:99–113
Chua F, Sly PD, Laurent GJ (2005) Pediatric lung disease: from proteinases to pulmonary fibrosis. Pediatr Pulmonol 39:392–401
Coombe DR, Kett WC (2005) Heparin sulfate-protein interactions: therapeutic potential through structure function insights. Cell Mol Life Sci 62:410–424
Cooper DN, Massa SM, Barondes SH (1991) Endogenous muscle lectin inhibits myoblast adhesion to laminin. J Cell Biol 115:1437–1448
Cornelison DD, Wilcox-Adelman SA, Goetinck PF, Rauvala H, Rapraeger AC, Olwin BB (2004) Essential and separable roles for Syndecan-3 and Syndecan-4 in skeletal muscle development and regeneration. Genes Dev 18:2231–2236
Cornelison DDW, Filla MS, Stanley HM, Rapraeger AC, Olwin BB (2001) Syndecan-3 and syndecan-4 specifically mark skeletal muscle satellite cells and are implicated in satellite cell maintenance and muscle regeneration. Dev Biol 239:79–94

Dellas C, Loskuttof DJ (2005) Historical analysis of PAI-I from its discovery to its potential role in cell motility and disease. Thromb Haemost 93:631–640

Dourdin N, Balcerzak D, Brustis JJ, Poussard JJ, Cottin P, Ducastaing A (1999) Potential m-calpain substrates during myoblast fusion. Exp Cell Res 246:433–442

Dourdin N, Brustis J-J, Balcerzak D, Elamrani N, Poussard S, Cottin P, Ducastaing A (1997) Myoblast fusion requires fibronectin degradation by exteriorized m-calpain. Exp Cell Res 235:385–394

Droguett R, Cabello-Verrugio C, Riquelme C, Brandan E (2006) Extracellular proteoglycans modify TGF-beta bio-availability attenuating its signaling during skeletal muscle differentiation. Matrix Biol 25:332–341

Echizenya M, Kondo S, Takahashi R, Oh J, Kawashima S, Kitayama H, Takahashi C, Noda M (2005) The membrane-anchored MMP-regulator RECK is a target of myogenic regulatory factors. Oncogene 24:5850–5857

Elson HF, Ingwall JS (1980) The cell substratum modulates skeletal muscle differentiation. J Supramol Struct 14:313–328

Fadic R, Mezzano V, Alvarez K, Cabrera D, Holmgren J, Brandan E (2006) Increase in decorin and biglycan in Duchenne Muscular Dystrophy: role of fibroblasts as cell source of these proteoglycans in the disease. J Cell Mol Med 10:758–769

Fiorotto ML, Lopez R, Oliver WT, Khan AS, Draghia-Akli R (2006) Transplacental transfer of a growth hormone-releasing hormone peptide from mother to fetus in the rat. DNA Cell Biol 25:429–437

Foster W, Li Y, Usas A, Somogyi G, Huard J (2003) Gamma interferon as an antifibrosis agent in skeletal muscle. J Orthop Res 21:798–804

Frieser M, Nockel H, Pausch F, Roder C, Hahn A, Deutzmann R, Sorokin LM (1997) Cloning of the mouse laminin alpha 4 cDNA. Expression in a subset of endothelium. Eur J Biochem 246:727–735

Fuentealba L, Carey DJ, Brandan E (1999) Antisense inhibition of syndecan-3 expression during skeletal muscle differentiation accelerates myogenesis through a basic fibroblast growth factor-dependent mechanism. J Biol Chem 274:37876–37884

Garcia AJ, Vega MD, Boettiger D (1999) Modulation of cell proliferation and differentiation through substrate-dependent changes in fibronectin conformation. Mol Biol Cell 10:785–798

Gatchalian CL, Schachner M, Sanes J (1989) Fibroblasts that proliferate near denervated synaptic sites in skeletal muscle synthesize the adhesive molecules tenascin(J1),N-CAM,fibronectin and a heparin sulfate proteoglycan. J Cell Biol 108:1873–1890

Gelse K, Poschl E, Aigner T (2003) Collagens–structure, function, and biosynthesis. Adv Drug Deliv Rev 55:1531–1546

Goetsch SC, Hawke TJ, Gallardo TD, Richardson JA, Garry DJ (2003) Transcriptional profiling and regulation of the extracellular matrix during muscle regeneration. Physiol Genomics 14:261–271

Gosselin LE, Williams JE, Personius K, Farkas GA (2007) A comparison of factors associated with collagen metabolism in different skeletal muscles from dystrophic (mdx) mice: impact of pirfenidone. Muscle Nerve 35:208–216

Grounds MD (1998) Age-associated changes in the response of skeletal muscle cells to exercise and regeneration. Ann NYAcad Sci 854:78–91

Grounds MD, Davies MJ (1996) Chemotaxis in myogenesis. Basic Appl Myology 6:469–483

Grounds MD, McGeachie JK, Davies MJ, Sorokin L, Maley MAL (1998) The expression of extracellular matrix during adult skeletal muscle regeneration: how the basement membrane, interstitium, and myogenic cells collaborate. Basic Appl Myology 8:129–141

Grounds MD, Sorokin L, White J (2005) Strength at the extracellular matrix-muscle interface. Scand J Med Sci Sports 15:381–391

Grounds MD, White J, Rosenthal N, Bogoyevitch MA (2002) The role of stem cells in skeletal and cardiac muscle repair. J Histochem Cytochem 50:589–610

Gulati AK, Reddi AH, Zalewski AA (1983a) Changes in the basement membrane zone component during skeletal muscle fiber degeneration and regeneration. J Cell Biol 97:957–962

Gulati AK, Zalewski AA, Reddi AH (1983b) An immunofluorescent study of the distribution of fibronectin and laminin during limb regeneration in the adult newt. Dev Biol 96:355–365

Gullberg D, Tiger C-F, Velling T (1999) Laminins during muscle development and in muscular dystrophies. Cell Mol Life Sci 56:442–460

Gutierrez Jea (2006) Changes in cell secreted and cell associated proteoglycan synthesis during conversion of myoblasts to osteoblasts in response to bone morphogenetic protein-2: role of decorin in cell response to BMP-2. J Cell Physiol 206:58–67

Hagios C, Brown-Luedi M, Chiquet-Ehrismann R (1999) Tenascin-Y, a component of distinctive connective tissues, supports muscle cell growth. Exp Cell Res 253:607–617

Hallmann R, Horn N, Selg M, Wendler O, Pausch F, Sorokin LM (2005) Expression and function of laminins in the embryonic and mature vasculature. Physiol Rev 85:979–1000

Hartley RS, Yablonka-Reuveni Z (1990) Long term maintenance of primary myogenic cultures on a reconstituted basement membrane. In Vitro Cell Dev Biol 26:955–961

Hascall VC, Majors AK, De La Motte CA, Evanko SP, Wang A, Drazba JA, Strong SA, Wight TN (2004) Intracellular hyaluronan: a new frontier for inflammation? Biochim Biophys Acta 1673:3–12

Hauschka SD, Konigsberg IR (1966) The influence of collagen on the development of muscle clones. Zoology 55:119–126

Henriquez JP, Casar JC, Fuentealba L, Carey DJ, Brandan E (2002) Extracellular matrix histone H1 binds to perlecan, is present in regenerating skeletal muscle and stimulates myoblast proliferation. J Cell Sci 115:2041–2051

Hewitson TD, Martic M, Kelynack KJ, Pagel CN, Mackie EJ, Becker GJ (2005) Thrombin is a profibrotic factor for rat renal fibroblasts in vitro. Nephron Exp Nephrol 101:e42–e49

Hill E, Boontheekul T, Mooney DJ (2006) Designing scaffolds to enhance transplanted myoblast survival and migration. Tissue Eng 12:1295–1304

Hodgetts S, Radley H, Davies M, Grounds MD (2006) Reduced necrosis of dystrophic muscle by depletion of host neutrophils, or blocking TNFalpha function with Etanercept in mdx mice. Neuromuscul Disord 16:591–602

Houzelstein D, Goncalves IR, Fadden AJ, Sidhu SS, Cooper DN, Drickamer K, Leffler H, Poirier F (2004) Phylogenetic analysis of the vertebrate galectin family. Mol Biol Evol 21:1177–1187

Huijbregts J, White JD, Grounds MD (2001) The absence of MyoD in regenerating skeletal muscle affects the expression pattern of basement membrane, interstitial matrix and integrin molecules, consistent with delayed myotube formation. Acta Histochem 103:379–396

Iba K, Albrechtsen R, Gilpin B, Frohlich C, Loechel F, Zolkiewska A, Ishiguro K, Kojima T, Liu W, Langford JK, Sanderson RD, Brakebusch C, Fassler R, Wewer UM (2000) The cysteine-rich domain of human ADAM 12 supports cell adhesion through syndecans and triggers signaling events that lead to beta1 integrin-dependent cell spreading. J Cell Biol 149:1143–1156

Ingber DE (2006) Cellular mechanotransduction: putting all the pieces together again. FASEB J 20:811–827

Jarvinen TA, Jarvinen TL, Kaariainen M, Kalimo H, Jarvinen M (2005) Muscle Injuries. Biology and treatment. Am J Sports Med 33:745–764

Jenniskens GJ, Veerkamp JH, Van Kuppevelt TH (2006) Heparan sulfates in skeletal muscle development and physiology. J Cell Physiol 206:283–294

Kami K, Senba E (2005) Galectin-1 is a novel factor that regulates myotube growth in regenerating skeletal muscles. Curr Drug Targets 6:395–405

Kanagawa M, Michele DE, Satz JS, Barresi R, Kusano H, Sasaki T, Timpl R, Henry MD, Campbell KP (2005) Disruption of perlecan binding and matrix assembly by post-translational or genetic disruption of dystroglycan function. FEBS Lett 579:4792–4796

Khan ZA, Farhangkhoee H, Mahon JL, Bere L, Gonder JR, Chan BM, Unival S, Chakrabarti S (2006) Endothelins: regulators of extracellular matrix protein production in diabetes. Exp Biol Med 231:1022–1029

Kherif S, Lafuma C, Dehaupas M, Lachkar S, Fournier J-G, Verdiere-Sahuque M, Fardeau M, Alameddine HS (1999) Expression of matrix metalloproteinases 2 and 9 in regenerating skeletal muscle: a study in experimentally injured and mdx muscles. Dev Biol 205:158–170

Kjaer M (2004) Role of Extracellular Matrix in Adaptation of Tendon and Skeletal Muscle to Mechanical Loading. Physiol Rev 84:649–698

Kricker JA, Towne CL, Firth SM, Herington AC, Upton Z (2003) Structural and functional evidence for the interaction of insulin-like growth factors (IGFs) and IGF binding proteins with vitronectin. Endocrinology 144:2807–2815

Kujawa MJ, Pechak DG, Fiszman MY, Caplan AI (1986) Hyaluronic acid bonded to cell culture surfaces inhibits the program of myogenesis. Dev Biol 113:10–16

Kumar A, Mohan S, Newton J, Rehage M, Tran K, Baylink D, Qin X (2005) Pregnancy-associated plasma protein-A regulates myoblast proliferation and differentiation through an insulin-like growth factor-dependent mechanism. J Biol Chem 280:37782–37789

Kurosu H, Yamamoto M, Clark JD, Pastor JV, Nandi A, Gurnani P, McGuinness OP, Chikuda H, Yamaguchi M, Kawaguchi H, Shimomura I, Takayama Y, Herz J, Kahn CR, Rosenblatt KP, Kuro-o M (2005) Suppression of aging in mice by the hormone Klotho. Science 309:1829–1833

Lafuste P, Sonnet C, Chazaud B, Dreyfus PA, Gherardi RK, Wewer UM, Authier FJ (2005) ADAM12 and {alpha}9{beta}1 Integrin Are Instrumental in Human Myogenic Cell Differentiation. Mol Biol Cell 16:861–870

Larsen M, Artym VV, Green JA, UYamada KM (2006) The matrix reorganized: extracellular matrix remodeling and integrin signaling. Curr Opin Cell Biol 18:463–471

Lechner BE, Lim JH, Mercado ML, Fallon JR (2006) Developmental regulation of biglycan expression in muscle and tendon. Muscle Nerve 34:347–55

Liu FT, Rabinovich GA (2005) Galectins as modulators of tumour progression. Nat Rev Cancer 5:29–41

Lluri G, Jaworski DM (2005) Regulation of TIMP-2, MT1-MMP, and MMP-2 expression during C2C12 differentiation. Muscle Nerve 32:492–499

Loechel F, Fox JW, Murphy G, Albrechtsen R, Wewer UM (2000) ADAM 12-S cleaves IGFBP-3 and IGFBP-5 and is inhibited by TIMP-3. Biochem Biophys Res Commun 278:511–515

Lopez-Alemany R, Suelves M, Munoz-Canoves P (2003) Plasmin generation dependent on alpha-enolase-type plasminogen receptor is required for myogenesis. Thromb Haemost 90:724–733

Mackie EJ, Pagel CN, Smith R, de Niese MR, Song SJ, Pike RN (2002) Protease-activated receptors: a means of converting extracellular proteolysis into intracellular signals. IUBMB Life 53:277–281

Mackie EJ, Tucker RP (1999) The tenascin-C knockout revisited. J Cell Sci 112:3847–3853

Maile LA, Busby WH, Sitko K, Capps BE, Sergent T, Badley-Clarke J, Ling Y, Clemmons DR (2006) The heparin binding domain of vitronectin is the region that is required to enhance insulin-like growth factor-I signaling. Mol Endocrinol 20:881–892

Majors AK, Austin RC, de la Motte CA, Pyeritz RE, Hascall VC, Kessler SP, Sen G, Strong SA (2003) Endoplasmic reticulum stress induces hyaluronan deposition and leukocyte adhesion. J Biol Chem 278:47223–47231

Maley MAL, Davies MJ, Grounds MD (1995) Extracellular matrix, growth factors, genetics: Their influence on cell proliferation and myotube formation in primary cultures of adult mouse skeletal muscle. Exp Cell Res 219:169–179

Martin PT (2006) Mechanisms of Disease: congenital muscular dystrophies-glycosylation takes center stage. Nat Clin Pract Neurol 2:222–230

Meddahi A, Bree F, Papy-Garcia D, Gautron J, Barritault D, Caruelle JP (2002) Pharmacological studies of RGTA(11), a heparan sulfate mimetic polymer, efficient on muscle regeneration. J Biomed Mater Res 62:525–531

Miura T, Kishioka Y, Wakamatsu J, Hattori A, Hennebry A, Berry CJ, Sharma M, Kambadur R, Nishimura T (2006) Decorin binds myostatin and modulates its activity to muscle cells. Biochem Biophys Res Commun 340:675–680

Moghadaszadeh B, Albrechtsen R, Guo L, Zaik M, Kawaguchi N, Borup RH, FKronquist P, Schroder HD, Davies KE, Voit T, Neielson FC, Engvall E, Wewer UM (2003) Compensation for dystrophin-deficiency: ADAM 12 overexpression in skeletal muscle results in increased alpha 7 integrin, utrophin, and associated glycoproteins. Hum Mol Genet 12:2467–2493

Molteni R, Fabbri M, Bender JR, Pardi R (2006) Pathophysiology of leukocyte-tissue interactions. Curr Opin Cell Biol 18:491–498

Myllyharju J, Kivirikko KI (2004) Collagens, modifying enzymes and their mutations in humans, flies and worms. Trends Genet 20:33–43

Nagamine Y, Medcalf RL, Munoz-Canoves P (2005) Transcriptional and posttranscriptional regulation of the plasminogen activator system. Thromb Haemost 93:661–675

Ocalan M, Goodman SL, Kuhl U, Hauschka SD, von der Mark K (1988) Laminin alters cell shape and stimulates motility and proliferation of murine skeletal myoblasts. Dev Biol 125:158–167

Ohtake Y, Tojo H, Seiki M (2006) Multifunctional roles of MT1-MMP in myofiber formation and morphostatic maintenance of skeletal muscle. J Cell Sci 119:3822–3832

Olwin BB, Rapraeger A (1992) Repression of myogenic differentiation by aFGF, bFGF, and K-FGF is dependent on cellular heparan sulfate. J Cell Biol 118:631–639

Overall CM, Blobel CP (2007) In search of partners: linking extracellular proteases to substrates. Nat Rev Mol Cell Biol 8:245–57

Papadimitriou JM, Robertson TA, Mitchell CA, Grounds MD (1990) The process of new plasmalemma formation in focally injured skeletal muscle fibres. J Struct Biol 103:(2) 124–134

Pescatori M, Broccolini A, Minetti C, Bertini E, Bruno C, D'Amico A, Bernardini C, Mirabella Ddagger Ddagger M, Silvestri G, Giglio V, Modoni A, Pedemonte M, Tasca G, Galluzzi G, Mercuri E, Tonali PA, Ricci E (2007) Gene expression profiling in the early phases of DMD: a constant molecular signature characterizes DMD muscle from early postnatal life throughout disease progression. FASEB J 21:1210–1226

Pike RN, Buckle AM, le Bonniec BF, Church FC (2005) Control of the coagulation system by serpins. FEBS J 272:4842–4851

Radley HG, Grounds MD (2006) Cromolyn administration (to block mast cell degranulation) reduces necrosis of dystrophic muscle in mdx mice. Neurobiol Disord 23:387–397

Rahmani M, Wong BW, Ang L, Cheung CC, Carthy JM, Walinski H, McManus BM (2006) Versican: signaling to transcriptional control pathways. Can J Physiol Pharmacol 84:77–92

Rapraeger AC, Krufka A, Olwin BB (1991) Requirement of heparan sulfate for bFGF-mediated fibroblast growth and myoblast differentiation. Science 252:1705–1708

Ricard-Blum S, Ruggiero F (2005) The collagen superfamily: from the extracellular matrix to the cell membrane. Pathol Biol (Paris) 53:430–442

Ringelmann B, Roder C, Hallmann R, Maley M, Davies M, Grounds M, Sorokin L (1999) Expression of Laminin $\alpha1$, $\alpha2$, $\alpha4$ and $\alpha5$ chains, fibronectin and tenascin-C in skeletal muscle of dystrophic 129ReJ dy/dy mice. Exp Cell Res 246:165–182

Roberts P, McGeachie JK (1990) Endothelial cell activation during angiogenesis in freely transplanted skeletal muscles in mice and its relationship to the onset of myogenesis. J Anat 169:197–207

Robertson TA, Papadimitriou JM, Grounds MD (1993) Fusion of myogenic cells to the newly sealed region of damaged myofibres in skeletal muscle regeneration. Neuropath Appl Neurobiol 19:350–358

Salimena MC, Lagrota-Candido J, Quirico-Santos Tl (2004) Gender dimorphism influences extracellular matrix expression and regeneration of muscular tissue in mdx dystrophic mice. Histochem Cell Biol 122:435–444

Sarasa-Renedo A, Chiquet M (2005) Mechanical signals regulating extracellular matrix gene expression in fibroblasts. Scand J Med Sci Sports 15:223–230

Sato K, Li Y, Foster W, Fukushima K, Badlani N, Adachi N, Usas A, Fu FH, Huard J (2003) Improvement of muscle healing through enhancement of muscle regeneration and prevention of fibrosis. Muscle Nerve 28:365–372

Shafat I, Vlodavsky I, Ilan N (2006) Characterization of mechanisms involved in secretion of active heparanase. J Biol Chem 281:23804–12811

Shavlakadze T, Grounds MD (2006) Of bears, frogs, meat, mice and men: insights into the complexity of factors affecting skeletal muscle atrophy/hypertrophy and myogenesis/adipogenesis. BioEssays 28: 994–1009

Shavlakadze T, Grounds MD (2003) Theraputic interventions for age-related muscle wasting: importance of innervation and exercise for preventing sarcopenia. In Rattan S, ed. Modulating aging and longevity. The Netherlands, Kluwer Academic. 139–166

Shefer G, Yablonka-Reuveni Z (2007) The ins and auts of satellite cell myogenesis; the role of the ruling growth factors. Schiaffino S, Partridge T, (eds) Skeletal muscle Repair and Regeneration. New York, springer. Chapter 6, pp 107–143.

Shen W, Li Y, Tang Y, Cummins J, Huard J (2005) NS-398, a cyclooxygenase-2-specific inhibitor, delays skeletal muscle healing by decreasing regeneration and promoting fibrosis. Am J Pathol 167:1105–1117

Shin H, Jo S, Mikos AG (2003) Biomimetic materials for tissue engineering. Biomaterials 24:4353–4364

Slater CR, Schiaffino S (2007) Innervation of regenerating muscle. In Schiaffino S, Partridge T, eds. Skeletal Muscle Repair and Regeneration. New York, Springer. Chapter 14, pp XXX

Smythe GM, Lai MC, Grounds MD, Rakoczy P (2002) Adeno-associated virus-mediated transfer of vascular endothelial growth factor in skeletal muscle prior to transplantation promotes revascularisation of the regenerating skeletal muscle. Tissue Engin 8:879–891

Sorokin LM, Maley MAL, Moch H, von der Mark H, von der Mark K, Cadalbert L, Karosi S, Davies MJ, McGeachie JK, Grounds MD (2000) Laminin α4 and integrin α6 are upregulated in regenerating dy/dy skeletal muscle: comparative expression of laminin and integrin isoforms in muscles regenerating after crush injury. Exp Cell Res 256:500–514

Stringer SE (2006) The role of heparan sulphate proteoglycans in angiogenesis. Biochem Soc Trans 34:451–453

Suelves M, Vidal B, Ruiz V, Baeza-Raja B, Doaz-Ramos A, Cuartas I, Lluis F, Parra M, Jardi M, Lopez-Alemany R, Serrano AL, Munoz-Canovez P (2005) The plasminogen activation system in skeletal muscle regeneration: antagonistic roles of urokinase-type plasminigen activator (uPA) and its inhibitor (PAI-1). Front Biosci 10:2978–2985

Tidball JG (2007) Inflammation in skeletal muscle regeneration. In Schiaffino.S, Partridge T, eds. Skeletal Muscle Repair and Regeneration. New York, Springer. Chapter 12, pp XXX

Tidball JG (1991) Force transmission across muscle cell membranes. J Biomech 24:43–52

Velleman SG, Liu X, Coy CS, McFarland DC (2004) Effects of syndecan-1 and glypican on muscle cell proliferation and differentiation: implications for possible functions during myogenesis. Poult Sci 83:1020–1027

Wang S, Voisin MB, Larbi KY, Dangerfield J, Scheiermann C, Tran M, Maxwell PH, Sorokin L, Nourshargh S (2006) Venular basement membranes contain specific matrix protein low expression regions that act as exit points for emigrating neutrophils. J Exp Med 203:1519–1532

Watt DJ, Jones GE, Goldring K (2004) The involvement of galectin-1 in skeletal muscle determination, differentiation and regeneration. Glycoconj J 19:615–619

Wewer U, Albrechrtsen R, Engvall E (2005) ADAM12: the long and the short of it. In Hooper NM, Lendeckel U, (eds.) The ADAM Family of Proteases. Proteases in Biology and Disease. Vol 4. Springer, The Netherlands pp 123–146.

Whitehead NP, Yeung EW, Allen DG (2006) Muscle damage in mdx (dystrophic) mice: role of calcium and reactive oxygen species. Clin Exp Pharmacol Physiol 33:657–662

Wiberg C, Heinegard D, Wenglen C, Timpl R, Morgelin M (2002) Biglycans organizes collagen VI into hexagonal-like networks resembling tissue structures. J Biol Chem 277:49120–49126

Yoshimura M (1985) Changes of hyaluronic acid synthesis during differentiation of myogenic cells and its relation to transformation of myoblasts by Rous sarcoma virus. Cell Differ 16:175–185

Zammit PS, Partridge TA, Yablonka-Reuveni Z (2006) The skeletal muscle satellite cell: the stem cell that came in from the cold. J Histochem Cytochem 54:1177–1191

Zanotti S, Negri T, Cappelletti C, Bernasconi P, Canioni E, Di Blasi C, Pegoraro E, Angelini C, Ciscato P, Prelle A, Mantegazza R, Morandi, Mora M (2005) Decorin and biglycan expression is differentially altered in several muscular dystrophies. Brain 128:2546–2555

Zimowska M, Szczepankowska D, Streminska W, Papy D, Tournaire MC, Gautron J, Barritault D, Moraczewski J, Martelly I (2001) Heparan sulfate mimetics modulate calpain activity during rat Soleus muscle regeneration. J Cell Physiol 188:178–187

CHAPTER 14

INNERVATION OF REGENERATING MUSCLE

CLARKE R. SLATER[1] AND STEFANO SCHIAFFINO[2]

[1] *Institute of Neuroscience, Faculty of Medical Sciences, University of Newcastle upon Tyne, Framlington Place, Newcastle upon Tyne, NE2 4HH, UK*

[2] *Department of Biomedical Sciences, University of Padova, Via G. Colombo 3, 35121 Padova, Italy*

1. INTRODUCTION

The regeneration of damaged or diseased muscles is only beneficial if the regenerated muscles become effectively innervated. Thus an important feature of adult vertebrate motor neurons is the ability of their axons to make new synaptic contacts with muscle. A number of cellular and molecular factors contribute to this ability and provide the basis for the often successful innervation of regenerated muscles. Nonetheless, the extent of that success in different circumstances varies greatly. In some cases, regenerated muscles may be as effectively innervated as they were before damage while in others, the results may be almost useless. The quality of the outcome depends critically on both the effectiveness of innervation of individual muscle fibres, and on the pattern of connectivity between the spinal motor pools and the muscles in question.

In the adult neuromuscular system, the properties of motor neurons are closely matched to those of the muscle fibres they innervate. At one level, this means that the muscles innervated by a given motor neuron are appropriate for the pattern of synaptic inputs to that neuron. For example, an 'extensor' motor neuron is only effective if it is connected to and activates an 'extensor', rather than a 'flexor', muscle. At another level, within the muscle, 'slow' motor neurons that are tonically active, and normally used for the maintenance of prolonged postural contractions, are only fully effective if they are connected to and activate muscle fibres that can sustain contraction over long periods of time. Similarly, 'fast' motor neurons, that fire in brief high frequency bursts, are only fully effective of connected to muscle fibres that can generate force rapidly. Thus the success of innervation of a

S. Schiaffino and T. Partridge (eds.), Skeletal Muscle Repair and Regeneration, 303–333.

regenerated muscle is not simply determined by how reliably its muscle fibres can be activated at the neuromuscular junction (NMJ), but also by the appropriateness of the resulting patterns of connectivity.

A crucial factor in determining the success of innervation of regenerated muscles is the nature and extent of the damage. In particular, if the basal lamina (BL) sheaths that surround each muscle fibre and each motor axon remain intact, innervation is very much more likely to be successful than if the continuity of the sheaths is broken. The importance of these sheaths is both in their role as a mechanical 'scaffold' for regeneration (Vracko and Benditt, 1972) and in their specific molecular composition (Sanes, 1989), which promotes regeneration and localised differentiation of both nerve and muscle.

In this chapter, the current understanding of the main factors that determine the success of innervation of regenerated muscles will be presented. Emphasis will be on the formation of functional NMJs and on how matching between the properties of motor axons and the muscle fibres they come to innervate is achieved. While the emphasis will be on studies of mammals, some particularly illuminating experiments made on frogs will also be described.

2. INNERVATION OF REGENERATING MUSCLES FOLLOWING MINIMAL NERVE DAMAGE

2.1 Muscle Degeneration Resulting from a Dystrophin Mutation

An important naturally occurring example of muscle degeneration that arises from abnormalities within the muscle fibre itself is that associated with the X-linked mutations of dystrophin that give rise to many forms of muscular dystrophy. Of these, the most susceptible to experimental analysis is that in the *mdx* mouse (Bulfield et al., 1984). This condition results from frame-shift mutations in the gene encoding dystrophin (Sicinski et al., 1989; Durbeej and Campbell, 2002). In contrast to the mutations in the analogous gene that give rise to Duchenne muscular dystrophy in humans, the consequences for the *mdx* mice are much less serious.

Following a round of muscle fibre degeneration and regeneration starting several weeks after birth and lasting 1–2 months, the degenerative phase subsides and the mice remain essentially normal (Torres and Duchen, 1987). The regenerated muscle fibres contract well in response to nerve stimulation, indicating that they become functionally innervated. Examination of the NMJs on the newly regenerated muscle fibres showed that they are composed of numerous spot like contacts, typically 3 μm in diameter, in contrast to the more continuous bands of normal mouse NMJs (Fig. 1) (Torres and Duchen, 1987; Lyons and Slater, 1991). However, the total area of synaptic contact is close to normal. Furthermore, the extent of postsynaptic folding is variable and generally less well developed than at NMJs in wild type mice (Torres and Duchen, 1987; Lyons and Slater, 1991; Nagel et al., 1990). Functional studies have shown that the NMJs on the regenerated fibres function as well as those

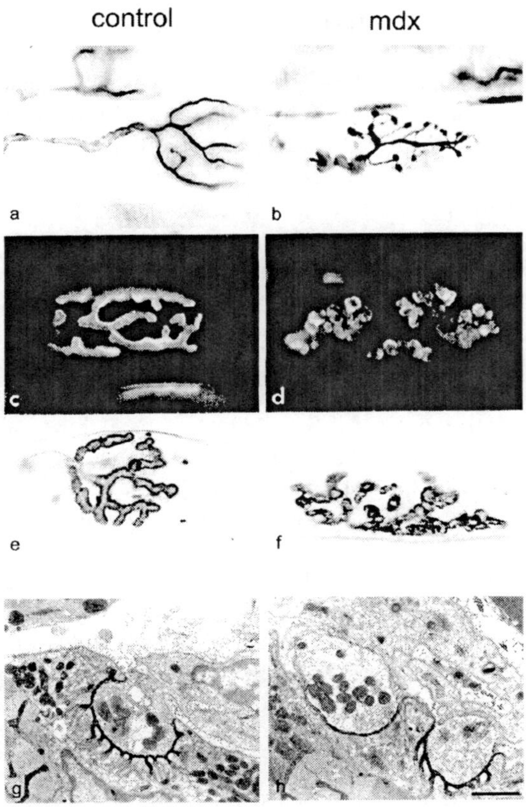

Figure 1. Neuromuscular junctions from epitrochleoanconeus muscles of control and *mdx* mice, aged 2 months. (a,b) nerve terminals impregnated with silver. Note that *mdx* terminals consist of numerous spot-like contacts. (c,d) acetylcholine receptors (AChRs) labelled with rhodamine-α-bungarotoxin. (e,f) acetylcholinesterase (AChE) activity. Note that the distribution of AChRs and AChE also forms small spots rather than the continuous bands present in the control. (g,h) Electron micrographs. Note that the extent of postsynaptic folding is reduced at some contacts in the mdx mice and that the extent of folding varies within the same NMJ. AChRs are labelled with biotin-α-bungarotoxin. Scale bar = 20 μm in a-f and 1 μm in g, h. From (Lyons and Slater, 1991)

in normal muscles (Nagel et al., 1990; Lyons and Slater, 1991). These findings establish that adult mouse motor neurons retain the capacity to form new synaptic contacts with regenerated muscle fibres.

2.2 Muscle Degeneration Induced by Natural Myotoxins

An experimental limitation of the *mdx* mouse for studies of the innervation of regenerating muscle is that the degeneration of the muscle fibres is asynchronous. This prevents more detailed analysis of the sequence of events leading to successful innervation. This limitation is overcome by experimental approaches that use chemical or mechanical treatments to induce muscle degeneration.

In some of the early studies of the innervation of regenerating muscles, crude snake venoms were used to damage the muscle. These venoms usually contain many toxic fractions, some of which also affect the nerve, potentially complicating the interpretation of the results. One of the earliest studies using a purified myotoxin was made with the 'depolarizing fraction' of the venom of the mamba, *Dendroaspis jamesoni* (Duchen et al., 1974). Injection of this fraction into the mouse leg causes localised muscle fibre breakdown with little immediate effect on the nerve terminal. During the process of degeneration, the muscle fibres break down completely. Nonetheless, the BL sheaths that surrounded the original muscle fibres survive as do the myogenic satellite cells within them. At the sites of the original NMJs, remnants of the BL that had extended into the space between the postsynaptic folds persist and mark those sites. Soon after the breakdown of the muscle fibres, new fibres begin to form within the original BL sheaths as a result of the proliferation and fusion of the surviving satellite cells. The new muscle fibres form within a week and are soon innervated by the surviving nerve terminals.

Within two weeks of damage, newly formed NMJs between the surviving axons and the regenerated muscle fibres can be identified. These differ in their detailed structure from those in normal muscles. In particular, the postsynaptic folds are much reduced in extent and the normally continuous regions of close contact between nerve and muscle are much shorter than normal. At later times, 8–10 weeks after damage, the regions of synaptic contact consist of numerous spot-like contacts roughly 3 μm in diameter instead of the more continuous bands typical of normal mouse NMJs. At the ultrastructural level, although postsynaptic folds are present, they are considerably less extensive than normal.

Although no detailed studies of neuromuscular transmission have been reported, the rapid growth of the regenerating muscle fibres suggests that effective transmission was restored soon after structural innervation was established. At all stages of muscle breakdown and regeneration cholinesterase activity, normally restricted to the NMJ, remains associated with the nerve terminals. It thus appears likely that in this example of innervation of regenerating muscle fibres, the nerve reoccupies the original synaptic site, but that the detailed conformation of the NMJ region is somewhat modified. Similar observations have been made after treatment of mouse muscles with the lethal toxin of *Clostridium sordellii* (Barbier et al., 2004).

The recovery of functional neuromuscular transmission in a similar example was described in a study was made using notexin, the myotoxic component of the venom of the Australian tiger snake *Notechis scutatus* (Grubb et al., 1991). The process of muscle breakdown was qualitatively similar to that described above and followed a similar time course, with new muscle fibres observed within less than a week of intoxication (Fig. 2A). Intracellular recordings from these muscle fibres showed that both spontaneous miniature endplate potentials (mEPPs) and evoked endplate potentials (EPPs) could be recorded 3–5 days after intoxication, when the first

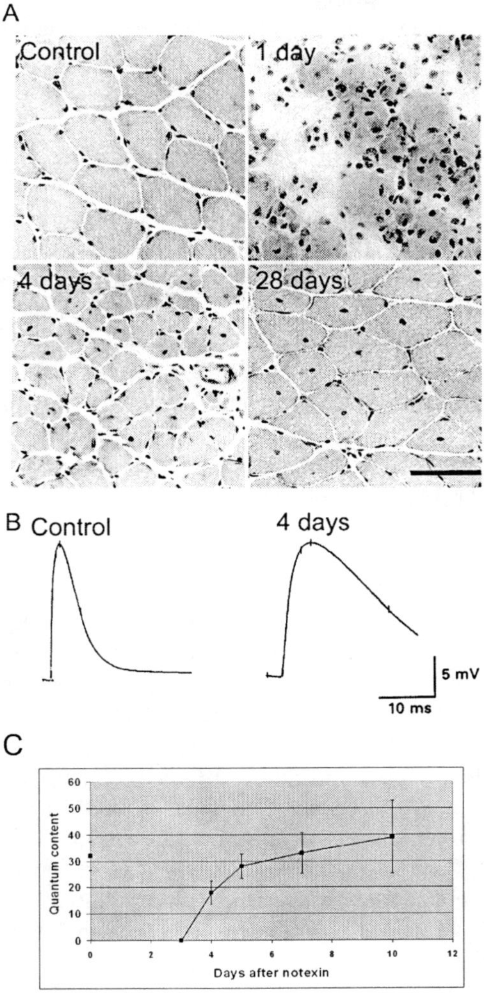

Figure 2. Functional innervation of rat soleus muscles regenerated after destruction by notexin. (A) Time course of muscle regeneration. Muscle fibres are destroyed within 1 day but have regenerated by 4 days after treatment. Scale bar = 50 μm. (B) Synaptic responses to nerve stimulation in control and regenerated muscles. Note that by 4 days after treatment, an endplate potential (EPP) on normal amplitude can be recorded. The relatively slow decay phase of the EPP on the regenerated fibres results primarily from the presence of AChRs containing a γ-subunit in place of the normal ϵ-subunit and from the reduced AChE activity. (C) Recovery of quantal ACh release from the nerve. The quantum content (the number of ACh quanta released from the nerve by a single nerve impulse) returns to the normal range within 1 week of treatment. From (Grubb et al., 1991)

nerve-evoked contractions were observed. The EPP amplitude, and the number of acetylcholine 'quanta' released by a single nerve action potential increased steadily to near normal values by 10 days after intoxication (Fig. 2B,C).

2.3 Muscle Degeneration Induced by a Local Anaesthetic

Innervation has also been studied in rat muscles regenerating after injection of bupivacaine, a local anaesthetic (Benoit and Belt, 1970; Jirmanova, 1975; Jirmanova and Thesleff, 1976; Jirmanova and Thesleff, 1972). In this situation, although the breakdown of treated muscle fibres is as complete and rapid as after snake toxins, the reconstruction of the postsynaptic apparatus is more complete than that described after intoxication with *Dendroaspis* myotoxin. The basis for this difference remains unknown. By examining the reconstruction of the postsynaptic apparatus in muscle that had been damaged with bupivacaine and then treated with botulinum toxin to block ACh release from the nerve, it was found that neither neuromuscular transmission nor muscle activity were required for the reconstruction of the postsynaptic apparatus to take place (Jirmanova and Thesleff, 1976).

2.4 Conclusions

Taken together, these studies show that when the nerve terminal remains intact new postsynaptic specializations can form on regenerated muscle fibres at sites where they come in contact with the surviving nerve terminal. The speed of this process is, if anything, faster than during normal development and independent of nerve or muscle activity. This suggests that some signals might persist in the region of the original NMJ that could promote the rapid differentiation of the surface of a regenerating muscle fibre into a postsynaptic domain. In each of the situations described above, the BL surrounding nerve and muscle cells, and the common 'synaptic' BL, remained intact. This raised the possibility that such signals might be associated with the BL. Additional studies (see below) have shown that signals associated with the synaptic BL do indeed play a key role in the successful innervation of regenerating muscle fibres.

3. INNERVATION OF INTACT MUSCLES AFTER NERVE DAMAGE

In many situations, damage to muscles is accompanied by damage to some or all of the motor axons that innervate them. If a motor axon is cut through, its distal portion breaks down by the process of Wallerian degeneration. The initial stages of muscle regeneration can take place in the absence of any innervation. However, the severed axon may eventually regenerate and re-establish functional contacts with muscle. An understanding of how the nerve responds to damage when the muscle is left intact provides an important foundation for understanding the innervation of regenerating muscles when both nerve and muscle have been damaged.

3.1 Reinnervation After Partial Nerve Section

When only some of the motor axons innervating a muscle degenerate, as a result of disease or trauma, the remaining axons often grow sprouts from their intramuscular portions that may innervate the denervated fibres (Son et al., 1996). A well-known clinical example of this behaviour occurs in spinal muscular atrophy when the death of some motor neurons leaves the muscle partially denervated (Coers and Woolf, 1959). The adaptive plasticity of the surviving motor neurons has many similarities with the response to blocking doses of botulinum toxin (Duchen, 1970). In both cases, it is likely that the axonal outgrowth is triggered by inactivity of the muscle. This effect of muscle activity may be mediated by growth factors released from the intact, but inactive, muscle fibres (Caroni et al., 1994; Aigner et al., 1995).

In most cases, reinnervation of denervated muscle fibres in partially denervated muscles occurs at the sites of the original NMJs. It appears that the growing axonal sprouts are guided to the denervated NMJs by processes of the 'denervated' terminal Schwann cells (Reynolds and Woolf, 1992; Son et al., 1996). These grow away from the denervated NMJs and are apparently attracted to NMJs with intact innervation (Love and Thompson, 1999). When they contact an intact NMJ, they provide a favourable substrate for the outgrowth of axonal sprouts, which are thus guided to the denervated NMJs.

An important feature of this form of adaptive plasticity is that it may lead to a mismatch of the properties of the motor neuron and the muscle fibres it comes to innervate. In normally innervated muscles, all the muscle fibres innervated by a single motor axon have very similar functional properties. However, these may differ substantially from those in other 'motor units'. When a motor neuron innervates denervated muscle fibres with inappropriate functional properties, the new innervation may bring about a change in the molecular and functional properties of the muscle fibres it takes over. How this functional transformation comes about is described in detail later in this chapter. That it happens at all indicates that the mechanisms that account for nerve-muscle matching operate less effectively in the adult than during normal development.

3.2 Reinnervation After Nerve Crush

An extreme form of nerve injury involves breaking the continuity of all the axons, either by crushing the nerve sufficiently to interrupt the axons or by cutting the nerve. Both methods result in degeneration of the axon distal to the site of injury. The degeneration of the axon is followed within days by the outgrowth of sprouts from the proximal stump. Whether these sprouts reinnervate the muscle, and if so how rapidly, depends very much on the type and extent of nerve damage. If the whole nerve is damaged by crushing, the BL sheaths surrounding each axon and its associated Schwann cells remain intact. These provide an excellent matrix and scaffold for the growth of the regenerating axons. Simply by growing within the surviving BL sheath, a regenerating axon may be directed accurately back to the same muscle fibres it previously innervated.

A particularly striking example of the accuracy of reinnervation after nerve crush has been described in the frog (Letinsky et al., 1976). In this case, the regenerating axons reoccupy precisely the sites on the muscle fibre membrane that formed the original postsynaptic apparatus. Furthermore, the specialised 'active zones' in the nerve terminal (Harlow et al., 2001), from which ACh quanta are released (Heuser et al., 1979), form exactly opposite the surviving postsynaptic folds, just as at the normal NMJ. This study supports the view that during reinnervation of an adult muscle, factors that survive nerve degeneration and are presumably located in the muscle fibre or the synaptic BL can induce the detailed differentiation of the regenerating presynaptic nerve terminal.

3.3 Reinnervation After Cutting the Nerve

If the BL sheaths of the nerve are disrupted by injury, reinnervation of the muscle is much less effective. While the axons may eventually grow back to the muscle they originally innervated, this process is usually slower and less accurate than when the BL sheaths remain intact (Gutmann and Young, 1944; Bennett et al., 1973). Nonetheless, effective functional reinnervation of the muscle may still occur. Of particular interest is the observation that when it does, the sites of the newly formed NMJs are often similar, and possibly identical, to those of the original ones (Miledi, 1960; Bennett et al., 1973). This is true even if the regenerating axons are forced to contact the muscle some distance from the normal zone of innervation. Analysis of the pattern of axon growth in such cases suggests that the axons are not specifically attracted to the sites of the original NMJs. However, if these sites are contacted by a growing axon, they readily form new synaptic contacts (Bennett et al., 1973). These studies thus suggest that the site of an original NMJ is one where reinnervation of a muscle can occur preferentially.

3.4 Reinnervation at Ectopic Sites

One group of experiments appears initially to contradict the conclusion just stated. These studies involved the formation of new NMJs at ectopic sites (i.e. ones where NMJs would normally not be found) on the denervated rat soleus muscle. In these experiments a 'foreign' nerve (i.e. one that normally innervates a different muscle) was placed over a region of the soleus muscle that normally has no NMJs. After a period of several weeks, in which the axons of the foreign nerve grew extensively over the surface of the muscle fibres, the normal innervation of the soleus was cut. Within less than a week, many new NMJs formed in the region of the foreign nerve implant and stimulation of the foreign nerve evoked contraction of the host muscle (Fig. 3) (Jansen et al., 1973; Lømo and Slater, 1978). These ectopic NMJs have all the main postsynaptic specialisations of normal NMJs, including accumulations of AChRs and AChE and extensive postsynaptic folds. This is a striking demonstration of the ability of an adult motor axon to form completely new synaptic contacts with an intact, similarly adult, muscle fibre.

In the context of Section 3.2, there are two important features of this situation. First, the soleus muscle is a better 'host' of ectopic innervation than other muscles

A

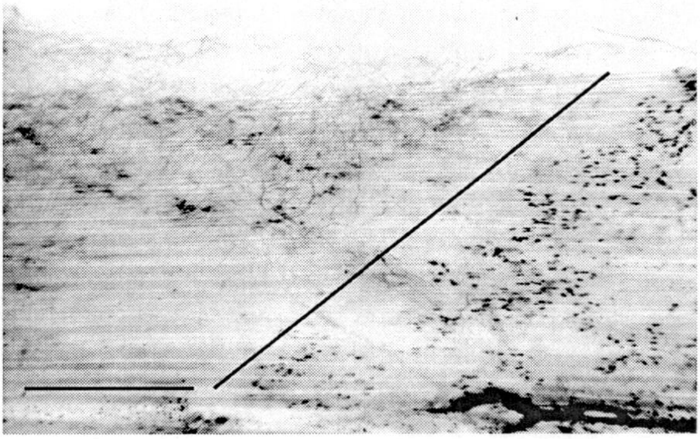

B

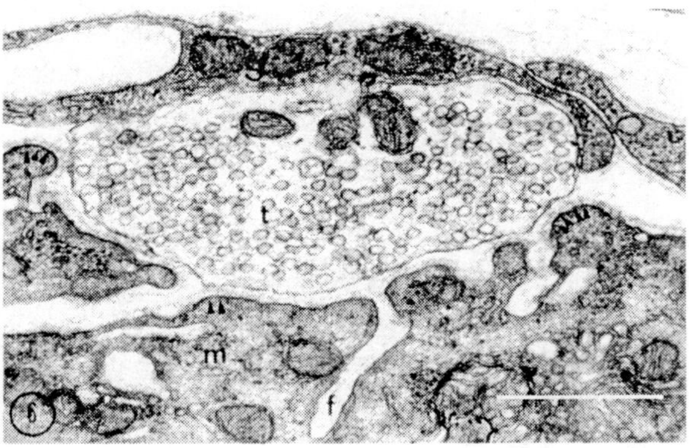

Figure 3. Distribution and structure of new NMJs formed at ectopic sites on adult rat soleus muscles. (A) Whole mount of rat soleus muscle 14 days after onset of ectopic NMJ formation. Nerves are labelled with methylene blue and the position of NMJs indicated by AChE activity. The nerves to the original NMJs (to the lower right of the line) have degenerated. Ectopic NMJs can be seen to the upper left of the line. AChE activity is present but still weaker that at the original NMJs. Scale bar = 3 mm. From (Lømo and Slater, 1978). (B) Ultrastructure of ectopic NMJ 3 weeks after initiation of ectopic NMJ formation. Synaptic vesicles in the nerve terminal and folds of the postsynaptic membrane ('f') are similar to those in normal mature NMJs. Scale bar = 1 μm. From (Korneliussen and Sommerschild, 1976)

(Taxt, 1983). This is almost certainly because it is composed largely of 'slow' muscle fibres, innervated by motor neurons with a tonic pattern of activity (Hennig and Lømo, 1985). Second, even in the soleus, it is not uncommon for some of the foreign axons to grow along the surface of the muscle fibres and to reinnervate the denervated original NMJs (Lømo and Slater, 1978). Thus, although these experiments demonstrate clearly the ability of motor axons to form new NMJs, they also underline the fact that previously formed NMJs are sites of preferential reinnervation.

3.5 Neural Induction of Postsynaptic Differentiation

The ability of an adult motor axon to make new synaptic contacts with a muscle is of obvious relevance to the process of innervation of regenerating muscles. An outstanding question raised by the phenomenon of ectopic NMJ formation is: How does the nerve induce postsynaptic specialisations in the muscle fibre? One clue was given by experiments in which the foreign nerve was cut 2 days after cutting the soleus nerve, i.e. after the muscle became fully inactive but before new ectopic NMJs had become functional. If such muscles were left inactive, many dispersed clusters of AChRs formed in the muscle fibres in the region of the implanted, but now degenerated, foreign nerve. However, if the muscles were kept active by direct stimulation, a small number of compact AChR clusters, each with an associated accumulation of AChE activity, persisted. These experiments suggested that during the 2-day period following transection of the soleus nerve, the foreign axons modified the surface of the muscle fibres in such a way as to leave a 'trace' that marked spots where postsynaptic differentiation would occur if the muscle became active. Studies that have revealed the likely molecular basis of this 'trace' are described below (Section 4).

3.6 Conclusions

The studies of reinnervation of intact muscles described above all suggest that the region of a mature NMJ has distinctive properties that allow it to become preferentially reinnervated after denervation. Further, they suggest that these properties might in some way originate from a relatively brief interaction between the nerve and the muscle at an early, prefunctional, stage of NMJ formation. One possible location of such signals is the synaptic region of BL. The following section describes experiments that have provided strong evidence in support of this suggestion and have gone on to identify a molecular signalling pathway that seems likely to underlie these effects.

4. THE ROLE OF THE BL AND ITS COMPONENTS IN MUSCLE REINNERVATION

Some of the experiments that have revealed most about how regenerating muscle becomes innervated have been made with the aim of determining the possible role of the BL in reconstruction of the NMJ. These experiments involved damaging the cellular components of the NMJ (nerve, Schwann cells and/or muscle) in ways that leave the BL intact. The initial damage was combined with treatments that allowed some, but not all, of the cellular components of the NMJ to regenerate. Most of these experiments were initially made on frogs, but the main findings have subsequently been observed in mammals.

4.1 Regenerating Nerve Differentiates when it Makes Contact with 'Bare' Synaptic BL

The possibility that the synaptic BL might induce differentiation of the nerve terminal was studied by damaging both nerve and muscle and then preventing muscle regeneration by localised X-irradiation (Sanes et al., 1978). When regenerating axons made contact with the synaptic BL, identified by its persisting AChE activity, they formed characteristic accumulations of synaptic vesicles (Fig. 4). These were precisely aligned with the sites of the openings of the postsynaptic folds which could be identified by the protrusions of BL that had occupied the interfold space. Furthermore, functional and molecular studies suggested that these terminals could release transmitter (Glicksman and Sanes, 1983; Dunaevsky and Connor, 1995). In the absence of any persisting muscle fibres, these observations suggest that factors in the synaptic BL can induce local differentiation of the regenerating nerve terminal.

4.2 AChRs Accumulate on Regenerating Muscle at Site of Original NMJ in Absence of Nerve or Schwann Cell

A complementary approach was used to see whether the synaptic BL might also be able to induce postsynaptic properties in regenerating muscle fibres (Burden et al., 1979; McMahan and Slater, 1984). Muscles, and the overlying nerve terminal and Schwann cells, were damaged and the nerve was cut so that the muscle fibres, but not the nerve, could regenerate (Fig. 5A). At the sites where the regenerating muscle fibres made contact with the synaptic BL, characteristic features of the normal postsynaptic region formed. These included clusters of AChRs, AChE, postsynaptic folds and accumulations of myonuclei (Fig. 5B,C). Since neither of the presynaptic cellular components of the NMJ was present during regeneration of the muscle fibre, it was concluded that the synaptic BL also contained factors that can induce differentiation of the muscle fibre.

The ability of synaptic BL to induce postsynaptic differentiation has also been observed in mammals (Womble, 1986; Strochlic et al., 2004; Brenner et al., 1992). There it has also been shown that myonuclei of the regenerating muscle fibres accumulate at the original synaptic site. As in normal muscles, these 5–10 nuclei express a distinctive pattern of genes encoding key molecules of the postsynaptic region, including both AChRs and Na_V1 channels (Jo and Burden, 1992; Goldman et al., 1991; Brenner et al., 1992; Stocksley et al., 2005; Lupa and Caldwell, 1994). Thus the synaptic BL can induce not only the aggregation of postsynaptic molecules, but also their preferential expression by underlying myonuclei.

4.3 Discovery and Properties of Agrin

Evidence of the importance of the synaptic BL as a key factor in the control of muscle innervation prompted efforts to identify the molecules involved. These

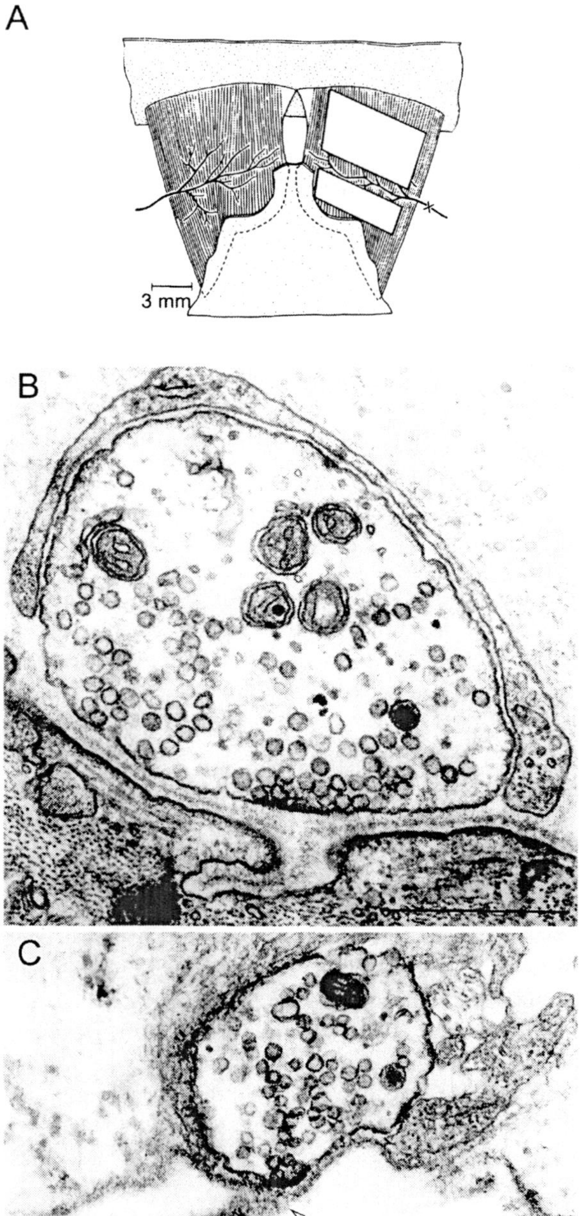

Figure 4. (Continued)

efforts have led to the identification of agrin, a protein made by and released from the motor nerve terminal that can induce essentially all the known molecular specialisations of the postsynaptic region (McMahan, 1990).

4.3.1 Discovery and characterisation of agrin

Agrin was initially isolated from extracts of the extracellular matrix of the electric organ of *Torpedo* (Nitkin et al., 1987). This organ is a modified, and very large, NMJ which has served as the source of a number of key components of the NMJ. Agrin is a heparansulfate proteoglycan. Different alternatively spliced forms are synthesised by motor neurons (Magill-Solc and McMahan, 1988; Hoch et al., 1993) and muscle fibres, as well as cells in the CNS. The motor neuron form is transported to, and released from, the nerve terminal. Following its release, it binds to the synaptic BL (Fig. 6) (Magill-Solc and McMahan, 1990). From this location, it activates a muscle specific kinase (MuSK), thereby triggering a complex signalling cascade that leads to postsynaptic differentiation (Glass et al., 1996; Moore et al., 2001). These effects of agrin are mimicked by soluble agrin applied ectopically to the surface of muscle fibres in vivo (Cohen et al., 1997; Meier et al., 1997).

Taken together, these findings suggest that agrin is at least one of the molecules that give the synaptic BL its distinctive ability to induce postsynaptic differentiation during NMJ formation or reconstruction. The unique importance of agrin in the process of NMJ formation was established by the creation of mice lacking agrin (Gautam et al., 1996). In these mice, NMJs failed to form, resulting in perinatal death. A similar phenotype was observed in mice lacking MuSK (Gautam et al., 1999) or its essential activator Dok-7 (Okada et al., 2006), further indicating the importance of the signalling pathway activated by agrin for NMJ formation.

4.3.2 Agrin as a possible axonal 'stop' signal

Less is known about the possible role of agrin as a mediator of presynaptic differentiation of motor nerve terminals. In mice deficient in agrin, MuSK or Dok-7, the motor axons fail to elaborate normal presynaptic terminals and instead grow extensively over the surface of the muscle (Gautam et al., 1996; Gautam et al., 1999; Okada et al., 2006). Furthermore, in vitro, motor neurite outgrowth is selectively inhibited

Figure 4. (Continued) Differentiation of regenerated frog nerve terminals at original synaptic sites in the absence of muscle. (A) Scheme of experiment. On one side of the flat *cutaneous pectoris* muscle, the muscle on either side of the band of innervation was removed. The fragments of muscle fibres in the remaining 'bridge' degenerate completely. The frog was X-irradiated to kill proliferating myogenic cells, thus preventing muscle regeneration. At the time of the initial operation, the nerve was crushed ('x'), causing its distal part, including the presynaptic terminals, to degenerate. Subsequently the axons regenerated within the surviving BL tubes. (B) Ultrastructure of normal frog NMJ, showing nerve terminal with synaptic vesicles and one postsynaptic fold with a projection of synaptic BL into it BL. (C) Regenerated axon terminal in contact with synaptic BL, indicated by a surviving projection of BL. Note the dense thickening of the nerve terminal representing a presynaptic active zone immediately opposite the BL projection. Modified from (Burden et al. 1979, McMahan and Wallace, 1989)

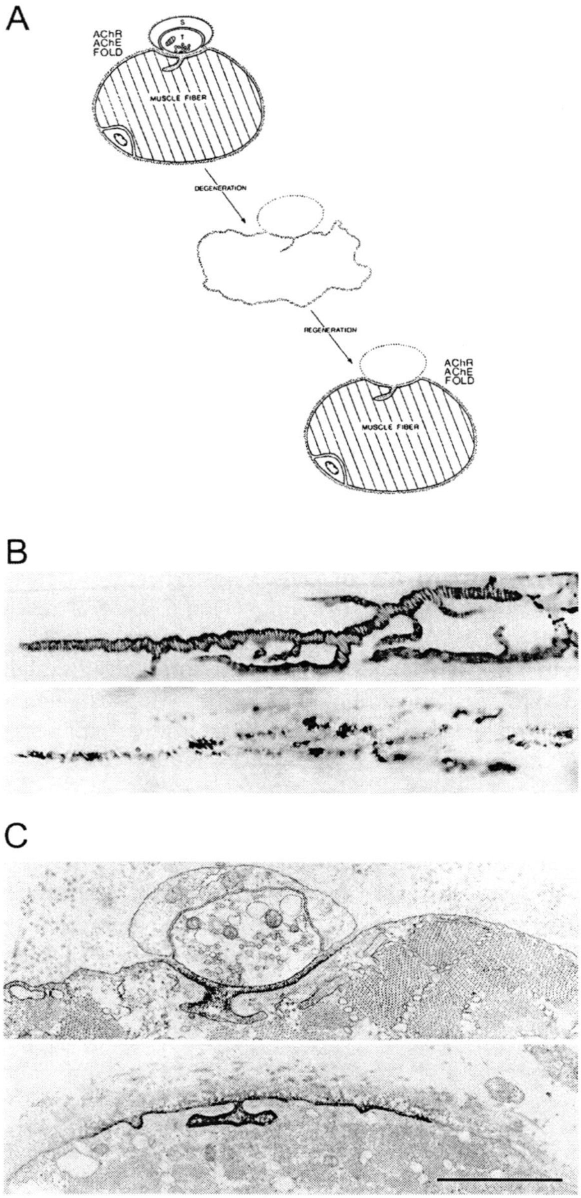

Figure 5. Aggregation of AChRs at sites of original NMJs on frog muscle that had regenerated in the absence of nerve or Schwann cells. (A) Scheme of experiment. The muscle was frozen to kill all muscle fibres and myogenic cells and the Schwann cells. At the same time the nerve was cut so as to prevent reinnervation. Degeneration of all damaged cells initially leaves the BL tubes that had surrounded nerve and muscle fibres completely empty. Subsequently the muscle fibres regenerate. (B) AChRs, labelled by biotin-α-bungarotoxin, at normal (top panel) and regenerated (bottom panel) synaptic sites. Although

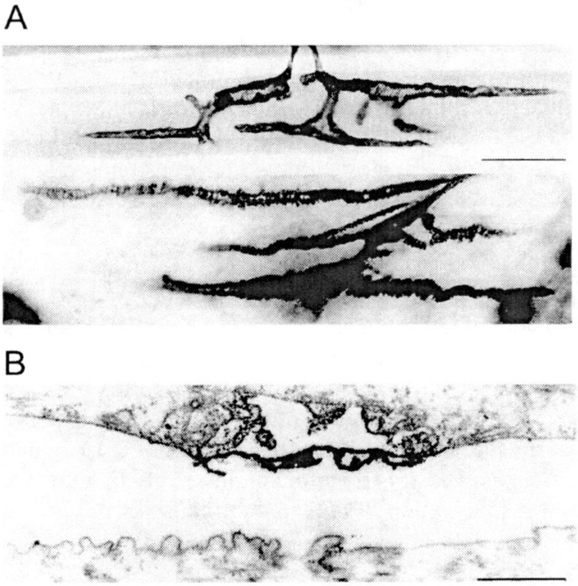

Figure 6. Distribution of agrin at intact and damaged frog NMJs. The *cutaneous pectoris* muscle was damaged by crushing, causing degeneration of all muscle fibres, axons and Schwann cells at the NMJ but leaving intact the BL. (A) Immunolabelling of agrin at normal (top panel) and damaged muscle, 23 days after crushing (bottom panel). Note strong immunolabelling of agrin at the original synaptic site on the BL of the damaged muscle. Scale bar40 μm. (B) Ultrastructural demonstration of agrin immunolabelling in the synaptic BL of an NMJ damaged by crushing 21 days previously. Note that agrin is most intense in the portion of BL that had extended between the nerve and the muscle. Scale bar = 1 μm. From (Reist et al., 1987)

by cell surface MuSK and agrin (Dimitropoulou and Bixby, 2005). These observations are consistent with agrin playing a key role as a component of the synaptic BL that promotes NMJ formation or reconstruction. However, the molecular signalling pathways that mediate agrin's effect on the nerve are not known.

4.4 Laminins as Possible Regulators of Nerve Differentiation

Although agrin is apparently the dominant molecule in the synaptic BL that influences NMJ formation, it is unlikely to be the only one. A further distinctive component of

Figure 5. (Continued) AChR labelling at the regenerated NMJ sites is discontinuous, it clearly outlines the original course of the nerve. (C) Ultrastructure of normal (top panel) and regenerated (bottom panel) synaptic sites. AChRs labelled by biotin-α-bungarotoxin. Where the regenerated muscle fibre makes contact with the original synaptic sites, indicated by the formation of folds, there is a high concentration of AChRs. Scale bar = 20 μm in B and 1.4 μm in C. From (McMahan and Slater, 1984)

the synaptic BL is laminin β2 (s-laminin) (Hunter et al., 1989a; Hunter et al., 1989b). Studies in vitro show that this form of laminin can act as a 'stop signal' for motor neurons (Porter et al., 1995). Further, there is evidence that an additional form of laminin in the synaptic cleft is linked to voltage-gated calcium channels in the presynaptic membrane of the *Torpedo* electroplaque (Sunderland et al., 2000). Such links, if present at the NMJ, might help to align molecules necessary for transmitter release from the nerve with the folds in the muscle fibre membrane.

4.5 Conclusions

There is much experimental evidence for the view that the synaptic BL contains molecules that can induce the differentiation of synaptic properties in both motor nerve terminals and muscle fibres. Thus the BL persists after many forms of nerve and muscle damage and can promote the reconstruction of the NMJ following nerve and/or muscle regeneration. Agrin has many of the properties expected of such molecules and is present in the synaptic BL. It is therefore highly likely that it plays a key role in ensuring the effective innervation of regenerating muscles.

5. INNERVATION OF MUSCLES AFTER DISRUPTION OF BOTH NERVE AND MUSCLE

The previous sections present evidence, largely from studies of the frog, that the synaptic BL contains molecules that promote the reconstruction of functional innervation of the regenerating muscle fibres. In this section we describe further examples of muscle regeneration in mammals in which both nerve and muscle were damaged. In some cases this was done in a way that left the BL grossly normal, as in some of the frog experiments cited above. In others, the architecture of the muscle was grossly disrupted, for example by mincing. Even then, some degree of regeneration and functionally innervation occurred. Unfortunately, there have been very few studies of the structure or function of the NMJs in these muscles, so the success of innervation can only be assessed from studies of nerve-evoked contraction.

5.1 Innervation of Muscles After Free Grafting

In a major series of studies, muscle regeneration was studied after 'free grafting' i.e. removing the muscle from the body and then returning after suturing the tendons to their original fixation points (see Carlson, this volume). In all these studies, whether made on rats or cats, the muscle fibres all degenerate as a result of ischemia. The extent and efficacy of subsequent regeneration depends very much on the success of revascularization and hence on the size of the muscle and the species. For muscles beyond a certain size, the central portion fails to become adequately revascularized and the muscle fibres within it fail to regenerate. In all free grafted muscles, the force developed following direct or nerve stimulation is less than normal, both in terms of total force and force per unit area (Faulkner et al., 1983; Carlson and Faulkner, 1988). This is likely due to in part to increased amounts of connective tissue and the failure of some muscle fibres to regenerate. However a key factor influencing the recovery of force output is the extent of reinnervation.

Following 'free-grafting' of rat soleus muscles into the opposite leg, the extent of reinnervation was varied. In some muscle fibres, which themselves appeared thick and healthy when studied 30–60 days after grafting, mature NMJs were seen (Fig. 7) (Schmalbruch, 1977). These had extensive folds and were occupied by nerve terminals. However, many other fibres remained thin and apparently uninnervated. Thus there was a close correlation between the presence of innervation and the state of maturation of the regenerated muscle fibres.

The importance of reinnervation as a determinant of the success of muscle regeneration was studied by varying the treatment of the nerve at the time of muscle grafting (cf. (Faulkner et al., 1983; Carlson and Faulkner, 1988)). The treatments used included leaving the nerve (and necessarily its vascular supply) intact, cutting the nerve but surgically joining the ends immediately, implanting the cut proximal end of the nerve into the muscle, and cutting the nerve and leaving it with no special connection to the muscle. As the degree of disconnection of the original nerve increases, the force output of the regenerated muscle increases. This was assumed, without extensive direct evidence, to be a result of decreasingly effective innervation of the regenerating muscle fibres. This would be consistent with an important role of the endoneurial sheaths in guiding regenerating axons back to the muscle (Gutmann and Young, 1944). In the absence of detailed anatomical and physiological studies, it is unclear to what extent the

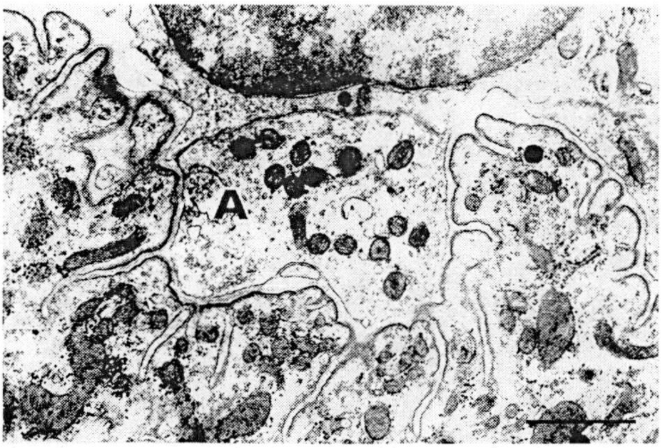

Figure 7. Neuromuscular junction formed in regenerated rat soleus muscle autografted 30 days previously. Note that the region of postsynaptic folding on the regenerated muscle extends well beyond the region of nerve-muscle contact. The increased density at the crests of the reconstructed folds shows the regions of high AChR density. Scale bar = 1 μm. from (Schmalbruch, 1977)

innervation of these regenerating muscles occurred at the sites of the original NMJs.

5.2 Innervation of Regenerating NMJ-free Segments of Muscle

A variant of the free grafting procedure was used to study the ability of regenerating axons to innervate regenerating muscles in the absence of the sites of the original NMJs. To this end the muscle (rat soleus) was transected on either side of the zone containing the original NMJs and the two segments joined before replacing the graft (Womble, 1986). Regenerating axons made functional contacts with the new muscle within three weeks. This confirms that while the original NMJs remain as sites of preferential innervation of regenerating muscles, effective innervation can be established at ectopic sites as on intact muscles.

5.3 Innervation of Muscles Regenerating After Mincing

One of the first systematic studies of muscle regeneration used possibly the most thorough method for damaging the muscle. This involves mincing the muscle into fragments a millimetre or so in length before replacing the minced tissue mass into its original bed. The history of this approach, pioneered by the Russian biologist Studitsky, is recounted in Carlson's article in this volume. In spite of the mincing procedure, the BL tubes surrounding the short lengths of minced muscle fibre remain and form scaffolds for subsequent muscle fibre regeneration.

When this procedure is used in rat leg muscles, new muscle fibres are present within two weeks of muscle damage (Bennett et al., 1974). Intracellular recordings from these fibres already detect the spontaneous and evoked electrical events (mEPPs and EPPs) that are the hallmark of functional innervation. The strength of neuromuscular transmission increased up to 60 days post operation when NMJs had morphological and functional properties similar to those in normal muscles. At this time, the NMJs appear similar to normal with the typical 'pretzel' shape, abundant AChE and deep postsynaptic folds (Fig. 8). These findings attest to the remarkable ability of the peripheral neuromuscular system to reconstitute itself, even after devastating damage.

5.4 Conclusions

The success of regeneration of damaged mammalian muscle depends critically on the integrity of the BL sheaths of both muscle and nerve. If those sheaths remain intact, muscle fibre regeneration is speedy and complete as is the innervation of the regenerated fibres. The greater the loss of integrity of the sheaths resulting from the primary muscle damage, the less well motor the axons are able to form functional NMJs with the regenerated muscle fibres.

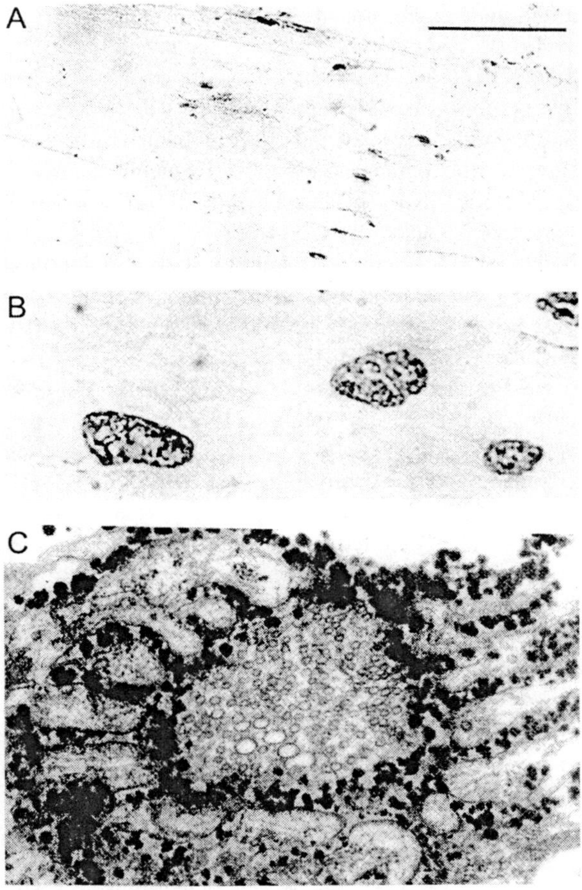

Figure 8. Neuromuscular junctions in regenerated rat leg muscles 30–60 days after mincing. (A) Distribution of NMJs (indicated by dark regions of high AChE activity) in central band of muscle. Scale bar = 130 μm. (B) Higher magnification showing typical 'pretzel' configuration of NMJs. Scale bar = 50 μm. (C) Ultrastructure of NMJs marked by dark granular deposits indicating high AChE activity. Note presence of abundant synaptic vesicles in the nerve terminal and folds of the postsynaptic membrane. Scale bar = 0.8 μm. From (Bennett et al., 1974)

6. NERVE-INDUCED MATURATION OF REGENERATING MUSCLE

6.1 Muscle Growth and Fiber Type Specification in Regenerating Muscle are Dependent on Innervation

While the very initial phases of muscle regeneration after injury can take place in the absence of neural influence, the subsequent growth and maturation of regenerating muscle fibers require the presence of the nerve. Nerve activity appears to affect directly protein turnover and gene expression within multinucleated regenerating myotubes (see below, Sections 6.3 and 6.4), and indirectly the proliferation and differentiation of mononucleated myogenic precursors such as satellite cells. Loss and dysfunction of myogenic precursor cells has been observed following denervation or muscle unloading induced by hind limb suspension, which is accompanied by reduced motor neuron activity, and this impairs the increase in myonuclei which is necessary for muscle growth (Mozdziak et al., 2001; Mitchell and Pavlath, 2004). It is likely that these indirect effects of nerve activity on myogenic precursors are mediated by the release of specific factors by the muscle fibers. For example, studies on cultured muscle cells have shown that growing myotubes release interleukin 4, which acts on mononucleated myoblasts inducing their fusion with the myotubes (see Chapter 7). However, the role of nerve activity on this process has not been explored.

Myosin isoforms have been used as markers to define the role of the nerve in controlling fiber type specification in regenerating muscle. Regenerating muscle fibers initially express developmental myosin heavy chain (MyHC) forms, such as embryonic and neonatal MyHCs, and later on adult fast and slow MyHC forms (Sartore et al., 1982; Whalen et al., 1990). The role of the nerve in controlling these MyHC switches is well illustrated in a widely used experimental model, i.e. the regeneration of the rat soleus after degeneration of muscle fibers induced by bupivacaine or notexin injection. In this model, miniature endplate potential (mEPPs) are detected as early as 3 days after injury (Jirmanova and Thesleff, 1972) and functional reinnervation, as determined by the ability of muscle fibers to generate an action potential in response to stimulation of the motor nerve, was present in many fibers at 4–5 days (Whalen et al., 1990; Grubb et al., 1991).

By day 3 after injury, the innervated and denervated soleus muscles are similarly composed of thin myotubes expressing embryonic and neonatal myosin heavy chains (MyHCs). The innervated muscle fibers then undergo a rapid increase in size and express predominantly MyHC-slow whereas the denervated muscle fibers remain atrophic and express predominantly fast-type MyHCs (Fig. 9).

Two consecutive switches in MyHC gene expression occur in a surprisingly short temporal window around day 4 and 5 after injury (Jerkovic et al., 1997), immediately after the formation of functional NMJs (see above, Section 2). The first switch from embryonic and neonatal to fast MyHC-2X and -2B is independent of neural influence since it is detected in both innervated and denervated muscle at day 4 (Fig. 10).

This initial embryonic/neonatal-to-fast MyHC transition reflects an intrinsic default program of MyHC gene expression that was first described in neonatally denervated skeletal muscle (Butler-Browne et al., 1982) and is in part controlled by thyroid hormones (d'Albis et al., 1987). The second fast-to-slow switch, detected at day 5, is dependent on innervation and is characterized by the transient up-regulation of MyHC-2A and persistent accumulation of MyHC-slow transcripts with a concomitant down-regulation of MyHC-2X and -2B transcripts (Jerkovic et al., 1997). The antithetical changes in fast and slow myosin transcripts taking

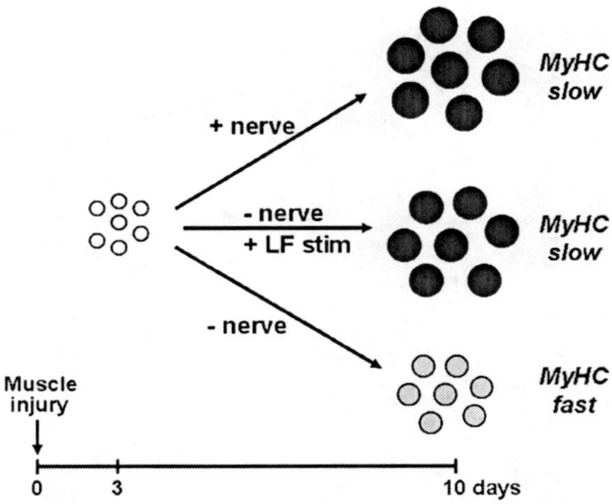

Figure 9. Effect of nerve activity on muscle fiber growth and fiber type specification in the regenerating rat soleus muscle. Bupivacaine (marcaine) injection causes muscle degeneration and subsequent regeneration. At day 3 after injury the regenerating soleus consists of small myotubes expressing embryonic myosin heavy chain (MyHC). During subsequent days in response to slow motor neuron activity the regenerating fibers grow in size and start to express MyHC-slow but in the absence of innervation muscle growth is blocked and fibers undergo a default differentiation leading to MyHC-fast gene expression. The effect of nerve activity can be reproduced by low frequency stimulation (LF stim) of the regenerating denervated muscle. Modified from (Schiaffino et al., 2006)

place in the same short time period point to a common regulatory mechanism mediated by the slow motor neurons innervating the soleus muscle. In the regenerating fast extensor digitorum longus (EDL) the switch to MyHC-slow does not occur and most fibers continue to express fast-type contractile proteins (Esser et al., 1993).

6.2 The Effect of Motor Neurons on Regenerating Muscle is Mediated by Specific Impulse Patterns

The switch to MyHC-slow induced by innervation in the regenerating rat soleus is due to the specific impulse patterns of slow motor neurons since it can be reproduced by direct electrical stimulation of denervated regenerating muscle using low frequency stimulation that resembles the natural firing pattern of slow motor neurons (Kalhovde et al., 2005). This stimulation pattern also induced significant increase in muscle fiber size, although denervated and stimulated fibers tended to be smaller than the innervated fibers (Fig. 11).

The regenerating muscle appears to have greater plasticity than intact adult muscle, as shown by the greater changes in contractile properties and fiber type composition observed after cross-transplantation and cross-reinnervation (Donovan and Faulkner, 1987). Whereas cross-reinnervation with a slow nerve or low

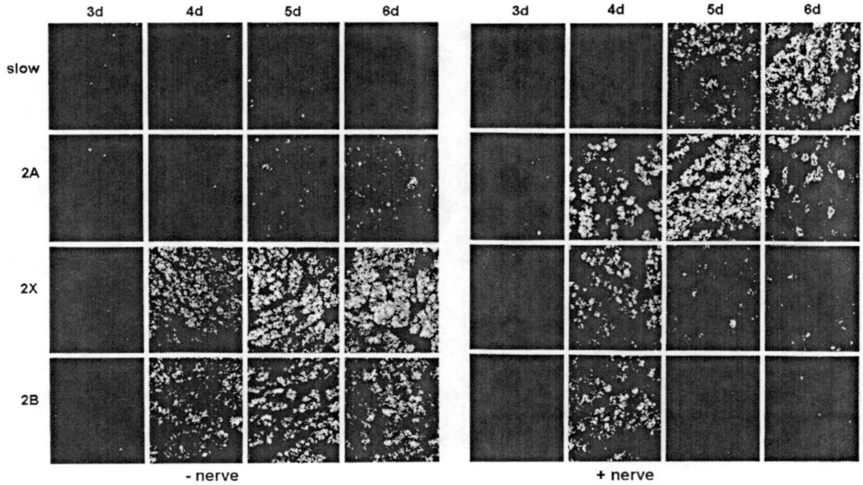

Figure 10. Nerve activity controls MyHC gene expression in regenerating rat soleus muscle. In situ hybridization reveals the distribution of MyHC-slow, -2A, -2X and -2B transcripts at day 3–6 after injury in the absence of the nerve (−nerve) or in the presence of the nerve (+nerve). Serial cryosections were hybridized with ^{35}S-labeled probes then processed for autoradiography and viewed by dark-field microscopy. White areas denote the presence of fibers expressing the specific MyHC mRNAs. No reactivity is seen at day 3 because at this stage regenerating myofibers express only embryonic and neonatal MyHCs. Modified from (Jerkovic et al., 1997)

frequency stimulation does not induce significant fast-to-slow fiber type conversion in adult rat muscles, MyHC-slow expression can be rapidly induced in the regenerating fast EDL when innervated by a slow nerve or stimulated with a low frequency pattern through the nerve (Erzen et al., 1999; Pette et al., 2002; Kalhovde et al., 2005). However, direct stimulation experiments have shown that fast and slow regenerating muscles do not respond in the same way to an identical stimulation pattern. The amount of MyHC-slow accumulated in the regenerating EDL is much lower than that induced in the regenerating soleus following the same low frequency stimulation, suggesting the existence of intrinsic differences between satellite cells of fast and slow muscles (Kalhovde et al., 2005).

6.3 Nerve-dependent Regulation of Fiber Growth in Regenerating Muscle

In vivo transfection of regenerating innervated or denervated soleus with constitutively active or dominant negative mutants of different signal transducers has been used to identify the intracellular signaling pathways that mediate the effects of nerve activity. As shown in Fig. 12, one would predict that the effect of nerve activity on muscle growth or fiber type specification could be blocked in muscle fibers transfected with appropriate dominant negative mutants or inhibitory peptides in

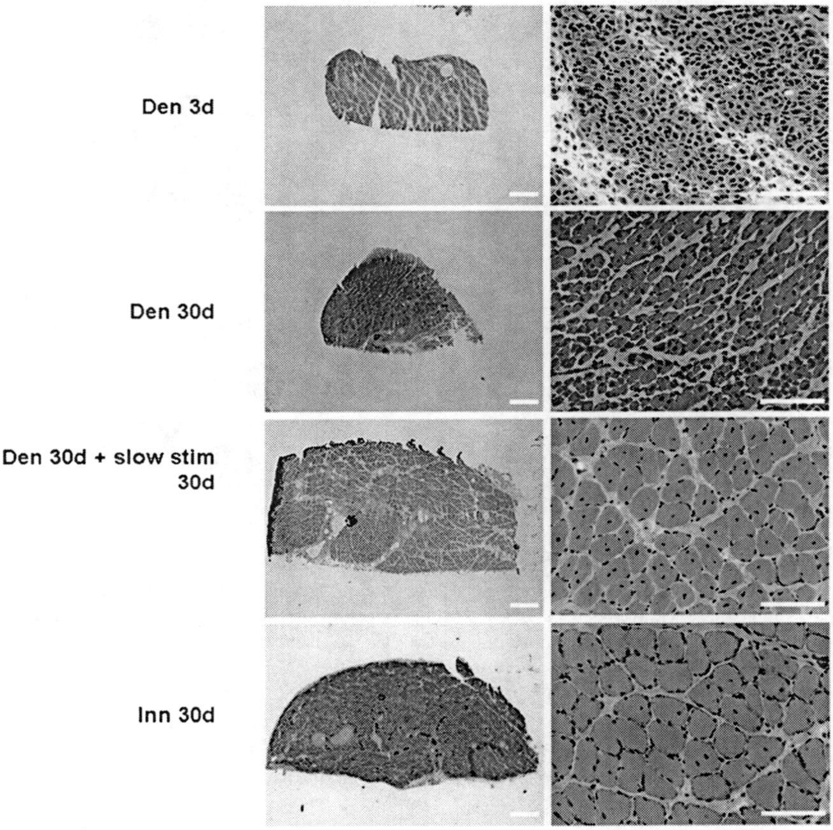

Figure 11. The effect of nerve activity on regenerating muscle growth is partly reproduced by direct electrical stimulation of denervated regenerating muscles. Cross-section of regenerating soleus muscles are shown at low (left panels) or higher magnification (right panels). Note the poor growth of denervated muscle seen at 3 days (Den 3d) and 30 days (Den 30d) after injury compared to the large increase in muscle fiber size in innervated muscle (Inn 30d) or in denervated muscle after stimulation with a low frequency pattern (Den 30d + slow stim 30d). Modified from (Kalhovde et al., 2005)

innervated muscle; on the other hand, transfection with constitutively active mutants could reproduce the effect of nerve activity in denervated muscle.

A major signaling pathway involved in muscle growth control in vitro and in vivo is the phosphoinositide 3-kinase (PI3K)-protein kinase B (PKB, also known as Akt) pathway (Rommel et al., 2001; Bodine et al., 2001). Constitutively active PI3K and PKB/Akt, as well as a Ras double mutant (RasV12C40) that activates selectively the phosphoinositide 3-kinase-PKB/Akt pathway, increase fiber size and prevent denervation atrophy when transfected in regenerating muscles but do not affect the fiber type profile (Murgia et al., 2000; Pallafacchina et al., 2002). The coexistence of hypertrophic muscle fibers overexpressing activated PKB/Akt with

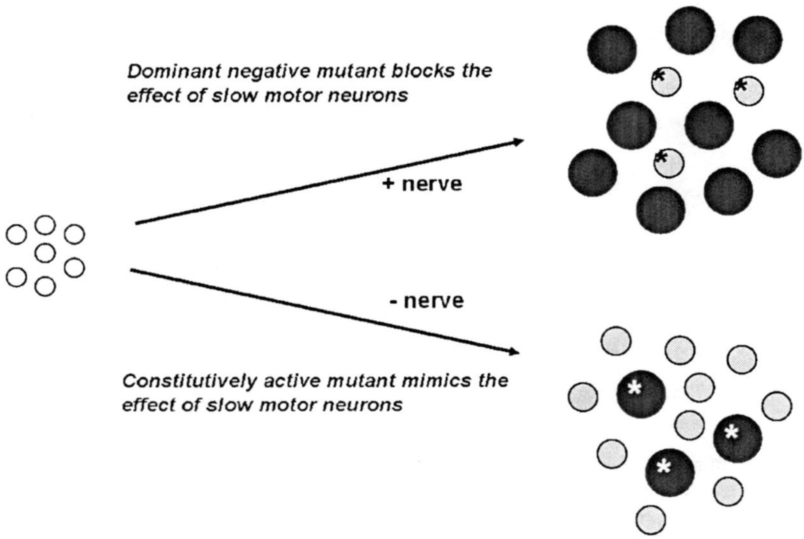

Figure 12. In vivo transfection in the regenerating rat soleus muscle allows to identify activity-dependent signaling pathways involved in fiber size and fiber type regulation. Transfection with a dominant negative mutant of a signal transducer blocks the effect of slow motor neuron activity in transfected fibers, labeled with asterisks, in the innervated regenerating soleus. On the other hand transfection of the regenerating denervated soleus with a constitutively active mutant induces muscle growth and upregulation of the MyHC-slow gene in transfected fibers. Modified from (Schiaffino et al., 2006)

normal-size untransfected fibers within the same muscle points to a cell-autonomous control of muscle growth by PKB/Akt (Fig. 13). This does not rule the possibility that PKB/Akt is also involved in the secretion from muscle fibers of factors acting on myogenic progenitor cells or other cell types. For example, activation of myogenic PKB/Akt signaling promotes increased secretion of VEGF (vascular endothelial growth factor) and capillary vessel formation (Takahashi et al., 2002). Two major downstream pathways mediate the hypertrophic effect of PKB/Akt in skeletal muscle fibers: the mammalian target of rapamycin (mTOR) kinase, that is activated by PKB/Akt and stimulates protein synthesis (Pallafacchina et al., 2002), and the transcription factors of the FoxO family, that are inhibited by PKB/Akt and stimulate protein degradation via the ubiquitin-proteasome pathway (Sandri et al., 2004).

6.4 Nerve-dependent Regulation of Fiber Type Specification in Regenerating Muscle

Two intracellular signaling pathways, the Ras-ERK and calcineurin-NFAT pathways, are involved in the nerve-dependent activation of the slow gene program in regenerating muscle. Transfection of regenerating denervated soleus with a Ras

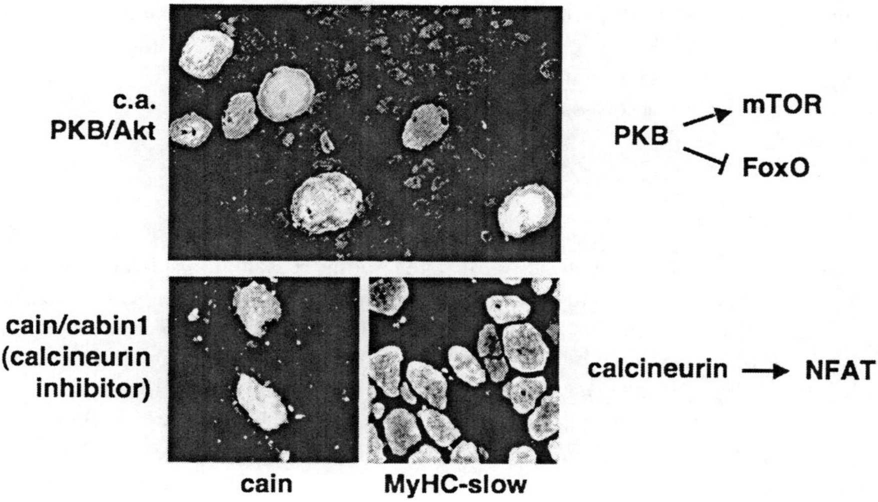

Figure 13. Signaling pathways involved in nerve activity-dependent regulation of muscle fiber size and fiber type regulation in regenerating muscle. *Upper panel*: Constitutively active (c.a.) PKB/Akt induces muscle fiber hypertrophy in regenerating skeletal muscle. Immunofluorescence analysis of a transverse section of regenerating denervated soleus muscle transfected with HA-tagged c.a. PKB and stained with anti-HA antibodies. Note that transfected fibers are much larger in size than surrounding untransfected fibers. PKB is known to stimulate protein synthesis by activating the mTOR (mammalian target of rapamycin) kinase and to block protein degradation through the ubiquitin-proteasome pathway by inhibiting the transcription factors of the FoxO gene family. (modified from Pallafacchina et al., 2002). *Lower panels*: The calcineurin inhibitor cain/cabin-1 prevents the up-regulation of MyHC-slow in regenerating soleus muscle. Serial cross sections of regenerating soleus muscle transfected with myc-tagged cain and stained with anti-myc (*left*) or anti-MyHC-slow (*right*) antibodies. Note that the two fibers expressing myc-tagged cain do not express MyHC-slow, unlike most surrounding untransfected fibers, but are similar in size. Calcineurin is known to induce the activation of the slow gene program by activating the transcription factors of the NFAT gene family. Modified from (Serrano et al., 2001)

double mutant (RasV12S35) that activates selectively the ERK1/2 pathway leads to expression of MyHC-slow without affecting muscle fiber size (Murgia et al., 2000). The physiological role of this pathway is supported by the finding that the effect of slow motor neurons is inhibited by a dominant-negative Ras mutant and that ERK activity is increased by innervation and low-frequency electrical stimulation (Murgia et al., 2000). The role of the calcineurin-NFAT pathway is supported by the finding that the up-regulation of MyHC-slow in regenerating rat soleus muscle is prevented by the calcineurin inhibitors cyclosporin A (CsA), FK506, and the calcineurin inhibitory protein domain from cain/cabin-1 (Serrano et al., 2001) (Fig. 13). Furthermore, the activation of MyHC-slow induced by direct electrostimulation of denervated regenerating muscle with a continuous low frequency impulse pattern is blocked by CsA, showing that calcineurin function in muscle fibers and not in motor neurons is responsible for the nerve-dependent specification of slow muscle fibers. The transcription factors of the NFAT gene family appear to mediate the effect of calcineurin in regenerating muscle since VIVIT, a specific peptide inhibitor of Cn-mediated NFAT activation, blocks the expression of MyHC-slow in regenerating soleus and a constitutively active NFATc1 mutant induces MyHC-slow expression in regenerating denervated soleus (McCullagh et al., 2004).

6.5 Conclusions

The extent of growth and maturation of regenerating muscle depends critically on the contractile activity normally induced by innervation. Activity in a general sense, and the specific patterns of activity associated with functionally distinct motor units, both play a role in regulating gene expression and shaping the functional properties of the regenerated muscle. These effects are facilitated because regenerating muscles are more plastic than adult muscles, presumably due to a progressive restriction in the transcriptional potential of muscle fibre nuclei. However, there are clear and significant limits to this plasticity. Thus regenerating fast and slow muscles respond differently to identical patterns of activity. This implies that regenerated muscles retain a 'memory' of their original type that is independent of patterned activation by the nerve, possibly due intrinsic differences between satellite cells from different muscles types.

7. GENERAL CONCLUSIONS

It is obvious that a regenerated muscle will only be useful if it is effectively innervated. The cellular and molecular events that account for the innervation of regenerating muscles have become increasingly well understood in recent years. In parallel, the impact of that innervation on muscle gene regulation, muscle growth and the functional diversification of regenerating muscle fibres is now better known. As a result, the older notion that the nerve released 'trophic' factors that controlled the global properties of muscles has been replaced by evidence that the pattern of use evoked by the nerve is a primary determinant of the functional properties of regenerated muscles. However, factors released by the nerve have a major role locally at the level of the NMJ in the differentiation of this specific domain of the muscle fibre.

The initial events of muscle regeneration can take place in the absence of the nerve. Thus the proliferation and fusion of myogenic satellite cells occur at much the same rate in muscles regenerating in the absence of the nerve as when the nerve is intact. This is hardly surprising since even in the most favourable situations the ability of the nerve to evoke muscle contraction does not develop until after the main events of myotube formation have occurred. However, in the absence of innervation, the functional differentiation of regenerated myotubes is arrested in a 'default' state similar to that of fast muscles. Muscle activity, normally driven by the nerve, is required for the emergence of 'slow' gene programme and for normal muscle fibre growth.

A key determinant of the success of both the regeneration and the subsequent innervation of muscles is the BL. The BL sheaths surrounding each muscle and nerve fibre provide a mechanical scaffold and supporting matrix that promotes orderly interactions between muscle and nerve cells during regeneration. If the BL is disrupted by the events causing muscle damage then both muscle regeneration and innervation are impaired. In addition to its general supporting role during regeneration, the BL acts as substrate for the attachment of important signalling molecules at the NMJ. The most important of these is agrin. By virtue of being bound to the synaptic BL, agrin survives the degeneration of both nerve and muscle cells, and serves to promote and coordinate rapid and effective reconstruction of NMJs on the regenerated muscle fibres.

For regenerated muscle fibres to have their full functionality, their properties must match those of the motor neurons that come to innervate them. It is likely that the processes that ensure this matching differ in some ways from those that operate during normal development. During development, each muscle fibre is initially innervated by numerous axons, only one of which survives in the adult. It is likely that the competitive process that selects the one survivor optimises the degree of functional matching between nerve and muscle cells. During the innervation of regenerated muscles in adults, most muscle fibres are never innervated by more than one axon. In this situation, functional matching comes about mainly by modification of the properties of the muscle fibres in response to the pattern of activity imposed by the nerve. While there are limits to this induced matching, it goes a long way to ensuring the effectiveness of the regenerated muscle. It is clear that the success of this process involves multiple signalling pathways that regulate different aspects of muscle fibre differentiation e.g. growth, contractile properties, electrical properties, endurance.

REFERENCES

Aigner L, Arber S, Kapfhammer JP, Laux T, Schneider C, Botteri F, Brenner HR, Caroni P (1995) Overexpression of the neural growth-associated protein GAP-43 induces nerve sprouting in the adult nervous system of transgenic mice. Cell 83:269–278

Barbier J, Popoff MR, Molgo J (2004) Degeneration and regeneration of murine skeletal neuromuscular junctions after intramuscular injection with a sublethal dose of Clostridium sordellii lethal toxin. Infect Immun 72:3120–3128

Bennett MR, Florin T, Woog R (1974) The formation of synapses in regenerating mammalian striated muscle. J Physiol 238:79–92

Bennett MR, McLachlan EM, Taylor RS (1973) The formation of synapses in reinnervated mammalian striated muscle. J Physiol 233:481–500

Benoit PW, Belt WD (1970) Destruction and regeneration of skeletal muscle after treatment with a local anaesthetic, bupivacaine (Marcaine). J Anat 107:547–556

Bodine SC, Stitt TN, Gonzalez M, Kline WO, Stover GL, Bauerlein R, Zlotchenko E, Scrimgeour A, Lawrence JC, Glass DJ, Yancopoulos GD (2001) Akt/mTOR pathway is a crucial regulator of skeletal muscle hypertrophy and can prevent muscle atrophy in vivo. Nat Cell Biol 3:1014–1019

Brenner HR, Herczeg A, Slater CR (1992) Synapse-specific expression of acetylcholine receptor genes and their products at original synaptic sites in rat soleus muscle fibres regenerating in the absence of innervation. Development 116:41–53

Bulfield G, Siller WG, Wight PA, Moore KJ (1984) X chromosome-linked muscular dystrophy (mdx) in the mouse. Proc Natl Acad Sci U S A 81:1189–1192

Burden SJ, Sargent PB, McMahan UJ (1979) Acetylcholine receptors in regenerating muscle accumulate at original synaptic sites in the absence of the nerve. J Cell Biol 82:412–425

Butler-Browne GS, Bugaisky LB, Cuenoud S, Schwartz K, Whalen RG (1982) Denervation of newborn rat muscle does not block the appearance of adult fast myosin heavy chain. Nature 299:830–833

Carlson BM, Faulkner JA (1988) Reinnervation of long-term denervated rat muscle freely grafted into an innervated limb. Exp Neurol 102:50–56

Caroni P, Schneider C, Kiefer MC, Zapf J (1994) Role of muscle insulin-like growth factors in nerve sprouting:suppression of terminal sprouting in paralyzed muscle by IGF-binding protein 4. J Cell Biol 125:893–902

Coers C, Woolf AL (1959) The innervation of muscle: a biopsy study. Oxford, Blackwell

Cohen I, Rimer M, Lømo T, McMahan UJ (1997) Agrin-induced postsynaptic-like apparatus in skeletal muscle fibers in vivo. Mol Cell Neurosci 9:237–253

d'Albis A, Weinman J, Mira JC, Janmot C, Couteaux R (1987) The regulator role of thyroid hormones in myogenesis. Analysis of isoforms of myosin in muscular regeneration. C R Acad Sci III 305:697–702

Dimitropoulou A, Bixby JL (2005) Motor neurite outgrowth is selectively inhibited by cell surface MuSK and agrin. Mol Cell Neurosci 28:292–302

Donovan CM, Faulkner JA (1987) Plasticity of skeletal muscle: regenerating fibers adapt more rapidly than surviving fibers. J Appl Physiol 62:2507–2511

Duchen LW (1970) Changes in motor innervation and cholinesterase localization induced by botulinum toxin in skeletal muscle of the mouse: differences between fast and slow muscles. J Neurol Neurosurg Psychiatry 33:40–54

Duchen LW, Excell BJ, Patel R, Smith B (1974) Changes in motor end-plates resulting from muscle fibre necrosis and regeneration. A light and electron microscopic study of the effects of the depolarizing fraction (cardiotoxin) of Dendroaspis jamesoni venom. J Neurol Sci 21:391–417

Dunaevsky A, Connor EA (1995) Long-term maintenance of presynaptic function in the absence of target muscle fibers. J Neurosci 15:6137–6144

Durbeej M, Campbell KP (2002) Muscular dystrophies involving the dystrophin-glycoprotein complex: an overview of current mouse models. Curr Opin Genet Dev 12:349–361

Erzen I, Primc M, Janmot C, Cvetko E, Sketelj J, d'Albis A (1999) Myosin heavy chain profiles in regenerated fast and slow muscles innervated by the same motor nerve become nearly identical. Histochem J 31:277–83

Esser K, Gunning P, Hardeman E (1993) Nerve-dependent and -independent patterns of mRNA expression in regenerating skeletal muscle. Dev Biol 159:173–83

Faulkner JA, Markley JM Jr, McCully KK, Watters CR, White TP (1983) Characteristics of cat skeletal muscles grafted with intact nerves or with anastomosed nerves. Exp Neurol 80:682–696

Gautam M, DeChiara TM, Glass DJ, Yancopoulos GD, Sanes JR (1999) Distinct phenotypes of mutant mice lacking agrin, MuSK, or rapsyn. Brain Res Dev Brain Res 114:171–178

Gautam M, Noakes PG, Moscoso L, Rupp F, Scheller RH, Merlie JP, Sanes JR (1996) Defective neuromuscular synaptogenesis in agrin-deficient mutant mice. Cell 85:525–535

Glass DJ, Bowen DC, Stitt TN, Radziejewski C, Bruno J, Ryan TE, Gies DR, Shah S, Mattsson K, Burden SJ, DiStefano PS, Valenzuela DM, DeChiara TM, Yancopoulos GD (1996) Agrin acts via a MuSK receptor complex. Cell 85:513–523

Glicksman MA, Sanes JR (1983) Differentiation of motor nerve terminals formed in the absence of muscle fibres. J Neurocytol 12:661–671

Goldman D, Carlson BM, Staple J (1991) Induction of adult-type nicotinic acetylcholine receptor gene expression in noninnervated regenerating muscle. Neuron 7:649–658

Grubb BD, Harris JB, Schofield IS (1991) Neuromuscular transmission at newly formed neuromuscular junctions in the regenerating soleus muscle of the rat. J Physiol 441:405–421

Gutmann E, Young JZ (1944) The reinnervation of muscle after various periods of atrophy. J Anat 78:15–43

Harlow ML, Ress D, Stoschek A, Marshall RM, McMahan UJ (2001) The architecture of active zone material at the frog's neuromuscular junction. Nature 409:479–484

Hennig R, Lømo T (1985) Firing patterns of motor units in normal rats. Nature 314:164–166

Heuser JE, Reese TS, Dennis MJ, Jan Y, Jan L, Evans L (1979) Synaptic vesicle exocytosis captured by quick freezing and correlated with quantal transmitter release. J Cell Biol 81:275–300

Hoch W, Ferns M, Campanelli JT, Hall ZW, Scheller RH (1993) Developmental regulation of highly active alternatively spliced forms of agrin. Neuron 11:479–490

Hunter DD, Porter BE, Bulock JW, Adams SP, Merlie JP, Sanes JR (1989a) Primary sequence of a motor neuron-selective adhesive site in the synaptic basal lamina protein S-laminin. Cell 59:905–913

Hunter DD, Shah V, Merlie JP, Sanes JR (1989b) A laminin-like adhesive protein concentrated in the synaptic cleft of the neuromuscular junction. Nature 338:229–234

Jansen JK, Lømo T, Nicolaysen K, Westgaard RH (1973) Hyperinnervation of skeletal muscle fibers: dependence on muscle activity. Science 181:559–561

Jerkovic R, Argentini C, Serrano-Sanchez A, Cordonnier C, Schiaffino S (1997) Early myosin switching induced by nerve activity in regenerating slow skeletal muscle. Cell Struct Funct 22:147–153

Jirmanova I (1975) Ultrastructure of motor end-plates during pharmacologically-induced degeneration and subsequent regeneration of skeletal muscle. J Neurocytol 4:141–155

Jirmanova I, Thesleff S (1972) Ultrastructural study of experimental muscle degeneration and regeneration in the adult rat. Z Zellforsch Mikrosk Anat 131:77–97

Jirmanova I, Thesleff S (1976) Motor end-plates in regenerating rat skeletal muscle exposed to botulinum toxin. Neuroscience 1:345–347

Jo SA, Burden SJ (1992) Synaptic basal lamina contains a signal for synapse-specific transcription. Development 115:673–680

Kalhovde JM, Jerkovic R, Sefland I, Cordonnier C, Calabria E, Schiaffino S, Lømo T (2005) 'Fast' and 'slow' muscle fibres in hindlimb muscles of adult rats regenerate from intrinsically different satellite cells. J Physiol 562:847–857

Korneliussen H, Sommerschild H (1976) Ultrastructure of the new neuromuscular junctions formed during reinnervation of rat soleus muscle by a 'foreign' nerve. Cell Tissue Res 167:439–452

Letinsky MS, Fischbeck KH, McMahan UJ (1976) Precision of reinnervation of original postsynaptic sites in frog muscle after a nerve crush. J Neurocytol 5:691–718

Lømo T, Slater CR (1978) Control of acetylcholine sensitivity and synapse formation by muscle activity. J Physiol 275:391–402

Love FM, Thompson WJ (1999) Glial cells promote muscle reinnervation by responding to activity-dependent postsynaptic signals. J Neurosci 19:10390–10396

Lupa MT, Caldwell JH (1994) Sodium channels aggregate at former synaptic sites in innervated and denervated regenerating muscles. J Cell Biol 124:139–147

Lyons PR, Slater CR (1991) Structure and function of the neuromuscular junction in young adult mdx mice. J Neurocytol 20:969–981

Magill-Solc C, McMahan UJ (1988) Motor neurons contain agrin-like molecules. J Cell Biol 107:1825–1833

Magill-Solc C, McMahan UJ (1990) Agrin-like molecules in motor neurons. J Physiol (Paris) 84:78–81

McCullagh KJ, Calabria E, Pallafacchina G, Ciciliot S, Serrano AL, Argentini C, Kalhovde JM, Lømo T, Schiaffino S (2004) NFAT is a nerve activity sensor in skeletal muscle and controls activity-dependent myosin switching. Proc Natl Acad Sci U S A 101:10590–10595

McMahan UJ (1990) The agrin hypothesis. Cold Spring Harb Symp Quant Biol 55:407–418

McMahan UJ, Slater CR (1984) The influence of basal lamina on the accumulation of acetylcholine receptors at synaptic sites in regenerating muscle. J Cell Biol 98:1453–1473

McMahan UJ, Wallace BG (1989) Molecules in basal lamina that direct the formation of synaptic specializations at neuromuscular junctions. Dev Neurosci 11:227–247

Meier T, Hauser DM, Chiquet M, Landmann L, Ruegg MA, Brenner HR (1997) Neural agrin induces ectopic postsynaptic specializations in innervated muscle fibers. J Neurosci 17:6534–6544

Miledi R (1960) Properties of regenerating neuromuscular synapses in the frog. J Physiol 154:190–205

Mitchell PO, Pavlath GK (2004) Skeletal muscle atrophy leads to loss and dysfunction of muscle precursor cells. Am J Physiol Cell Physiol 287:1753–1762

Moore C, Leu M, Muller U, Brenner HR (2001) Induction of multiple signaling loops by MuSK during neuromuscular synapse formation. Proc Natl Acad Sci U S A 98:14655–14660

Mozdziak PE, Pulvermacher PM, Schultz E (2001) Muscle regeneration during hindlimb unloading results in a reduction in muscle size after reloading. J Appl Physiol 91:183–190

Murgia M, Serrano AL, Calabria E, Pallafacchina G, Lømo T, Schiaffino S (2000) Ras is involved in nerve-activity-dependent regulation of muscle genes. Nat Cell Biol 2:142–147

Nagel A, Lehmann-Horn F, Engel AG (1990) Neuromuscular transmission in the mdx mouse. Muscle Nerve 13:742–749

Nitkin RM, Smith MA, Magill C, Fallon JR, Yao YM, Wallace BG, McMahan UJ (1987) Identification of agrin, a synaptic organizing protein from Torpedo electric organ. J Cell Biol 105:2471–2478

Okada K, Inoue A, Okada M, Murata Y, Kakuta S, Jigami T, Kubo S, Shiraishi H, Eguchi K, Motomura M, Akiyama T, Iwakura Y, Higuchi O, Yamanashi Y (2006) The muscle protein Dok-7 is essential for neuromuscular synaptogenesis. Science 312:1802–1805

Pallafacchina G, Calabria E, Serrano AL, Kalhovde JM, Schiaffino S (2002) A protein kinase B-dependent and rapamycin-sensitive pathway controls skeletal muscle growth but not fiber type specification. Proc Natl Acad Sci U S A 99:9213–9218

Pette D, Sketelj J, Skorjanc D, Leisner E, Traub I, Bajrovic F (2002) Partial fast-to-slow conversion of regenerating rat fast-twitch muscle by chronic low-frequency stimulation. J Muscle Res Cell Motil 23:215–221

Porter BE, Weis J, Sanes JR (1995) A motoneuron-selective stop signal in the synaptic protein S-laminin. Neuron 14:549–559

Reist NE, Magill C, McMahan UJ (1987) Agrin-like molecules at synaptic sites in normal, denervated, and damaged skeletal muscles. J Cell Biol 105:2457–2469

Reynolds ML, Woolf CJ (1992) Terminal Schwann cells elaborate extensive processes following denervation of the motor endplate. J Neurocytol 21:50–66

Rommel C, Bodine SC, Clarke BA, Rossman R, Nunez L, Stitt TN, Yancopoulos GD, Glass DJ (2001) Mediation of IGF-1-induced skeletal myotube hypertrophy by PI(3)K/Akt/mTOR and PI(3)K/Akt/GSK3 pathways. Nat Cell Biol 3:1009–1013

Sandri M, Sandri C, Gilbert A, Skurk C, Calabria E, Picard A, Walsh K, Schiaffino S, Lecker SH, Goldberg AL (2004) Foxo transcription factors induce the atrophy-related ubiquitin ligase atrogin-1 and cause skeletal muscle atrophy. Cell 117:399–412

Sanes JR (1989) Extracellular matrix molecules that influence neural development. Annu Rev Neurosci 12:491–516

Sanes JR, Marshall LM, McMahan UJ (1978) Reinnervation of muscle fiber basal lamina after removal of myofibers. Differentiation of regenerating axons at original synaptic sites. J Cell Biol 78:176–198

Sartore S, Gorza L, Schiaffino S (1982) Fetal myosin heavy chains in regenerating muscle. Nature 298:294–296

Schiaffino S, Sandri M, Murgia M (2006) Signalling pathways controlling muscle fiber size and type in response to nerve activity. In: Bottinelli R, Reggiani C (eds) Skeletal muscle plasticity in health and disease. From genes to whole muscle. Berlin, Springer, pp 91–119

Schmalbruch H (1977) Regeneration of soleus muscles of rat autografted in toto as studied by electron microscopy. Cell Tissue Res 177:159–180

Serrano AL, Murgia M, Pallafacchina G, Calabria E, Coniglio P, Lømo T, Schiaffino S (2001) Calcineurin controls nerve activity-dependent specification of slow skeletal muscle fibers but not muscle growth. Proc Natl Acad Sci U S A 98:13108–13113

Sicinski P, Geng Y, Ryder-Cook AS, Barnard EA, Darlison MG, Barnard PJ (1989) The molecular basis of muscular dystrophy in the mdx mouse: a point mutation. Science 244:1578–1580

Son YJ, Trachtenberg JT, Thompson WJ (1996) Schwann cells induce and guide sprouting and reinnervation of neuromuscular junctions. Trends Neurosci 19:280–285

Stocksley MA, Awad SS, Young C, Lightowlers RN, Brenner HR, Slater CR (2005) Accumulation of Na(V)1 mRNAs at differentiating postsynaptic sites in rat soleus muscles. Mol Cell Neurosci 28:694–702

Strochlic L, Cartaud A, Mejat A, Grailhe R, Schaeffer L, Changeux JP, Cartaud J (2004) 14-3-3 gamma associates with muscle specific kinase and regulates synaptic gene transcription at vertebrate neuromuscular synapse. Proc Natl Acad Sci U S A 101:18189–18194

Sunderland WJ, Son YJ, Miner JH, Sanes JR, Carlson SS (2000) The presynaptic calcium channel is part of a transmembrane complex linking a synaptic laminin (alpha4beta2gamma1) with non-erythroid spectrin. J Neurosci 20:1009–1019

Takahashi A, Kureishi Y, Yang J, Luo Z, Guo K, Mukhopadhyay D, Ivashchenko Y, Branellec D, Walsh K (2002) Myogenic Akt signaling regulates blood vessel recruitment during myofiber growth. Mol Cell Biol 22:4803–4814

Taxt T (1983) Cross-innervation of fast and slow-twitch muscles by motor axons of the sural nerve in the mouse. Acta Physiol Scand 117:331–341

Torres LF, Duchen LW (1987) The mutant mdx: inherited myopathy in the mouse. Morphological studies of nerves, muscles and end-plates. Brain 110(Pt 2):269–299

Vracko R, Benditt EP (1972) Basal lamina: the scaffold for orderly cell replacement. Observations on regeneration of injured skeletal muscle fibers and capillaries. J Cell Biol 55:406–419

Whalen RG, Harris JB, Butler-Browne GS, Sesodia S (1990) Expression of myosin isoforms during notexin-induced regeneration of rat soleus muscles. Dev Biol 141:24–40

Womble MD (1986) The clustering of acetylcholine receptors and formation of neuromuscular junctions in regenerating mammalian muscle grafts. Am J Anat 176:191–205

CHAPTER 15

BOOSTING MUSCLE REGENERATION

TOMMASO NASTASI AND NADIA ROSENTHAL
Mouse Biology Unit, European Molecular Biology Laboratory, Campus "A. Buzzati-Traverso", Via Ramarini 32, 00015 Monterotondo Scalo (Rome) ITALY

1. SUMMARY

Promoting efficient regeneration of damaged or degenerating human organs remains one of the most challenging aspirations of medicine. The problem is greatest in tissues such as the nervous system or the heart that are traditionally compromised in their regenerative capacity, hindered in their full recovery by the relatively poor response of progenitor cells to injury. By contrast, over six hundred muscles that compose the human body maintain a rich resource of resident satellite cell progenitors, which are activated to increase and rebuild muscle tissue during exercise or upon injury. Yet even the robust regenerative response of adult skeletal muscle is insufficient to maintain its form and function in ageing, where frailty is a major cause of morbidity. This deficit is exaggerated in muscle pathologies such as muscular dystrophies and myopathies, which progressively compromise skeletal muscle functional capacity, resulting in tissue atrophy through persistent protein degradation and activation of apoptotic and necrotic pathways.

The difficulty in countering muscle degeneration during ageing or disease derives in part from its heterogeneous composition. Adult skeletal muscle can adapt its phenotypic properties to fluctuating stimuli, including physical activity, injury, stimulation by motor neurons, oxygen and nutrient supply, and changes in hormone levels. To meet these functional demands, each muscle contains a multiplicity of fiber types that vary in composition and physiological properties. This heterogeneity is specified during development, but remains relatively plastic in response to changing environmental and activity demands (Russell et al., 2000), through efficient synthetic (anabolic) and degradative (catabolic) mechanisms that trigger the appropriate hypertrophic or hypotrophic/atrophic responses.

S. Schiaffino and T. Partridge (eds.), Skeletal Muscle Repair and Regeneration, 335–358.

The ability to remodel fiber structures and regulate fiber density, size and type in response to environmental and/or physiological demands is crucial for the integrity and functioning of the muscular system (Glass, 2005; Guttridge, 2004). Excessive and sustained anabolic hypertrophic conditions can lead to chronic hypertrophy, which has less serious consequences on skeletal muscle than on cardiac muscle, but has nevertheless implications on human health (Heineke and Molkentin, 2006; Hesselink and van Baak, 2004; Nader, 2005). More commonly, alteration in muscle fiber composition represents a major component in the muscle degeneration associated with massive catabolic conditions in the wake of cancer, heart failure or infectious disease, leading to pathological decrease of muscle mass in age-related sarcopenia, and other cases of muscle wasting and atrophy (Schakman and Thissen, 2006; Shavlakadze and Grounds, 2006).

In this chapter we will discuss the most recent achievements in the elucidation of mechanisms underlying muscle growth, plasticity and repair. Behind muscle homeostasis lies a complex and interconnected network of signalling pathways that dynamically modulate the balance between anabolic and catabolic states, and control the activation and subsequent differentiation of satellite cell progenitors. Understanding the molecular basis of these pathways is critical to the design of therapies to control, ameliorate or cure age-related, metabolic and genetic disruptions of muscle function. We will review the discovery of molecular targets that modulate muscle homeostasis, and will review their relative efficacy as tools in boosting muscle regenerative potential, focusing on the action of two proteins, Insulin-like Growth Factor 1 (IGF-1) and Myostatin (MSTN), which play key roles in boosting muscle capacity to repair and regenerate. Modulation of these two muscle regulators in murine models and in some human myopathies has shown considerable potential for increasing muscle mass and function, relieving some of the phenotypes of muscle pathologies and ameliorating the loss of plasticity in ageing muscles. The tremendous social impact that therapeutic agents could have in the treatment and or prevention of a large group of diseased states affecting muscle function cannot be overestimated.

2. CONTROL OF SKELETAL MUSCLE MASS

Formation and patterning of limb and body muscles are achieved through a precisely orchestrated series of events that occur during embryonic and fetal development (Buckingham et al., 2003; Buckingham et al., 2003; Christ and Brand-Saberi, 2002; Hollway and Currie, 2003). After birth, skeletal muscle tissue undergoes postnatal growth based on residual proliferation/differentiation of neonatal myoblasts as well as on the stimulation of genetic programs controlling muscle hypertrophy; the number of nuclei per fiber increases with fiber size and density of the contractile matrix (Cossu and Biressi, 2005; Tajbakhsh, 2005). The muscle growth that is seen in the early postnatal stages and adult transition follows a specific program of muscle maturation which replaces developmental isoforms of several muscle proteins with their adult counterparts, and leads to establishment of efficient specialized muscle structures, such as the neuromuscular junctions or the sarcomeres and their membrane costameric connections (Allen and Leinwand, 2001; Eizema et al., 2007; Geiger et al., 2006; Strbenc et al., 2006). Aberrant deviation from such muscle growth program is a peculiar feature of several muscle congenital diseases. In addition to the intrinsic developmental program of muscle formation, it is generally accepted that loading, i.e. an increased demand for muscle activity, is a physiological prerequisite during the early postnatal stages, and accounts in large part for the striking increase in muscle mass occurring in this period (Bodine, 2006; Buckingham, 2006; Shavlakadze and Grounds, 2006).

2.1 Mechanisms of Muscle Hypertrophy

Adult skeletal muscle tissue responds to nutritional, hormonal or mechanical stimuli by increasing strength of contraction and/or mass (Shavlakadze and Grounds, 2006). This adaptation can occur in different ways. An increase in myofibrillar density correlates with modulation of contraction patterns and expression of specific protein isoforms, which in turn will allow for more efficient cell metabolism and contraction properties. Alternatively, activation of muscle precursors will increase the number of fibers in a muscle. These effects *per se*, also when coupled to increased number of nuclei and/or increased myofibrillar content per fiber, do not necessarily lead to fiber hypertrophy (Shavlakadze and Grounds, 2006). Increased force loading on the muscle can be accompanied, instead, by increases in fiber diameter and muscle mass, a prerequisite of the hypertrophic response (Fig. 1). No less relevant is the function exerted by fiber innervations to accommodate for a change in loading demands (Paul and Rosenthal, 2002; Pette, 2001), an effect that can be mimicked experimentally, leading to change in type, size and physiological properties of a muscle fiber.

The plasticity of the systems involved in muscle hypertrophy, and their complexity underscore how robust the processes controlling these events must be. In fact, the molecular players of the hypertrophic response form interconnected networks, which ultimately modulate muscle fiber size and muscle mass in response to hormonal or mechanical stimuli. Ultimately, the signals activated by increased loading or by hormonal clues result in a strong anabolic response. The most extensively documented pathway involves the PI3K/Akt/mTOR signaling cascade (Bodine, 2006; Guttridge, 2004) (Fig. 3). Simplistically, upon stimulation of this pathway Akt phosphorylates and activates the mammalian Target Of Rapamycin (mTOR) (Bodine, 2006; Bodine et al., 2001b; Reynolds et al., 2002) enhancing protein synthesis (Glass, 2005), and phosphorylates and inhibits Glycogen Synthase Kinase 3β (GSK3β), which has been implicated in cell survival as well as regulation of protein synthesis (Bodine, 2006; Glass, 2005). The fact that rapamycin, which blocks mTOR-dependent signaling, is able to suppress the exercise-induced hypertrophy confirms the importance of this pathway in the control of muscle fiber size (Bodine et al., 2001b). In reality the situation is far more complex, involving different isoforms of each member of the PI3K/Akt/mTOR cascade, which can

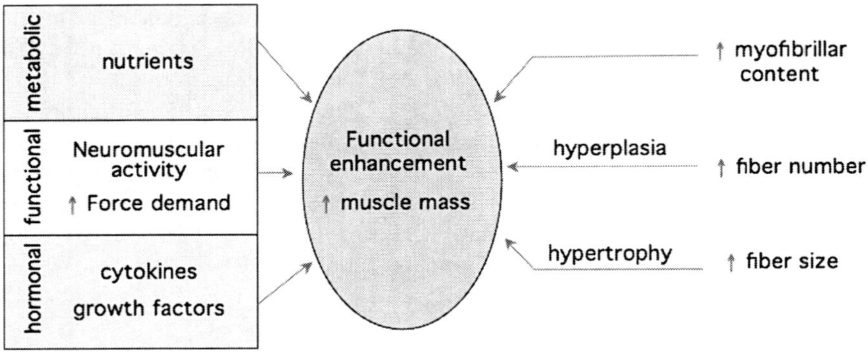

Figure 1. Signalling pathways involved in muscle hypertrophy

combine in several modes according to the tissue of expression and threshold of activation (Cho et al., 2001; Nader, 2005; Wullschleger et al., 2006). Moreover, the specific relevance of GSK3 in muscle hypertrophy has been recently put into question by McManus and coworkers (McManus et al., 2005) who generated knock-in animals carrying GSK3 genes mutated at the Ser residues phosphorylated by Akt. Few significant effects were seen, other than impaired Glycogen Synthase (GS) activation in muscles upon insulin stimulation. Whether these mice present with other more subtle defects arising from deficient Akt-dependent GSK3 inhibition awaits further investigation.

Other pathways that participate in muscle hypertrophy are controlled by members of the Mitogen Associated Protein Kinase (MAPK), namely Extracellular signal Regulated Kinase 1/2 (ERK1/2), c-jun N-terminal Kinase (JNK) and p38 MAPK, or by the calcium-dependent phosphatase calcineurin (Bassel-Duby and Olson, 2003; Bassel-Duby and Olson, 2006; Glass, 2005; Shavlakadze and Grounds, 2006) (Fig. 3). The relevance of these pathways is underscored by their evolutionary conservation, in the regulation of human skeletal muscle metabolism (Sakamoto et al., 2004).

2.2 Mechanisms of Muscle Atrophy

Muscles also adapt to reduction of nutrients or activity by increasing lysosome-dependent or cytosolic degradative pathways that are physiologically responsible for the turnover of muscle proteins. In addition, new pathways are activated to reduce the muscle mass and to promote a shift towards atrophy (Glass, 2005; Guttridge, 2004).

At least three degradative systems are active in muscle fibers to modulate protein turnover, in order of increasing complexity (Fig. 2): lysosomal degradation; ubiquitin-mediated proteasomal systems (Ciechanover, 2005; Reinstein and Ciechanover, 2006) and a recently uncovered pathway where ubiquitin-mediated degradation is coupled to the proteolytic activity of calpain-3 directed

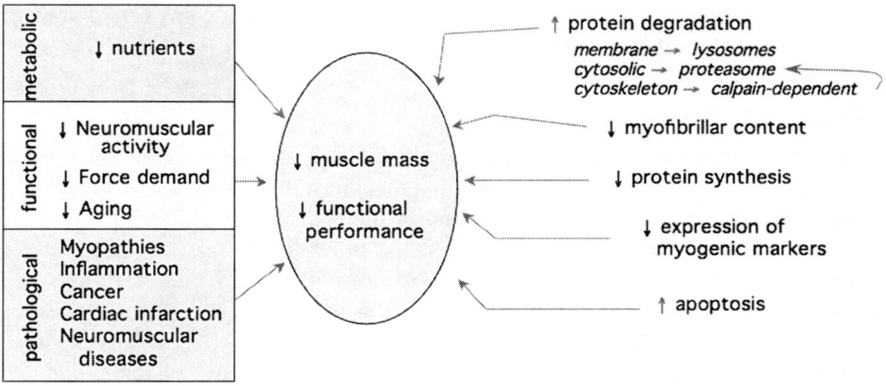

Figure 2. Signalling pathways involved in muscle atrophy

towards cytoskeleton and sarcomeric components (Duguez et al., 2006; Kramerova et al., 2005; Kramerova et al., 2006). Several proteins involved in selective degradation of muscle-specific substrates have been described (Bodine et al., 2001a; Gomes et al., 2001; Jagoe and Goldberg, 2001; Nastasi et al., 2004), and the importance played by degradative pathways in muscle atrophic states is well established (Attaix et al., 2005; Cao et al., 2005; Glass, 2005; Kandarian and Jackman, 2006).

A balance between anabolic and catabolic signals is clearly critical for the efficient maintenance of muscle fiber homeostasis. Recent reports describe an important pathway for controlling muscle size that involves activation of Akt and Akt-dependent inactivation of FOXO transcription factors, members of the forkhead family. This in turn leads to inactivation of MuRF1 and MAFbx E3 ubiquitin-ligases, involved in several models of muscle atrophy (Sandri et al., 2004; Stitt et al., 2004). Thus the PI3K/Akt pathway is able to stimulate muscle growth by stimulating hypertrophy, on one hand, and by inhibiting muscle atrophy on the other. However, the PI3K/Akt pathway alone is not likely to be responsible for all modes of fiber hypertrophy, as other pathways are also able to promote these changes. Conversely, stimulation of this one pathway cannot counteract all modes of muscle atrophy, such as the one induced by pro-inflammatory cytokines. Indeed, muscle atrophy induced by inflammation seems to be mediated by NF-kB-dependent and Akt-independent pathways (Dehoux et al., 2007; Mourkioti et al., 2006; Mourkioti and Rosenthal, 2005; Spate and Schulze, 2004).

2.3 Pathological Changes in Muscle Mass

Several pathological conditions can lead to massive increase (hypertrophy) or loss of muscle mass (muscle wasting and cachexia). These conditions, whether congenital or acquired, ultimately lead to inability of the muscle to compensate for loss of mass or to maintain the hypertrophic state.

2.3.1 *Anabolic conditions*

Skeletal muscle hypertrophy induced by increased loading upon exercise is balanced by the energy needed for this activity, by the physiological damage repair and protective mechanisms of muscle tissue and by the general self-limiting and conservational mechanisms of the whole body (Close et al., 2005). However, after exercise muscles grow by increasing mass and activity, which requires more nutrients and blood supply and more efficient nerve-muscle connections, as well as by regenerating damaged muscle fibers (Close et al., 2005). Exercise, especially resistance and training exercise, is known to trigger variations in local and systemic concentrations of selected hormones (Kraemer and Ratamess, 2005), whereas boosting muscle growth by means of drugs, such as anabolic steroids, produces secondary pathologies arising from administration even at doses lower than those illicitly used to achieve better muscle performance (Hartgens and Kuipers, 2004; Herbst and Bhasin, 2004; Kraemer and Ratamess, 2005).

2.3.2 *Catabolic conditions*

Independent of the pathways that initially trigger loss of muscle mass, defined as muscle wasting or atrophy, the common end point is the shift of the balance between protein synthesis and degradation towards a catabolic state (Glass, 2005; Guttridge, 2004). Loss of muscle mass can occur under many different conditions: it can originate from genetic defects affecting the body hormonal, neuronal or muscular systems; it occurs in conditions of starvation or forced disuse of muscles; it presents as secondary effects of other pathological conditions, such as congestive heart failure (cardiac cachexia), cancer (cachexia), diabetes, immune disease; and it is one of the most debilitating aspects of the ageing process (Kandarian and Jackman, 2006; Schakman and Thissen, 2006). Most importantly, muscle loss, when associated with these other conditions, significantly lowers the probability of a positive prognosis; under such conditions, loss of more than 40% of muscle mass is *per se* a life threatening condition (Schakman and Thissen, 2006).

It is likely that in absence of sufficient neuro-hormonal stimulation, or repair of neuromuscular damage, muscle tissue cannot sustain physiological homeostasis and succumbs to catabolic degradation. The importance of hypertrophy as the engine driving muscles away from atrophic decline is underscored by the observation that Akt1/Akt2 double knock-out mice show marked muscle atrophy (Peng et al., 2003), combined with evidence of decreased Akt/mTOR activation in at least two models of muscle atrophy, namely cardiac cachexia (Delafontaine and Akao, 2006; Song et al., 2005b) and hindlimb suspension (Reynolds et al., 2002). A major goal of clinical research is to identify targets of molecules that can control and re-establish the balance between anabolic and catabolic functions, in conditions where this balance is lost.

Considering its numerous origins, muscle wasting is clearly a more complex multifactorial phenomenon that can compromise tissue integrity by interfering with mechanisms that repair damage, protect from inflammatory-toxic insults, regenerate new fibers, modulate fiber size and myofibrillar density or upset the balance between protein synthesis vs. degradation. During muscle atrophy each of these homeostatic mechanisms can fail either singly or in combination (Schakman and Thissen, 2006; Shavlakadze and Grounds, 2006). Although ubiquitin-proteasome mediated protein degradation is currently a promising target for therapies (Attaix et al., 2005; Glass, 2005; Jagoe and Goldberg, 2001; Tisdale, 2005), the variety of modes and intensity of muscle atrophy constitute a major determinant in the outcome of any potential treatment.

3. REGENERATING MUSCLE

In designing therapies that can cure or ameliorate conditions of muscle damage or atrophy, it is important to take into account the limitations of intrinsic regenerative potential of adult and senescent muscle tissue (Buckingham, 2006; Zammit et al., 2006). It has been shown, for example, that age-related sarcopenia is associated with reactivation of myogenic factors and embryonic myosin isoform expression, but also with incomplete differentiation of the regenerating fibers (Edstrom and Ulfhake, 2005). Other studies have revealed that ageing muscles retain the ability to regenerate, but lack the appropriate stimuli to activate endogenous progenitor cells (Luo et al., 2005; Wagers and Conboy, 2005). The field is currently focused on how the intrinsic deficits that lead to muscle degenerative states can be modulated. Developing methods to supply or increase activated muscle precursors is crucial for the treatment of congenital muscle disease as well. As discussed elsewhere in this volume, several potential candidate cell types, if properly activated, could be useful in the development of cell-mediated therapies for muscle pathologies. We will comment in this chapter on some of the genetic pathways that impact upon the regenerative capacity of muscle that may be alternative or supplementary targets of therapeutic intervention.

3.1 Muscle Boosters

An ideal muscle booster would be a molecule able to maintain muscle mass, suppress muscle loss and stimulate muscle regeneration. To date two factors, IGF-1 and MSTN, have been implicated in the control of these functions, each targeting different intracellular pathways.

3.1.1 Insulin-like growth factor 1

Insulin-like Growth Factor 1 (IGF-1), also known as Somatomedin C, is a member of insulin-related bioactive peptides. Expressed in almost every tissue, IGF-1 participates in body growth, together with growth hormone (GH) that induces its secretion from the liver. Its role during development and in the early postnatal phases is

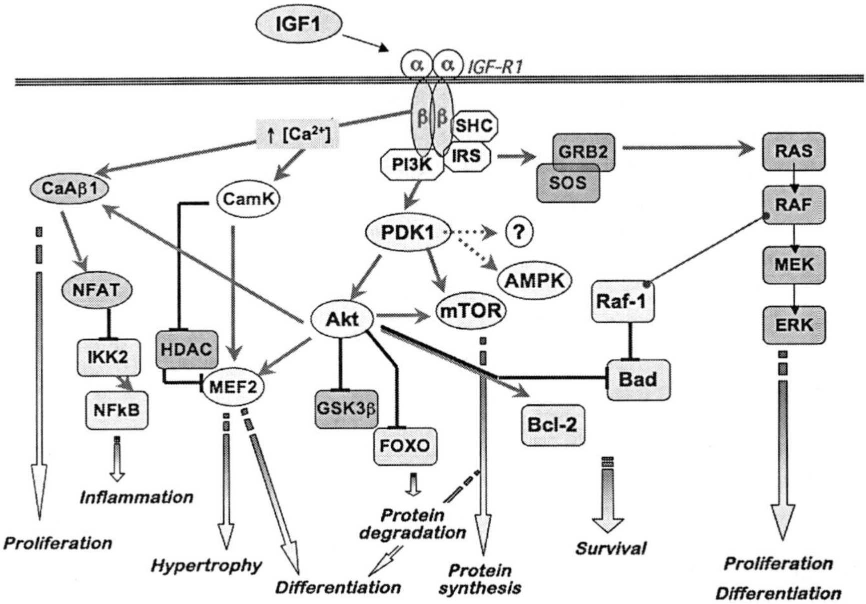

Figure 3. Modes of action of IGF-1

crucial as deficiency of this factor, either as primary defect or in response to reduced production/response to GH levels, is the leading cause of a subset of growth pathologies.

Unlike the insulin gene, the single-copy IGF-1 gene locus encodes multiple proteins with variable N- and C-terminal amino acid sequences, generated by use of variable transcriptional start sites and alternative splicing events (Winn et al., 2002). Although the IGF-1 gene is highly conserved in numerous species, its relatively large size (over 70 kb), and its complex transcript production have complicated its analysis. IGF-1 is synthesized as a pre-pro-hormone, which undergoes at least two processing events: cleavage of the signal peptide and of the C-terminal peptide. These events generate the mature 70aa-long active peptide, which is responsible for mediating the action of GH (Etherton, 2004; Le Roith et al., 2001a; Le Roith et al., 2001b).

Although circulating IGF-1 is provided mainly by the liver, its paracrine actions through expression of different isoforms in extra-hepatic tissues may be the most important contributors to growth (Fig. 3). Processing of these isoforms, either intracellularly or extracellularly, is necessary to produce mature active IGF-1 peptide, a single-chain protein of 70 amino acids that differs from insulin by retention of the C-domain, by a short extension of the A-domain to include a novel domain D, and by the presence of variable C-terminal Extension peptides (E-peptides). Data from transgenic animals expressing various IGF-1 isoforms in skeletal muscle suggest that specific responses elicited by local production of the factor *in vivo* may be distinct from the response elicited by exogenously provided IGF-1 (Barton, 2006; Shavlakadze et al., 2005).

In muscle, IGF-1 is highly upregulated upon differentiation (McLellan et al., 2006), stimulating the expression of differentiation markers (Musaro and Rosenthal, 1999; Noguchi, 2005; Tiffin et al., 2004; Xu and Wu, 2000) and down regulating the expression of embryonic muscle protein isoforms (Rodgers, 2005). Thus IGF-1 actively contributes to the muscle differentiation program.

IGF-1 function is also critical in modulating the balance between anabolic and catabolic signals. IGF-1 mediated signal transduction activates intracellular PI3K activity, resulting in increased phosphorylation and activation of the Akt/mTOR regulated pathways: these events lead to muscle hypertrophy (Bassel-Duby and Olson, 2006; Glass, 2005; Rommel et al., 2001) (Fig. 3). However, in the hypertrophic muscles of animals expressing a full length native IGF-1 transgene (mIGF-1), chronic Akt phosphorylation was not observed and was supplanted by PDK1 activation, indicating that the paracrine hypertrophic action of IGF-1 might employ alternative pathways to activate mTOR (Shavlakadze et al., 2005; Song et al., 2005a; Song et al., 2005b).

IGF-1 dramatically suppresses protein degradation as well. Muscle overexpression of mIGF-1 in transgenic mice with left ventricular dysfunction prevented muscular atrophy and inhibited skeletal muscle Foxo transcription factors, ultimately leading to decreased expression of at least two E3-ubiquitin ligases involved in muscle atrophy and to reduced activity of the ubiquitin-mediated proteasomal degradation system (Sandri et al., 2004; Schulze et al., 2005; Stitt et al., 2004). The same transgene also ameliorated the effects of angiotensin II-mediated muscle wasting (Song et al., 2005b).

Expression of IGF-1 in muscle has the potential to ameliorate tissue deterioration caused by very diverse mechanisms. Crossing an mIGF-1 transgene into the *mdx* animal model has proved effective in reducing the dystrophic phenotype (Barton et al., 2002; Noguchi, 2005). In addition to its direct effects on muscle atrophy, IGF-1 has a neurotrophic effect on neurons, and promotes elongation and branching (Rabinovsky, 2004). This function extends also to peripheral motor neurons, and muscle expression of IGF-1 can rescue not only the muscular atrophy but also some of the neuronal defects in a mouse model of Amyotrophic Lateral Sclerosis (ALS) (Dobrowolny et al., 2005).

Although the evidence for IGF-1 pleiotropic actions in the counteraction of muscle wasting is supported by these studies, we are still far from understanding how the IGF-1 system works and how it might be tweaked in a clinical setting. This is largely because the IGF-1 system is extremely complex. The peptide exerts its actions through the IGF-1 Receptor, but has lower affinity for Insulin Receptor as well; activation of these receptors transduces alternative or overlapping intracellular signaling events, mediated by different combinations or levels of receptor activation, phosphorylation of one or more members of the Insulin Receptor Substrate family, and activation of additional intracellular kinases. In addition to this scenario, IGF-1 forms complexes with a group of more than six different IGF Binding Proteins (IGFBP), likely responsible for its stability in the extracellular milieu, or for its presentation in an active conformation.

In addition to the direct effects of IGF-1 stimulation, specific isoforms of IGF-1 may be effective in stimulating the regeneration of ageing muscle by activating locally residing myogenic cells or by increased recruitment of circulating multipotent precursors (Musaro et al., 2004). This is of particular relevance for the development of future IGF-1-based therapies and for the amelioration of muscle loss associated with ageing. In ageing muscle, in fact, muscle wasting is associated with the inability to trigger Notch-dependent activation of resident regenerating satellite cells (Luo et al., 2005; Wagers and Conboy, 2005) coupled with reduced activity and with alteration of the neuro-hormonal environment and of physiological hypertrophic signals. The atrophy associated with ageing and its potential treatment could have significant impact on general health conditions (Edstrom and Ulfhake, 2005; Harridge, 2003; Musaro et al., 2001).

3.1.2 Myostatin

Myostatin (MSTN), also called Growth and Differentiation Factor 8 (GDF8), is a member of the TGFβ superfamily of growth factors. Identified relatively recently, this protein has quickly captured the interest of the scientific community because of its multifaceted functions on muscle development and regeneration (Joulia-Ekaza and Cabello, 2006).

MSTN, like other members of the TGFβ family, functions as ligand of a class of transmembrane receptors, which can trigger the activation of intracellular signaling carried out by the Smad family of transcriptional regulators or by other Smad-independent regulators (Lee, 2004; Moustakas and Heldin, 2005). Regulation of the ligand-receptor binding as well as of the Smad signaling network is extremely complex and still under investigation; however it is widely accepted that this system plays a pivotal role in a broad range of events, ranging from cell fate decision to control of tumor cell progression (Chen and Meng, 2004; Fogarty et al., 2005; Hubmacher et al., 2006; Izzi and Attisano, 2006; Kluppel and Wrana, 2005; Miyazono et al., 2004; Tian and Meng, 2006).

MSTN is synthesized as a pre-pro-peptide, which undergoes cleavage of the signal peptide to yield a secretable pro-peptide form. Correct processing of the inhibitory N-terminal pro-peptide and dimerization of the C-terminal active peptide are two requirements for binding to the serine/threonine kinase Activin type II Receptor (ActRIIB) and for the activity of the factor (Joulia-Ekaza and Cabello, 2006).

MSTN knockout animals show a marked increase of muscle mass, due to both increase in muscle fiber number and size (McPherron et al., 1997). On the other hand, overexpression of the inhibitory N-terminal pro-domain or of a cleavage-deficient dominant negative mutant of MSTN, seemed to lead to muscle hypertrophy rather than increased fiber number. One explanation might be that these overexpression studies used promoters that become active upon differentiation, thereby hindering the effects of MSTN blockade in undifferentiated myoblasts

(Joulia-Ekaza and Cabello, 2006). Indeed, experiments performed *in vitro*, either expressing MSTN or blocking its activity, provide rigorous evidence for a role of this protein in regulating cell cycle. In a proliferative context, MSTN induced cell cycle arrest of myoblasts in G0/G1 and G2 phases, by stimulating accumulation of the cyclin-CDK inhibitor p21 and by reducing the levels of phosphorylated inactive retinoblastoma (Rb) tumor suppressor protein (Joulia et al., 2003). In line with these data, MSTN reduced satellite cell activation and migration in *ex vivo* experiments, confirming the inhibitory role of MSTN on activation of muscle precursors (McCroskery et al., 2005; Walsh and Celeste, 2005).

When MSTN effects are studied upon induction of differentiation, the results are coherent with a block of this program, as indicated by reduced levels of MyoD and myogenin, lack of increased expression of terminal differentiation markers and reduced expression of p21, which under irreversible G0/G1 arrest is upregulated in a MyoD-dependent manner. Accordingly, when MSTN is blocked, the fusion index of cultured myoblasts is greatly enhanced (McCroskery et al., 2005). Thus, MSTN seems to be a master tuner controlling the programs of muscle proliferation and differentiation, and exerts its myostatic activity by blocking proliferation in conjunction with preventing progression of the differentiation program (Joulia et al., 2003).

In a pathological context, MSTN is upregulated during processes of muscle atrophy. Muscle loss following leg immobilization or HIV infection is associated to increased levels of MSTN (Carlson et al., 1999; Gonzalez-Cadavid et al., 1998; Morley et al., 2006). These effects are reversed after a cycle of exercise, which reestablish normal levels of MSTN, in mice as well as in humans (Jones et al., 2004; Wehling et al., 2000). However the role of MSTN on muscle loss may be dependent on the peculiar pattern of progression of a specific model of muscle atrophy; in fact, after hind limb suspension, muscle atrophy was even augmented, rather than diminished, in MSTN knockout animals compared to wild type controls (McMahon et al., 2003).

Recently, emphasis has been given to the use of neutralizing antibodies, which block MSTN signaling, to the point that clinical trials using this type of therapy are currently ongoing for the treatment of several muscle pathologies (Holzbaur et al., 2006; Walsh and Celeste, 2005). Use of MSTN neutralizing antibodies has been successful in the treatment of the mouse model for Duchenne Muscular Dystrophy (the *mdx* mutant) (McCroskery et al., 2005; Patel et al., 2005), in the suppression of muscle atrophy in a ALS mouse model (Holzbaur et al., 2006) and in the treatment of dexamethasone-induced muscle atrophy (Gilson et al., 2007). MSTN neutralizing antibodies did not produce significant side effects other than improving the atrophic phenotype, which is promising from a therapeutical perspective (Walsh and Celeste, 2005); however, while both hyperplasia and hypertrophy caused increased muscle mass in MSTN knock-out animals, treatment with neutralizing antibodies induces only hypertrophy, which again raises the argument that some of the effects attributed to MSTN reflect a more complex action. Nevertheless, the beneficial effects of blocking MSTN activity include increased fiber size and strength, increased regeneration and decreased inflammation and fibrosis of the damaged tissue. Whether MSTN has functions on tissues other than muscle also remains unresolved. In MSTN -/- animals increased myogenesis is coupled to decreased fat tissue, suggesting a role for this protein in muscle vs. fat cell fate decision (Lin et al., 2002).

In addition to all these effects, blockade of MSTN function also represses protein degradation, by down-regulating the expression of ubiquitin ligases and other proteases (Gilson et al., 2007). Intriguingly, MSTN controls the expression levels of the ubiquitin ligases MuRF1 and Atrogin-1/MAFbox, by reducing Akt activity and thereby enhancing FoxO1 transcription of those genes; a mechanism that mirrors the IGF-1-PI3K-Akt pathway (Gilson et al., 2007; McFarlane et al., 2006). Whether IGF-1 and MSTN crosstalk at key nodes of the intracellular signaling network is not known, however they seem to oppositely swing the balance of the muscle fiber towards primarily anabolic or catabolic condition. It is therefore not surprising that marked levels of muscle atrophy have been associated with increased MSTN and decreased IGF-1 levels, in a mouse model of liver cirrhosis (Dasarathy et al., 2004). Further investigation is needed to determine whether simultaneous and localized modulation of the action of these two factors would be effective in treating muscle wasting conditions.

3.1.3 NF-κB

The diverse molecular players in muscle atrophy include proinflammatory cytokines, which have been implicated in the loss of muscle mass and function (Spate and Schulze, 2004). Among the pathways controlling inflammation, those activating transcription factor NF-κB play a major pleiotropic role in the modulation of immune, inflammatory, cell survival and proliferative responses (Karin et al., 2004). In unstimulated cells, NF-κB is kept inactive via association with inhibitory proteins, the IκBs. Phosphorylation of IκB is mediated by the IκB kinase (IKK) complex, which contains two catalytic subunits (IKK1/α and IKK2/β) (Zandi et al., 1997) and a regulatory subunit termed NEMO/IKKγ (Yamaoka et al., 1998). Upon a variety of stimuli, including TNFα, IL-1 and other growth factors, IκBa is phosphorylated and degraded through the ubiquitination pathway, rendering NF-κB free to accumulate in the nucleus (Ghosh and Karin, 2002). Once NF-κB is in the nucleus, it binds to κB binding sites and interacts with transcriptional co-activators to regulate gene expression (Li, 2002).

The idea that NF-κB may be responsible for mediating the skeletal muscle loss first arose from studies in muscle tissue cultures, when it was shown to influence myogenic growth and differentiation (Guttridge et al., 1999; Li, 2003), down-regulate MyoD expression (Guttridge et al., 2000; Langen et al., 2004) and induce muscle atrophy in mice (Cai et al., 2004; Mourkioti et al., 2006). Recently, the *in vivo* contribution of NF-κB to skeletal muscle development and regeneration has been demonstrated (Mourkioti et al., 2006). Deletion of IKK2, a critical modulator of NF-κB pathway, increased muscle strength and led to maintenance of both muscle mass and function in response either to denervation of the sciatic nerve, which causes degeneration of the leg muscles, or to induction of muscle regeneration after cardiotoxin administration. Moreover, it was shown that IGF-1 overexpression acts additively with NF-κB inhibition, resulting in enhanced protection against muscle atrophy. Interestingly, this improvement seems to be mediated by preventing induction of two important ubiquitin ligases (MuRF1 and Atrogin-1/MAFbox) involved in muscle atrophy (Mourkioti et al., 2006).

Although these data clearly demonstrate that local NF-κB inhibition raises new possibilities for clinical applications against muscle degenerative diseases, a systemic blockade of this pathway would be fatal since NF-κB signaling is a key mediator of vital inflammatory responses (Bonizzi and Karin, 2004; Li, 2002). Further research is needed in order to evaluate the exact mechanisms whereby this pathway acts in various tissues and under different conditions, and to devise methods to target skeletal muscles specifically, without affecting the immune response.

3.2 Other Potential Candidates

3.2.1 Calcineurin

Calcineurin plays an important role in a wide variety of physiological and pathological processes, including the immune response, neuronal plasticity and cardiac development and hypertrophy (Schulz and Yutzey, 2004). In response to calcium increase, calcineurin induces dephosphorylation, nuclear translocation and activation of the NFAT transcription factors, a process sensitive to the action of the immunosuppressive drug Cyclosporin-A (CsA).

The involvement of calcineurin in skeletal muscle hypertrophy is controversial (Musaro et al., 2001; Musaro and Rosenthal, 1999; Rommel et al., 2001), and confounded by the complexity of calcineurin isoforms. In skeletal muscle, the Cn/NFAT pathway mediates myotube differentiation, enhances myoblast recruitment, controls muscle fiber type specification and ameliorates injury to dystrophic muscles (Friday et al., 2000; Horsley et al., 2001; Horsley et al., 2003; Naya et al., 2000; Parsons et al., 2003; Stupka et al., 2006).

We have recently characterized a naturally occurring splicing variant of the Calcineurin A catalytic subunit (CnAb1) in which the autoinhibitory domain that controls enzyme activation is replaced with a unique C-terminal region (E. Lara-Pezzi and N. Rosenthal, unpublished results). Interestingly, increased CnAb1 expression was noted in response to the muscle-restricted overexpression of mIGF-1, which preserved muscle integrity and enhanced the response to chronic degeneration in a mouse model of amyotrophic lateral sclerosis (ALS) (Dobrowolny et al., 2005). The CnAb1 enzyme is constitutively active and dephosphorylates its NFAT target in a cyclosporin-insensitive manner. CnAb1 is highly expressed in proliferating myoblasts and human tumors, where it inhibits FoxO transcription factors independently of its phosphatase activity. In myoblasts, CnAb1 knockdown activates FoxO-regulated genes, reduces proliferation and induces myoblast differentiation. Supplemental CnAb1 transgene expression in skeletal muscle leads to enhanced regeneration, reduced scar formation and accelerated resolution of inflammation. This unique mode of action distinguishes the CnAb1 isoform as a candidate for interventional strategies in muscle wasting.

3.2.2 Calpains

Calpains are a class of calcium-dependent proteases that target several muscle specific proteins (Burkard et al., 2005; Otani et al., 2006; Wei et al., 2006). They are involved in Limb Girdle Muscular Dystrophy type 2A (Kramerova et al., 2007; Kramerova et al., 2005; Kramerova et al., 2006), where their role in establishment and maintenance of correct myofibrillar apparatus has been amply elucidated (Kramerova et al., 2005; Kramerova et al., 2006).

However, their levels increase also in muscle degenerative disorders while their inhibition has proved useful to suppress some of the wasting phenotype associated to muscle dystrophy and sepsis in mice, identifying this class of molecules as potential target for therapy (Burdi et al., 2006; Jones et al., 2004; Wei et al., 2005).

3.2.3 Steroids

Anabolic steroids include a number of testosterone-related molecules that exert muscle growth effects (Herbst and Bhasin, 2004). The muscle growth promoting activity of this class of molecules has been exploited in the past for therapeutic purposes (Muscaritoli et al., 2006), but the uncertainty about long-term effects and the abuse of these drugs among athletes has raised serious concerns (Bahrke and Yesalis, 2004; Hartgens and Kuipers, 2004; Muscaritoli et al., 2006).

3.2.4 β2-agonists

β2-adrenergic agonists represent another class of molecules that stimulate muscle growth, to the point of being used for therapy of pathologies associated with muscle loss. β2-agonists include Clenbuterol, which is known to boost muscle growth and counteract muscle atrophy (Agrawal et al., 2003; Dodd and Koesterer, 2002; Yimlamai et al., 2005). The mechanisms by which β2-agonists stimulate muscle growth seem extremely diverse: they suppress protein degradation (Yimlamai et al., 2005), stimulate isoform replacement of muscle-specific proteins (Arai et al., 2006), and stimulate motoneuron growth and activity (Zeman et al., 2004). Interestingly, the effects of β2-agonists are reproducible in organisms as evolutionarily distinct as mammals and fishes (Salem et al., 2006). The anabolic effects of Clenbuterol are suppressed by Rapamycin, a potent inhibitor of mTOR, suggesting that at least some of these effects are also mediated by the Akt/mTOR pathway (Kline et al., 2007). However, the side effects associated with use of β2-agonists as muscle boosters have raised serious concern both for human health and for the meat industry (Burniston et al., 2002; Sleeper et al., 2002). It is likely that many of the reported side effects are due to excessive dosage, obscuring the anabolic effects of the compounds that could be dissociated from their myotoxicity by administration of lower doses (Burniston et al., 2006; Burniston et al., 2007). Nevertheless it remains to be seen whether chronic use of such molecules would be beneficial to enhance muscle performance and to treat muscle disease.

4. CLINICAL APPLICATION OF MUSCLE BOOSTERS

4.1 IGF-1 Based Therapies

Stimulation of IGF-1-dependent pathways has mainly focused on the use of recombinant forms of the active 70 a mature peptide. Several drugs have been developed for the treatment of IGF-1 deficiencies, namely Increlex™ (Tercica, Brisbane, CA, USA), Myotrophin™ (Cephalon, Frazer, PA, USA) and Somatokine™/iPLEX™ (InsMed, Glen Allen, VA, USA), also known as mecasermin rinfabate. Primarily developed for treating IGF-1 deficiencies, these drugs have recently been approved in clinical trials for the treatment of several myopathies. They supply the patients with a recombinant version of human IGF-1 (rhIGF-1), with the exception of Somatokine™/iPLEX™ which delivers rhIGF-1 associated with its binding protein IGFBP3 for improved stability in the circulation.

Use of IGF-1-based drugs has proved useful to increase IGF-1 levels in patients suffering of Growth Hormone insensitivity and IGF-1 deficiency, indicating side effects including cranial hypertension, hipbone problems and hypoglycemia (Kemp and Thrailkill, 2006); reduction in adverse effects coupled to increased stability have been reported after use of Somatokine™/iPLEX™, which has recently received FDA approval as orphan drug (Kemp, 2006). Studies in mice have also confirmed that mecasermin rinfabate, i.e. the IGF-1 associated to IGFBP3, has beneficial effects on levels of muscle protein synthesis during starvation (Svanberg et al., 2000).

A 2-year phase III clinical trial for treatment of ALS with Myotrophin™ has completed the recruiting phase (NCT00035815); while Somatokine™/iPLEX™ is being tested in phase I and II clinical trial for the treatment of muscle wasting associated to Myotonic Dystrophy type I, a study that should be completed by 2010 (NCT00233519).

4.2 Myostatin as Target for Therapies

The MSTN blocking antibody MYO029 (Wyeth, Madison, NJ, USA) is one of the most promising therapeutic drugs designed to inhibit MSTN activity and lessen muscle wasting. Phase I and II clinical trials have recently completed enrolling patients affected by Becker, Fascioscapulohumeral and Limb-Girdle Muscular Dystrophies (NCT00104078), in the wake of encouraging results obtained in animal models (Bogdanovich et al., 2005; Chakkalakal et al., 2005; Ohsawa et al., 2006; Parsons et al., 2006). Another inhibitor of MSTN, a soluble form of the activin Receptor IIB (ACVR2B), developed by Wyeth and Metamorphix industries, induced an even more robust hypertrophic response in mice when compared to the effects of the blocking antibody MYO029 (Lee et al., 2005).

4.3 Drug-based Therapies

As discussed above, the use of anabolic compounds, such as steroids and β2-agonists, has proven successful in limiting the muscle wasting effects caused by inactivity, cancer, muscle diseases and HIV infections. As with any drug, it is important to consider the long-term consequences and dosage side effects that may hamper the potential of such compounds to successfully treat chronic muscle degenerative conditions.

The ever growing list of the molecular players involved in muscle degeneration continue to present new prospects for therapeutic targets. Of particular interest are the proteins involved in the different degradative pathways such as the E3-ubiquitin ligases Atrogin-1/MAFbox and MuRF1, but also other proteolytic enzymes such as the calpains. In fact, for this latter group of proteins, inhibitors have already been identified and tested in animal models of muscle dystrophy and muscle wasting, proving, at least in theory, that specifically blocking the degradative processes could be beneficial for certain aspects of these diseases (Burdi et al., 2006; Fareed et al., 2006). The use of calpain inhibitors will have to be further elucidated, particularly in light of evidence that calpains may play an important role in aspects of muscle regeneration that may be blocked by the use of these inhibitors (Duguez et al., 2006; Kramerova et al., 2005; Raynaud et al., 2004). In general, fighting the catabolic aspects of muscle degeneration should ideally be coupled with stimulation of physiological hypertrophic and regenerative pathways (Tisdale, 2005).

5. LESSONS FROM SPORTS AND SOCIETY

Elucidation of the mechanisms underlying muscle regeneration is critical for the understanding of debilitating myopathic conditions, and for the development of therapeutic programs to rescue muscle loss, to boost muscle mass and performance that accompany them. However, a scientific commitment to develop such strategies has a broad impact on other aspects of social health. Growth deficit disorders and the muscle deterioration which occurs during ageing could benefit from better knowledge of the mechanisms controlling muscle growth and better drugs to modulate such mechanisms (Chien and Karsenty, 2005; Close et al., 2005; Harridge, 2003; Karakelides and Sreekumaran Nair, 2005). These aspects of muscle research are increasingly important, particularly in developed societies; according to the latest U.S. Health Reports, treatment and management of chronic pathologies affecting elderly people will absorb a great part of financial social security expenses (National Center for Health Statistics, 2006).

Athletics is another field of human activities where scientific and the health-related goals converge for a better good. Sport is not only dominated by competitions where the best athlete is selected, but it has become a practical need in developed societies where the heavy use of transportation and easy access to resources has curtailed the daily requirement for physical activity. Awareness of the beneficial effects of exercising has also led to an increased interest in the improvement of personal and athletic performance, and in sport-related research to explore and assess the use of substances stimulating muscle performance (Bahrke and Yesalis, 2004; Hartgens and Kuipers, 2004). The modulation of molecular targets to achieve improved muscle performance has now permeated the professional athletic scene, raising concerns in the medical and sport sciences regarding the chronic use of certain anabolic compounds and their long-term effects, but has also prompted the Medical Commission of the International Olympic Committee (IOC) and the World Anti-Doping Agency (WADA) to modify their definition of doping, by including also "the nontherapeutic use of cells, genes, genetic elements, or of modulation of gene expression, having the capacity to enhance athletic performance" (World Anti-Doping Agency, 2003; World Anti-Doping Agency, 2006). Despite the unfortunate abuse of muscle boosting substances, the sport world nevertheless provides valuable health information on the effects of methods and drugs that are in use or could be used for therapeutic purposes. With the ever-growing list of molecules involved in regulating muscle mass provided by basic and clinical research, sport agencies will need tighter control of the use and abuse of these information.

From the ethical point of view, the athlete's world is at a key checkpoint: redefining doping from the point of view of genetic variations. If modulating the action of genetic modifiers, which physiologically contribute to muscle mass and performance, is to be considered an illicit enhancement of personal performance, where do naturally occurring mutations stand? Cases of athletes bearing mutations in the erythropoietin gene, or the recent report of an over-muscular child born with genetic MSTN deficiency (Haisma and de Hon, 2006) are examples of enhanced skeletal muscle performance through spontaneous genetic variation and highlight the potential for societal intervention or abuse in the athletic arena. These ethical concerns notwithstanding, the fortuitous benefits of these interventions for patients where boosting muscle regeneration could be a lifesaving intervention should not be underestimated.

ACKNOWLEDGEMENTS

We thank Foteini Mourkioti for suggestions and critical reading of the manuscript.

REFERENCES

Agrawal S, Thakur P, Katoch SS (2003) Beta adrenoceptor agonists, clenbuterol, and isoproterenol retard denervation atrophy in rat gastrocnemius muscle: use of 3-methylhistidine as a marker of myofibrillar degeneration. Jpn J Physiol 53:229–237

Allen DL, Leinwand LA (2001) Postnatal myosin heavy chain isoform expression in normal mice and mice null for IIb or IId myosin heavy chains. Dev Biol 229:383–395

Arai C, Ohnuki Y, Umeki D, Hirashita A, Saeki Y (2006) Effects of clenbuterol and cyclosporin A on the myosin heavy chain mRNA level and the muscle mass in rat masseter. J Physiol Sci 56:205–209

Attaix D, Ventadour S, Codran A, Bechet D, Taillandier D, Combaret L (2005) The ubiquitin-proteasome system and skeletal muscle wasting. Essays Biochem 41:173–186

Bahrke MS, Yesalis CE (2004) Abuse of anabolic androgenic steroids and related substances in sport and exercise. Curr Opin Pharmacol 4:614–620

Barton ER (2006) Viral expression of insulin-like growth factor-I isoforms promotes different responses in skeletal muscle. J Appl Physiol 100:1778–1784

Barton ER, Morris L, Musaro A, Rosenthal N, Sweeney HL (2002) Muscle-specific expression of insulin-like growth factor I counters muscle decline in mdx mice. J Cell Biol 157:137–148

Bassel-Duby R, Olson EN (2003) Role of calcineurin in striated muscle: development, adaptation, and disease. Biochem Biophys Res Commun 311:1133–1141

Bassel-Duby R, Olson EN (2006) Signaling pathways in skeletal muscle remodeling. Annu Rev Biochem 75:19–37

Bodine SC (2006) mTOR Signaling and the Molecular Adaptation to Resistance Exercise. Med Sci Sports Exerc 38:1950–1957

Bodine SC, Latres E, Baumhueter S, Lai VK, Nunez L, Clarke BA, Poueymirou WT, Panaro FJ, Na E, Dharmarajan K, Pan ZQ, Valenzuela DM, DeChiara TM, Stitt TN, Yancopoulos GD, Glass DJ (2001a) Identification of ubiquitin ligases required for skeletal muscle atrophy. Science 294:1704–1708

Bodine SC, Stitt TN, Gonzalez M, Kline WO, Stover GL, Bauerlein R, Zlotchenko E, Scrimgeour A, Lawrence JC, Glass DJ, Yancopoulos GD (2001b) Akt/mTOR pathway is a crucial regulator of skeletal muscle hypertrophy and can prevent muscle atrophy in vivo. Nat Cell Biol 3:1014–1019

Bogdanovich S, Perkins KJ, Krag TO, Whittemore LA, Khurana TS (2005) Myostatin propeptide-mediated amelioration of dystrophic pathophysiology. Faseb J 19:543–549

Bonizzi G, Karin M (2004) The two NF-kappaB activation pathways and their role in innate and adaptive immunity. Trends Immunol 25:280–288

Buckingham M (2003) How the community effect orchestrates muscle differentiation. Bioessays 25:13–16

Buckingham M (2006) Myogenic progenitor cells and skeletal myogenesis in vertebrates. Curr Opin Genet Dev 16:525–532

Buckingham M, Bajard L, Chang T, Daubas P, Hadchouel J, Meilhac S, Montarras D, Rocancourt D, Relaix F (2003) The formation of skeletal muscle: from somite to limb. J Anat 202:59–68

Burdi R, Didonna MP, Pignol B, Nico B, Mangieri D, Rolland JF, Camerino C, Zallone A, Ferro P, Andreetta F, Confalonieri P, De Luca A (2006) First evaluation of the potential effectiveness in muscular dystrophy of a novel chimeric compound, BN 82270, acting as calpain-inhibitor and anti-oxidant. Neuromuscul Disord 16:237–248

Burkard N, Becher J, Heindl C, Neyses L, Schuh K, Ritter O (2005) Targeted proteolysis sustains calcineurin activation. Circulation 111:1045–1053

Burniston JG, Clark WA, Tan LB, Goldspink DF (2006) Dose-dependent separation of the hypertrophic and myotoxic effects of the beta(2)-adrenergic receptor agonist clenbuterol in rat striated muscles. Muscle Nerve 33:655–663

Burniston JG, McLean L, Beynon RJ, Goldspink DF (2007) Anabolic effects of a non-myotoxic dose of the beta2-adrenergic receptor agonist clenbuterol on rat plantaris muscle. Muscle Nerve 35:217–223

Burniston JG, Ng Y, Clark WA, Colyer J, Tan LB, Goldspink DF (2002) Myotoxic effects of clenbuterol in the rat heart and soleus muscle. J Appl Physiol 93:1824–1832

Cai D, Frantz JD, Tawa NE, Jr., Melendez PA, Oh BC, Lidov HG, Hasselgren PO, Frontera WR, Lee J, Glass DJ, Shoelson SE (2004) IKKbeta/NF-kappaB activation causes severe muscle wasting in mice. Cell 119:285–298

Cao PR, Kim HJ, Lecker SH (2005) Ubiquitin-protein ligases in muscle wasting. Int J Biochem Cell Biol 37:2088–2097

Carlson CJ, Booth FW, Gordon SE (1999) Skeletal muscle myostatin mRNA expression is fiber-type specific and increases during hindlimb unloading. Am J Physiol 277:R601–R606

Chakkalakal JV, Thompson J, Parks RJ, Jasmin BJ (2005) Molecular, cellular, and pharmacological therapies for Duchenne/Becker muscular dystrophies. Faseb J 19:880–891

Chen YG, Meng AM (2004) Negative regulation of TGF-beta signaling in development. Cell Res 14:441–449

Chien KR, Karsenty G (2005) Longevity and lineages: toward the integrative biology of degenerative diseases in heart, muscle, and bone. Cell 120:533–544

Cho H, Thorvaldsen JL, Chu Q, Feng F, Birnbaum MJ (2001) Akt1/PKBalpha is required for normal growth but dispensable for maintenance of glucose homeostasis in mice. J Biol Chem 276:38349–38352
Christ B, Brand-Saberi B (2002) Limb muscle development. Int J Dev Biol 46:905–914
Ciechanover A (2005) Intracellular protein degradation: from a vague idea thru the lysosome and the ubiquitin-proteasome system and onto human diseases and drug targeting. Cell Death Differ 12:1178–1190
Close GL, Kayani A, Vasilaki A, McArdle A (2005) Skeletal muscle damage with exercise and aging. Sports Med 35:413–427
Cossu G, Biressi S (2005) Satellite cells, myoblasts and other occasional myogenic progenitors: possible origin, phenotypic features and role in muscle regeneration. Semin Cell Dev Biol 16:623–631
Dasarathy S, Dodig M, Muc SM, Kalhan SC, McCullough AJ (2004) Skeletal muscle atrophy is associated with an increased expression of myostatin and impaired satellite cell function in the portacaval anastamosis rat. Am J Physiol Gastrointest Liver Physiol 287:G1124–G1130
Dehoux M, Gobier C, Lause P, Bertrand L, Ketelslegers JM, Thissen JP (2007) IGF-I does not prevent myotube atrophy caused by proinflammatory cytokines despite activation of Akt/Foxo and GSK3{beta} pathways and inhibition of Atrogin-1 mRNA. Am J Physiol Endocrinol Metab 292:E145–E150
Delafontaine P, Akao M (2006) Angiotensin II as candidate of cardiac cachexia. Curr Opin Clin Nutr Metab Care 9:220–224
Dobrowolny G, Giacinti C, Pelosi L, Nicoletti C, Winn N, Barberi L, Molinaro M, Rosenthal N, Musaro A (2005) Muscle expression of a local Igf-1 isoform protects motor neurons in an ALS mouse model. J Cell Biol 168:193–199
Dodd SL, Koesterer TJ (2002) Clenbuterol attenuates muscle atrophy and dysfunction in hindlimb-suspended rats. Aviat Space Environ Med 73:635–639
Duguez S, Bartoli M, Richard I (2006) Calpain 3: a key regulator of the sarcomere? Febs J 273:3427–3436
Edstrom E, Ulfhake B (2005) Sarcopenia is not due to lack of regenerative drive in senescent skeletal muscle. Aging Cell 4:65–77
Eizema K, van der Wal DE, van den Burg MM, de Jonge HW, Everts ME (2007) Differential Expression of Calcineurin and SR Ca2+ Handling Proteins in Equine Muscle Fibers During Early Postnatal Growth. J Histochem Cytochem 55:247–254
Etherton TD (2004) Somatotropic function: the somatomedin hypothesis revisited. J Anim Sci 82 E-Suppl:E239–E244
Fareed MU, Evenson AR, Wei W, Menconi M, Poylin V, Petkova V, Pignol B, Hasselgren PO (2006) Treatment of rats with calpain inhibitors prevents sepsis-induced muscle proteolysis independent of atrogin-1/MAFbx and MuRF1 expression. Am J Physiol Regul Integr Comp Physiol 290:R1589–1597
Fogarty MP, Kessler JD, Wechsler-Reya RJ (2005) Morphing into cancer: the role of developmental signaling pathways in brain tumor formation. J Neurobiol 64:458–475
Friday BB, Horsley V, Pavlath GK (2000) Calcineurin activity is required for the initiation of skeletal muscle differentiation. J Cell Biol 149:657–666
Geiger PC, Bailey JP, Mantilla CB, Zhan WZ, Sieck GC (2006) Mechanisms underlying myosin heavy chain expression during development of the rat diaphragm muscle. J Appl Physiol 101:1546–1555
Ghosh S, Karin M (2002) Missing pieces in the NF-kappaB puzzle. Cell 109 Suppl:S81–S96
Gilson H, Schakman O, Combaret L, Lause P, Grobet L, Attaix D, Ketelslegers JM, Thissen JP (2007) Myostatin Gene Deletion Prevents Glucocorticoid-Induced Muscle Atrophy. Endocrinology 148:452–460
Glass DJ (2005) Skeletal muscle hypertrophy and atrophy signaling pathways. Int J Biochem Cell Biol 37:1974–1984
Gomes MD, Lecker SH, Jagoe RT, Navon A, Goldberg AL (2001) Atrogin-1, a muscle-specific F-box protein highly expressed during muscle atrophy. Proc Natl Acad Sci U S A 98:14440–14445
Gonzalez-Cadavid NF, Taylor WE, Yarasheski K, Sinha-Hikim I, Ma K, Ezzat S, Shen R, Lalani R, Asa S, Mamita M, Nair G, Arver S, Bhasin S (1998) Organization of the human myostatin gene and expression in healthy men and HIV-infected men with muscle wasting. Proc Natl Acad Sci U S A 95:14938–14943

Guttridge DC (2004) Signaling pathways weigh in on decisions to make or break skeletal muscle. Curr Opin Clin Nutr Metab Care 7:443–450

Guttridge DC, Albanese C, Reuther JY, Pestell RG, Baldwin AS Jr (1999) NF-kappaB controls cell growth and differentiation through transcriptional regulation of cyclin D1. Mol Cell Biol 19:5785–5799

Guttridge DC, Mayo MW, Madrid LV, Wang CY, Baldwin AS Jr (2000) NF-kappaB-induced loss of MyoD messenger RNA: possible role in muscle decay and cachexia. Science 289:2363–2366

Haisma HJ, de Hon O (2006) Gene doping. Int J Sports Med 27:257–266

Harridge SD (2003) Ageing and local growth factors in muscle. Scand J Med Sci Sports 13:34–39

Hartgens F, Kuipers H (2004) Effects of androgenic-anabolic steroids in athletes. Sports Med 34:513–554

Heineke J, Molkentin JD (2006) Regulation of cardiac hypertrophy by intracellular signalling pathways. Nat Rev Mol Cell Biol 7:589–600

Herbst KL, Bhasin S (2004) Testosterone action on skeletal muscle. Curr Opin Clin Nutr Metab Care 7:271–277

Hesselink MK, van Baak MA (2004) Physical activity and health, novel concepts and new targets: report from the 12th Conference of the International Research Group on the Biochemistry of Exercise. Proc Nutr Soc 63:189–197

Hollway GE, Currie PD (2003) Myotome meanderings. Cellular morphogenesis and the making of muscle. EMBO Rep 4:855–860

Holzbaur EL, Howland DS, Weber N, Wallace K, She Y, Kwak S, Tchistiakova LA, Murphy E, Hinson J, Karim R, Tan XY, Kelley P, McGill KC, Williams G, Hobbs C, Doherty P, Zaleska MM, Pangalos MN, Walsh FS (2006) Myostatin inhibition slows muscle atrophy in rodent models of amyotrophic lateral sclerosis. Neurobiol Dis 23:697–707

Horsley V, Friday BB, Matteson S, Kegley KM, Gephart J, Pavlath GK (2001) Regulation of the growth of multinucleated muscle cells by an NFATC2-dependent pathway. J Cell Biol 153:329–338

Horsley V, Jansen KM, Mills ST, Pavlath GK (2003) IL-4 acts as a myoblast recruitment factor during mammalian muscle growth. Cell 113:483–494

Hubmacher D, Tiedemann K, Reinhardt DP (2006) Fibrillins: from biogenesis of microfibrils to signaling functions. Curr Top Dev Biol 75:93–123

Izzi L, Attisano L (2006) Ubiquitin-dependent regulation of TGFbeta signaling in cancer. Neoplasia 8:677–688

Jagoe RT, Goldberg AL (2001) What do we really know about the ubiquitin-proteasome pathway in muscle atrophy? Curr Opin Clin Nutr Metab Care 4:183–190

Jones SW, Hill RJ, Krasney PA, O'Conner B, Peirce N, Greenhaff PL (2004) Disuse atrophy and exercise rehabilitation in humans profoundly affects the expression of genes associated with the regulation of skeletal muscle mass. Faseb J 18:1025–1027

Joulia D, Bernardi H, Garandel V, Rabenoelina F, Vernus B, Cabello G (2003) Mechanisms involved in the inhibition of myoblast proliferation and differentiation by myostatin. Exp Cell Res 286:263–275

Joulia-Ekaza D, Cabello G (2006) Myostatin regulation of muscle development: molecular basis, natural mutations, physiopathological aspects. Exp Cell Res 312:2401–2414

Kandarian SC, Jackman RW (2006) Intracellular signaling during skeletal muscle atrophy. Muscle Nerve 33:155–165

Karakelides H, Sreekumaran Nair K (2005) Sarcopenia of aging and its metabolic impact. Curr Top Dev Biol 68:123–148

Karin M, Yamamoto Y, Wang QM (2004) The IKK NF-kappa B system: a treasure trove for drug development. Nat Rev Drug Discov 3:17–26

Kemp SF, Fowlkes JL, Thrailkill KM (2006) Efficacy and safety of mecasermin rinfabate. Expert Opin Biol Ther 6:533–538

Kemp SF, Thrailkill KM (2006) Investigational agents for the treatment of growth hormone-insensitivity syndrome. Expert Opin Investig Drugs 15:409–415

Kline WO, Panaro FJ, Yang H, Bodine SC (2007) Rapamycin Inhibits the Growth and Muscle Sparing Effects of Clenbuterol. J Appl Physiol 102:740–747

Kluppel M, Wrana JL (2005) Turning it up a Notch: cross-talk between TGF beta and Notch signaling. Bioessays 27:115–118

Kraemer WJ, Ratamess NA (2005) Hormonal responses and adaptations to resistance exercise and training. Sports Med 35:339–361

Kramerova I, Beckmann JS, Spencer MJ (2007) Molecular and cellular basis of calpainopathy (limb girdle muscular dystrophy type 2A). Biochimica et Biophysica Acta (BBA) - Molecular Basis of Disease 1772:128–144

Kramerova I, Kudryashova E, Venkatraman G, Spencer MJ (2005) Calpain 3 participates in sarcomere remodeling by acting upstream of the ubiquitin-proteasome pathway. Hum Mol Genet 14:2125–2134

Kramerova I, Kudryashova E, Wu B, Spencer MJ (2006) Regulation of the M-Cadherin-{beta}-Catenin Complex by Calpain 3 during Terminal Stages of Myogenic Differentiation. Mol Cell Biol 26:8437–8447

Langen RC, Van Der Velden JL, Schols AM, Kelders MC, Wouters EF, Janssen-Heininger YM (2004) Tumor necrosis factor-alpha inhibits myogenic differentiation through MyoD protein destabilization. Faseb J 18:227–237

Le Roith D, Bondy C, Yakar S, Liu JL, Butler A (2001a) The somatomedin hypothesis: 2001. Endocr Rev 22:53–74

Le Roith D, Scavo L, Butler A (2001b) What is the role of circulating IGF-I? Trends Endocrinol Metab 12:48–52

Lee SJ (2004) Regulation of muscle mass by myostatin. Annu Rev Cell Dev Biol 20:61–86

Lee SJ, Reed LA, Davies MV, Girgenrath S, Goad ME, Tomkinson KN, Wright JF, Barker C, Ehrmantraut G, Holmstrom J, Trowell B, Gertz B, Jiang MS, Sebald SM, Matzuk M, Li E, Liang LF, Quattlebaum E, Stotish RL, Wolfman NM (2005) Regulation of muscle growth by multiple ligands signaling through activin type II receptors. Proc Natl Acad Sci U S A 102:18117–18122

Li Q, Verma IM (2002) NF-kB regulation in the immune system. Nat Rev Immunol 2:725–734

Li YP (2003) TNF-alpha is a mitogen in skeletal muscle. Am J Physiol Cell Physiol 285:C370–C376

Lin J, Arnold HB, Della-Fera MA, Azain MJ, Hartzell DL, Baile CA (2002) Myostatin knockout in mice increases myogenesis and decreases adipogenesis. Biochem Biophys Res Commun 291:701–706

Luo D, Renault VM, Rando TA (2005) The regulation of Notch signaling in muscle stem cell activation and postnatal myogenesis. Semin Cell Dev Biol 16:612–622

McCroskery S, Thomas M, Platt L, Hennebry A, Nishimura T, McLeay L, Sharma M, Kambadur R (2005) Improved muscle healing through enhanced regeneration and reduced fibrosis in myostatin-null mice. J Cell Sci 118:3531–3541

McFarlane C, Plummer E, Thomas M, Hennebry A, Ashby M, Ling N, Smith H, Sharma M, Kambadur R (2006) Myostatin induces cachexia by activating the ubiquitin proteolytic system through an NF-kappaB-independent, FoxO1-dependent mechanism. J Cell Physiol 209:501–514

McLellan AS, Kealey T, Langlands K (2006) An E box in the exon 1 promoter regulates insulin-like growth factor-I expression in differentiating muscle cells. Am J Physiol Cell Physiol 291:C300–C307

McMahon CD, Popovic L, Oldham JM, Jeanplong F, Smith HK, Kambadur R, Sharma M, Maxwell L, Bass JJ (2003) Myostatin-deficient mice lose more skeletal muscle mass than wild-type controls during hindlimb suspension. Am J Physiol Endocrinol Metab 285:E82–E87

McManus EJ, Sakamoto K, Armit LJ, Ronaldson L, Shpiro N, Marquez R, Alessi DR (2005) Role that phosphorylation of GSK3 plays in insulin and Wnt signalling defined by knockin analysis. Embo J 24:1571–1583

McPherron AC, Lawler AM, Lee SJ (1997) Regulation of skeletal muscle mass in mice by a new TGF-beta superfamily member. Nature 387:83–90

Miyazono K, Maeda S, Imamura T (2004) Coordinate regulation of cell growth and differentiation by TGF-beta superfamily and Runx proteins. Oncogene 23:4232–4237

Morley JE, Thomas DR, Wilson MM (2006) Cachexia: pathophysiology and clinical relevance. Am J Clin Nutr 83:735–743

Mourkioti F, Kratsios P, Luedde T, Song YH, Delafontaine P, Adami R, Parente V, Bottinelli R, Pasparakis M, Rosenthal N (2006) Targeted ablation of IKK2 improves skeletal muscle strength, maintains mass, and promotes regeneration. J Clin Invest 116:2945–2954

Mourkioti F, Rosenthal N (2005) IGF-1, inflammation and stem cells: interactions during muscle regeneration. Trends Immunol 26:535–542

Moustakas A, Heldin CH (2005) Non-Smad TGF-beta signals. J Cell Sci 118:3573–3584

Musaro A, Giacinti C, Borsellino G, Dobrowolny G, Pelosi L, Cairns L, Ottolenghi S, Cossu G, Bernardi G, Battistini L, Molinaro M, Rosenthal N (2004) Stem cell-mediated muscle regeneration is enhanced by local isoform of insulin-like growth factor 1. Proc Natl Acad Sci U S A 101:1206–1210

Musaro A, McCullagh K, Paul A, Houghton L, Dobrowolny G, Molinaro M, Barton ER, Sweeney HL, Rosenthal N (2001) Localized Igf-1 transgene expression sustains hypertrophy and regeneration in senescent skeletal muscle. Nat Genet 27:195–200

Musaro A, Rosenthal N (1999) Maturation of the myogenic program is induced by postmitotic expression of insulin-like growth factor I. Mol Cell Biol 19:3115–3124

Muscaritoli M, Bossola M, Aversa Z, Bellantone R, Rossi Fanelli F (2006) Prevention and treatment of cancer cachexia: new insights into an old problem. Eur J Cancer 42:31–41

Nader GA (2005) Molecular determinants of skeletal muscle mass: getting the "AKT" together. Int J Biochem Cell Biol 37:1985–1996

Nastasi T, Bongiovanni A, Campos Y, Mann L, Toy JN, Bostrom J, Rottier R, Hahn C, Conaway JW, Harris AJ, D'Azzo A (2004) Ozz-E3, a muscle-specific ubiquitin ligase, regulates beta-catenin degradation during myogenesis. Dev Cell 6:269–282

National Center for Health Statistics (2006) Health, United States, 2006. In: Hyatsville MD (ed) With Chartbook on Trends in the Helath of Americans, N. N. C. f. H. Statistics

Naya FJ, Mercer B, Shelton J, Richardson JA, Williams RS, Olson EN (2000) Stimulation of slow skeletal muscle fiber gene expression by calcineurin in vivo. J Biol Chem 275:4545–4548

Noguchi S (2005) The biological function of insulin-like growth factor-I in myogenesis and its therapeutic effect on muscular dystrophy. Acta Myol 24:115–118

Ohsawa Y, Hagiwara H, Nakatani M, Yasue A, Moriyama K, Murakami T, Tsuchida K, Noji S, Sunada Y (2006) Muscular atrophy of caveolin-3-deficient mice is rescued by myostatin inhibition. J Clin Invest 116:2924–2934

Otani K, Polonsky KS, Holloszy JO, Han DH (2006) Inhibition of calpain results in impaired contraction-stimulated GLUT4 translocation in skeletal muscle. Am J Physiol Endocrinol Metab 291:E544–E548

Parsons SA, Millay DP, Sargent MA, McNally EM, Molkentin JD (2006) Age-dependent effect of myostatin blockade on disease severity in a murine model of limb-girdle muscular dystrophy. Am J Pathol 168:1975–1985

Parsons SA, Wilkins BJ, Bueno OF, Molkentin JD (2003) Altered skeletal muscle phenotypes in calcineurin Aalpha and Abeta gene-targeted mice. Mol Cell Biol 23:4331–4343

Patel K, Macharia R, Amthor H (2005) Molecular mechanisms involving IGF-1 and myostatin to induce muscle hypertrophy as a therapeutic strategy for Duchenne muscular dystrophy. Acta Myol 24:230–241

Paul AC, Rosenthal N (2002) Different modes of hypertrophy in skeletal muscle fibers. J Cell Biol 156:751–760

Peng XD, Xu PZ, Chen ML, Hahn-Windgassen A, Skeen J, Jacobs J, Sundararajan D, Chen WS, Crawford SE, Coleman KG, Hay N (2003) Dwarfism, impaired skin development, skeletal muscle atrophy, delayed bone development, and impeded adipogenesis in mice lacking Akt1 and Akt2. Genes Dev 17:1352–1365

Pette D (2001) Historical Perspectives: plasticity of mammalian skeletal muscle. J Appl Physiol 90:1119–1124

Rabinovsky ED (2004) The multifunctional role of IGF-1 in peripheral nerve regeneration. Neurol Res 26:204–210

Raynaud F, Carnac G, Marcilhac A, Benyamin Y (2004) m-Calpain implication in cell cycle during muscle precursor cell activation. Exp Cell Res 298:48–57

Reinstein E, Ciechanover A (2006) Narrative review: protein degradation and human diseases: the ubiquitin connection. Ann Intern Med 145:676–684

Reynolds TH, Bodine SC, Lawrence JC Jr (2002) Control of Ser2448 phosphorylation in the mammalian target of rapamycin by insulin and skeletal muscle load. J Biol Chem 277:17657–17662

Rodgers BD (2005) Insulin-like growth factor-I downregulates embryonic myosin heavy chain (eMyHC) in myoblast nuclei. Growth Horm IGF Res 15:377–383

Rommel C, Bodine SC, Clarke BA, Rossman R, Nunez L, Stitt TN, Yancopoulos GD, Glass DJ (2001) Mediation of IGF-1-induced skeletal myotube hypertrophy by PI(3)K/Akt/mTOR and PI(3)K/Akt/GSK3 pathways. Nat Cell Biol 3:1009–1013

Russell B, Motlagh D, Ashley WW (2000) Form follows function: how muscle shape is regulated by work. J Appl Physiol 88:1127–1132

Sakamoto K, Arnolds DE, Ekberg I, Thorell A, Goodyear LJ (2004) Exercise regulates Akt and glycogen synthase kinase-3 activities in human skeletal muscle. Biochem Biophys Res Commun 319:419–425

Salem M, Levesque H, Moon TW, Rexroad CE, Yao J (2006) Anabolic effects of feeding beta2-adrenergic agonists on rainbow trout muscle proteases and proteins. Comp Biochem Physiol A Mol Integr Physiol 144:145–154

Sandri M, Sandri C, Gilbert A, Skurk C, Calabria E, Picard A, Walsh K, Schiaffino S, Lecker SH, Goldberg AL (2004) Foxo transcription factors induce the atrophy-related ubiquitin ligase atrogin-1 and cause skeletal muscle atrophy. Cell 117:399–412

Schakman O, Thissen JP (2006) Gene therapy with anabolic growth factors to prevent muscle atrophy. Curr Opin Clin Nutr Metab Care 9:207–213

Schulz RA, Yutzey KE (2004) Calcineurin signaling and NFAT activation in cardiovascular and skeletal muscle development. Dev Biol 266:1–16

Schulze PC, Fang J, Kassik KA, Gannon J, Cupesi M, MacGillivray C, Lee RT, Rosenthal N (2005) Transgenic overexpression of locally acting insulin-like growth factor-1 inhibits ubiquitin-mediated muscle atrophy in chronic left-ventricular dysfunction. Circ Res 97:418–426

Shavlakadze T, Grounds M (2006) Of bears, frogs, meat, mice and men: complexity of factors affecting skeletal muscle mass and fat. Bioessays 28:994–1009

Shavlakadze T, Winn N, Rosenthal N, Grounds MD (2005) Reconciling data from transgenic mice that overexpress IGF-I specifically in skeletal muscle. Growth Horm IGF Res 15:4–18

Sleeper MM, Kearns CF, McKeever KH (2002) Chronic clenbuterol administration negatively alters cardiac function. Med Sci Sports Exerc 34:643–650

Song YH, Godard M, Li Y, Richmond SR, Rosenthal N, Delafontaine P (2005a) Insulin-like growth factor I-mediated skeletal muscle hypertrophy is characterized by increased mTOR-p70S6K signaling without increased Akt phosphorylation. J Investig Med 53:135–142

Song YH, Li Y, Du J, Mitch WE, Rosenthal N, Delafontaine P (2005b) Muscle-specific expression of IGF-1 blocks angiotensin II-induced skeletal muscle wasting. J Clin Invest 115:451–458

Spate U, Schulze PC (2004) Proinflammatory cytokines and skeletal muscle. Curr Opin Clin Nutr Metab Care 7:265–269

Stitt TN, Drujan D, Clarke BA, Panaro F, Timofeyva Y, Kline WO, Gonzalez M, Yancopoulos GD, Glass DJ (2004) The IGF-1/PI3K/Akt pathway prevents expression of muscle atrophy-induced ubiquitin ligases by inhibiting FOXO transcription factors. Mol Cell 14:395–403

Strbenc M, Smerdu V, Pogacnik A, Fazarinc G (2006) Myosin heavy chain isoform transitions in canine skeletal muscles during postnatal growth. J Anat 209:149–163

Stupka N, Plant DR, Schertzer JD, Emerson TM, Bassel-Duby R, Olson EN, Lynch GS (2006) Activated calcineurin ameliorates contraction-induced injury to skeletal muscles of mdx dystrophic mice. J Physiol 575:645–656

Svanberg E, Ohlsson C, Kimball SR, Lundholm K (2000) rhIGF-I/IGFBP-3 complex, but not free rhIGF-I, supports muscle protein biosynthesis in rats during semistarvation. Eur J Clin Invest 30:438–446

Tajbakhsh S (2005) Skeletal muscle stem and progenitor cells: reconciling genetics and lineage. Exp Cell Res 306:364–372

Tian T, Meng AM (2006) Nodal signals pattern vertebrate embryos. Cell Mol Life Sci 63:672–685

Tiffin N, Adi S, Stokoe D, Wu NY, Rosenthal SM (2004) Akt phosphorylation is not sufficient for insulin-like growth factor-stimulated myogenin expression but must be accompanied by down-regulation of mitogen-activated protein kinase/extracellular signal-regulated kinase phosphorylation. Endocrinology 145:4991–4996

Tisdale MJ (2005) The ubiquitin-proteasome pathway as a therapeutic target for muscle wasting. J Support Oncol 3:209–217

Wagers AJ, Conboy IM (2005) Cellular and molecular signatures of muscle regeneration: current concepts and controversies in adult myogenesis. Cell 122:659–667

Walsh FS, Celeste AJ (2005) Myostatin: a modulator of skeletal-muscle stem cells. Biochem Soc Trans 33:1513–1517

Wehling M, Cai B, Tidball JG (2000) Modulation of myostatin expression during modified muscle use. Faseb J 14:103–110

Wei W, Fareed MU, Evenson A, Menconi MJ, Yang H, Petkova V, Hasselgren PO (2005) Sepsis stimulates calpain activity in skeletal muscle by decreasing calpastatin activity but does not activate caspase-3. Am J Physiol Regul Integr Comp Physiol 288:R580–R590

Wei W, Yang H, Cao P, Menconi M, Chamberlain C, Petkova V, Hasselgren PO (2006) Degradation of C/EBPbeta in cultured myotubes is calpain-dependent. J Cell Physiol 208:386–398

Winn N, Paul A, Musaro A, Rosenthal N (2002) Insulin-like growth factor isoforms in skeletal muscle aging, regeneration, disease. Cold Spring Harb Symp Quant Biol 67:507–518

World Anti-Doping Agency (2003) The World Anti-Doping Code, W. A.-D. A. WADA

World Anti-Doping Agency (2006) The 2007 Prohibited List. International Standard, W. A.-D. A. WADA

Wullschleger S, Loewith R, Hall MN (2006) TOR signaling in growth and metabolism. Cell 124:471–484

Xu Q, Wu Z (2000) The insulin-like growth factor-phosphatidylinositol 3-kinase-Akt signaling pathway regulates myogenin expression in normal myogenic cells but not in rhabdomyosarcoma-derived RD cells. J Biol Chem 275:36750–36757

Yamaoka S, Courtois G, Bessia C, Whiteside ST, Weil R, Agou F, Kirk HE, Kay RJ, Israel A (1998) Complementation cloning of NEMO, a component of the IkappaB kinase complex essential for NF-kappaB activation. Cell 93:1231–1240

Yimlamai T, Dodd SL, Borst SE, Park S (2005) Clenbuterol induces muscle-specific attenuation of atrophy through effects on the ubiquitin-proteasome pathway. J Appl Physiol 99:71–80

Zammit PS, Partridge TA, Yablonka-Reuveni Z (2006) The skeletal muscle satellite cell: the stem cell that came in from the cold. J Histochem Cytochem 54:1177–1191

Zandi E, Rothwarf DM, Delhase M, Hayakawa M, Karin M (1997) The IkappaB kinase complex (IKK) contains two kinase subunits, IKKalpha and IKKbeta, necessary for IkappaB phosphorylation and NF-kappaB activation. Cell 91:243–252

Zeman RJ, Peng H, Etlinger JD (2004) Clenbuterol retards loss of motor function in motor neuron degeneration mice. Exp Neurol 187:460–467

CHAPTER 16

AGE-DEPENDENT CHANGES IN SKELETAL MUSCLE REGENERATION

ANDREW S. BRACK[1] AND THOMAS A. RANDO[1,2]

[1] *Department of Neurology and Neurological Sciences, Stanford University School of Medicine, Stanford, CA, 94305, USA*

[2] *GRECC and Neurology Service, VA Palo Alto Health Care System, Palo Alto, CA 94304, USA*

1. INTRODUCTION TO TISSUE AGING

The aging of multicellular organisms is characterized by structural and functional changes at all levels, ranging from the molecular to the organismal. These changes are generally manifested as a gradual decline in physiological function associated with characteristic histological and morphological changes. Whereas steady-state functional changes may be subtle in some tissues, the reduced ability to respond to stress is a general property of aged tissues and is often quite pronounced. Normal regenerative responses are impaired, typically with less rapid and ultimately less effective repair. Tissues such as epithelia of skin and gut, skeletal muscle, and liver that display a rapid and complete wound healing response in young individuals of the species begin to display impaired healing with age (Fry et al., 1984; Martin et al., 1998; Ashcroft et al., 2002; Li et al., 2004). Instead of restoring the tissue to the pre-injury state, the wound healing response in elderly individuals instead is characterized by ineffective regenerative processes, such as fibrogenesis and adipogenesis (Everitt et al., 1985; Sadeh, 1988; Abrass et al., 1995; Gagliano et al., 2002). The result is that the healed tissue is less capable of normal physiological function and even less capable of subsequent effective regeneration.

The focus of this chapter is the effects of aging on the process of muscle regeneration. We will first discuss briefly some of the characteristics of aged skeletal muscle. We will then review the observable changes in the regenerative response in aged muscle compared to young muscle. Finally, we will discuss the cellular and molecular bases of those changes, addressing separately the properties

S. Schiaffino and T. Partridge (eds.), Skeletal Muscle Repair and Regeneration, 359–374.

of the cells that mediate that regenerative response and the properties of the "aged environment", both locally and systemically, in which the muscle stem and progenitor cells must carry out the complex process of tissue restoration. There are many important processes that also change with age, that are associated with tissue repair in general, and that are likely to affect muscle regeneration, such as revascularization, inflammation, and neural reinnervation. Although any complete description of the effects of aging on muscle regeneration would necessarily include a detailed examination of each of these processes, they are not the focus of this chapter and are mentioned only briefly.

2. THE PHENOTYPE OF AGED MUSCLE

It is necessary to begin any review of age-related changes in tissue repair and regeneration with a review of the basic changes that occur in that tissue under normal conditions with age. In skeletal muscle, the physiological, biochemical, and anatomical changes that are part of the normal aging process lead to generally smaller and weaker muscles (Fiatarone et al., 1990; Sayer et al., 2006). We will review briefly the general changes that characterize aged muscle that are responsible for significant morbidity and mortality in the elderly due to a loss of strength and mobility (Fiatarone et al., 1990; Sayer et al., 2006).

2.1 Physiological Changes

Aged skeletal muscle exhibits a decline in many functional properties such as power, strength and endurance (Fiatarone et al., 1990; Sayer et al., 2006) The reasons for this are multifactorial. There is a general overall decrease in strength in mammalian muscle with age that is due in part to an overall reduction in size of the muscles (see "Anatomical changes" below) but also due to intrinsically less functional muscle fibers that display a decrease in specific force and maximal power with age (Brooks and Faulkner, 1988; Brooks and Faulkner, 1991). Other changes, such as aberrant neural firing patterns and contractile properties due to the denervation that occur during aging, also explain some aspects of the altered physiology of aged muscle (Larsson and Ansved, 1995).

2.2 Anatomical Changes

2.2.1 Muscle fibers

During aging there is a loss of muscle mass. There are many reports of muscle fiber atrophy during aging (Coggan et al., 1992; Lexell, 1995; Dow et al., 2005; Brack et al., 2005), however the extent of atrophy varies among muscle types and species. In humans it has been reported that the number of fibers within specific large muscles declines during aging (Lexell et al., 1988). Therefore both atrophy and loss of individual muscle fibers would lead to a decline in muscle mass of the aged individual.

Muscles are characterized by individual fibers of different metabolic and contractile properties. During aging there is a shift in contractile properties within a muscle: fast, glycolytic fibers switch to a slower more oxidative phenotype. However, like fiber atrophy, fiber type shifts vary among muscles and species (Li and Larsson, 1996; Balagopal et al., 2001; Snow et al., 2005).

The post-mitotic nuclei that reside within the syncytia of adult muscle fibers undergo gradual replacement over time (Darr and Schultz, 1987). In aged muscle, it appears that there is a loss of fiber myonuclei which precedes the decrease in fiber size, resulting in a temporary increase in nuclear domain size (Brack et al., 2005). It is possible that this temporary increase in domain size is the stimulus for muscle atrophy, which then returns the domain size to its optimal set-point.

Other phenotypic changes that have been described in aged muscle include single fibers having a branched appearance, or fiber splitting (Blaivas and Carlson, 1991; Charge et al., 2002), and chains of centrally located nuclei (Brack et al., 2005). Whether these are features of aging per se, or of an ongoing degeneration/regeneration process induced by the increased injury susceptibility (Brooks and Faulkner, 1996) is unknown.

2.2.2 *Satellite cells*

The most abundant and well characterized skeletal muscle stem cell is the satellite cell (Mauro, 1961). The satellite cell is closely apposed to the muscle fiber when quiescent; when activated, satellite cells migrate away from the fiber and extend cytoplasmic projections (Carlson et al., 2001). These cells are responsible for the impressive regenerative potential of skeletal muscle (Collins and Partridge, 2005), even though in the adult they represent a very small percentage of the nuclear content of the tissue (Zammit et al., 2002). Light-microscopic and ultrastructural properties of satellite cells have been described for young and old animals (Snow, 1977; Carlson et al., 2001). By morphological criteria examined, satellite cells in old muscles were indistinguishable from those in young muscle (Snow, 1977).

Different studies comparing the number of satellite cells in aged muscles to that in young muscle have yielded divergent results ranging from a marked decline, to little change, to a relative and absolute increase (Snow, 1977; Gibson and Schultz, 1983; Conboy et al., 2003; Sajko et al., 2004; Brack et al., 2005; Shefer et al., 2006). However, it is difficult to compare directly the results of most studies because of differences of species examined, muscles surveyed, ages used for comparison, and techniques used to gauge satellite cell number. Clearly, a decline in the number of progenitor cells in the muscles of very old animals could account in part for any age-related decline in tissue regenerative potential. However, the decline in regenerative potential with age is more severe than can be accounted for on the basis of even the largest estimates of a decline in satellite cell number, suggesting that the functionality of satellite cells in aged muscle is impaired, a suggestion supported by both in vitro and in vivo data (Bockhold et al., 1998; Shefer et al., 2006). The age-related changes of satellite cell functionality are considered in detail in subsequent sections.

2.2.3 *Extracellular matrix*

The extracellular matrix of muscle includes both the interstitial connective tissue and the muscle fiber basal lamina which ensheaths satellite cells. Muscle tissue from aged rodents or humans has more connective tissue than from younger individuals of each species (Goldspink et al., 1994; Lexell, 1995) There is also an increase in the basal lamina surrounding satellite cells in aged muscle (Snow, 1977).

2.2.4 *Non-muscle cellular components*

Skeletal muscle, like all tissues, is composed of primary parenchymal cells and other cellular constituents that are essential for normal function. The vascular supply of skeletal muscle branches to form a capillary network in which several capillaries are associated with each muscle fiber. In aged muscle there are fewer capillaries per fiber compared to young muscle (Coggan et al., 1992; Porter et al., 1995; Ryan et al., 2006). Fibroblasts represent a major cellular component of the interstitial space. Although the number and characteristics of interstitial fibroblasts in aged muscle compared to young muscle have not been carefully studied, the increased deposition of fibrous connective tissue in aged muscle could reflect changes of fibroblast number or activity. Finally, there are hematopoietic cellular components that migrate from the blood into muscle tissue normally. During muscle regeneration in response to injury, a key component of that infiltration in terms of effective regenerative responses is the macrophage (Tidball, 2005). With age, macrophages exhibit impaired functional characteristics during the wound healing response of skeletal muscle (Cannon, 1998).

3. MUSCLE REGENERATION IN AGED MAMMALS

The characteristics and mechanisms of muscle regeneration in adult and aged animals have been studied primarily in response to acute physical (e.g. crush, cut), chemical (e.g. cardiotoxin, bupivicaine, notexin), and cryo-injury. These methods, although distinct, all yield reproducible injuries that are useful for analyzing the regeneration process. After the initial insult to the muscle, the injured fibers degenerate. The normally quiescent satellite cells then activate, enter the cell cycle and begin to proliferate in order to generate a sufficient number of progenitors to repair or replace the damaged fibers. In so doing, the proliferating myogenic cells exit the cell cycle, terminally differentiate and fuse to injured myofibers or to each other to form nascent myotubes (Hawke and Garry, 2001; Charge and Rudnicki, 2004; Zammit et al., 2006).

In aged muscle, there are detectable changes in virtually every aspect of the regenerative process (Sadeh, 1988; Carlson and Faulkner, 1989; McGeachie and Grounds, 1995; Carlson et al., 2001; Conboy et al., 2003). There is a delay in satellite cell activation and less proliferation, resulting in fewer myogenic progenitors. Myogenic cells in old muscle do exit the cell cycle and form nascent muscle fibers. However this occurs later than in young muscle and the nascent fibers formed are generally fewer in number and of smaller caliber during the early regenerative response. Therefore, restoration of normal muscle architecture is impaired. In contrast, there is an increased accumulation of connective tissue between the muscle fibers in the old muscle tissue.

4. CELLULAR AND MOLECULAR BASIS OF THE IMPAIRED REGENERATIVE RESPONSE OF AGED SKELETAL MUSCLE

During regeneration of aged muscle, satellite cell activation and proliferation are markedly reduced compared with young muscle, resulting in fewer satellite cell progeny and consistent with an age-related decline in the functionality of satellite cells and their progeny (Schultz and Lipton, 1982; Bockhold et al., 1998; Conboy and Rando, 2005). However, this does not necessarily indicate there is an intrinsic defect in the satellite cell. The functionality of the satellite cell could be impaired due to altered signals within the local and systemic environment.

Satellite cell function is certainly regulated at many levels, all of which may be subject to age-related changes. There are cell-intrinsic regulatory pathways of activation, proliferation and lineage progression, and these could be subject to both genetic and epigenetic modifications leading to intrinsic changes in satellite cells with age. To the extent that such changes are epigenetic, they would potentially be reversible. All of the intrinsic pathways are likely to be regulated by both local soluble and insoluble factors within the stem cell niche and by systemic factors. These cell-extrinsic influences on satellite cell function may also change with age, perhaps negatively affecting the capacity of these stem cells to participate in effective tissue repair. To the extent that such external influences could be regulated, such age-dependent inhibitory effects could also, theoretically, represent reversible age-related changes of satellite cell functionality. These two aspects of the aging of satellite cell functionality, cell-intrinsic and cell-extrinsic, will be considered separately.

4.1 Cell-intrinsic

When studying the intrinsic cellular aspects of aging, there has to be a distinction from the aging of the cell-external environment that occurs in vivo. Therefore, the most direct tests of intrinsic aging are those in which isolated cells are transferred to identical environments, either in vitro or in vivo. Furthermore, as noted above, it is important to distinguish whether any intrinsic difference that is measured is stable and irreversible (such as cumulative genetic mutations) or potentially reversible (such as epigenetic regulation of gene expression).

4.1.1 Molecular and biochemical properties

Gene expression studies from isolated satellite cells or their progeny have demonstrated that the transcriptional readout changes with age (Bortoli et al., 2003). Whether these alterations are due to irreversible genetic change or reversible epigenetic effects remains to be determined. In satellite cells in resting muscle or from freshly isolated single fibers, there are conflicting findings whether the myogenic regulatory factors change during aging (Dedkov et al., 2003; Brack et al., 2005). The discrepancy may be due to species differences, the ages compared or whether the fibers examined are undergoing low level injury.

One cell-intrinsic change that occurs with age in virtually all dividing somatic cells is telomere shortening (Bekaert et al., 2004). This process has been associated with phenotypic changes in aging tissues as the shortening of telomere length below a critical level can result in cell death or senescence instead of proliferation (Campisi, 2001). Although skeletal muscle is a predominantly postmitotic tissue, telomere shortening in the satellite cell compartment could theoretically impact the regenerative response of old muscle. Satellite cells isolated from muscles of human patients do not show a decrease in minimum or mean telomere length during aging (Decary et al., 1997), but a correlation between proliferation potential and telomere length has been observed (Decary et al., 2000; Mouly et al., 2005). Comparable studies have not been done in rodents, but rodents have much longer telomeres than do humans so their satellite cells are less likely to display critical telomere shortening. Furthermore, whether cell senescence occurs in vivo in normal mammalian aging or is a consequence of cell culturing conditions has yet to be clarified. Since age-related impairment of skeletal muscle regeneration extends across species, it is currently not clear to what extent telomere shortening is an important cell-intrinsic change contributing to the aging phenotype.

4.1.2 In vitro functional studies

In addition to direct biochemical and molecular assays of freshly isolated satellite cells from young and old members of different species, in vitro functional assays have been used to characterize age-related changes. The following sections review the studies that have tested for age-related changes of different aspects of satellite cell functionality in vitro.

4.1.3 Activation and proliferation

Satellite cells from young and old muscle have been tested for their ability to activate and proliferate in vitro. Old myogenic cells show a delay in responding to an activating stimulus (Schultz and Lipton, 1982; Barani et al., 2003) and therefore give rise to fewer progeny compared to young satellite cells (Bockhold et al., 1998; Conboy et al., 2003; Shefer et al., 2006). The elevated expression of negative cell cycle regulators in satellite cell progeny from old muscle is consistent with the impaired proliferation observed both in vitro and in vivo (Machida and Booth, 2004). However, the results are more equivocal when myogenic precursors derived from satellite cells of old animals are passaged over time and then analyzed for cell proliferation. Progenitors from old muscle that have been maintained in culture after many passages proliferate as well as progenitors from young muscle (Barani et al., 2003; Shefer et al., 2006). Whether the difference observed during initial growth in vitro is a cell-intrinsic change in the aged satellite cells that is reversible, or whether a similar subpopulation of cells are selected for over time from both the young and old satellite cells, remains to be determined.

4.1.4 Myogenic lineage progression

The progeny of the satellite cells progress in their lineage from an early progenitor to a fusion-competent myoblast. The signals that regulate this are complex. Notch signaling is clearly an important mediator of the proliferative expansion of transient amplifying cells in the transition from the quiescent satellite cell to fusion-competent myoblasts (Conboy and Rando, 2002). Notch activation is important for efficient cell cycle progression and the G1-S-phase transition (Conboy et al., 2003; Sarmento et al., 2005). Inhibiting Notch activation in regenerating young muscle inhibits the regenerative process. In aged muscle, an impairment of Notch signaling, related to a failure of expression of the Notch ligand Delta in response to stimuli that induce Delta expression in young muscle, results in reduced proliferative expansion of progenitors and ineffective regeneration (Conboy et al., 2003). Forced activation of Notch signaling in injured muscle of aged mice restores regenerative potential to that tissue (Conboy et al., 2003).

We have found that Wnt signaling plays an equally critical role in the lineage progression of satellite cell progeny (Brack et al., in press). Following proliferative expansion of the transient amplifying cells, active Wnt signaling in those cells decreases the proliferative capacity and promotes the progression of those cells to become fusion-competent myoblasts expressing high levels of desmin. It is not known whether the transition from transit amplifying cell to committed myoblast is affected by aging, but it is clear that effective regeneration depends on a tightly controlled temporal switch between Notch and Wnt signaling to control myogenic lineage progression. Satellite cells from old muscle do progress down the myogenic lineage in vitro as there are numerous reports of myoblasts expressing later lineage markers such as desmin and myogenin (Bockhold et al., 1998; Charge et al., 2002; Shefer et al., 2006), but the reduced number of such cells may be due to impaired lineage progression as well as impaired activation and would be consistent with an alteration in the balance between Notch and Wnt signaling. Determining whether there is a specific defect in myogenic lineage progression of aged satellite cell progeny will require an analysis of individual cells to determine whether, once they have begun proliferating, they proceed along the myogenic lineage as readily as cells from young tissue.

4.1.5 Alternate lineage pathways (fibrogenic, adipogenic) or cell fate (apoptosis, senescence)

It is clear that satellite cells, when placed into specific conditions in vitro, can be induced to express markers of alternate mesenchymal lineage pathways, including osteogenic, chondrogenic, and adipogenic lineages (Asakura et al., 2001; Shefer et al., 2004). Whether it is interesting to consider the increased fibrosis and adipogenesis seen in aged muscle, both uninjured and regenerated (Everitt et al., 1985; Sadeh, 1988; Goldspink et al., 1994), as possibly being related to this potential of myogenic progenitors to adopt alternate fates.

It is interesting that Wnt signaling, while being important in the regulation of myogenic lineage progression, may also be important for lineage determination (Taylor-Jones et al., 2002; Vertino et al., 2005). Both in vitro and in vivo, mice lacking Wnt 10 b displayed myogenic cells that had an enhanced adipogenic profile. This co-expression in the muscle cells would presumably lead to impaired muscle regeneration and muscle function. After muscle injury there was an accumulation of fat deposits in the muscle (Vertino et al., 2005).

Conversely, we have found that increased Wnt signaling, apparently via the canonical β-catenin pathway (Logan and Nusse, 2004), can induce myogenic progenitors to diverge from the myogenic lineage and adopt a fibrogenic phenotype (Brack et al., 2007). This is particularly relevant to impaired regeneration of old muscle because this altered lineage progression is associated with enhanced Wnt signaling in satellite cell progeny isolated from aged muscle after injury compared with that seen in young muscle. Given the pleiotropic effects of Wnts and the number of different Wnts and Wnt receptors, it is possible that specific combinations have both positive and negative affects on maintaining the myogenic lineage. Also, as noted above, it may be that the specific timing is critical as to whether a single Wnt may have positive or negative effects on cell fate in aged muscle progenitors and regeneration in vivo.

In addition to the possibility of satellite cell progeny adopting an alternate lineage phenotype and thus precluding their effective participation in muscle repair, the increased propensity of aged satellite cells to undergo either apoptosis or senescence compared with younger cells would also be consistent with an impaired regenerative response in aged muscle being due to an intrinsic property of the stem cells. Satellite cells isolated from old muscle and plated in culture for 3 days were more susceptible to an exposure of apoptotic-inducing agents than young cells (Jejurikar et al., 2006). If satellite cells are more susceptible to apoptosis in vivo, this combined would have detrimental effects on regeneration in aging muscle.

4.1.6 Myogenic differentiation

Aged muscle does give rise to multinucleated muscle fibers during regeneration, although the regenerating fibers are smaller and therefore differentiation may be impaired. For the limited number of satellite cells that do activate, proliferate, and progress down the myogenic lineage pathway in aged muscle, their ability to differentiate into multinucleated myotubes has been compared to that of young cells. From single fiber cultures, fewer myogenic cells isolated from old muscle expressed myosin heavy chain after 2 days in differentiating conditions, consistent with impaired differentiation (Charge et al., 2002). Furthermore, this defect was independent of the density of the culture. In support, there was less total myosin protein in aged myogenic precursors when differentiated for 2 days (Lees et al., 2006). In contrast, when young and old cells were left in culture for longer periods and then induced to differentiate, no difference between the young and old cells was observed (Conboy et al., 2003; Shefer et al., 2006). These data suggest that there are cell-intrinsic changes with age that render older satellite cells less effective in undergoing myogenic differentiation, but that these changes are in fact reversible when the cells are placed in an environment that promotes the processes of growth and differentiation necessary for effective regeneration.

4.2 Cell-extrinsic

Not only is there unequivocal evidence of changes of the systemic milieu with age that are likely to influence the functionality of all cells including satellite cells (Sidorenko et al., 1986; Hirayama et al., 1993), but the reversibility of some of the cell-intrinsic changes noted above also supports the notion that satellite cell functionality is profoundly influenced by the environment in which they reside at rest and activate in response to various stimuli. Direct evidence of the affect of the environment on the functionality of aged satellite cells comes from two different kinds of heterochronic studies. The first involves classic heterochronic transplantation experiments in rats in which whole muscles were isolated from young or old donors and then transplanted into young or old hosts (Carlson and Faulkner, 1989). These studies demonstrated that engraftment success was improved if the old muscle was transplanted into a young host compared to an old host, whereas the regeneration of young muscle transplanted in an old host was impaired relative to that in a young host. These results strongly supported the hypothesis that factors within the tissue itself in which muscle stem cells were proliferating and differentiating were more important than the source of the cells (i.e. from young or old donor) in determining the effectiveness of the regenerative response.

More recent heterochronic experiments that suggest that the host environment is critical to the success of regeneration mediated by aged satellite cells involve parabiotic studies (Conboy et al., 2005). In these studies, parabiotic pairings of young and old mice were followed by muscle injury to either or both mice and an analysis of the success of the regenerative response. Consistent with the heterochronic transplantation results, it was found that satellite cells in aged mice that had been paired with young mice showed marked improvements in functionality both in vivo and in vitro suggesting that exposure to the young systemic environment could reverse some of the age-related cellular changes. Likewise, satellite cells in young mice that had been paired with old mice showed a decline in functionality (Conboy et al., 2005). These results suggest that there is some combination of positive influence of the young systemic environment and a negative effect of the old systemic environment on stem cell functionality.

Together, these results make a very strong case for needing to understand the effects of aging not only the stem cells themselves but also on the mechanisms by which local and systemic factors may influence the phenotype and behavior of stem cells. The environment is complex, ranging from the components of the stem cell niche that are in direct contact with satellite cells (matrix components, muscle fiber membrane constituents), autocrine and paracrine factors both at rest and following injury, and finally circulating factors. How each of these may change with age and, consequently, influence stem cell functionality, will be considered separately.

4.2.1 Satellite cell niche

The stem cell niche is the microenvironment that sustains the characteristic of the stem cell, the ability to maintain quiescence until required to activate and self-renew under conditions of damage and regeneration. The niche of the satellite cell involves the myofiber with which it is closely associated on its "basal" surface without any intervening basal lamina, and an overlying basal lamina that separates the "apical" surface of the satellite cell from other fibers and interstitial cells. Other constituents in the niche could include infiltrating inflammatory cells and local vascular components, particularly adjacent capillaries (Fujino et al., 2005). Is the satellite cell niche altered during aging? The myofiber is altered in many ways during aging, as noted above. This could influence both the physical connection with the satellite cell, but also paracrine factors released from the fiber (discussed below). There is an increase in connective tissue between the fibers and surrounding the fibers in aged compared to young muscle (Goldspink et al., 1994). The basal lamina encapsulating the fiber is also thicker (Snow, 1977; Goldspink et al., 1994). Clearly, age-related alterations of the composition and deposition of extracellular matrix components could influence every aspect of satellite cell function. Recently, it was reported that mesenchymal stem cells can alter their fate depending on the stiffness of the matrix on which the cells reside (Engler et al., 2006). Applying this principle to satellite cells in aging, it is conceivable that with more connective tissue and a thicker, and thus potentially "stiffer" basal lamina, the cells may be less able to maintain their myogenic fate in the aged niche. Reduced vascularization during aging would not only reduce blood flow to the muscle, it could also reduce the availability of systemic (circulating) factors necessary for satellite cell function (discussed below).

4.2.2 Autocrine/paracrine factors

Extracts from crushed muscle contain mitogens, such as hepatocyte growth factor (HGF) and fibroblast growth factor (FGF), which can enhance the activation and proliferation of aged satellite cells in vitro (Bischoff, 1986; Chen and Quinn, 1992; Yablonka-Reuveni et al., 1999; Sheehan et al., 2000; Shefer et al., 2006). Another source of paracrine factors is fibroblasts, which can secrete soluble factors that act on myoblasts (Quinn et al., 1990; Chen and Quinn, 1992). There are also factors released from invading macrophages (Robertson et al., 1993), such as TGFβ, which can affect myogenic cell function (Pampusch et al., 1990). Therefore locally secreted factors can modulate satellite cell function, and any changes of these factors with age would be important in terms of understanding declining satellite cell functionality (Vandenburgh et al., 1984). Indeed, muscle extracts from old mice were less mitogenic to both young and old myogenic progenitors than were extracts from young muscle, consistent with an age-dependent change in paracrine regulation (Mezzogiorno et al., 1993).

It is equally plausible that age-related loss of satellite cell functionality is due to an increase in inhibitory factors (Vandenburgh et al., 1984; Brack et al., 2007). In cell culture, myoblasts from old muscle had elevated TGFβ signaling (Beggs et al., 2004). TGFβ impairs proliferation of myoblasts (Pampusch et al., 1990) and can increase fibrosis (Li et al., 2004). Whether elevated TGFβ signaling can explain the impaired regeneration and elevated fibrosis in vivo in aged muscle has not been determined.

4.2.3 Systemic factors of host environment

In addition to local changes in the muscle, the systemic milieu is another potential source of growth promoting and inhibitory factors that could alter satellite cell function and mediate muscle regeneration during aging. The findings from the heterochronic parabiotic studies described above clearly suggest that systemic factors play a major role in modifying the activity of satellite cells in response to injury (Conboy et al., 2005). The identity of the positive and negative regulators that change with age remains to be determined, but both proteinaceous and non-proteinaceous constituents need to be considered. Given the vast complexity of the composition of serum, it is likely that systemic influences on stem cell function will be multifactorial and that the net effect will be derived from the integration of the many factors present and their changes with age.

Most protein growth factors that are known to regulate myogenic progenitor proliferation in vitro are present in the circulation. The primary source of systemic IGF-1 is the liver (Froesch and Zapf, 1985). IGF-1 can stimulate myoblast cell proliferation in cell culture (Florini and Magri, 1989). In aged humans, reduced serum levels of IGF-1 have been observed (Reeves et al., 2000), making systemic IGF-1 levels one plausible candidate for environment-mediated impaired muscle regeneration. We have observed that Wnt signaling is increased in aged satellite cells, an effect that is mediated by serum (Brack et al., 2007). The increased Wnt activity resulted in a myogenic-to-fibrogenic conversion of progenitors and an increase in fibrosis. Inhibition of Wnt signaling in the aged animal (both systemically and intramuscularly) reduced fibrosis and improved muscle regeneration. Therefore Wnt proteins may act systemically and may be altered during aging.

Steroid hormones are another category of circulating factors that need to be considered as systemic modulators of stem cell function. Androgens have been found to play a role in proliferation and differentiation of satellite cells (Chen et al., 2005). The mechanism of action may be direct, by binding to androgen receptors expressed on satellite cells (Doumit et al., 1996), or indirect, for example by increasing sensitivity to IGF-1 mediated pathways (Thompson et al., 1989). Decreased levels of androgens have been reported in aged men (Snyder, 2001), raising the possibility that this could account in part for declining satellite cell functionality.

5. CONCLUSIONS

Understanding the characteristics of tissue aging and the genetic, biochemical, and cellular mechanisms that underlie those changes is a fundamental challenge of biology. Theories of the mechanisms of the aging process abound, and there is considerable evidence in support of widely divergent hypotheses with no unifying theory. As such, when trying to understand the mechanisms of age-related decline in a process as complex as muscle regeneration, one is left with traditional descriptive biology and then explorations of mechanisms only to the extent that the normal mechanisms are understood, superimposed on the complexity of how aging affects every aspect of the individual, from single molecules to the whole organism. That there is an age-related decline in muscle regenerative potential is clear. That the decline can be attributed to known changes in biochemical process and cellular function is also clear and reflects the multifactorial nature of those changes. Future research will only add to the complexity of that mechanistic understanding, but may also reveal multiple different targets for therapeutic intervention whereby the regeneration of aged muscle maybe enhanced. This could lead to more rapid recovery of function and/or mobility, particularly under conditions of relatively rapid loss of tissue (e.g. injury, post-operative bed rest) when restoration of muscle can profoundly reduce morbidity and enhance quality of life.

REFERENCES

Abrass CK, Adcox MJ, Raugi GJ (1995) Aging-associated changes in renal extracellular matrix. Am J Pathol 146:742–752

Asakura A, Komaki M, Rudnicki M (2001) Muscle satellite cells are multipotential stem cells that exhibit myogenic, osteogenic, and adipogenic differentiation. Differentiation 68:245–253

Ashcroft GS, Mills SJ, Ashworth JJ (2002) Ageing and wound healing. Biogerontology 3:337–345

Balagopal P, Schimke JC, Ades P, Adey D, Nair KS (2001) Age effect on transcript levels and synthesis rate of muscle MHC and response to resistance exercise. Am J Physiol Endocrinol Metab 280:E203–E208

Barani AE, Durieux AC, Sabido O, Freyssenet D (2003) Age-related changes in the mitotic and metabolic characteristics of muscle-derived cells. J Appl Physiol 95:2089–2098

Beggs ML, Nagarajan R, Taylor-Jones JM, Nolen G, Macnicol M, Peterson CA (2004) Alterations in the TGFbeta signaling pathway in myogenic progenitors with age. Aging Cell 3:353–361

Bekaert S, Derradji H, Baatout S (2004) Telomere biology in mammalian germ cells and during development. Dev Biol 274:15–30

Bischoff R (1986) A satellite cell mitogen from crushed adult muscle. Dev Biol 115:140–147

Blaivas M, Carlson BM (1991) Muscle fiber branching–difference between grafts in old and young rats. Mech Ageing Dev 60:43–53

Bockhold KJ, Rosenblatt JD, Partridge TA (1998) Aging normal and dystrophic mouse muscle: analysis of myogenicity in cultures of living single fibers. Muscle Nerve 21:173–183

Bortoli S, Renault V, Eveno E, Auffray C, Butler-Browne G, Pietu G (2003) Gene expression profiling of human satellite cells during muscular aging using cDNA arrays. *Gene* 321:145–154

Brack AS, Bildsoe H, Hughes SM (2005) Evidence that satellite cell decrement contributes to preferential decline in nuclear number from large fibres during murine age-related muscle atrophy. J Cell Sci 118:4813–4821

Brack AS, Conboy IM, Conboy MJ, Shen J, Rando TA. A temporal switch from Notch to Wnt signaling in muscle stem cells is necessary for normal adult myogenesis. Cell Stem Cell, in press.

Brack AS, Conboy MJ, Lee M, Roy S, Kuo CJ, Keller C, Rando TA (2007) Increased Wnt signaling during aging alters myogenic stem cell fate and increases fibrosis. Science 317:807–810.

Brooks SV, Faulkner JA (1988) Contractile properties of skeletal muscles from young, adult and aged mice. J Physiol 404:71–82

Brooks SV, Faulkner JA (1991) Maximum and sustained power of extensor digitorum longus muscles from young, adult, and old mice. J Gerontol 46:B28–B33

Brooks SV, Faulkner JA (1996) The magnitude of the initial injury induced by stretches of maximally activated muscle fibres of mice and rats increases in old age. J Physiol 497 (Pt 2):573–580

Campisi J (2001) From cells to organisms: can we learn about aging from cells in culture? Exp Gerontol 36:607–618

Cannon JG (1998) Intrinsic and extrinsic factors in muscle aging. Ann N Y Acad Sci 854:72–77

Carlson BM, Dedkov EI, Borisov AB, Faulkner JA (2001) Skeletal muscle regeneration in very old rats. J Gerontol A Biol Sci Med Sci 56:B224–B233

Carlson BM, Faulkner JA (1989) Muscle transplantation between young and old rats: age of host determines recovery. Am J Physiol 256:C1262–C1266

Charge SB, Brack AS, Hughes SM (2002) Aging-related satellite cell differentiation defect occurs prematurely after Ski-induced muscle hypertrophy. Am J Physiol Cell Physiol 283:C1228–C1241

Charge SB, Rudnicki MA (2004) Cellular and molecular regulation of muscle regeneration. Physiol Rev. 84:209–238

Chen G, Quinn LS (1992) Partial characterization of skeletal myoblast mitogens in mouse crushed muscle extract. J Cell Physiol 153:563–574

Chen Y, Zajac JD, MacLean HE (2005) Androgen regulation of satellite cell function. J Endocrinol 186:21–31

Coggan AR, Spina RJ, King DS, Rogers MA, Brown M, Nemeth PM, Holloszy JO (1992) Histochemical and enzymatic comparison of the gastrocnemius muscle of young and elderly men and women. J Gerontol 47:B71–B76

Collins CA, Partridge TA (2005) Self-renewal of the adult skeletal muscle satellite cell. Cell Cycle 4:1338–1341

Conboy IM, Conboy MJ, Smythe GM, Rando TA (2003) Notch-mediated restoration of regenerative potential to aged muscle. Science 302:1575–1577

Conboy IM, Conboy MJ, Wagers AJ, Girma ER, Weissman IL, Rando TA (2005) Rejuvenation of aged progenitor cells by exposure to a young systemic environment. Nature 433:760–764

Conboy IM, Rando TA (2002) The regulation of Notch signaling controls satellite cell activation and cell fate determination in postnatal myogenesis. Dev Cell 3:397–409

Conboy IM, Rando TA (2005) Aging, stem cells and tissue regeneration: lessons from muscle. Cell Cycle 4:407–410

Darr KC, Schultz E (1987) Exercise-induced satellite cell activation in growing and mature skeletal muscle. J Appl Physiol 63:1816–1821

Decary S, Hamida CB, Mouly V, Barbet JP, Hentati F, Butler-Browne GS (2000) Shorter telomeres in dystrophic muscle consistent with extensive regeneration in young children. Neuromuscul Disord 10:113–120

Decary S, Mouly V, Hamida CB, Sautet A, Barbet JP, Butler-Browne GS (1997) Replicative potential and telomere length in human skeletal muscle: implications for satellite cell-mediated gene therapy. Hum Gene Ther 8:1429–1438

Dedkov EI, Kostrominova TY, Borisov AB, Carlson BM (2003) MyoD and myogenin protein expression in skeletal muscles of senile rats. Cell Tissue Res 311:401–416

Doumit ME, Cook DR, Merkel RA (1996) Testosterone up-regulates androgen receptors and decreases differentiation of porcine myogenic satellite cells in vitro. Endocrinology 137:1385–1394

Dow DE, Dennis RG, Faulkner JA (2005) Electrical stimulation attenuates denervation and age-related atrophy in extensor digitorum longus muscles of old rats. *J. Gerontol.* A Biol Sci Med Sci 60:416–424

Engler AJ, Sen S, Sweeney HL, Discher DE (2006) Matrix elasticity directs stem cell lineage specification. Cell 126:677–689

Everitt AV, Shorey CD, Ficarra MA (1985) Skeletal muscle aging in the hind limb of the old male Wistar rat: inhibitory effect of hypophysectomy and food restriction. Arch Gerontol Geriatr 4:101–115

Fiatarone MA, Marks EC, Ryan ND, Meredith CN, Lipsitz LA, Evans WJ (1990) High-intensity strength training in nonagenarians. Effects on skeletal muscle. JAMA 263:3029–3034

Florini JR, Magri KA (1989) Effects of growth factors on myogenic differentiation. Am J Physiol 256:C701–C711

Froesch ER, Zapf J (1985) Insulin-like growth factors and insulin: comparative aspects. Diabetologia 28:485–493

Fry M, Silber J, Loeb LA, Martin GM (1984) Delayed and reduced cell replication and diminishing levels of DNA polymerase-alpha in regenerating liver of aging mice. J Cell Physiol 118:225–232

Fujino H, Kohzuki H, Takeda I, Kiyooka T, Miyasaka T, Mohri S, Shimizu J, Kajiya F (2005) Regression of capillary network in atrophied soleus muscle induced by hindlimb unweighting. J Appl Physiol 98:1407–1413

Gagliano N, Arosio B, Grizzi F, Masson S, Tagliabue J, Dioguardi N, Vergani C, Annoni G (2002) Reduced collagenolytic activity of matrix metalloproteinases and development of liver fibrosis in the aging rat. Mech Ageing Dev 123:413–425

Gibson MC, Schultz E (1983) Age-related differences in absolute numbers of skeletal muscle satellite cells. Muscle Nerve 6:574–580

Goldspink G, Fernandes K, Williams PE, Wells DJ (1994) Age-related changes in collagen gene expression in the muscles of mdx dystrophic and normal mice. *Neuromuscul.* Disord 4:183–191

Hawke TJ, Garry DJ (2001) Myogenic satellite cells: physiology to molecular biology. J Appl Physiol 91:534–551

Hirayama R, Takemura K, Nihei Z, Ichikawa W, Takagi Y, Mishima Y, Utsuyama M, Hirokawa K (1993) Differential effect of host microenvironment and systemic humoral factors on the implantation and the growth rate of metastatic tumor in parabiotic mice constructed between young and old mice. Mech Ageing Dev 71:213–221

Jejurikar SS, Henkelman EA, Cederna PS, Marcelo CL, Urbanchek MG, Kuzon WM, Jr (2006) Aging increases the susceptibility of skeletal muscle derived satellite cells to apoptosis. Exp Gerontol. 41:828–836

Larsson L, Ansved T (1995) Effects of ageing on the motor unit. Prog Neurobiol 45:397–458

Lees SJ, Rathbone CR, Booth FW (2006) Age-associated decrease in muscle precursor cell differentiation. Am J Physiol Cell Physiol 290:C609–C615.

Lexell J (1995) Human aging, muscle mass, and fiber type composition. J Gerontol A Biol Sci Med Sci 50 Spec No:11–16.

Lexell J, Taylor CC, Sjostrom M (1988) What is the cause of the ageing atrophy? Total number, size and proportion of different fiber types studied in whole vastus lateralis muscle from 15- to 83-year-old men. J Neurol Sci 84:275–294

Li X, Larsson L (1996) Maximum shortening velocity and myosin isoforms in single muscle fibers from young and old rats. Am J Physiol 270:C352–C360

Li Y, Foster W, Deasy BM, Chan Y, Prisk V, Tang Y, Cummins J, Huard J (2004) Transforming growth factor-beta1 induces the differentiation of myogenic cells into fibrotic cells in injured skeletal muscle: a key event in muscle fibrogenesis. Am J Pathol 164:1007–1019

Logan CY, Nusse R (2004) The Wnt signaling pathway in development and disease. Annu Rev Cell Dev Biol 20:781–810

Machida S, Booth FW (2004) Insulin-like growth factor 1 and muscle growth: implication for satellite cell proliferation. Proc Nutr Soc 63:337–340

Martin K, Potten CS, Roberts SA, Kirkwood TB (1998) Altered stem cell regeneration in irradiated intestinal crypts of senescent mice. J Cell Sci 111 (Pt 16):2297–2303

Mauro A (1961) Satellite cell of skeletal muscle fibers. J Biophys Biochem Cytol 9:493–495

McGeachie JK, Grounds MD (1995) Retarded myogenic cell replication in regenerating skeletal muscles of old mice: an autoradiographic study in young and old BALBc and SJL/J mice. Cell Tissue Res 280:277–282

Mezzogiorno A, Coletta M, Zani BM, Cossu G, Molinaro M (1993) Paracrine stimulation of senescent satellite cell proliferation by factors released by muscle or myotubes from young mice. Mech Ageing Dev 70:35–44

Mouly V, Aamiri A, Bigot A, Cooper RN, Di DS, Furling D, Gidaro T, Jacquemin V, Mamchaoui K, Negroni E, Perie S, Renault V, Silva-Barbosa SD, Butler-Browne GS (2005) The mitotic clock in skeletal muscle regeneration, disease and cell mediated gene therapy. Acta Physiol Scand 184:3–15

Pampusch MS, Hembree JR, Hathaway MR, Dayton WR (1990) Effect of transforming growth factor beta on proliferation of L6 and embryonic porcine myogenic cells. J Cell Physiol 143:524–528

Porter MM, Vandervoort AA, Lexell J (1995) Aging of human muscle: structure, function and adaptability. *Scand.* J Med Sci Sports 5:129–142

Quinn LS, Ong LD, Roeder RA (1990) Paracrine control of myoblast proliferation and differentiation by fibroblasts. Dev Biol 140:8–19

Reeves I, Abribat T, Laramee P, Jasmin G, Brazeau P (2000) Age-related serum levels of insulin-like growth factor-I, -II and IGF-binding protein-3 following myocardial infarction. Growth Horm IGF Res 10:78–84

Robertson TA, Maley MA, Grounds MD, Papadimitriou JM (1993) The role of macrophages in skeletal muscle regeneration with particular reference to chemotaxis. Exp Cell Res 207:321–331

Ryan NA, Zwetsloot KA, Westerkamp LM, Hickner RC, Pofahl WE, Gavin TP (2006) Lower skeletal muscle capillarization and VEGF expression in aged vs. young men. J Appl Physiol 100:178–185

Sadeh M (1988) Effects of aging on skeletal muscle regeneration. J Neurol Sci 87:67–74

Sajko S, Kubinova L, Cvetko E, Kreft M, Wernig A, Erzen I (2004) Frequency of M-cadherin-stained satellite cells declines in human muscles during aging. J Histochem Cytochem 52:179–185

Sarmento LM, Huang H, Limon A, Gordon W, Fernandes J, Tavares MJ, Miele L, Cardoso AA, Classon M, Carlesso N (2005) Notch1 modulates timing of G1-S progression by inducing SKP2 transcription and p27 Kip1 degradation. J Exp Med 202:157–168

Sayer AA, Syddall H.E, Martin HJ, Dennison EM, Roberts HC, Cooper C (2006) Is grip strength associated with health-related quality of life? Findings from the Hertfordshire Cohort Study. Age Ageing 35:409–415

Schultz E, Lipton BH (1982) Skeletal muscle satellite cells: changes in proliferation potential as a function of age. Mech Ageing Dev 20:377–383

Sheehan SM, Tatsumi R, Temm-Grove CJ, Allen RE (2000) HGF is an autocrine growth factor for skeletal muscle satellite cells in vitro. Muscle Nerve 23:239–245

Shefer G, Van de Mark DP, Richardson JB, Yablonka-Reuveni Z (2006) Satellite-cell pool size does matter: defining the myogenic potency of aging skeletal muscle. Dev Biol 294:50–66.

Shefer G, Wleklinski-Lee M, Yablonka-Reuveni Z (2004) Skeletal muscle satellite cells can spontaneously enter an alternative mesenchymal pathway. J Cell Sci 117:5393–5404

Sidorenko AV, Gubrii IB, Andrianova LF, Macsijuk TV, Butenko GM (1986) Functional rearrangement of lymphohemopoietic system in heterochronically parabiosed mice. *Mech. Ageing Dev.* 36:41–56.

Snow LM, McLoon LK, Thompson LV (2005) Adult and developmental myosin heavy chain isoforms in soleus muscle of aging Fischer Brown Norway rat. Anat Rec A Discov Mol Cell Evol Biol 286:866–873

Snow MH (1977) The effects of aging on satellite cells in skeletal muscles of mice and rats. Cell Tissue Res 185:399–408

Snyder PJ (2001) Effects of age on testicular function and consequences of testosterone treatment. J Clin Endocrinol Metab 86:2369–2372

Taylor-Jones JM, McGehee RE, Rando TA, Lecka-Czernik B, Lipschitz DA, Peterson CA (2002) Activation of an adipogenic program in adult myoblasts with age. Mech Ageing Dev 123:649–661

Thompson SH, Boxhorn LK, Kong WY, Allen RE (1989) Trenbolone alters the responsiveness of skeletal muscle satellite cells to fibroblast growth factor and insulin-like growth factor I. Endocrinology 124:2110–2117

Tidball JG (2005) Inflammatory processes in muscle injury and repair. *Am. J. Physiol Regul. Integr.* Comp Physiol 288:R345–R353

Vandenburgh HH, Sheff MF, Zacks SI (1984) Soluble age-related factors from skeletal muscle which influence muscle development. Exp Cell Res 153:389–401

Vertino AM, Taylor-Jones JM, Longo KA, Bearden ED, Lane TF, McGehee RE, Jr, MacDougald OA, Peterson CA (2005) Wnt10b deficiency promotes coexpression of myogenic and adipogenic programs in myoblasts. Mol Biol Cell 16:2039–2048

Yablonka-Reuveni Z, Seger R, Rivera AJ (1999) Fibroblast growth factor promotes recruitment of skeletal muscle satellite cells in young and old rats. J Histochem Cytochem 47:23–42

Zammit PS, Heslop L, Hudon V, Rosenblatt JD, Tajbakhsh S, Buckingham ME, Beauchamp JR, Partridge TA (2002) Kinetics of myoblast proliferation show that resident satellite cells are competent to fully regenerate skeletal muscle fibers. Exp Cell Res 281:39–49

Zammit PS, Partridge TA, Yablonka-Reuveni Z (2006) The skeletal muscle satellite cell: the stem cell that came in from the cold. J Histochem Cytochem 54:1177–1191

INDEX

CPSIA information can be obtained at www.ICGtesting.com
Printed in the USA
LVOW082142221112

308450LV00005B/94/P

9 781402 067679